DATE DUE

Methods in Enzymology

Volume 288

CHEMOKINE RECEPTORS

METHODS IN ENZYMOLOGY

EDITORS-IN-CHIEF

John N. Abelson   Melvin I. Simon

DIVISION OF BIOLOGY
CALIFORNIA INSTITUTE OF TECHNOLOGY
PASADENA, CALIFORNIA

FOUNDING EDITORS

Sidney P. Colowick and Nathan O. Kaplan

Methods in Enzymology

Volume 288

Chemokine Receptors

EDITED BY

Richard Horuk

DEPARTMENT OF IMMUNOLOGY
BERLEX BIOSCIENCES
RICHMOND, CALIFORNIA

ACADEMIC PRESS
San Diego London Boston New York Sydney Tokyo Toronto

This book is printed on acid-free paper. ♾

Academic Press
15 East 26th Street, 15th Floor, New York, New York 10010, USA
http://www.apnet.com

Academic Press Limited
24-28 Oval Road, London NW1 7DX, UK
http://www.hbuk.co.uk/ap/

International Standard Book Number: 0-12-182189-7

PRINTED IN THE UNITED STATES OF AMERICA
97 98 99 00 01 02 MM 9 8 7 6 5 4 3 2 1

Table of Contents

Contributors to Volume 288 ix

Preface xiii

Volumes in Series xv

Section I. Chemokine Receptors

1. C-X-C Chemokine Receptor Desensitization Mediated through Ligand-Enhanced Receptor Phosphorylation on Serine Residues — Ann Richmond, Susan Mueller, John R. White, and Wayne Schraw — 3

2. Generation of Monoclonal Antibodies to Chemokine Receptors — Anan Chuntharapai and K. Jin Kim — 15

3. Chemokine Receptors in Developing Human Brain — Meredith Halks-Miller, Joseph Hesselgesser, Ilona J. Miko, and Richard Horuk — 27

4. Expression of Chemokine Receptors in Insect Cells Using Baculovirus Vectors — Zi-xuan Wang, Ying-hua Cen, Hai-hong Guo, Jian-guo Du, and Stephen C. Peiper — 38

5. Chimeric Chemokine Receptors for Analysis of Structure–Function Relationships — Stephen C. Peiper, Zhao-hai Lu, Tian-yuan Zhang, and Zi-xuan Wang — 56

6. Molecular Approaches to Identifying Ligand Binding and Signaling Domains of C-C Chemokine Receptors — Felipe S. Monteclaro, Hidenori Arai, and Israel F. Charo — 70

7. Characterization of Functional Activity of Chemokine Receptors Using the Cytosensor Microphysiometer — Simon Pitchford, Margaret Hirst, H. Garrett Wada, Samuel D. H. Chan, Véronique E. Timmermans, and Gillian M. K. Humphries — 84

8. Calcium Flux Assay of Chemokine Receptor Expression in *Xenopus* Oocytes — Philip M. Murphy — 108

9. Cell–Cell Fusion Assay to Study Role of Chemokine Receptors in Human Immunodeficiency Virus Type 1 Entry — Joseph Rucker, Benjamin J. Doranz, Aimee L. Edinger, Deborah Long, Joanne F. Berson, and Robert W. Doms 118

10. Iodination of Chemokines for Use in Receptor Binding Analysis — Gregory L. Bennett and Richard Horuk 134

11. Expression of Chemokine Receptors by Endothelial Cells: Detection by Intravital Microscopy Using Chemokine-Coated Fluorescent Microspheres — Victor H. Fingar, Hai-hong Guo, Zhao-hai Lu, and Stephen C. Peiper 148

Section II. Chemokines in Disease

12. Neutralization of Interleukin-8 in *in Vivo* Models of Lung and Pleural Injury — V. Courtney Broaddus and Caroline A. Hébert 161

13. Murine Experimental Autoimmune Encephalomyelitis: A Model of Immune-Mediated Inflammation and Multiple Sclerosis — Andrzej R. Glabinski, Marie Tani, Vincent K. Tuohy, and Richard M. Ransohoff 182

14. *In Vitro* and *in Vivo* Systems to Assess Role of C-X-C Chemokines in Regulation of Angiogenesis — Douglas A. Arenberg, Peter J. Polverini, Steven L. Kunkel, Armen Shanafelt, and Robert M. Strieter 190

15. Role of Chemokines in Antibacterial Host Defense — Theodore J. Standiford, Steven L. Kunkel, and Robert M. Strieter 220

16. Animal Models of Asthma: Role of Chemokines — David A. Griffiths-Johnson, Paul D. Collins, Peter J. Jose, and Timothy J. Williams 241

17. Identification and Structural Characterization of Chemokines in Lesional Skin Material of Patients with Inflammatory Skin Disease — Jens-Michael Schröder 266

Section III. Signal Transduction of Chemokines

18. Calcium Mobilization Assays — Shaun R. McColl and Paul H. Naccache 301

19. G-Protein Activation by Chemokines — Suzanne K. Beckner 309

20. Adenylate Cyclase Assays to Measure Chemokine Receptor Function — RICHARD HORUK — 326

21. Analysis of Signal Transduction Following Lymphocyte Activation by Chemokines — KEVIN B. BACON — 340

22. Calcium Mobilization and Phosphoinositide Turnover as Measure of Chemokine Receptor Function in Lymphocytes — KEVIN B. BACON — 362

AUTHOR INDEX 385

SUBJECT INDEX 415

Contributors to Volume 288

Article numbers are in parentheses following the names of contributors.
Affiliations listed are current.

HIDENORI ARAI (6), *Gladstone Institute of Cardiovascular Disease, University of California, San Francisco, California 94141*

DOUGLAS A. ARENBERG (14), *University of Michigan, Ann Arbor, Michigan 48109*

KEVIN B. BACON (21, 22), *Department of Immunobiology, DNAX Research Institute, Palo Alto, California 94304-1104*

SUZANNE K. BECKNER (19), *Westat, Rockville, Maryland 20850*

GREGORY L. BENNETT (10), *Department of Bioanalytical Technology, Genentech, Inc., South San Francisco, California 94080*

JOANNE F. BERSON (9), *Department of Pathology and Laboratory Medicine, University of Pennsylvania, Philadelphia, Pennsylvania 19104*

V. COURTNEY BROADDUS (12), *Lung Biology Center, Department of Medicine, University of California, San Francisco, San Francisco, California 94143-0854*

SAMUEL D. H. CHAN (7), *Molecular Devices Corporation, Sunnyvale, California 94089*

ISRAEL F. CHARO (6), *Gladstone Institute of Cardiovascular Disease, University of California, San Francisco, California 94141*

YING-HUA CHEN (4), *Departments of Biochemistry and Molecular Biology, Pathology and Laboratory Medicine, Internal Medicine, and Surgery, Henry Vogt Cancer Research Institute of James Graham Brown Cancer Center, University of Louisville, Louisville, Kentucky 40292*

ANAN CHUNTHARAPAI (2), *Department of Antibody Techniques, Genentech, Inc., South San Francisco, California 94080*

PAUL D. COLLINS (16), *Division of Applied Pharmacology, Imperial College School of Medicine at the National Heart and Lung Institute, London SW3 6LY, United Kingdom*

ROBERT W. DOMS (9), *Department of Pathology and Laboratory Medicine, University of Pennsylvania, Philadelphia, Pennsylvania 19104*

BENJAMIN J. DORANZ (9), *Department of Pathology and Laboratory Medicine, University of Pennsylvania, Philadelphia, Pennsylvania 19104*

JIAN-GUO DU (4), *Departments of Biochemistry and Molecular Biology, Pathology and Laboratory Medicine, Internal Medicine, and Surgery, Henry Vogt Cancer Research Institute of James Graham Brown Cancer Center, University of Louisville, Louisville, Kentucky 40292*

AIMEE L. EDINGER (9), *Department of Pathology and Laboratory Medicine, University of Pennsylvania, Philadelphia, Pennsylvania 19104*

VICTOR H. FINGAR (11), *Department of Surgery, University of Louisville, Louisville, Kentucky 40292*

ANDRZEJ R. GLABINSKI (13), *Department of Neurology, Medical University of Lodz, Lodz, Poland*

DAVID A. GRIFFITHS-JOHNSON (16), *Division of Applied Pharmacology, Imperial College School of Medicine at the National Heart and Lung Institute, London SW3 6LY, United Kingdom*

HAI-HONG GUO (4, 11), *Department of Biochemistry and Molecular Biology, University of Louisville, Louisville, Kentucky 40292*

MEREDITH HALKS-MILLER (3), *Department of Experimental Pathology, Berlex Biosciences, Richmond, California 94804*

Caroline A. Hébert (12), *Genentech, Inc., South San Francisco, California 94080-4990*

Joseph Hesselgesser (3), *Department of Immunology, Berlex Biosciences, Richmond, California 94804*

Margaret Hirst (7), *Molecular Devices Corporation, Sunnyvale, California 94089*

Richard Horuk (3, 10, 20), *Department of Immunology, Berlex Biosciences, Richmond, California 94804*

Gillian M. K. Humphries (7), *Molecular Devices Corporation, Sunnyvale, California 94089*

Peter J. Jose (16), *Division of Applied Pharmacology, Imperial College School of Medicine at the National Heart and Lung Institute, London SW3 6LY, United Kingdom*

K. Jin Kim (2), *Department of Antibody Techniques, Genentech, Inc., South San Francisco, California 94080*

Steven L. Kunkel (14, 15), *Department of Pathology, University of Michigan Medical Center, Ann Arbor, Michigan 48109*

Deborah Long (9), *Department of Pathology and Laboratory Medicine, University of Pennsylvania, Philadelphia, Pennsylvania 19104*

Zhao-hai Lu (5, 11), *Department of Biochemistry and Molecular Biology, University of Louisville, Louisville, Kentucky 40292*

Shaun R. McColl (18), *Department of Microbiology and Immunology, The University of Adelaide, Adelaide, South Australia 5005*

Ilona J. Miko (3), *Department of Experimental Pathology, Berlex Biosciences, Richmond, California 94804*

Felipe S. Monteclaro (6), *Gladstone Institute of Cardiovascular Disease, University of California, San Francisco, California 94141*

Susan Mueller (1), *Department of Cell Biology, Vanderbilt University School of Medicine, Nashville, Tennessee 37232-2175*

Philip M. Murphy (8), *Laboratory of Host Defenses, National Institute of Allergy and Infectious Diseases, National Institutes of Health, Bethesda, Maryland 20892*

Paul H. Naccache (18), *Centre de Rechérche en Rhumatologie et Inflammation, Centre de Rechérche du CHUL and Faculty of Medicine, Université Laval, Québec, Canada G1V 4G2*

Stephen C. Peiper (4, 5, 11), *Departments of Biochemistry and Molecular Biology, Pathology and Laboratory Medicine, University of Louisville, Louisville, Kentucky 40292*

Simon Pitchford (7), *Molecular Devices Corporation, Sunnyvale, California 94089*

Peter J. Polverini (14), *University of Michigan, Ann Arbor, Michigan 48109*

Richard M. Ransohoff (13), *The Research Institute and Mellen Center for Multiple Sclerosis Treatment and Research, Cleveland Clinic Foundation, Cleveland, Ohio 44195*

Ann Richmond (1), *Department of Cell Biology, Vanderbilt University School of Medicine, Nashville, Tennessee 37232-2175, and Department of Veterans Affairs Medical Center, Nashville, Tennessee 37232-2175*

Joseph Rucker (9), *Department of Pathology and Laboratory Medicine, University of Pennsylvania, Philadelphia, Pennsylvania 19104*

Wayne Schraw (1), *Department of Cell Biology, Vanderbilt University School of Medicine, Nashville, Tennessee 37232-2175*

Jens-Michael Schröder (17), *Department of Dermatology, University of Kiel, D-24105 Kiel, Germany*

Armen Shanafelt (14), *Institute of Molecular Biologicals and Institute of Research Technologies, Bayer Corporation, West Haven, Connecticut 06516*

Theodore J. Standiford (15), *Division of Pulmonary and Critical Care Medicine, University of Michigan Medical Center, Ann Arbor, Michigan 48109*

Robert M. Strieter (14, 15), *Division of Pulmonary and Critical Care Medicine, University of Michigan Medical Center, Ann Arbor, Michigan 48109*

MARIE TANI (13), *Department of Neurosciences, The Research Institute, Cleveland Clinic Foundation, Cleveland, Ohio 44195*

VÉRONIQUE E. TIMMERMANS (7), *Molecular Devices Corporation, Sunnyvale, California 94089*

VINCENT K. TUOHY (13), *Department of Immunology, The Research Institute, Cleveland Clinic Foundation, Cleveland, Ohio 44195*

H. GARRETT WADA (7), *Molecular Devices Corporation, Sunnyvale, California 94089*

ZI-XUAN WANG (4, 5), *Departments of Biochemistry and Molecular Biology, Pathology and Laboratory Medicine, Internal Medicine, and Surgery, Henry Vogt Cancer Research Institute of James Graham Brown Cancer Center University of Louisville, Louisville, Kentucky 40292*

JOHN R. WHITE (1), *Department of Immunology, SmithKline Beecham Pharmaceuticals, King of Prussia, Pennsylvania 19406-0939*

TIMOTHY J. WILLIAMS (16), *Division of Applied Pharmacology, Imperial College School of Medicine at the National Heart and Lung Institute, London SW3 6LY, United Kingdom*

TIAN-YUAN ZHANG (5), *Departments of Biochemistry and Molecular Biology, Pathology and Laboratory Medicine, Internal Medicine, and Surgery, Henry Vogt Cancer Research Institute of James Graham Brown Cancer Center, University of Louisville, Louisville, Kentucky 40292*

Preface

Chemokines play an important role in inducing the directed migration of blood leukocytes in the body. When this process goes awry and immune cells turn on and attack their own tissues, autoimmune diseases such as rheumatoid arthritis and multiple sclerosis result. In light of these proinflammatory properties, chemokines and their cellular receptors have become therapeutic targets for drug intervention by major pharmaceutical companies. In addition, chemokine receptors have been in the scientific spotlight recently because of the finding that they are coreceptors, along with CD4, for pathogenic organisms such as HIV-1, which use them to gain entry into, and infect, mammalian cells. These and other related findings have placed chemokines in the limelight and exposed them to intense scrutiny by an increasingly broad population of the scientific community.

Given this increased interest in chemokines there was a real need for a practical bench guide that gives detailed protocols and methods that can be used by researchers and advanced students as a step-by-step guide for studying these molecules. With this in mind I assembled a series of comprehensive articles from acknowledged experts in the chemokine field. They are presented in Volumes 287 and 288 of *Methods in Enzymology.* Volume 287 deals with methods in chemokine research; Volume 288 covers chemokine receptor protocols. These volumes provide a detailed compendium of laboratory methods that will appeal both to the novice and to the more experienced researcher wanting to enter this field.

I would like to express my sincere thanks to all the contributing authors for their outstanding efforts and patience during the production of this work. Also, I would like to thank Shirley Light of Academic Press for providing guidance, encouragement, and advice in the preparation of these volumes.

RICHARD HORUK

METHODS IN ENZYMOLOGY

VOLUME I. Preparation and Assay of Enzymes
Edited by SIDNEY P. COLOWICK AND NATHAN O. KAPLAN

VOLUME II. Preparation and Assay of Enzymes
Edited by SIDNEY P. COLOWICK AND NATHAN O. KAPLAN

VOLUME III. Preparation and Assay of Substrates
Edited by SIDNEY P. COLOWICK AND NATHAN O. KAPLAN

VOLUME IV. Special Techniques for the Enzymologist
Edited by SIDNEY P. COLOWICK AND NATHAN O. KAPLAN

VOLUME V. Preparation and Assay of Enzymes
Edited by SIDNEY P. COLOWICK AND NATHAN O. KAPLAN

VOLUME VI. Preparation and Assay of Enzymes (*Continued*)
Preparation and Assay of Substrates
Special Techniques
Edited by SIDNEY P. COLOWICK AND NATHAN O. KAPLAN

VOLUME VII. Cumulative Subject Index
Edited by SIDNEY P. COLOWICK AND NATHAN O. KAPLAN

VOLUME VIII. Complex Carbohydrates
Edited by ELIZABETH F. NEUFELD AND VICTOR GINSBURG

VOLUME IX. Carbohydrate Metabolism
Edited by WILLIS A. WOOD

VOLUME X. Oxidation and Phosphorylation
Edited by RONALD W. ESTABROOK AND MAYNARD E. PULLMAN

VOLUME XI. Enzyme Structure
Edited by C. H. W. HIRS

VOLUME XII. Nucleic Acids (Parts A and B)
Edited by LAWRENCE GROSSMAN AND KIVIE MOLDAVE

VOLUME XIII. Citric Acid Cycle
Edited by J. M. LOWENSTEIN

VOLUME XIV. Lipids
Edited by J. M. LOWENSTEIN

VOLUME XV. Steroids and Terpenoids
Edited by RAYMOND B. CLAYTON

VOLUME XVI. Fast Reactions
Edited by KENNETH KUSTIN

VOLUME XVII. Metabolism of Amino Acids and Amines (Parts A and B)
Edited by HERBERT TABOR AND CELIA WHITE TABOR

VOLUME XVIII. Vitamins and Coenzymes (Parts A, B, and C)
Edited by DONALD B. MCCORMICK AND LEMUEL D. WRIGHT

VOLUME XIX. Proteolytic Enzymes
Edited by GERTRUDE E. PERLMANN AND LASZLO LORAND

VOLUME XX. Nucleic Acids and Protein Synthesis (Part C)
Edited by KIVIE MOLDAVE AND LAWRENCE GROSSMAN

VOLUME XXI. Nucleic Acids (Part D)
Edited by LAWRENCE GROSSMAN AND KIVIE MOLDAVE

VOLUME XXII. Enzyme Purification and Related Techniques
Edited by WILLIAM B. JAKOBY

VOLUME XXIII. Photosynthesis (Part A)
Edited by ANTHONY SAN PIETRO

VOLUME XXIV. Photosynthesis and Nitrogen Fixation (Part B)
Edited by ANTHONY SAN PIETRO

VOLUME XXV. Enzyme Structure (Part B)
Edited by C. H. W. HIRS AND SERGE N. TIMASHEFF

VOLUME XXVI. Enzyme Structure (Part C)
Edited by C. H. W. HIRS AND SERGE N. TIMASHEFF

VOLUME XXVII. Enzyme Structure (Part D)
Edited by C. H. W. HIRS AND SERGE N. TIMASHEFF

VOLUME XXVIII. Complex Carbohydrates (Part B)
Edited by VICTOR GINSBURG

VOLUME XXIX. Nucleic Acids and Protein Synthesis (Part E)
Edited by LAWRENCE GROSSMAN AND KIVIE MOLDAVE

VOLUME XXX. Nucleic Acids and Protein Synthesis (Part F)
Edited by KIVIE MOLDAVE AND LAWRENCE GROSSMAN

VOLUME XXXI. Biomembranes (Part A)
Edited by SIDNEY FLEISCHER AND LESTER PACKER

VOLUME XXXII. Biomembranes (Part B)
Edited by SIDNEY FLEISCHER AND LESTER PACKER

VOLUME XXXIII. Cumulative Subject Index Volumes I–XXX
Edited by MARTHA G. DENNIS AND EDWARD A. DENNIS

VOLUME XXXIV. Affinity Techniques (Enzyme Purification: Part B)
Edited by WILLIAM B. JAKOBY AND MEIR WILCHEK

VOLUME XXXV. Lipids (Part B)
Edited by JOHN M. LOWENSTEIN

VOLUME XXXVI. Hormone Action (Part A: Steroid Hormones)
Edited by BERT W. O'MALLEY AND JOEL G. HARDMAN

VOLUME XXXVII. Hormone Action (Part B: Peptide Hormones)
Edited by BERT W. O'MALLEY AND JOEL G. HARDMAN

VOLUME XXXVIII. Hormone Action (Part C: Cyclic Nucleotides)
Edited by JOEL G. HARDMAN AND BERT W. O'MALLEY

VOLUME XXXIX. Hormone Action (Part D: Isolated Cells, Tissues, and Organ Systems)
Edited by JOEL G. HARDMAN AND BERT W. O'MALLEY

VOLUME XL. Hormone Action (Part E: Nuclear Structure and Function)
Edited by BERT W. O'MALLEY AND JOEL G. HARDMAN

VOLUME XLI. Carbohydrate Metabolism (Part B)
Edited by W. A. WOOD

VOLUME XLII. Carbohydrate Metabolism (Part C)
Edited by W. A. WOOD

VOLUME XLIII. Antibiotics
Edited by JOHN H. HASH

VOLUME XLIV. Immobilized Enzymes
Edited by KLAUS MOSBACH

VOLUME XLV. Proteolytic Enzymes (Part B)
Edited by LASZLO LORAND

VOLUME XLVI. Affinity Labeling
Edited by WILLIAM B. JAKOBY AND MEIR WILCHEK

VOLUME XLVII. Enzyme Structure (Part E)
Edited by C. H. W. HIRS AND SERGE N. TIMASHEFF

VOLUME XLVIII. Enzyme Structure (Part F)
Edited by C. H. W. HIRS AND SERGE N. TIMASHEFF

VOLUME XLIX. Enzyme Structure (Part G)
Edited by C. H. W. HIRS AND SERGE N. TIMASHEFF

VOLUME L. Complex Carbohydrates (Part C)
Edited by VICTOR GINSBURG

VOLUME LI. Purine and Pyrimidine Nucleotide Metabolism
Edited by PATRICIA A. HOFFEE AND MARY ELLEN JONES

VOLUME LII. Biomembranes (Part C: Biological Oxidations)
Edited by SIDNEY FLEISCHER AND LESTER PACKER

VOLUME LIII. Biomembranes (Part D: Biological Oxidations)
Edited by SIDNEY FLEISCHER AND LESTER PACKER

VOLUME LIV. Biomembranes (Part E: Biological Oxidations)
Edited by SIDNEY FLEISCHER AND LESTER PACKER

VOLUME LV. Biomembranes (Part F: Bioenergetics)
Edited by SIDNEY FLEISCHER AND LESTER PACKER

VOLUME LVI. Biomembranes (Part G: Bioenergetics)
Edited by SIDNEY FLEISCHER AND LESTER PACKER

VOLUME LVII. Bioluminescence and Chemiluminescence
Edited by MARLENE A. DELUCA

VOLUME LVIII. Cell Culture
Edited by WILLIAM B. JAKOBY AND IRA PASTAN

VOLUME LIX. Nucleic Acids and Protein Synthesis (Part G)
Edited by KIVIE MOLDAVE AND LAWRENCE GROSSMAN

VOLUME LX. Nucleic Acids and Protein Synthesis (Part H)
Edited by KIVIE MOLDAVE AND LAWRENCE GROSSMAN

VOLUME 61. Enzyme Structure (Part H)
Edited by C. H. W. HIRS AND SERGE N. TIMASHEFF

VOLUME 62. Vitamins and Coenzymes (Part D)
Edited by DONALD B. MCCORMICK AND LEMUEL D. WRIGHT

VOLUME 63. Enzyme Kinetics and Mechanism (Part A: Initial Rate and Inhibitor Methods)
Edited by DANIEL L. PURICH

VOLUME 64. Enzyme Kinetics and Mechanism (Part B: Isotopic Probes and Complex Enzyme Systems)
Edited by DANIEL L. PURICH

VOLUME 65. Nucleic Acids (Part I)
Edited by LAWRENCE GROSSMAN AND KIVIE MOLDAVE

VOLUME 66. Vitamins and Coenzymes (Part E)
Edited by DONALD B. MCCORMICK AND LEMUEL D. WRIGHT

VOLUME 67. Vitamins and Coenzymes (Part F)
Edited by DONALD B. MCCORMICK AND LEMUEL D. WRIGHT

VOLUME 68. Recombinant DNA
Edited by RAY WU

VOLUME 69. Photosynthesis and Nitrogen Fixation (Part C)
Edited by ANTHONY SAN PIETRO

VOLUME 70. Immunochemical Techniques (Part A)
Edited by HELEN VAN VUNAKIS AND JOHN J. LANGONE

VOLUME 71. Lipids (Part C)
Edited by JOHN M. LOWENSTEIN

VOLUME 72. Lipids (Part D)
Edited by JOHN M. LOWENSTEIN

VOLUME 73. Immunochemical Techniques (Part B)
Edited by JOHN J. LANGONE AND HELEN VAN VUNAKIS

VOLUME 74. Immunochemical Techniques (Part C)
Edited by JOHN J. LANGONE AND HELEN VAN VUNAKIS

VOLUME 75. Cumulative Subject Index Volumes XXXI, XXXII, XXXIV–LX
Edited by EDWARD A. DENNIS AND MARTHA G. DENNIS

VOLUME 76. Hemoglobins
Edited by ERALDO ANTONINI, LUIGI ROSSI-BERNARDI, AND EMILIA CHIANCONE

VOLUME 77. Detoxication and Drug Metabolism
Edited by WILLIAM B. JAKOBY

VOLUME 78. Interferons (Part A)
Edited by SIDNEY PESTKA

VOLUME 79. Interferons (Part B)
Edited by SIDNEY PESTKA

VOLUME 80. Proteolytic Enzymes (Part C)
Edited by LASZLO LORAND

VOLUME 81. Biomembranes (Part H: Visual Pigments and Purple Membranes, I)
Edited by LESTER PACKER

VOLUME 82. Structural and Contractile Proteins (Part A: Extracellular Matrix)
Edited by LEON W. CUNNINGHAM AND DIXIE W. FREDERIKSEN

VOLUME 83. Complex Carbohydrates (Part D)
Edited by VICTOR GINSBURG

VOLUME 84. Immunochemical Techniques (Part D: Selected Immunoassays)
Edited by JOHN J. LANGONE AND HELEN VAN VUNAKIS

VOLUME 85. Structural and Contractile Proteins (Part B: The Contractile Apparatus and the Cytoskeleton)
Edited by DIXIE W. FREDERIKSEN AND LEON W. CUNNINGHAM

VOLUME 86. Prostaglandins and Arachidonate Metabolites
Edited by WILLIAM E. M. LANDS AND WILLIAM L. SMITH

VOLUME 87. Enzyme Kinetics and Mechanism (Part C: Intermediates, Stereochemistry, and Rate Studies)
Edited by DANIEL L. PURICH

VOLUME 88. Biomembranes (Part I: Visual Pigments and Purple Membranes, II)
Edited by LESTER PACKER

VOLUME 89. Carbohydrate Metabolism (Part D)
Edited by WILLIS A. WOOD

VOLUME 90. Carbohydrate Metabolism (Part E)
Edited by WILLIS A. WOOD

VOLUME 91. Enzyme Structure (Part I)
Edited by C. H. W. HIRS AND SERGE N. TIMASHEFF

VOLUME 92. Immunochemical Techniques (Part E: Monoclonal Antibodies and General Immunoassay Methods)
Edited by JOHN J. LANGONE AND HELEN VAN VUNAKIS

VOLUME 93. Immunochemical Techniques (Part F: Conventional Antibodies, Fc Receptors, and Cytotoxicity)
Edited by JOHN J. LANGONE AND HELEN VAN VUNAKIS

VOLUME 94. Polyamines
Edited by HERBERT TABOR AND CELIA WHITE TABOR

VOLUME 95. Cumulative Subject Index Volumes 61–74, 76–80
Edited by EDWARD A. DENNIS AND MARTHA G. DENNIS

VOLUME 96. Biomembranes [Part J: Membrane Biogenesis: Assembly and Targeting (General Methods; Eukaryotes)]
Edited by SIDNEY FLEISCHER AND BECCA FLEISCHER

VOLUME 97. Biomembranes [Part K: Membrane Biogenesis: Assembly and Targeting (Prokaryotes, Mitochondria, and Chloroplasts)]
Edited by SIDNEY FLEISCHER AND BECCA FLEISCHER

VOLUME 98. Biomembranes (Part L: Membrane Biogenesis: Processing and Recycling)
Edited by SIDNEY FLEISCHER AND BECCA FLEISCHER

VOLUME 99. Hormone Action (Part F: Protein Kinases)
Edited by JACKIE D. CORBIN AND JOEL G. HARDMAN

VOLUME 100. Recombinant DNA (Part B)
Edited by RAY WU, LAWRENCE GROSSMAN, AND KIVIE MOLDAVE

VOLUME 101. Recombinant DNA (Part C)
Edited by RAY WU, LAWRENCE GROSSMAN, AND KIVIE MOLDAVE

VOLUME 102. Hormone Action (Part G: Calmodulin and Calcium-Binding Proteins)
Edited by ANTHONY R. MEANS AND BERT W. O'MALLEY

VOLUME 103. Hormone Action (Part H: Neuroendocrine Peptides)
Edited by P. MICHAEL CONN

VOLUME 104. Enzyme Purification and Related Techniques (Part C)
Edited by WILLIAM B. JAKOBY

VOLUME 105. Oxygen Radicals in Biological Systems
Edited by LESTER PACKER

VOLUME 106. Posttranslational Modifications (Part A)
Edited by FINN WOLD AND KIVIE MOLDAVE

VOLUME 107. Posttranslational Modifications (Part B)
Edited by FINN WOLD AND KIVIE MOLDAVE

VOLUME 108. Immunochemical Techniques (Part G: Separation and Characterization of Lymphoid Cells)
Edited by GIOVANNI DI SABATO, JOHN J. LANGONE, AND HELEN VAN VUNAKIS

VOLUME 109. Hormone Action (Part I: Peptide Hormones)
Edited by LUTZ BIRNBAUMER AND BERT W. O'MALLEY

VOLUME 110. Steroids and Isoprenoids (Part A)
Edited by JOHN H. LAW AND HANS C. RILLING

VOLUME 111. Steroids and Isoprenoids (Part B)
Edited by JOHN H. LAW AND HANS C. RILLING

VOLUME 112. Drug and Enzyme Targeting (Part A)
Edited by KENNETH J. WIDDER AND RALPH GREEN

VOLUME 113. Glutamate, Glutamine, Glutathione, and Related Compounds
Edited by ALTON MEISTER

VOLUME 114. Diffraction Methods for Biological Macromolecules (Part A)
Edited by HAROLD W. WYCKOFF, C. H. W. HIRS, AND SERGE N. TIMASHEFF

VOLUME 115. Diffraction Methods for Biological Macromolecules (Part B)
Edited by HAROLD W. WYCKOFF, C. H. W. HIRS, AND SERGE N. TIMASHEFF

VOLUME 116. Immunochemical Techniques (Part H: Effectors and Mediators of Lymphoid Cell Functions)
Edited by GIOVANNI DI SABATO, JOHN J. LANGONE, AND HELEN VAN VUNAKIS

VOLUME 117. Enzyme Structure (Part J)
Edited by C. H. W. HIRS AND SERGE N. TIMASHEFF

VOLUME 118. Plant Molecular Biology
Edited by ARTHUR WEISSBACH AND HERBERT WEISSBACH

VOLUME 119. Interferons (Part C)
Edited by SIDNEY PESTKA

VOLUME 120. Cumulative Subject Index Volumes 81–94, 96–101

VOLUME 121. Immunochemical Techniques (Part I: Hybridoma Technology and Monoclonal Antibodies)
Edited by JOHN J. LANGONE AND HELEN VAN VUNAKIS

VOLUME 122. Vitamins and Coenzymes (Part G)
Edited by FRANK CHYTIL AND DONALD B. MCCORMICK

VOLUME 123. Vitamins and Coenzymes (Part H)
Edited by FRANK CHYTIL AND DONALD B. MCCORMICK

VOLUME 124. Hormone Action (Part J: Neuroendocrine Peptides)
Edited by P. MICHAEL CONN

VOLUME 125. Biomembranes (Part M: Transport in Bacteria, Mitochondria, and Chloroplasts: General Approaches and Transport Systems)
Edited by SIDNEY FLEISCHER AND BECCA FLEISCHER

VOLUME 126. Biomembranes (Part N: Transport in Bacteria, Mitochondria, and Chloroplasts: Protonmotive Force)
Edited by SIDNEY FLEISCHER AND BECCA FLEISCHER

VOLUME 127. Biomembranes (Part O: Protons and Water: Structure and Translocation)
Edited by LESTER PACKER

VOLUME 128. Plasma Lipoproteins (Part A: Preparation, Structure, and Molecular Biology)
Edited by JERE P. SEGREST AND JOHN J. ALBERS

VOLUME 129. Plasma Lipoproteins (Part B: Characterization, Cell Biology, and Metabolism)
Edited by JOHN J. ALBERS AND JERE P. SEGREST

VOLUME 130. Enzyme Structure (Part K)
Edited by C. H. W. HIRS AND SERGE N. TIMASHEFF

VOLUME 131. Enzyme Structure (Part L)
Edited by C. H. W. HIRS AND SERGE N. TIMASHEFF

VOLUME 132. Immunochemical Techniques (Part J: Phagocytosis and Cell-Mediated Cytotoxicity)
Edited by GIOVANNI DI SABATO AND JOHANNES EVERSE

VOLUME 133. Bioluminescence and Chemiluminescence (Part B)
Edited by MARLENE DELUCA AND WILLIAM D. MCELROY

VOLUME 134. Structural and Contractile Proteins (Part C: The Contractile Apparatus and the Cytoskeleton)
Edited by RICHARD B. VALLEE

VOLUME 135. Immobilized Enzymes and Cells (Part B)
Edited by KLAUS MOSBACH

VOLUME 136. Immobilized Enzymes and Cells (Part C)
Edited by KLAUS MOSBACH

VOLUME 137. Immobilized Enzymes and Cells (Part D)
Edited by KLAUS MOSBACH

VOLUME 138. Complex Carbohydrates (Part E)
Edited by VICTOR GINSBURG

VOLUME 139. Cellular Regulators (Part A: Calcium- and Calmodulin-Binding Proteins)
Edited by ANTHONY R. MEANS AND P. MICHAEL CONN

VOLUME 140. Cumulative Subject Index Volumes 102–119, 121–134

VOLUME 141. Cellular Regulators (Part B: Calcium and Lipids)
Edited by P. MICHAEL CONN AND ANTHONY R. MEANS

VOLUME 142. Metabolism of Aromatic Amino Acids and Amines
Edited by SEYMOUR KAUFMAN

VOLUME 143. Sulfur and Sulfur Amino Acids
Edited by WILLIAM B. JAKOBY AND OWEN GRIFFITH

VOLUME 144. Structural and Contractile Proteins (Part D: Extracellular Matrix)
Edited by LEON W. CUNNINGHAM

VOLUME 145. Structural and Contractile Proteins (Part E: Extracellular Matrix)
Edited by LEON W. CUNNINGHAM

VOLUME 146. Peptide Growth Factors (Part A)
Edited by DAVID BARNES AND DAVID A. SIRBASKU

VOLUME 147. Peptide Growth Factors (Part B)
Edited by DAVID BARNES AND DAVID A. SIRBASKU

VOLUME 148. Plant Cell Membranes
Edited by LESTER PACKER AND ROLAND DOUCE

VOLUME 149. Drug and Enzyme Targeting (Part B)
Edited by RALPH GREEN AND KENNETH J. WIDDER

VOLUME 150. Immunochemical Techniques (Part K: *In Vitro* Models of B and T Cell Functions and Lymphoid Cell Receptors)
Edited by GIOVANNI DI SABATO

VOLUME 151. Molecular Genetics of Mammalian Cells
Edited by MICHAEL M. GOTTESMAN

VOLUME 152. Guide to Molecular Cloning Techniques
Edited by SHELBY L. BERGER AND ALAN R. KIMMEL

VOLUME 153. Recombinant DNA (Part D)
Edited by RAY WU AND LAWRENCE GROSSMAN

VOLUME 154. Recombinant DNA (Part E)
Edited by RAY WU AND LAWRENCE GROSSMAN

VOLUME 155. Recombinant DNA (Part F)
Edited by RAY WU

VOLUME 156. Biomembranes (Part P: ATP-Driven Pumps and Related Transport: The Na,K-Pump)
Edited by SIDNEY FLEISCHER AND BECCA FLEISCHER

VOLUME 157. Biomembranes (Part Q: ATP-Driven Pumps and Related Transport: Calcium, Proton, and Potassium Pumps)
Edited by SIDNEY FLEISCHER AND BECCA FLEISCHER

VOLUME 158. Metalloproteins (Part A)
Edited by JAMES F. RIORDAN AND BERT L. VALLEE

Volume 159. Initiation and Termination of Cyclic Nucleotide Action
Edited by Jackie D. Corbin and Roger A. Johnson

Volume 160. Biomass (Part A: Cellulose and Hemicellulose)
Edited by Willis A. Wood and Scott T. Kellogg

Volume 161. Biomass (Part B: Lignin, Pectin, and Chitin)
Edited by Willis A. Wood and Scott T. Kellogg

Volume 162. Immunochemical Techniques (Part L: Chemotaxis and Inflammation)
Edited by Giovanni Di Sabato

Volume 163. Immunochemical Techniques (Part M: Chemotaxis and Inflammation)
Edited by Giovanni Di Sabato

Volume 164. Ribosomes
Edited by Harry F. Noller, Jr., and Kivie Moldave

Volume 165. Microbial Toxins: Tools for Enzymology
Edited by Sidney Harshman

Volume 166. Branched-Chain Amino Acids
Edited by Robert Harris and John R. Sokatch

Volume 167. Cyanobacteria
Edited by Lester Packer and Alexander N. Glazer

Volume 168. Hormone Action (Part K: Neuroendocrine Peptides)
Edited by P. Michael Conn

Volume 169. Platelets: Receptors, Adhesion, Secretion (Part A)
Edited by Jacek Hawiger

Volume 170. Nucleosomes
Edited by Paul M. Wassarman and Roger D. Kornberg

Volume 171. Biomembranes (Part R: Transport Theory: Cells and Model Membranes)
Edited by Sidney Fleischer and Becca Fleischer

Volume 172. Biomembranes (Part S: Transport: Membrane Isolation and Characterization)
Edited by Sidney Fleischer and Becca Fleischer

Volume 173. Biomembranes [Part T: Cellular and Subcellular Transport: Eukaryotic (Nonepithelial) Cells]
Edited by Sidney Fleischer and Becca Fleischer

Volume 174. Biomembranes [Part U: Cellular and Subcellular Transport: Eukaryotic (Nonepithelial) Cells]
Edited by Sidney Fleischer and Becca Fleischer

Volume 175. Cumulative Subject Index Volumes 135–139, 141–167

VOLUME 176. Nuclear Magnetic Resonance (Part A: Spectral Techniques and Dynamics)
Edited by NORMAN J. OPPENHEIMER AND THOMAS L. JAMES

VOLUME 177. Nuclear Magnetic Resonance (Part B: Structure and Mechanism)
Edited by NORMAN J. OPPENHEIMER AND THOMAS L. JAMES

VOLUME 178. Antibodies, Antigens, and Molecular Mimicry
Edited by JOHN J. LANGONE

VOLUME 179. Complex Carbohydrates (Part F)
Edited by VICTOR GINSBURG

VOLUME 180. RNA Processing (Part A: General Methods)
Edited by JAMES E. DAHLBERG AND JOHN N. ABELSON

VOLUME 181. RNA Processing (Part B: Specific Methods)
Edited by JAMES E. DAHLBERG AND JOHN N. ABELSON

VOLUME 182. Guide to Protein Purification
Edited by MURRAY P. DEUTSCHER

VOLUME 183. Molecular Evolution: Computer Analysis of Protein and Nucleic Acid Sequences
Edited by RUSSELL F. DOOLITTLE

VOLUME 184. Avidin–Biotin Technology
Edited by MEIR WILCHEK AND EDWARD A. BAYER

VOLUME 185. Gene Expression Technology
Edited by DAVID V. GOEDDEL

VOLUME 186. Oxygen Radicals in Biological Systems (Part B: Oxygen Radicals and Antioxidants)
Edited by LESTER PACKER AND ALEXANDER N. GLAZER

VOLUME 187. Arachidonate Related Lipid Mediators
Edited by ROBERT C. MURPHY AND FRANK A. FITZPATRICK

VOLUME 188. Hydrocarbons and Methylotrophy
Edited by MARY E. LIDSTROM

VOLUME 189. Retinoids (Part A: Molecular and Metabolic Aspects)
Edited by LESTER PACKER

VOLUME 190. Retinoids (Part B: Cell Differentiation and Clinical Applications)
Edited by LESTER PACKER

VOLUME 191. Biomembranes (Part V: Cellular and Subcellular Transport: Epithelial Cells)
Edited by SIDNEY FLEISCHER AND BECCA FLEISCHER

VOLUME 192. Biomembranes (Part W: Cellular and Subcellular Transport: Epithelial Cells)
Edited by SIDNEY FLEISCHER AND BECCA FLEISCHER

VOLUME 193. Mass Spectrometry
Edited by JAMES A. MCCLOSKEY

VOLUME 194. Guide to Yeast Genetics and Molecular Biology
Edited by CHRISTINE GUTHRIE AND GERALD R. FINK

VOLUME 195. Adenylyl Cyclase, G Proteins, and Guanylyl Cyclase
Edited by ROGER A. JOHNSON AND JACKIE D. CORBIN

VOLUME 196. Molecular Motors and the Cytoskeleton
Edited by RICHARD B. VALLEE

VOLUME 197. Phospholipases
Edited by EDWARD A. DENNIS

VOLUME 198. Peptide Growth Factors (Part C)
Edited by DAVID BARNES, J. P. MATHER, AND GORDON H. SATO

VOLUME 199. Cumulative Subject Index Volumes 168–174, 176–194

VOLUME 200. Protein Phosphorylation (Part A: Protein Kinases: Assays, Purification, Antibodies, Functional Analysis, Cloning, and Expression)
Edited by TONY HUNTER AND BARTHOLOMEW M. SEFTON

VOLUME 201. Protein Phosphorylation (Part B: Analysis of Protein Phosphorylation, Protein Kinase Inhibitors, and Protein Phosphatases)
Edited by TONY HUNTER AND BARTHOLOMEW M. SEFTON

VOLUME 202. Molecular Design and Modeling: Concepts and Applications (Part A: Proteins, Peptides, and Enzymes)
Edited by JOHN J. LANGONE

VOLUME 203. Molecular Design and Modeling: Concepts and Applications (Part B: Antibodies and Antigens, Nucleic Acids, Polysaccharides, and Drugs)
Edited by JOHN J. LANGONE

VOLUME 204. Bacterial Genetic Systems
Edited by JEFFREY H. MILLER

VOLUME 205. Metallobiochemistry (Part B: Metallothionein and Related Molecules)
Edited by JAMES F. RIORDAN AND BERT L. VALLEE

VOLUME 206. Cytochrome P450
Edited by MICHAEL R. WATERMAN AND ERIC F. JOHNSON

VOLUME 207. Ion Channels
Edited by BERNARDO RUDY AND LINDA E. IVERSON

VOLUME 208. Protein–DNA Interactions
Edited by ROBERT T. SAUER

VOLUME 209. Phospholipid Biosynthesis
Edited by EDWARD A. DENNIS AND DENNIS E. VANCE

VOLUME 210. Numerical Computer Methods
Edited by LUDWIG BRAND AND MICHAEL L. JOHNSON

VOLUME 211. DNA Structures (Part A: Synthesis and Physical Analysis of DNA)
Edited by DAVID M. J. LILLEY AND JAMES E. DAHLBERG

VOLUME 212. DNA Structures (Part B: Chemical and Electrophoretic Analysis of DNA)
Edited by DAVID M. J. LILLEY AND JAMES E. DAHLBERG

VOLUME 213. Carotenoids (Part A: Chemistry, Separation, Quantitation, and Antioxidation)
Edited by LESTER PACKER

VOLUME 214. Carotenoids (Part B: Metabolism, Genetics, and Biosynthesis)
Edited by LESTER PACKER

VOLUME 215. Platelets: Receptors, Adhesion, Secretion (Part B)
Edited by JACEK J. HAWIGER

VOLUME 216. Recombinant DNA (Part G)
Edited by RAY WU

VOLUME 217. Recombinant DNA (Part H)
Edited by RAY WU

VOLUME 218. Recombinant DNA (Part I)
Edited by RAY WU

VOLUME 219. Reconstitution of Intracellular Transport
Edited by JAMES E. ROTHMAN

VOLUME 220. Membrane Fusion Techniques (Part A)
Edited by NEJAT DÜZGÜNEŞ

VOLUME 221. Membrane Fusion Techniques (Part B)
Edited by NEJAT DÜZGÜNEŞ

VOLUME 222. Proteolytic Enzymes in Coagulation, Fibrinolysis, and Complement Activation (Part A: Mammalian Blood Coagulation Factors and Inhibitors)
Edited by LASZLO LORAND AND KENNETH G. MANN

VOLUME 223. Proteolytic Enzymes in Coagulation, Fibrinolysis, and Complement Activation (Part B: Complement Activation, Fibrinolysis, and Nonmammalian Blood Coagulation Factors)
Edited by LASZLO LORAND AND KENNETH G. MANN

VOLUME 224. Molecular Evolution: Producing the Biochemical Data
Edited by ELIZABETH ANNE ZIMMER, THOMAS J. WHITE, REBECCA L. CANN, AND ALLAN C. WILSON

VOLUME 225. Guide to Techniques in Mouse Development
Edited by PAUL M. WASSARMAN AND MELVIN L. DEPAMPHILIS

VOLUME 226. Metallobiochemistry (Part C: Spectroscopic and Physical Methods for Probing Metal Ion Environments in Metalloenzymes and Metalloproteins)
Edited by JAMES F. RIORDAN AND BERT L. VALLEE

VOLUME 227. Metallobiochemistry (Part D: Physical and Spectroscopic Methods for Probing Metal Ion Environments in Metalloproteins)
Edited by JAMES F. RIORDAN AND BERT L. VALLEE

VOLUME 228. Aqueous Two-Phase Systems
Edited by HARRY WALTER AND GÖTE JOHANSSON

VOLUME 229. Cumulative Subject Index Volumes 195–198, 200–227

VOLUME 230. Guide to Techniques in Glycobiology
Edited by WILLIAM J. LENNARZ AND GERALD W. HART

VOLUME 231. Hemoglobins (Part B: Biochemical and Analytical Methods)
Edited by JOHANNES EVERSE, KIM D. VANDEGRIFF, AND ROBERT M. WINSLOW

VOLUME 232. Hemoglobins (Part C: Biophysical Methods)
Edited by JOHANNES EVERSE, KIM D. VANDEGRIFF, AND ROBERT M. WINSLOW

VOLUME 233. Oxygen Radicals in Biological Systems (Part C)
Edited by LESTER PACKER

VOLUME 234. Oxygen Radicals in Biological Systems (Part D)
Edited by LESTER PACKER

VOLUME 235. Bacterial Pathogenesis (Part A: Identification and Regulation of Virulence Factors)
Edited by VIRGINIA L. CLARK AND PATRIK M. BAVOIL

VOLUME 236. Bacterial Pathogenesis (Part B: Integration of Pathogenic Bacteria with Host Cells)
Edited by VIRGINIA L. CLARK AND PATRIK M. BAVOIL

VOLUME 237. Heterotrimeric G Proteins
Edited by RAVI IYENGAR

VOLUME 238. Heterotrimeric G-Protein Effectors
Edited by RAVI IYENGAR

VOLUME 239. Nuclear Magnetic Resonance (Part C)
Edited by THOMAS L. JAMES AND NORMAN J. OPPENHEIMER

VOLUME 240. Numerical Computer Methods (Part B)
Edited by MICHAEL L. JOHNSON AND LUDWIG BRAND

VOLUME 241. Retroviral Proteases
Edited by LAWRENCE C. KUO AND JULES A. SHAFER

VOLUME 242. Neoglycoconjugates (Part A)
Edited by Y. C. LEE AND REIKO T. LEE

VOLUME 243. Inorganic Microbial Sulfur Metabolism
Edited by HARRY D. PECK, JR., AND JEAN LEGALL

VOLUME 244. Proteolytic Enzymes: Serine and Cysteine Peptidases
Edited by ALAN J. BARRETT

VOLUME 245. Extracellular Matrix Components
Edited by E. RUOSLAHTI AND E. ENGVALL

VOLUME 246. Biochemical Spectroscopy
Edited by KENNETH SAUER

VOLUME 247. Neoglycoconjugates (Part B: Biomedical Applications)
Edited by Y. C. LEE AND REIKO T. LEE

VOLUME 248. Proteolytic Enzymes: Aspartic and Metallo Peptidases
Edited by ALAN J. BARRETT

VOLUME 249. Enzyme Kinetics and Mechanism (Part D: Developments in Enzyme Dynamics)
Edited by DANIEL L. PURICH

VOLUME 250. Lipid Modifications of Proteins
Edited by PATRICK J. CASEY AND JANICE E. BUSS

VOLUME 251. Biothiols (Part A: Monothiols and Dithiols, Protein Thiols, and Thiyl Radicals)
Edited by LESTER PACKER

VOLUME 252. Biothiols (Part B: Glutathione and Thioredoxin; Thiols in Signal Transduction and Gene Regulation)
Edited by LESTER PACKER

VOLUME 253. Adhesion of Microbial Pathogens
Edited by RON J. DOYLE AND ITZHAK OFEK

VOLUME 254. Oncogene Techniques
Edited by PETER K. VOGT AND INDER M. VERMA

VOLUME 255. Small GTPases and Their Regulators (Part A: Ras Family)
Edited by W. E. BALCH, CHANNING J. DER, AND ALAN HALL

VOLUME 256. Small GTPases and Their Regulators (Part B: Rho Family)
Edited by W. E. BALCH, CHANNING J. DER, AND ALAN HALL

VOLUME 257. Small GTPases and Their Regulators (Part C: Proteins Involved in Transport)
Edited by W. E. BALCH, CHANNING J. DER, AND ALAN HALL

VOLUME 258. Redox-Active Amino Acids in Biology
Edited by JUDITH P. KLINMAN

VOLUME 259. Energetics of Biological Macromolecules
Edited by MICHAEL L. JOHNSON AND GARY K. ACKERS

VOLUME 260. Mitochondrial Biogenesis and Genetics (Part A)
Edited by GIUSEPPE M. ATTARDI AND ANNE CHOMYN

VOLUME 261. Nuclear Magnetic Resonance and Nucleic Acids
Edited by THOMAS L. JAMES

VOLUME 262. DNA Replication
Edited by JUDITH L. CAMPBELL

VOLUME 263. Plasma Lipoproteins (Part C: Quantitation)
Edited by WILLIAM A. BRADLEY, SANDRA H. GIANTURCO, AND JERE P. SEGREST

VOLUME 264. Mitochondrial Biogenesis and Genetics (Part B)
Edited by GIUSEPPE M. ATTARDI AND ANNE CHOMYN

VOLUME 265. Cumulative Subject Index Volumes 228, 230–262

VOLUME 266. Computer Methods for Macromolecular Sequence Analysis
Edited by RUSSELL F. DOOLITTLE

VOLUME 267. Combinatorial Chemistry
Edited by JOHN N. ABELSON

VOLUME 268. Nitric Oxide (Part A: Sources and Detection of NO; NO Synthase)
Edited by LESTER PACKER

VOLUME 269. Nitric Oxide (Part B: Physiological and Pathological Processes)
Edited by LESTER PACKER

VOLUME 270. High Resolution Separation and Analysis of Biological Macromolecules (Part A: Fundamentals)
Edited by BARRY L. KARGER AND WILLIAM S. HANCOCK

VOLUME 271. High Resolution Separation and Analysis of Biological Macromolecules (Part B: Applications)
Edited by BARRY L. KARGER AND WILLIAM S. HANCOCK

VOLUME 272. Cytochrome P450 (Part B)
Edited by ERIC F. JOHNSON AND MICHAEL R. WATERMAN

VOLUME 273. RNA Polymerase and Associated Factors (Part A)
Edited by SANKAR ADHYA

VOLUME 274. RNA Polymerase and Associated Factors (Part B)
Edited by SANKAR ADHYA

VOLUME 275. Viral Polymerases and Related Proteins
Edited by LAWRENCE C. KUO, DAVID B. OLSEN, AND STEVEN S. CARROLL

VOLUME 276. Macromolecular Crystallography (Part A)
Edited by CHARLES W. CARTER, JR., AND ROBERT M. SWEET

VOLUME 277. Macromolecular Crystallography (Part B)
Edited by CHARLES W. CARTER, JR., AND ROBERT M. SWEET

VOLUME 278. Fluorescence Spectroscopy
Edited by LUDWIG BRAND AND MICHAEL L. JOHNSON

VOLUME 279. Vitamins and Coenzymes, Part I
Edited by DONALD B. MCCORMICK, JOHN W. SUTTIE, AND CONRAD WAGNER

VOLUME 280. Vitamins and Coenzymes, Part J
Edited by DONALD B. MCCORMICK, JOHN W. SUTTIE, AND CONRAD WAGNER

VOLUME 281. Vitamins and Coenzymes, Part K
Edited by DONALD B. MCCORMICK, JOHN W. SUTTIE, AND CONRAD WAGNER

VOLUME 282. Vitamins and Coenzymes, Part L
Edited by DONALD B. MCCORMICK, JOHN W. SUTTIE, AND CONRAD WAGNER

VOLUME 283. Cell Cycle Control
Edited by WILLIAM G. DUNPHY

VOLUME 284. Lipases (Part A: Biotechnology)
Edited by BYRON RUBIN AND EDWARD A. DENNIS

VOLUME 285. Cumulative Subject Index Volumes 263, 264, 266–289 (in preparation)

VOLUME 286. Lipases (Part B: Enzyme Characterization and Utilization)
Edited by BYRON RUBIN AND EDWARD A. DENNIS

VOLUME 287. Chemokines
Edited by RICHARD HORUK

VOLUME 288. Chemokine Receptors
Edited by RICHARD HORUK

VOLUME 289. Solid Phase Peptide Synthesis
Edited by GREGG B. FIELDS

VOLUME 290. Molecular Chaperones (in preparation)
Edited by GEORGE H. LORIMER AND THOMAS O. BALDWIN

Section I

Chemokine Receptors

[1] C-X-C Chemokine Receptor Desensitization Mediated through Ligand-Enhanced Receptor Phosphorylation on Serine Residues

By ANN RICHMOND, SUSAN MUELLER, JOHN R. WHITE, and WAYNE SCHRAW

Introduction

The binding of ligand to most seven-transmembrane (STM) G-protein-coupled receptors leads to changes in the coupling of G proteins to the receptor.[1,2] A number of downstream signal transduction pathways become activated, including calcium mobilization,[3,4] phospholipase C,[5] phosphatidylinositol (PI) 3-kinase,[6] mitogen-activated protein (MAP) kinase,[7–10] and serine/threonine and tyrosine kinases.[9,11–15] In addition to these positive signaling events, other kinases are activated that play a role in downregulating the receptor-mediated response to the ligand.[16,17] In particular, the

[1] A. Levitzki, *Science* **241,** 800 (1988).
[2] S. S. G. Ferguson, L. Menard, L. S. Barak, W. J. Koch, A. M. Colapietro, and M. G. Caron, *J. Biol. Chem.* **270,** 24782 (1995).
[3] M. Baggiolini, B. Dewald, and B. Moser, *Adv. Immunol.* **55,** 97 (1994).
[4] T. Geiser, B. Dewald, M. U. Ehrengruber, I. Clark-Lewis, and M. Baggiolini, *J. Biol. Chem.* **268,** 15419 (1993).
[5] M. C. Galas and T. K. Harden, *Eur. J. Pharmacol.: Mol. Pharmacol.* **291,** 175 (1995).
[6] J. Ding, C. J. Vlahos, R. Liu, R. F. Brown, and J. A. Badwey, *J. Biol. Chem.* **270,** 11684 (1995).
[7] P. Crespo, T. G. Cachero, N. Z. Xu, and J. S. Gutkind, *J. Biol. Chem.* **270,** 25259 (1995).
[8] T. Van Biesen, B. E. Hawes, J. R. Raymond, L. M. Luttrell, W. J. Koch, and R. J. Lefkowitz, *J. Biol. Chem.* **271,** 1266 (1996).
[9] S. A. Jones, B. Moser, and M. Thelen, *FEBS Lett.* **364,** 211 (1995).
[10] C. Knall, S. Young, J. A. Nick, A. M. Buhl, G. S. Worthen, and G. L. Johnson, *J. Biol. Chem.* **271,** 2832 (1996).
[11] S. G. Mueller, W. P. Schraw, and A. Richmond, *J. Biol. Chem.* **269,** 1973 (1994).
[12] S. G. Mueller, W. P. Schraw, and A. Richmond, *J. Biol. Chem.* **270,** 10439 (1995).
[13] L. M. Luttrell, B. E. Hawes, T. van Biesen, D. K. Luttress, T. J. Lansing, and R. J. Lefkowitz, *J. Biol. Chem.* **271,** 19443 (1996).
[14] Q. C. Cheng, J. H. Han, H. G. Thomas, E. Balentien, and A. Richmond, *J. Immunol.* **148,** 451 (1992).
[15] Y.-H. Chen, J. Pouyssegur, S. A. Courtneidge, and E. Van Obberghen-Schilling, *J. Biol. Chem.* **269,** 27372 (1994).
[16] M. Schondorf, F. Bidlingmaier, and A. A. von Rueker, *Biochem. Biophys. Res. Commun.* **197,** 549 (1993).
[17] S. Sozzani, M. Molino, M. Locati, W. Luini, C. Cerletti, A. Vecchi, and A. Mantovani, *J. Immunol.* **150,** 1544 (1993).

0076-6879/97 $25.00

release of free $\beta\gamma$ subunits from the tripartite G protein, subsequent to ligand binding, is associated with the activation of a kinase that phosphorylates the receptor, and this event is thought to facilitate sequestration and downregulation of the receptor.[2,18–23] A number of these G-protein-activated receptor kinases (GRKs) have been characterized and shown to phosphorylate various STM receptors.[24] In general, receptor phosphorylation by GRKs occurs along the carboxyl tail of the receptors.[18,24,25] For some of the STM receptors, both serine and threonine residues become phosphorylated with ligand binding.[26] After the receptor is phosphorylated on certain residues, it loses its ability to respond to a second ligand binding event. This process is known as desensitization.[18,24,25] To determine how the binding of a C-X-C chemokine affects the CXCR2 receptor, we used assays to follow the phosphorylation of CXCR2 in response to ligand binding. After demonstrating that ligand binding to CXCR2 results in the phosphorylation of CXCR2 on serine residues, we developed a series of CXCR2 mutants with serine to alanine substitutions, or truncations at specific serine residues along the carboxyl tail of the receptor, to further characterize the role of individual serine residues in signaling by the CXCR2 receptor.

The wild-type CXCR2 cDNA and various receptor mutants are transfected into fibroblast or other cell types that normally do not express C-X-C chemokine receptors. Stable clones of the transfected cells that express the wild-type or mutant receptors are generated, and receptor phosphorylation, desensitization, and degradation are monitored after ligand activation of the receptor. Comparisons are made between ligand activation of receptor phosphorylation and phorbol ester activation through induction of protein kinase C activity. Receptor desensitization is monitored

[18] J. Pitcher, M. J. Lohse, J. Codina, M. G. Caron, and R. J. Lefkowitz, *Biochemistry* **31,** 3193 (1992).

[19] G. Pei, P. Samama, M. Lohse, M. Wang, J. Codina, and R. J. Lefkowitz, *Proc. Natl. Acad. Sci. U.S.A.* **91,** 2699 (1994).

[20] R. M. Richardson, H. Ali, E. D. Tomhave, B. Haribabu, and R. Snyderman, *J. Biol. Chem.* **270,** 27829 (1995).

[21] V. V. Gurevich, S. B. Dion, J. J. Onorato, J. Ptasienski, C. M. Kim, R. Sterne-Marr, M. M. Hosey, and J. L. Benovic, *J. Biol. Chem.* **270,** 720 (1995).

[22] S. S. G. Ferguson, W. E. Downey III, A.-M. Colapietro, L. S. Barak, L. Menard, and M. G. Caron, *Science* **271,** 363 (1996).

[23] M. Oppermann, N. J. Freedman, W. Alexander, and R. J. Lefkowitz, *J. Biol. Chem.* **271,** 13266 (1996).

[24] R. J. Lefkowitz, *Cell (Cambridge, Mass.)* **74,** 409 (1993).

[25] J. A. Pitcher, J. Inglese, J. B. Higgins, J. L. Arriza, P. J. Casey, C. Kim, J. L. Benovic, M. M. Kwatra, M. G. Caron, and R. J. Lefkowitz, *Science* **257,** 1264 (1992).

[26] R. M. Richardson, R. A. DuBose, H. Ali, E. D. Tomhave, D. Haribabu, and R. Synderman, *Biochemistry* **34,** 14193 (1995).

by following the release of intracellular free calcium using Fura-2 following ligand stimulation, or ligand restimulation. For these studies, multiple clones of each mutant are examined, and the ligand binding for each clone is characterized by Scatchard analysis.

Methods

Generation of Truncated CXCR2 Mutants

Polymerase chain reaction (PCR) strategies are employed to generate truncated CXCR2 mutants such that stop codons are introduced at Ser-331, Ser-342, or Ser-352. Amplification by PCR is conducted on the cDNA encoding the entire open reading frame for the CXCR2, which has been subcloned into BlueScript (Stratagene, La Jolla, CA). The primer pair for each reaction includes a common primer for the 5′ end of the open reading frame. Unique primers that would introduce the desired stop codons are used for the 3′ end and are as follows: Ser-331 (331T), gcgaagcttttagatcaagccatgtatagc; Ser-342 (342T), gcgaagcttttaaggcctgctgtctttggg; and Ser-352 (352T), gcgaagcttttaagtgtgccctgaagaaga. These primers also include a *Hin*dIII restriction site for future subcloning strategies. The PCR is conducted using 10 ng of the CXCR2/BlueScript plasmid as the template. AmpliTaq (Perkin-Elmer, Norwalk, CT) serves as the polymerase. Typically 30 cycles are completed using an annealing temperature of 48°. The generated PCR fragments are isolated, subcloned into BlueScript, and then sequenced to ensure that the stop codons have been introduced and that the PCR has not generated additional mutations within the fragment. Once the sequences are confirmed, the cDNAs for the truncated receptors are subcloned into the mammalian expression vector pRc/CMV (Invitrogen, San Diego, CA) and subsequently transfected into the cell line of choice, and stable G418-resistant clones are selected.

Site-Directed Mutagenesis of CXCR2

Mutagenesis of specific serine to alanine residues is conducted using the pALTER site-directed mutagenesis system (Promega, Madison, WI). The cDNA for the wild-type CXCR2 is subcloned into the pALTER-1 plasmid. Single-stranded DNA is isolated for the site-directed mutagenesis. The following mutations are synthesized using the indicated primers: Ser-342-Ala (S342A), agacagcaggccggcctttgttggc; Ser-346-Ala (S346A), tcctttgttggcgcctcttcagggcac; Ser-347-Ala (S347A), ctttgttggctcagcttcagggcaca; Ser-348-Ala (S348A), gttggctcttctgccgggcacacttcc; Ser-346/7/8-Ala (3A), tcctttgttggcgccgctgcagggcacactt. The primers used for the mutagenesis are designed to include novel restriction sites, and thus receptor mutants are initially screened by restriction analysis and subsequently sequenced to ensure that the desired mutation has been incorporated into the cDNA sequence. A Ser-342/6/7/8-Ala mutant is generated by using the Ser-346/7/8-Ala mutated sequence as the template and subjecting it to another round of mutagenesis with the Ser-342-Ala primer. Once the mutations are confirmed, the cDNAs encoding the open reading frame for the CXCR2 mutants are subcloned into the pRc/CMV expression vector and transfected into the cell line selected for study; again, stable G418-resistant clones selected.

Generation of Stable Transfectants

The transfected cell line [human placental cell line 3ASubE (ATTC, Rockville, MD), human 293 fibroblasts, or Chinese hamster ovarian fibroblasts (CHO cells)] is maintained in 5% fetal bovine serum/Dulbecco's modified Eagle's medium (FBS/DMEM; GIBCO, Grand Island, NY) at 37° with 5% (v/v) CO_2. Cells (30–40% confluence) are transfected by the $Ca_3(PO_4)_2$ precipitation method with 20 μg of the plasmid construct. Following glycerol shock, stable transfectants are selected in the presence of 400 μg/ml G418. Clonally selected stable transfectants are generated for all receptor mutants. The stable transfectants are analyzed for receptor expression either by Western blot analysis or indirect immunohistochemical staining, then subsequently for [^{125}I]-MGSA (melanoma growth stimulating activity) binding activity using our standard binding assay.[11] Multiple clones are generated for each mutation and analyzed for binding, phosphorylation, and calcium flux.

Radioiodinated Melanoma Growth Stimulating Activity Binding Assay

The chemokine MGSA (1 μg) can be iodinated using the chloramine-T method yielding a specific activity of approximately 100 μCi/μg. Alternatively, [^{125}I]-MGSA is available from Dupont–New England Nuclear (Boston, MA) at a specific activity of 272 μCi/μg. Stable transfectants are plated at a density of 10^5 cells per well in 24-well plates (Costar, Cambridge, MA). The binding assay is conducted at 4°, 48 hr after the initial plating as previously described.[11] The [^{125}I]-MGSA reagent (15,000–30,000 cpm/well) is diluted in binding buffer (0.1% ovalbumin/DMEM), in the absence or presence of unlabeled MGSA (1–256 ng/ml). After 200 μl of diluted [^{125}I]-MGSA is added per well, the plates are rocked at 4° for 4 hr. Wells are washed on ice three times with 1.0 ml ice-cold binding buffer. The bound [^{125}I]-MGSA is eluted with two washes of 0.1 *N* NaOH/1% sodium dodecyl sulfate (SDS) and counted in a gamma counter (Beckman, Columbia, MD, Gamma 5500). This assay should be repeated in triplicate for each of the representative clones.

In Vivo Phosphorylation Assay

To characterize the phosphorylation of the CXCR2 receptor in response to ligand binding, the *in vivo* phosphorylation assay can be performed on cultured cells expressing the transfected wild-type or mutant receptors.[11] Confluent cell cultures (35-mm plates) are washed twice in 2 ml of serum-free DMEM and incubated in the latter for 40 hr at 37°. The medium is replaced with 2 ml phosphate-free minimal essential medium (MEM; Life Technologies, Gaithersburg, MD), and the cells are incubated for 3 hr more at 37°. After phosphate starvation, cells are incubated in 1 ml phosphate-free MEM containing 250 μCi/ml ortho[^{32}P]phosphate (9000 Ci/mmol) (Amersham, Arlington Heights, IL) for 3 hr at 37°. MGSA (5 n*M*), 12-*O*-tetradecanoylphorbol 13-acetate (TPA, 400 n*M*), or the appropriate vehicle is added directly to the ortho[^{32}P]phosphate-containing medium, and cells are incubated for 10 min at 37°. After this incubation, the ortho[^{32}P]phosphate-containing medium is removed, plates are put on ice, and 1 ml of Triton X-100 lysis buffer [10 m*M* Tris, 150 m*M* NaCl, 1 m*M* EDTA, 1 m*M* phenylmethylsulfonyl fluoride (PMSF), 10 μg/ml aprotinin, 10 μg/ml leupeptin, and 1% (w/v) Triton X-100] is added directly to the plate. Cells are scraped, and the lysate is transferred to a microtube and centrifuged at 15,000 *g* at 4° in a microfuge for 15 min. The clarified lysates are transferred to a fresh tube, trichloroacetic acid (TCA)-precipitable counts are determined, and protein concentration is estimated (BCA, Pierce, Rockford, IL).

The CXCR2 is typically immunoprecipitated from 5×10^6 TCA-precipitable counts or 10–25 μg protein/ml buffer. Lysates are incubated with 5 μg of affinity-purified anti-NH_2-terminal peptide antibodies specific for the CXCR2, at 4° with rocking for 2 hr, followed by precipitation with 30 μl of a 1:1 dilution of protein A/G agarose (Pierce) for 1 hr at 4°. The immunoprecipitates are washed four times with 1 ml ice-cold lysis buffer. Washed pellets are denatured in 40 μl of 2× Laemmli sample buffer containing 0.5% SDS and 10% (v/v) 2-mercaptoethanol. Samples are electrophoresed through a 9% SDS–polyacrylamide gel and either dried and analyzed by autoradiography or transblotted onto a polyvinylidene difluoride (PVDF) membrane (Bio-Rad, Richmond, CA) and subjected to autoradiography and Western blot analysis.

Western Blot Analysis

The expression and phosphorylation of the CXCR2 receptor or mutant receptors can be followed by Western analysis of whole cell lysates according to our established procedures.[11] The PVDF membranes (Bio-Rad) are blocked for 1 hr at room temperature in 5% (w/v) milk powder in

Tris-buffered saline (TBS; 10 mM Tris-HCl, pH 7.5, 150 mM NaCl), then subsequently incubated overnight with the above-mentioned affinity purified anti-NH_2-terminal peptide antibodies (2 μg/ml) in 5% milk powder/TBS. Blots are rinsed with four 5-min washes in TBS, then incubated in the presence of goat anti-rabbit immunoglobulin G (IgG) whole molecule conjugated with alkaline phosphatase (Sigma, St. Louis, MO) at a 1:2000 dilution in 5% milk powder/TBS. Blots are washed as described above and developed at pH 9.5 using bromochloroindolyl phosphate (Sigma) and nitro blue tetrazolium (Sigma).[12]

Receptor Degradation Studies

Confluent cultures (60-mm dishes) of the transfectants are placed in serum-free DMEM and treated with MGSA (50 nM), TPA (400 nM), or the appropriate vehicle control for 2 hr at 37°. Plates are rinsed on ice with TBS, then scraped in 300 μl of the above-mentioned Triton X-100 lysis buffer. The cell lysates are clarified by centrifugation at 4° for 15 min, the supernatants are transferred to a fresh tube, and protein estimates are performed (BCA, Pierce). Twenty-five micrograms of protein are loaded per lane and electrophoresed through a 9% SDS–polyacrylamide gel, transblotted onto a nitrocellulose membrane, and then analyzed as described above.[11]

[^{35}S]Methionine/Cysteine Labeling of 3ASubE P-3 Cells

Cells expressing receptor and cells expressing only the neomycin selection marker are grown to confluence in 35-mm plates. Cells are rinsed twice with phosphate-buffered saline and then incubated with cysteine- and methionine-free MEM (Life Technologies, Gaithersburg, MD) for 1 hr at 37°, after which the culture medium is replaced with cysteine/methionine-free MEM containing 100 μCi/ml [^{35}S]cysteine/methionine (Tran^{35}S-label, 1000 Ci/mmol; ICN, Costa Mesa, CA). Cells are labeled for 6 hr at 37°. Cells are rinsed, and fresh medium containing unlabeled cysteine/methionine is added to the cells. Cells are either untreated or treated with MGSA (50 nM) or TPA (400 nM) for 2 hr. Triton X-100 extracts are prepared, and the CXCR2 is immunoprecipitated from an equal number of TCA-precipitable counts (2×10^7 cpm), electrophoresed through a 9% SDS–polyacrylamide gel, dried, and exposed to autoradiographic film.

Calcium Fluorimetry

Transfected cells expressing truncation and Ser/Ala mutants are grown until confluent. Cells are released by a short exposure (1–2 min) to Versine (trypsin–EDTA) and washed once in culture medium containing 5% fetal calf serum (FCS). Cells are then washed a second time in Krebs–Ringer solution (118 mM NaCl, 4.56 mM KCl, 25 mM $NaHCO_3$, 1.03 mM KH_2PO_4, 11.1 mM glucose, and 5 mM HEPES) minus Ca^{2+} and Mg^{2+}. Cells are resuspended at 2×10^6 cells/ml and incubated with Fura-2 for 30 min (2 μM final concentration) at 37°. After 30 min the volume of buffer is doubled with the Krebs–Ringer (minus Ca^{2+} and Mg^{2+}) solution, and the cells are incubated for 10 min at 37°. Cells are then centrifuged (300 g, 6 min) and washed once (50 ml) in Krebs–Ringer solution containing Ca^{2+} and Mg^{2+} (1 mM). The cells are finally adjusted to 1×10^6 cells/ml. Calcium mobilization experiments are performed using a single scanning spectrofluorometer. Data are collected using an IBM Model PS-II computer and analyzed using the software program, Igor, which uses the following equation to determine free Ca^{2+}:

$$Ca^{2+}\ (\mathrm{n}M) = 224(F - F_{\mathrm{min}})/(F_{\mathrm{max}} - F)$$

where F_{max} is the maximum fluorescence (in the presence of 1 mM free Ca^{2+}) and F_{min} is the minimum fluorescence in the presence of EGTA. The constant, 224, is the dissociation (K_d) constant between Fura-2 and Ca^{2+}.

Generally, cells (2 ml) are allowed to reach 37° for 5 min prior to stimulation with C-X-C chemokine at the indicated concentration. The fluorescence is monitored continually for the specified time. In the cross-desensitization experiments, cells (6 ml total, 1×10^6 cells/ml, 37°) are stimulated with 20 nM chemokine (stimulated) or buffer (control) for 5 min before being washed three times (15 ml) with Krebs–Ringer buffer plus Ca^{2+} and Mg^{2+}. Cells are finally resuspended at (1×10^6 cells/ml) and kept on ice until needed. Cells are allowed to warm to 37° for 5 min before the second stimulus of chemokine (5 nM) as previously described.

Phosphoamino Acid Analysis

Phosphoamino acid analysis is performed as described by Boyle *et al.*[27] Briefly, the receptor is immunoprecipitated from ortho[^{32}P]phosphate-labeled 3ASubE P-3 cells, which have been treated with MGSA or vehicle, and the precipitate is electrophoresed through a 7.5% SDS–polyacrylamide gel, transferred to an Immobilon membrane (Millipore, Bedford, MA), and subjected to autoradiography. After autoradiography, the bands corresponding to the CXCR2 are cut from the membrane, incubated in the presence of 6 M HCl for 1 hr at 110°, lyophilized, and then electrophoresed through a cellulose thin-layer plate in two dimensions. The first dimension is in pH 1.9 buffer, and the second dimension is in pH 3.5 buffer. Phosphoserine, phosphothreonine, and phosphotyrosine standards are included in

[27] W. J. Boyle, P. van der Geer, and T. Hunter, *Methods Enzymol.* **201,** 110 (1991).

the loading buffer of the samples. After electrophoresis, the plates are stained with ninhydrin to detect the position of the phosphoamino acid standards and then subjected to autoradiography.

GTPγ³⁵S Binding Assay

Membranes for the GTPγ^{35}S binding assay are prepared using a modified version of a protocol for the isolation of fibroblast membranes.[28] The GTPγ^{35}S binding assay is performed as described by Gierschik *et al.*[29] Typically, 2 μg of crude membranes is used per reaction. The reaction is initiated by the addition of membranes to the reactive mixture, and the reaction is allowed to proceed at 37° for 10 min, or as indicated. Nonspecific binding (<15% of total binding) is estimated in the presence of 100 μM GTP. Reactions are terminated by the addition of 0.9 ml of ice-cold 50 mM Tris-HCl, pH 7.4, 5 mM $MgCl_2$, and 1 mM EDTA (wash buffer) to each tube. The contents are transferred to a vacuum-filtration apparatus containing a 0.45-μm Gelman (Ann Arbor, MI) A/E fiberglass filter. After filtration, the filter is washed three times with 3 ml of ice-cold wash buffer and dried. The amount of GTPγ^{35}S bound is determined by scintillation counting (Beckman, LS 3801).

Chemotaxis

Although the 3ASubE placental cells are quite efficient for monitoring signal transduction in response to MGSA binding to the CXCR2, these cells do not readily undergo chemotaxis in response to gradients of chemokines. Therefore, we have transfected the wild-type and mutant receptors into the human 293 fibroblast cell line, selected stable polyclonal cell lines expressing this receptor, confirmed that the polyclonal lines specifically bind the MGSA/growth regulated protein (GRO) ligand, and subsequently monitored chemotaxis. For 293 fibroblasts expressing the CXCR2, the protocols of Ben-Baruch *et al.*[30] can be used, with modification. The 96-well chemotaxis chamber (Neuroprobe, Cabin John, MD) is used, and the lower compartment of the chamber is loaded with 360-μl aliquots of 1 mg/ml ovalbumin in DMEM or MGSA/GRO diluted in ovalbumin/DMEM (chemotaxis buffer). The polycarbonate membrane is the 10 μm pore size that has been coated on both sides with 20 μg/ml human collagen type IV for 2 hr at 37° and stored overnight at 4°. The cells are removed by trypsiniza-

[28] P. H. Howe and E. B. Leof, *Biochem. J.* **261,** 879 (1989).

[29] P. Gierschik, R. Moghtader, C. Straub, K. Dieterich, and K. H. Jacobs, *Eur. J. Biochem.* **197,** 725 (1991).

[30] A. Ben-Baruch, L. L. Xu, P. R. Young, K. Bengali, J. J. Oppenheim, and J. M. Wang, *J. Biol. Chem.* **270,** 22123 (1995).

tion, incubated in 10% FBS in DMEM 1.5 hr at 37° to allow restoration of receptor, washed in chemotaxis buffer, and then placed into the upper chamber in the ovalbumin medium. The chambers are incubated for 4.5–6 hr at 37° in humidified air with 5% (v/v) CO_2, after which the filter is removed, washed, fixed, and stained with a Diff-Quick kit. Quantitation of relative chemotactic index of cells on the lower filter surface was by densitometric scanning of the stained filter using an Epson ES 1200C scanner (Seiko Epson Corp., Torrance, CA) and Adobe Photoshop software (Adobe Systems Inc., San Jose, CA). Digitized images are subsequently quantitated using the PhosphorImager (Molecular Dynamics, Sunnyvale, CA). Volumes are integrated and normalized to a value of 1.

Receptor Sequestration

Receptor sequestration in cells expressing wild-type or mutant receptors can be monitored using a modification of the protocols described by Chuntharapai and Kim.[31] Briefly, confluent 24-well plate cultures of each cell type are pretreated with MGSA/GRO (100 n*M*) for varying times (10 min to 2 hr) at 4° or at 37°. After the initial incubation, the cultures are washed three times with binding buffer (1 mg/ml ovalbumin in DMEM) at 4°. The presence of the receptor on the cell surface is monitored by indirect immunodetection. After first blocking nonspecific antibody binding sites with a solution of goat anti-mouse IgG (1–5 μg/ml), the cells are incubated with 2 μg of rabbit anti-amino-terminal CXCR2 receptor antibodies (415) per well of cells,[11] washed three times with ovalbumin/DMEM, and then incubated for 30–60 min at 4° with 2×10^5 cpm of goat anti-rabbit ^{125}I-labeled antibody (ICN, 300 μCi/ml). Nonspecific binding of antibodies is monitored using the rabbit antibody to the carboxyl tail of the CXCR2 followed by goat anti-rabbit ^{125}I-labeled antibody. Alternatively, nonspecific binding can be determined by blocking the binding of the N-terminal antipeptide antibody to CXCR2 with peptides to which the antibody was raised. The nonspecific binding is washed away with three 5-min washes of binding buffer, cells are lysed in SDS/NaOH (1%/0.1 *N*), and radioactivity is determined by gamma counting.

Comments

A number of methods have been used to examine phosphorylation sites on receptors: mass spectrophotometric analysis of the *in vivo* phosphorylated receptors, direct *in vitro* phosphorylation assays using purified receptor, and analysis of loss of ligand-induced receptor phosphorylation *in vivo* using cell lines expressing receptor mutants where nonphosphorylatable

[31] A. Chuntharapai and K. J. Kim, *J. Immunol.* **15,** 2587 (1995).

Full length CXCR2

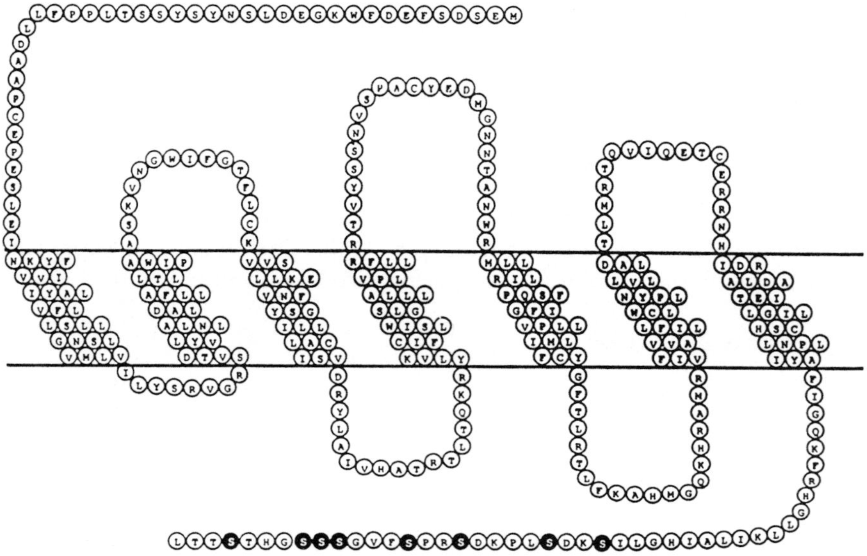

Serine-342 truncation mutation

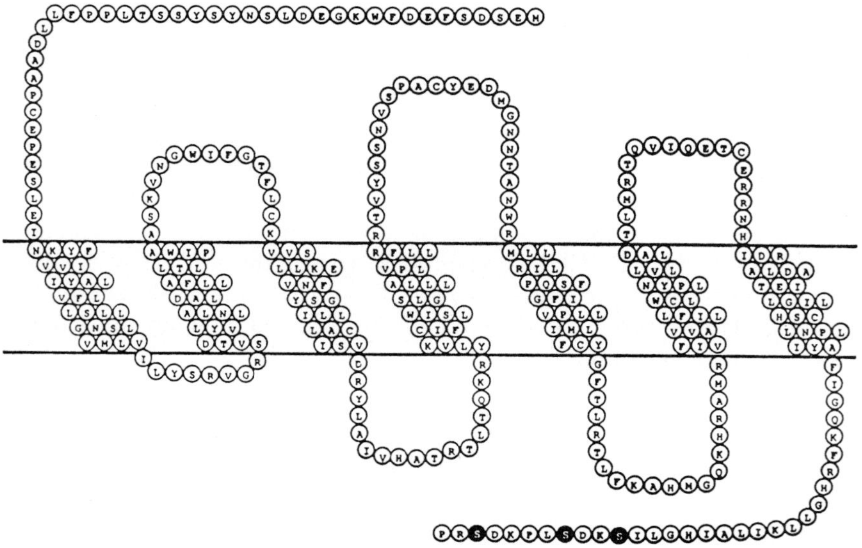

FIG. 1. Diagram showing the placement of the truncation and Ser/Ala mutations in the CXCR2. The serine residues in the carboxyl tail of the receptor that were mutated to alanine are indicated by blackened circles.

Serine-352 truncation mutation

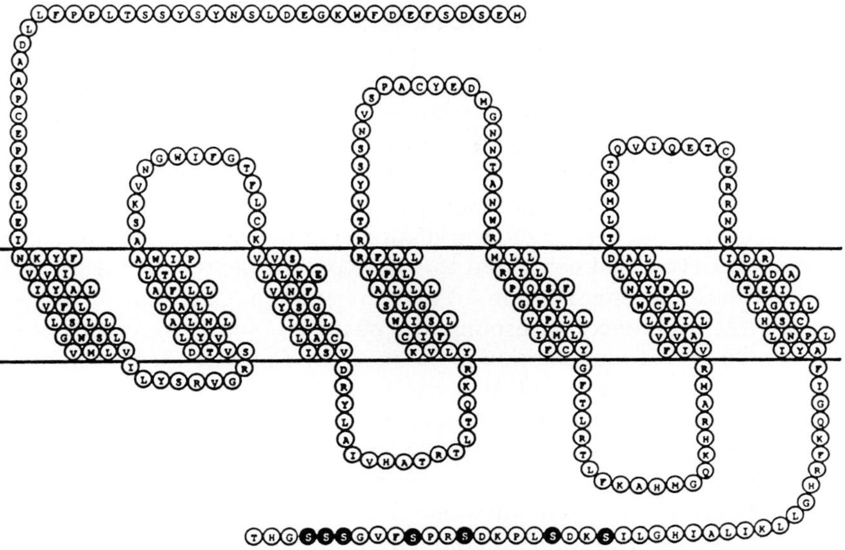

Serine-331 truncation mutation

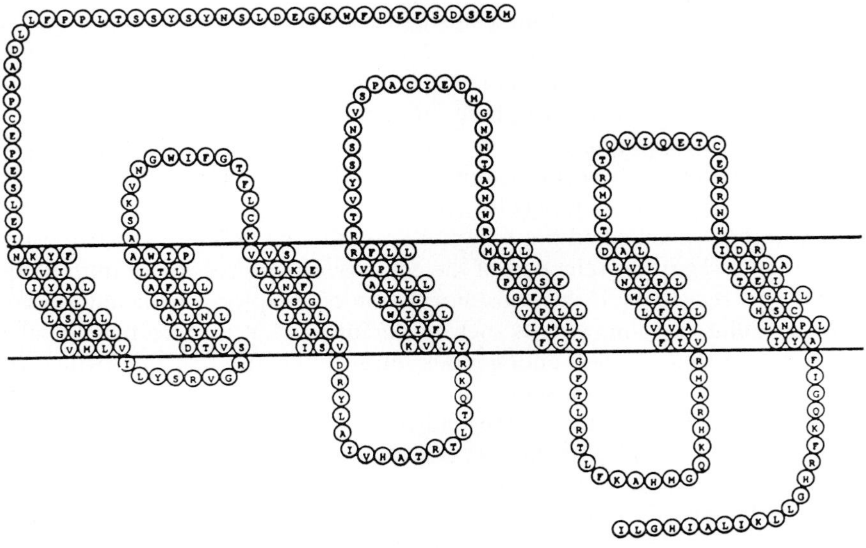

FIG. 1. (*continued*)

amino acids replace the serine, threonine, or tyrosine residues believed to be involved in the phosphorylation event. This third approach is the one we have used here to characterize the ligand-induced receptor phosphorylation of the CXCR2 chemokine receptor. This approach has the advantage of allowing examination of biological consequences of loss of the phosphorylation event, particularly relating to effects on downstream signaling events. Although selection and characterization of stable clones expressing these mutant receptors are somewhat tedious, once selected and characterized, the clones are quite useful to study alterations in calcium flux, receptor sequestration, and receptor degradation. Furthermore, affinity for ligand can be accessed and compared to wild-type receptors with regard to the role of these phosphorylation events in regulation. Care must be taken to verify that differences in response are observed in multiple clones of each mutant, and that the mutation in the receptor does not alter the affinity of the receptor for the ligand. It is also important to determine the number of receptors expressed per mutant clone as compared to wild type to avoid attributing gain or loss of function to an artifact resulting from differential expression of receptor at the cell membrane. This is particularly important for STM receptors, where often only 1 or 2 of every 50 clones selected will actually express the receptor at the cell surface. Screening of selected clones by Western blot will often show receptor expression in the other 48 clones, but ligand binding studies will be negative.

Using the methods outlined above, we have determined that phosphorylation of serine residues between 342 and 352 is crucial for ligand-mediated desensitization of the CXCR2 (see Fig. 1). In contrast to the findings with CXCR1 expression in basophilic leukemia cells, where both serine and threonine phosphorylation events were believed to be involved in receptor desensitization, only serine residues of CXCR2 appear to be phosphorylated in 3ASubE placental cells. However, these apparent differences may be cell type specific and not receptor specific. This brings up an important point with regard to selection of the cell type for expression of mutant and wild-type receptors. If the phosphorylation of receptor is believed to affect intracellular signaling events such as calcium flux, it is important to select a cell type that has low endogenous intracellular free calcium, which can be readily increased in response to signals that stimulate release of calcium from intracellular stores. If the end point of phosphorylation is believed to affect cell motility or chemotaxis, then the transfected cell should be one that can show chemotaxis toward the ligand. To this end, we found that although the release of intracellular free calcium could be readily followed in 3ASubE clones, these cells were not readily adaptable to the chemotaxis assay, and for those studies additional stable clones of mutant and wild-type receptor were selected in human 293 kidney fibroblasts. When the 342T mutant exhibiting truncation at Ser-342 in the carboxyl tail was expressed in 293 cells, the cells continue to exhibit chemotaxis toward an MGSA gradient, though this mutant exhibits little desensitization after ligand binding. Moreover, this mutant receptor is poorly sequestered following ligand stimulation.[32] These data suggest that desensitization is not a required event for chemotaxis.

Ben-Baruch *et al.* have shown that carboxyl tail truncation mutants retain the ability to undergo chemotaxis if as few as eight amino acids remain in the carboxyl tail.[30] Since this would eliminate the serine residues believed to be involved in ligand-induced receptor phosphorylation, the data from our laboratory in combination with their data demonstrate that receptor phosphorylation and desensitization are not directly involved in chemotaxis. We cannot rule out the possibility that there may be long-term effects of failure to desensitize that could ultimately affect chemotaxis.

Acknowledgments

This work was sponsored by the an Associate Career Scientist Award from the Department of Veterans Affairs (A. R.), a grant from the National Cancer Institute (CA34590), the Medical Research Council of Canada (S. M.), and the Skin Disease Center Grant (AR41943).

[32] S. G. Mueller, J. R. White, W. P. Schraw, V. Lam, and A. Richmond, *J. Biol. Chem.* **272,** 8207 (1997).

[2] Generation of Monoclonal Antibodies to Chemokine Receptors

By ANAN CHUNTHARAPAI and K. JIN KIM

Introduction

A family of chemokines has been identified that contains more than 20 chemokines with the ability to attract and activate leukocytes and play important roles as mediators of inflammation.[1,2] These chemokines can be divided into three subfamilies according to the position of the first two cysteines (C-X-C, C-C, and C chemokines). These chemokines often exhibit pleiotropic effects and bind to more than one receptor. Therefore, monoclonal antibodies (MAbs) that are specific to each receptor and/or that

[1] J. J. Oppenheim, C. O. Zachariae, N. Mukaida, and K. Matsushima, *Annu. Rev. Immunol.* **9,** 617 (1991).

[2] M. Baggiolini, B. Dewald, and B. Moser, *Adv. Immunol.* **55,** 97 (1994).

0076-6879/97 $25.00

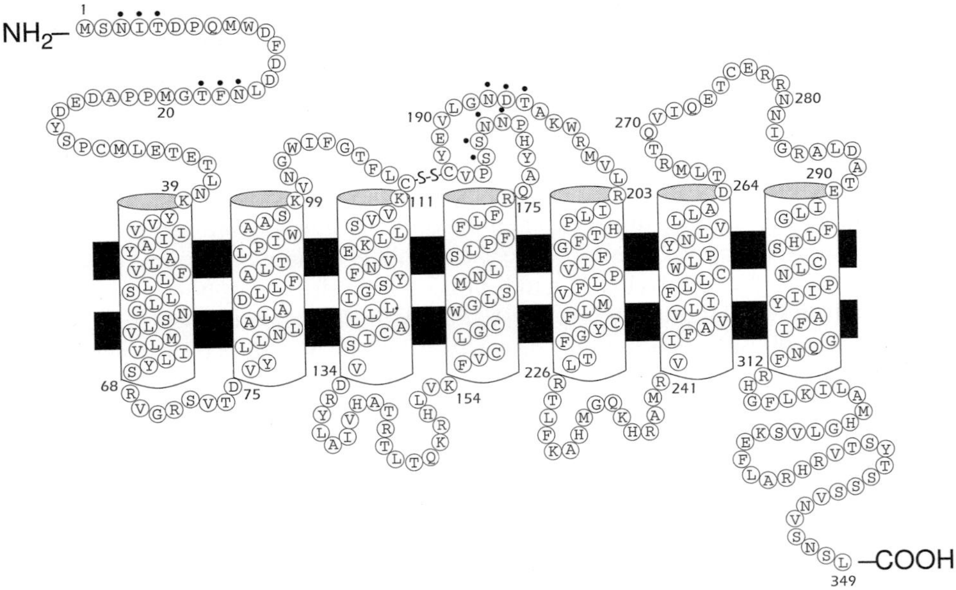

FIG. 1. Model of the secondary structure of human interleukin-8 receptor A. Dots represent potential glycosylation sites.

block the ligand binding to a particular receptor would be extremely useful for understanding the biology of each ligand and the distribution and regulation of its receptor expression.

Although much has been learned since the first report of monoclonal antibodies by Kohler and Milstein,[3] the generation of MAbs to chemokine receptors has remained a challenge. This may be due in part to several reasons. First, these 350- to 360-amino acid chemokine receptors consist of seven-transmembrane (STM) domains, with the extracellular domains (ECDs) of chemokine receptors being relatively short as shown in Fig. 1. The N-terminal portion of chemokine receptors is most exposed and consists of approximately 30–60 amino acid residues, whereas the other three extracellular loops consist of only 15–30 amino acids. Second, there are difficulties obtaining a pure source of antigen as an immunogen. Normal cells express only a limited number (10^3–10^5) of particular chemokine receptors while expressing many other receptors. Third, it is rather difficult to purify STM receptors as a source of antigen. In spite of the difficulties

[3] G. Köhler and C. Milstein, *Nature* (*London*) **256,** 495 (1975).

described above, we[4–6] and others[7,8] have been able to generate blocking MAbs to chemokine receptors. Because interleukin-8 (IL-8) is one of the best characterized chemokines and we have had the most experience with the IL-8 receptor, we describe here a method for generating MAbs to chemokine receptors using IL-8 receptor A (IL-8RA) as a prototype.

Immunogens

Synthetic Peptides as Immunogens

Synthetic peptides have often been used to generate MAbs when the native protein is not available. Using synthetic peptides covering the hydrophilic regions of the protein, one may obtain MAbs that can recognize the native proteins. We have synthesized seven peptides (residues 2–19, 12–31, 99–110, 176–186, 187–203, 265–277, and 277–291) covering the four extracellular domains of IL-8RA. We first tested the activities of antisera raised to each peptide conjugated to horse serum albumin for their ability to bind to 293 transfected cells expressing IL-8RA (293–IL-8RA) as well as neutrophils. Only peptide 2–19 induced antibodies recognizing the IL-8RA on cells (Fig. 2).

Because the N-terminal extracellular portion of the chemokine receptors is the largest and the most exposed, the N-terminal peptides should have a good chance to generate MAbs recognizing the receptor. We and others[4,5,7,9] have successfully raised polyclonal and monoclonal antibodies by immunizing with N-terminal peptides. Peptides are synthesized via solid-phase methodology[10] on either an ABI (Foster City, CA) Model 430 peptide synthesized using *t*-Boc (*tert*-butyloxycarbonyl) chemistry or a Milligen 9050 and ABI 431 using Fmoc (9-fluorenylmethoxycarbonyl) chemistry. Crude peptides are purified by high-performance liquid chromatography (HPLC) and analyzed by mass spectrometry. Peptides are conjugated to horse serum albumin using *m*-maleimidobenzoyl-*N*-hydroxysulfosuccinimide ester (Pierce, Rockford, IL) according to the manufacturer's instructions.

[4] A. Chuntharapai, J. Lee, J. Burnier, W. Wood, C. Hébert, and K. J. Kim, *J. Immunol.* **152,** 1783 (1994).

[5] A. Chuntharapai, J. Lee, C. A. Hébert, and K. J. Kim, *J. Immunol.* **153,** 5682 (1994).

[6] A. Chuntharapai and K. J. Kim, *J. Immunol.* **155,** 2587 (1995).

[7] J. M. Quan, T. R. Martin, G. B. Rosenberg, D. C. Foster, T. Whitmore, and R. B. Goodman, *Biochem. Biophys. Res. Commun.* **219,** 405 (1996).

[8] M. E. Nichols, P. Rubinstein, J. Barnwell, S. Rodriguez de Cordoba, and R. E. Rosenfield, *J. Exp. Med.* **166,** 776 (1987).

[9] M. E. W. Hammond, G. R. Lapointe, P. H. Feucht, S. Hilt, C. A. Gallegos, C. A. Gordon, M. A. Giedlin, G. Mullenbach, and P. Tekamp-Olson, *J. Immunol.* **155,** 1428 (1995).

[10] G. Barany and R. B. Merrifield, *Peptides* **2,** 1 (1980).

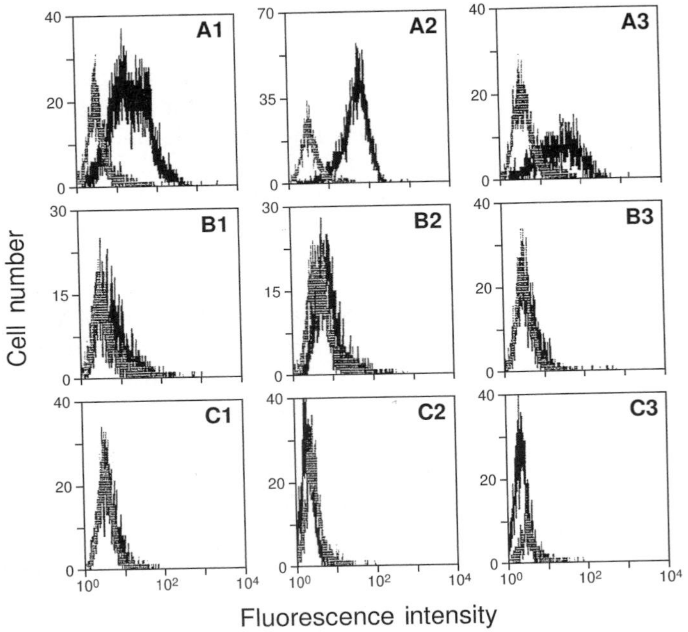

FIG. 2. Flow cytometric analysis of sera from mice immunized with IL-8RA synthetic peptides. Human neutrophils were incubated with various dilutions of sera for 30 min on ice. The MAbs bound to cells were detected by FITC-labeled goat anti-mouse Ig for 10 min. 1, 2, and 3 indicate serum dilutions of 1:100, 1:1000, and 1:10,000, respectively. (A) Peptide 2–19. (B) Peptide 12–31. Other peptides (99–110, 176–187, 187–203, 265–277, and 277–291) show a similar fluorescence staining profile (data not shown). (C) Normal mouse serum.

Transfected Cells Expressing Particular Receptor as Immunogens

Because normal cells express many closely related receptors and these cells need to be purified from other cell types for immunization, we routinely generate transfected cells expressing a high number of the particular receptor(s). Our experience suggests that the most potent blocking MAbs are obtained using transfected cells expressing high receptor numbers as shown in the flow diagram (Fig. 3).

Stably transfected 293 cells expressing IL-8RA (293–IL-8RA) or IL-8RB (293-27) are generated using clones described previously.[11,12] Briefly, 293 cells (a human fetal kidney cell line) are cotransfected with

[11] W. E. Holmes, J. Lee, W.-J. Kuang, G. C. Rice, and W. I. Wood, *Science* **253,** 1278 (1991).
[12] J. Lee, R. Horuk, G. C. Rice, G. L. Bennett, T. Camerato, and W. I. Wood, *J. Biol. Chem.* **267,** 16283 (1992).

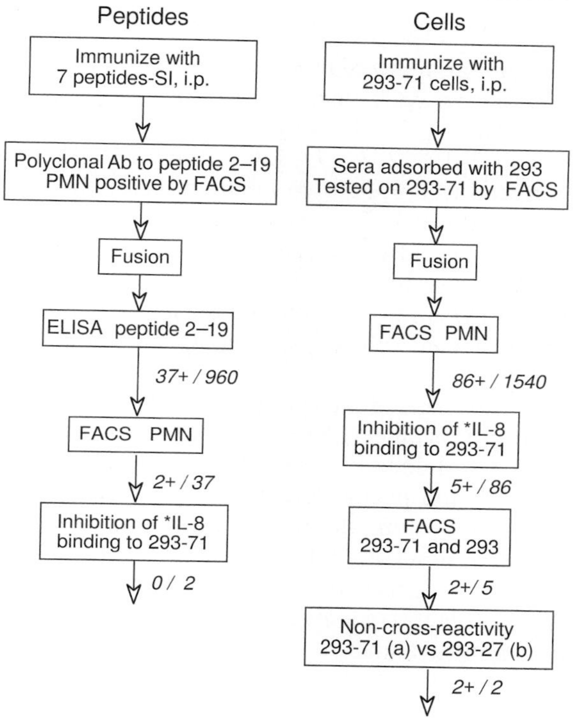

FIG. 3. Flowchart for the generation of MAbs to IL-8RA. Because only the group of mice immunized with peptide 2–19 showed positive staining on human neutrophils, only these mice were used to generate MAbs to IL-8RA. Numbers by arrows indicate the number of samples selected/the number of samples screened.

pRK5B.IL-8r1.1 (IL-8RA) or pRK5.8rr.27.1-1 (IL-8RB) plasmid and pSVENeoBal 6 plasmid using a calcium phosphate precipitation method as described previously.[13] Transfected cells are selected in Ham's F12/Dulbecco's modified Eagle's medium (DMEM) (v/v, 50:50) containing 10% fetal calf serum (FCS), 2 m*M* L-glutamine, 100 μg/ml penicillin, 100 μg/ml streptomycin, and 800 μg/ml of G418 (GIBCO, Grand Island, NY). For the IL-8RA study, clone 293–IL-8RA is chosen from 30 G418-resistant clones isolated from PRK5B.IL-8r1.1 transfection and evaluated for their ability to bind to [^{125}I]-IL-8. For the IL-8RB study, clone 293–IL-8RB is chosen from 30 G418-resistant clones isolated from pRK5.8rr.27.1-1 transfection and evaluated for [^{125}I]-IL-8 binding.

[13] C. Gorman, *in* "DNA Cloning: A Practical Approach" (D. M. Glover, ed.), Vol. 2, p. 143. IRL Press, Oxford, 1984.

Method of Immunization

BALB/c mice are immunized intraperitoneally (i.p.) or in the rear food pads with 10 μg of synthetic peptides, which cover various portions of extracellular domains of IL-8RA (e.g., peptide sequences corresponding to IL-8RA residues 2–19, 12–31, 99–110, 176–186, 187–203, 265–277, and 277–291), conjugated to horse serum albumin resuspended in monophosphoryl lipid A/trehalose dicorynomycolate (Ribi Immunochemical Research, Hamilton, MT). Individual peptides are used to immunize separate groups of mice. Mice are boosted several times with the same amount of peptide every 2 weeks until it is possible to detect antibodies binding to neutrophils or transfected cells expressing IL-8RA by flow cytometry, as shown in Fig. 3.

Mice are hyperimmunized intraperitoneally with 10^6 293–IL-8RA cells in phosphate-buffered saline (PBS) every 2 weeks until there are high antibody titers as determined by flow cytometry. Periodically we test antibody titers after the nonspecific activities are removed by incubating the sera with 293 cells and testing the remaining binding activities to 293 cells expressing IL-8RA by flow cytometry. When titers above 2000 are obtained, as determined by the reciprocal of the dilution that gives the half-maximum response by flow cytometry, we give a final prefusion boost with the same amount of antigen intravenously.

Fusion

Three days after the final boost of mice with transfected cells or peptides conjugated to a carrier, popliteal lymphoid cells or spleen cells of the immunized mice are fused with mouse myeloma PE×63Ag8U.1,[14] a nonsecreting clone of the myeloma P3×63Ag8, using 35% (v/v) polyethylene glycol. Briefly, myeloma cells grown to log phase and lymphoid cells are washed twice in prewarmed serum-free DMEM by centrifugation at 200 *g*. After washing, lymphoid cells and myeloma cells are mixed at a 10:1 ratio and pelleted together by centrifugation at 500 *g* for 7 min at room temperature. The supernatant is removed by aspiration, and the remaining pellet is loosened by gently raking against a round-bottom microtiter plate. One milliliter of prewarmed 35% polyethylene glycol in DMEM is added to the cell pellet over a 1-min period. After the cell pellet is exposed to polyethylene glycol for another 7 min at room temperature, 20 ml of prewarmed DMEM is added to the cells over the next 4 min in drops.

[14] D. E. Yelton, A. B. Diamond, S. P. Kwan, and M. D. Scharff, *Curr. Top. Microbiol. Immunol.* **81,** 1 (1978).

After centrifugation at 200 *g* for 7 min at room temperature, the cell pellet is resuspended in prewarmed super DMEM [DMEM plus 10% FCS, 10% NCTC-109 (BioWittaker, Wakersville, MD), 100 m*M* pyruvate, 100 U/ml insulin, 100 m*M* oxaloacetic acid, 2 m*M* glutamine, 1% nonessential amino acids (GIBCO), 100 U/ml penicillin, and 100 μg/ml streptomycin] at 2×10^6 lymphoid cells/ml. One hundred microliters of cell suspension is added to the flat-bottom microtiter wells. For the next 3 days 100 μl of super DMEM containing 100 μM hypoxanthine, 0.4 μM aminopterin, and 16 μM thymidine (1× HAT, Sigma, St. Louis, MO) is added to each well. Ten days after the fusion, culture supernatants are screened for the presence of MAbs to IL-8RA by enzyme-linked immunosorbent assay (ELISA) and/or fluorescence activated cell sorting (FACS) analysis.

Strategy for Screening Positive Hybridomas

When all tools are available, we recommend the use of different forms of antigens to screen MAbs to avoid nonspecific binding and to obtain biologically relevant MAbs. For example, when mice are immunized by peptides, the MAbs are initially screened for their abilities to bind to the peptides by ELISA followed by their abilities to bind to neutrophils or transfected cells by flow cytometry (Fig. 3). In contrast, when mice are immunized with 293–IL-8RA cells, we screen MAbs for their abilities to bind to human neutrophils by flow cytometry and to bind to the mixture of all IL-8RA peptides by ELISA. Positive MAbs are further analyzed for their abilities to bind to 293–IL-8RA cells but not to 293–IL-8RB cells by flow cytometry.

Enzyme-Linked Immunosorbent Assay

As a simple first screening method, we determine the ability of the hybridoma culture supernatants to bind to the IL-8RA peptide mixture in ELISA as described.[15] Nunc 96-well immunoplates (Flow Lab, McLean, VA) are coated with 50 μl/well of peptide (2 μg/ml) in carbonate buffer (pH 9.6) overnight at 4°. Nonspecific binding sites are blocked with PBS containing 2% (w/v) bovine serum albumin (BSA) for 1 hr. Plates are then incubated with 50 μl/well of culture supernatant for 1 hr followed by the addition of 50 μl/well of horseradish peroxidase-conjugated goat anti-mouse immunoglobulin (Ig). The bound enzyme is detected by the addition of *p*-nitrophenyl phosphate substrate in carbonate buffer plus 1 m*M* of $MgCl_2$ (Sigma 104 phosphate substrate). The color reaction is measured at 492 nm with an ELISA plate reader (Titertrek multiscan, Flow Lab). Be-

[15] K. J. Kim, M. Alphonso, C. H. Schmelzer, and D. Lowe, *J. Immunol. Methods* **156,** 9 (1992).

tween each step plates are washed three times in PBS containing 0.05% Tween 20.

Flow Cytometry

Initially all the culture supernatants (~2000 wells) are individually screened for their ability to bind to neutrophils. Selected hybridomas are further tested for their abilities to bind to the particular IL-8R but not to the nonspecific human cell surface antigens by flow cytometric analysis of binding to 293 cells and 293–IL-8R cells. Human neutrophils are prepared by using Mono-Poly Resolving medium (Flow Lab) according to the vendor's directions. Neutrophils or 293–IL-8RA cells are washed twice in cell sorter buffer (CSB; PBS containing 1% FCS and 0.02% NaN_3) at 300 *g* for 5 min. Twenty-five microliters of cell suspension (4×10^6 cells/ml) in CSB is added to U-bottom microtiter wells, mixed with 100 μl of culture supernatant or purified MAbs (10 μg/ml) in CSB, and incubated for 30 min on ice. After washing, cells are incubated with 100 μl of fluorescein isothiocyanate (FITC)-conjugated goat anti-mouse IgG for 30 min at 4°. Cells are washed twice in CSB, resuspended in 150 μl of CSB, and analyzed with a FACScan (Becton Dickinson, Mountain View, CA).

Selection of Blocking Monoclonal Antibodies

One can select antagonistic MAbs to a particular receptor by determining their abilities to inhibit bioactivities of the relevant ligand. In our study, neutrophils that can respond to IL-8 express many chemokine receptors including IL-8RA and IL-8RB. Therefore, neutrophils are not the best system to screen blocking MAbs specific for IL-8RA. However, 293–IL-8RA cells do not transduce the IL-8 signal. As there is no good bioassay specific for IL-8RA, the hybridomas potentially secreting antagonistic MAbs are screened using the IL-8 receptor binding assay described below.

Ligand Receptor Binding Assay

Transfected cells (50 μl, 4×10^6 cells/ml), resuspended in Hanks' balanced saline solution (HBSS) medium containing 0.5% BSA and 25 m*M* HEPES buffer, are incubated for 1 hr at 4° with 50 μl of hybridoma culture supernatant or purified MAbs in PBS. Cells are washed twice in the medium and resuspended to 1×10^6 cells/ml. One hundred microliters of cells is incubated for 1 hr at 4° with 100 μl of ^{125}I-IL-8. The unbound ^{125}I-IL-8 is removed by centrifuging the mixture over 0.5 ml of PBS containing 20% sucrose and 0.5% BSA at 1500 rpm for 5 min. The ^{125}I-IL-8 bound to the

TABLE I
CHARACTERISTICS OF MONOCLONAL ANTIBODIES TO IL-8RA

MAb	Antigen	Isotype	ED_{50} (pM)[a]	K_d (pM)[b]	Blocking[c]
4C8	Peptide 2–19	IgG_1	44	3,260	No
6E9	Peptide 2–19	IgG_1	563	17,000	No
2A4	293–IL-8RA	IgG_1	281	440	Yes
9H1	293–IL-8RA	IgG_1	125	88	Yes

[a] ED_{50} values were determined by the concentration of MAbs that gave 50% of the maximum binding to peptide covering the amino acid residues of IL-8RA.
[b] K_d values were determined by Scatchard plot analysis by competitive inhibition of ^{125}I-labeled MAb binding to 293–IL-8RA with various concentrations of unlabeled MAb.
[c] Blocking activities were determined by the abilities of the MAbs to inhibit ^{125}I-IL-8 binding to 293–IL-8RA.

cell pellets is counted using a gamma counter. The percentage of binding of ^{125}I-IL-8 to neutrophils or transfected cells in the presence of MAb is calculated by dividing the specific ^{125}I-IL-8 binding in the presence of MAb by the total specific ^{125}I-IL-8 binding. Specific ^{125}I-IL-8 binding is calculated by subtracting the nonspecific ^{125}I-IL-8 binding that is determined in the presence of 0.4 μM IL-8. The level of nonspecific ^{125}I-IL-8 binding is approximately 10% of the total ^{125}I-IL-8 binding.

Determinations of Monoclonal Antibody Affinities

To assess the quality of MAbs generated, we routinely compare the affinities of the MAbs by an ELISA as described by Van Heyningen.[16] The affinities of selected final MAbs are determined using a conventional radioimmunoprecipitation assay followed by Scatchard analysis.

Relative Affinities by Enzyme-Linked Immunosorbent Assay

The ELISA plates are coated with 2 μg/ml of peptide 2–19, blocked with BSA, and incubated with various concentrations of MAbs (0.005 to 2 μg/ml). The level of MAb bound to the plate is detected as described above. By comparing the MAb concentrations that gave the half-maximum binding to the peptide (ED_{50}), we have ranked the relative affinities of these MAbs (Table I).

Radioimmunoprecipitation Assay

The affinities of MAbs to IL-8RA on cells are determined by competitive inhibition of the binding of ^{125}I-labeled MAb to 293–IL-8RA transfected

[16] V. Van Heyningen, *Methods Enzymol.* **121,** 472 (1993).

cells with various concentrations of unlabeled MAbs. ^{125}I-Labeled MAbs are prepared using chloramine-T as described previously.[17] Fifty microliters of 293–IL-8RA (4 × 10^{6} cells/ml), resuspended in HBSS containing 0.5% BSA, is incubated with 100 μl of a fixed concentration of ^{125}I-labeled MAb and 50 μl of various concentrations of unlabeled MAbs for 1 hr at 4°. The unbound ^{125}I-labeled MAb is removed by centrifugation of the mixture over 0.5 ml of PBS containing 20% (w/v) sucrose and 0.5% BSA (w/v) at 1500 rpm for 5 min. The ^{125}I-labeled MAb bound to the cell pellet is counted using a gamma counter. The affinity of each MAb is determined by Scatchard analysis.[18]

Specific Comments

Comparison of Affinities of Monoclonal Antibodies to Interleukin-8 Receptor A versus Immunization Method

The relative affinities of MAbs to peptide 2–19 are determined by measuring the antibody concentration required to give 50% of the maximum binding (ED_{50}) to this peptide by ELISA as described by Van Heyningen[16] (Table I). The ED_{50} values of the peptide-derived MAbs 4C8 and 6E9 are 44 and 563 p*M*, respectively, whereas ED_{50} values of cell-derived MAbs 2A4 and 9H1 are 281 and 125 p*M*, respectively. The ranking of the MAbs in order of decreasing affinity is 4C8, 9H1, 2A4, and 6E9. It is not surprising that MAb 4C8, generated by immunizing peptide 2–19, has the highest affinity to this peptide. However, determination of the affinities of these MAbs for the IL-8RA expressed on transfected cells using Scatchard plot analysis shows that the cell-derived MAbs have the higher affinities for IL-8RA, 7- to 196-fold higher than peptide-derived MAbs (Table I). The dissociation constants of peptide-derived MAbs 4C8 and 6E9 are 3.26 and 17 n*M*, respectively, whereas those of cell-derived MAbs 2A4 and 9H1 are 0.44 and 0.088 n*M*, respectively. These observations suggest that high-affinity MAbs can be obtained with the cell membrane expressed receptor and that the higher affinities of cell-derived MAbs to IL-8RA may contribute to their blocking activities as described below.

Comparison of Blocking Activities of Monoclonal Antibodies versus Immunization Method

The blocking activities of MAbs obtained by immunizing cells or peptides have been compared (Fig. 4). The best cell-derived anti-IL-8RA MAbs

[17] E. Harlow and D. Lane, eds., "Antibodies," p. 328. Cold Spring Harbor Press, Cold Spring Harbor, N.Y., 1988.

[18] P. J. Munson and D. Rodbard, *Anal. Biochem.* **107,** 220 (1980).

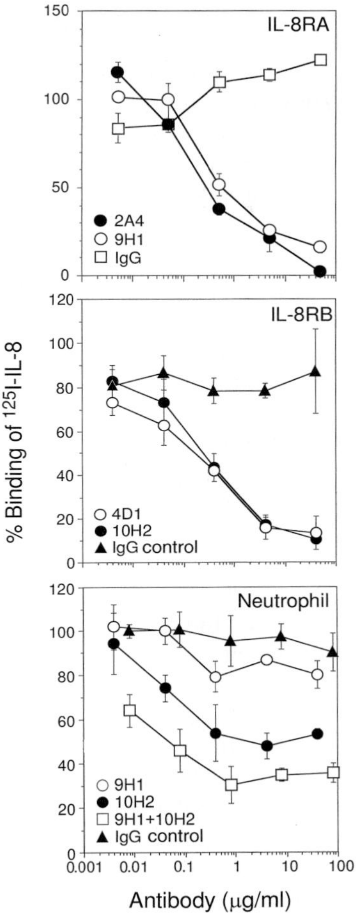

FIG. 4. Inhibition of ^{125}I-IL-8 binding to neutrophils or transfected 293 cells expressing IL-8RA and IL-8RB by MAbs. (*Top*) MAbs to IL-8RA (2A4 and 9H1) inhibited the binding of ^{125}I-IL-8 to 293–IL-8RA. (*Middle*) MAbs to IL-8RB (4D1 and 10H2) inhibited the binding of ^{125}I-IL-8 to 293–IL-8RB. (*Bottom*) ^{125}I-IL-8 binding to neutrophils was inhibited by anti-IL-8RA MAb (9H1) and/or anti-IL-8RB MAb (10H2).

2A4 and 9H1 show potent blocking of ^{125}I-IL-8 binding to the 293–IL-8RA cells. MAbs 2A4 and 9H1, at 50 μg/ml, show 98 and 85% inhibition of ^{125}I-IL-8 binding, respectively. In contrast, the best peptide-derived anti-IL-8RA MAbs 4C8 and 6E9 show no inhibition of ^{125}I-IL-8 binding (data not shown). We further investigated whether cell-derived MAbs 2A4 and 9H1 could block IL-8 binding to IL-8RA on human neutrophils. Because 10 μg/ml of these MAbs show almost 80% inhibition of ^{125}I-IL-8 binding to 293–IL-8RA cells, which express approximately 5 to 10 times more IL-8RA than human neutrophils, we determine the blocking abilities of these MAbs using 10 μg/ml. MAb 9H1 is able to block 20% of the ^{125}I-IL-8 binding to human neutrophils. A similar observation is made with anti-IL-8RB MAb 10H2, which is raised by immunizing 293–IL-8RB cells. MAb 10H2 is able to block approximately 90% of ^{125}I-IL-8 binding to 293–IL-8RB cells and is able to block approximately 50% of ^{125}I-IL-8 binding to neutrophils as shown in Fig. 4. From these results, we conclude that the most potent blocking MAbs can be obtained by immunizing with cells expressing particular chemokine receptors rather than by immunizing with synthetic peptides.

Quan *et al.*[7] have been able to generate blocking MAbs by immunizing with long N-terminal peptides (1–40 residues), whereas we were unable to generate blocking MAbs by immunizing with a short N-terminal peptide (residues 2–19). Thus, it is possible to generate blocking MAbs using long peptides, which potentially mimic the native conformation better.

General Comments

Some residues and motifs of the chemokine receptors are highly conserved among members of the chemokine family and show cross-reactivity with various ligands. For example, two interleukin-8 receptors, IL-8RA and IL-8RB, share 77% amino acid identity. However the homology of these two receptors is only 29% in the N-terminal regions that were shown to be involved in ligand binding.[19,20] Our results with polyclonal antibodies and MAbs suggest that the N-terminal portion of IL-8R is the most immunogenic region of the receptors. Therefore, it is reasonable to try to generate MAbs by immunizing with N-terminal synthetic peptides of chemokine receptors. This has been demonstrated by our laboratory[4] and by Quan *et al.*,[7] although we were able to generate the most potent blocking monoclonal antibodies by immunizing with cells expressing IL-8RA and IL-8RB rather than with peptides. The MAbs that we generated, using residues 2–19 of the IL-8RA peptide as an immunogen, were not blocking MAbs, although the main epitope recognized by our blocking MAbs, raised using transfected cells as an immunogen, was located within residues 2–14. In contrast, MAbs

[19] R. B. Gayle III, P. R. Sleath, S. Srinivason, C. W. Birks, K. S. Weerawarna, D. P. Cerretti, C. W. Birks, K. S. Weerawarna, D. P. Cerretti, C. J. Kozlosky, N. Nelson, T. Vanden Bos, and M. P. Beckmann, *J. Biol. Chem.* **268,** 7283 (1993).

[20] C. A. Hebert, A. Chuntharapai, M. Smith, T. Colby, J. Kim, and R. Horuk, *J. Biol. Chem.* **268,** 18549 (1993).

generated by Quan *et al.*[7] using the peptide covering the entire N-terminal ECD (1–40 amino acids) were blocking MAbs. Because the success by Quan *et al.*[7] may be due in part to the possibility that a longer peptide mimics the native conformation of the receptor, we suggest the use of a longer peptide as an immunogen. In any event, MAbs generated by immunizing with peptide 2–19 were able to recognize the native IL-8RA on cells and were very useful in initial receptor expression studies. Thus, when the sequence of a chemokine receptor is known and transfected cells expressing the particular receptor is not available, it is worthwhile to generate polyclonal and/or MAbs to peptides to facilitate the initial study.

Whether peptides or cells are used as immunogens, it is important to use different forms of antigens during the initial screening stage to select MAbs for the desired biological specificity. Furthermore, we found that the multiple immunization schemes were necessary to obtain high affinity MAbs to chemokine receptors. One explanation for this is that the ECD of the chemokine receptor are very small, which may result in low immunogenicity. Lastly, although we do not discuss phage antibodies in this chapter, an alternative method of generating monoclonal antibodies to chemokine receptors would be via construction of a combinatorial antibody library using phage display methods.[21]

[21] G. Winter, A. D. Griffiths, R. E. Hawkins, and H. R. Hoogenboom, *Annu. Rev. Immunol.* **12,** 433 (1994).

[3] Chemokine Receptors in Developing Human Brain

By MEREDITH HALKS-MILLER, JOSEPH HESSELGESSER, ILONA J. MIKO, and RICHARD HORUK

Introduction

Chemokines are a family of structurally related cytokines that control immunoregulatory and inflammatory processes by governing the trafficking and physiological activation of leukocytes.[1] The chemokines, which include interleukin-8 (IL-8), melanoma growth stimulating activity (MGSA), RANTES, monocyte chemotactic protein (MCP-1), and macrophage inflammatory protein (MIP-1), mediate their effects by specific binding to

[1] T. Schall, *in* "The Cytokine Handbook" (A. Thompson, ed.), p. 419. Academic Press, San Diego, 1994.

0076-6879/97 $25.00

high-affinity receptors expressed on the surface of target cells.[2] So far, 10 different chemokine receptors have been cloned.[3] These receptors are characterized by a heptahelical structure and belong to a superfamily of serpentine receptors. Most, but not all, are coupled to guanine nucleotide binding proteins (G proteins).

Although chemokines primarily chemoattract and activate immune cells, a number of reports show they are also expressed in the brain. Several chemokines including MCP-1 and RANTES have been implicated in inflammation in the central nervous system (CNS). In an experimental autoimmune encephalomyelitis (EAE) model of multiple sclerosis (MS) in the mouse, RANTES and MCP-1 mRNA expression levels in astrocytes seem to correlate with the onset of the disease, disappearing at its resolution.[4,5] The transient expression of these chemokines by astrocytes seems to precede both the influx of inflammatory cells and demyelination. MCP-1 and RANTES are chemotactic for monocytes and T cells, respectively,[1] and potentiate their adhesion to endothelial cells. There is also direct experimental evidence that chemokines are involved in modulating normal homeostatic mechanisms in the CNS; for example, IL-8 has been shown to enhance the survival of rat hippocampal neurons *in vitro.*[6]

In addition to the expression of chemokines in the CNS, it has been shown that chemokine receptors are present in the cytoplasm of selected neurons in normal adult human brain.[7] The IL-8 receptor B (CXCR2) has also been localized by immunocytochemistry to the fragmented dendrites in the senile plaques of Alzheimer's disease.[7] The role(s) that chemokine receptors may play in brain development, injury, or homeostasis is not well understood. To address the question of chemokine receptor function in the CNS, we first established a method of culturing neurons and glia from fetal human brain and determined that cells from these explants expressed at least one chemokine receptor, CXCR2, by immunocytochemistry. These immunochemical techniques were later applied to cultures of hNT cells[8] (human neuronal cells derived from an embryonal cell carcinoma line after treatment with retinoic acid) as well as to paraffin-embedded sections of human fetal brain obtained from samples collected after either autopsy or abortion. The results of these investigations show that CXCR2 is present in fetal brain cells grown *in vitro* and that it is also found in intact fetal

[2] R. Horuk, *Trends Pharmacol. Sci.* **15,** 159 (1994).
[3] C. R. Mackay, *J. Exp. Med.* **184,** 799 (1996).
[4] R. Godiska, D. Chantry, G. N. Dietsch, and P. W. Gray, *J. Neuroimmunol.* **58,** 167 (1995).
[5] R. M. Ransohoff *et al., FASEB J.* **7,** 592 (1993).
[6] D. M. Araujo and C. W. Cotman, *Brain Res.* **600,** 49 (1993).
[7] R. Horuk *et al., J. Immunol.* **158,** 2882 (1997).
[8] S. J. Pleasure, C. Page, and V. M.-Y. Lee, *J. Neurosci.* **12,** 1802 (1992).

brain from at least as early as gestational week 20. Followup studies showed that several different chemokine receptors are also present on hNT neurons.

Methods and Discussion

Human Fetal Brain Cultures

Human fetal brains are cultured to establish an *in vitro* model in which the development of brain chemokine receptors can be studied. Our goal is to grow mixed cultures of cortical neurons and glia in a defined culture system that can be made as reproducible as possible. To this end, explants are established using brain cells derived from aborted human fetuses between 17 and 22 weeks of gestation. (Brains from younger fetuses, although theoretically more desirable, are more difficult to obtain.) The cells are grown under several different culture conditions and evaluated morphologically with both immunofluorescence and immunohistochemistry.

Fetal brains are obtained from Advanced Bioscience Resources (ABR, Alameda, CA) 3 to 4 hr after abortion. The brains are shipped on ice in a nutrient medium [Iscove's minimal essential medium (IMEM) with high glucose (4.5 g/liter), 2 m*M* glutamine, 10% (v/v) fetal bovine serum (FBS, non-heat-inactivated), and gentamicin (50 μg/ml)] and are dissected immediately on receipt. Although the brains are often in fragments of about $4 \times 3 \times 1$ cm, it is not usually difficult to identify cerebral cortex and other large structures such as brain stem or thalamus. Smaller subdivisions are much more difficult to define with certainty. A dissecting microscope is of little use in clarifying subregions because brain tissues at this age are visually very homogeneous. In some samples, however, hippocampal tissue can be identified by its association with the choroid plexus. In these instances, hippocampal tissues are isolated and cultured separately from cortex. Both hippocampal and cortical neurons grow equally well in culture.

Under a dissecting microscope the meninges and any clotted blood are removed from the brain and discarded. Brain cortical tissues (volume of about 3 ml) are selected and transferred to a 15-ml sterile culture tube, where they are rinsed in 5 ml of nutrient medium (IMEM–high glucose, 10% FBS, with 50 U/ml penicillin, 50 μg/ml streptomycin, and 2 μg/ml insulin). The brain tissues are then resuspended in an equal volume of nutrient medium and gently triturated using a plastic 5-ml pipette. The resulting suspension of cells is filtered either through a nylon mesh bag of 210 μm pore diameter (fabric from Tetco, Briarcliff Manor, NY) or a stainless steel mesh cup of the same pore size. The cells are rinsed through with more nutrient medium, and a sterile glass rod (5 mm diameter) is used to press gently against the bag or steel mesh to facilitate cell passage.

The cell suspension is then further filtered by gravity through a nylon mesh filtration cup (100 μm pore diameter, Falcon) directly into a 50-ml centrifuge tube. The final filtrate is composed primarily of single cells. A 100-μl aliquot of the cells is diluted 1:9 with more nutrient medium and then 1:1 with a 0.4% solution of trypan blue (Sigma, St. Louis, MO). Microscopic examination reveals about 50% cell viability. The number of viable cells is estimated by counting two fields on a hemocytometer. The bulk of the cells are centrifuged at 200 *g* for 7 min. (Using this relative low gravitational force allows the lighter debris particles to remain in the supernatant after spinning.) The cell pellet is then resuspended in a volume of nutrient medium calculated to yield a concentration of 10^5 viable cells in 400 ml. A typical preparation yields a total of 30×10^6 to 50×10^6 viable cells per milliliter of initial wet brain cortical volume, or 150×10^6 cells in a standard procedure. Dissection of the hippocampus yields about 25×10^6 cells in toto.

To facilitate multiple immunohistochemical studies, the cells are plated onto either 4-, 8-, or 16-well chamber slides (Nalge Nunc, Rochester, NY). We find that plating the cells at a concentration of 10^5 cells in a volume of 400 μl per 8-well chamber is optimal. Half that number of cells is used in 16-well chambers and double that number in the 4-well chambers. Cells plated at lower densities do not grow as well or elaborate as many processes. We note that on glass slides, cells tend to grow better at the periphery of each well, close to the plastic walls of each chamber. Because the distance from the walls to the center of each well is smaller in the 8- and 16-well chambers, cells plated onto these smaller-size wells tend to grow more evently than in the 4-well chambers where cell growth in the center of the wells is sparse. Some cells are cultured on plastic slides; surprisingly, however, these slides do not appear to promote better growth and development than the 8- or 16-well glass slides. Glass slides are therefore used in all subsequent studies of primary brain cells because they are much easier to handle, coverslip, and view microscopically.

After plating, the cells are allowed to attach overnight. The next day about one-third of the medium is withdrawn and replaced with fresh nutrient medium without antibiotics. This step is needed to remove some of the cellular debris that has accumulated from the 50% of nonviable cells that are explanted along with the viable ones. (It is important not to remove all of the medium when feeding because fetal neurons are very easily damaged after exposure to air.) The cells are then allowed to grow undisturbed for 4 days. At the end of this time (day 5 *in vitro*), discrete clusters of cells with neuronal morphology have formed and have begun to send out radial processes. Glial background cells are minimal. Subsequently, the cultures are fed with an exchange of half of the nutrient medium three times per week.

Cells grown in IMEM with FBS develop a prominent layer of astroglial cells which gradually increase with time. In most explantations the glia have proliferated so heavily that by day 15 *in vitro* neuronal survival is greatly compromised. Therefore, our data from serum-fed cultures are all obtained from cultures terminated on either days 12 or 13 *in vitro,* before glial cell proliferation became too great.

To avoid the problems associated with glial proliferation, which is strongly associated with exposure to serum-containing medium, we have experimented with a serum-free medium designated as N-B27 (Neurobasal medium, supplemented 1:50 with B27, both from GIBCO-BRL, Grand Island, NY) that is developed[9] to promote neuronal (principally hippocampal) growth and maturation *in vitro.* Cells that are gradually switched to N-B27 by replacement of half of the feeding medium beginning on day 5 do not show significant glial cell proliferation over the course of the study and display more robust cable formation than those cells growing in serum (Fig. 1). The neuronal clusters in N-B27-fed explants continue to maintain integrity and elaborate thick, interconnecting neuronal processes up to day 20 *in vitro* (Fig. 2, see color insert).

When cells are switched to N-B27 medium after 8 days of growth in serum-containing IMEM, the glial outgrowth stops progressing; however, existing glial cells do not die back, and neuronal survival and cable formation are not as robust as in sister cultures that have been switched to N-B27 earlier. We conclude that switching from a serum-containing to a serum-free medium after 4 to 5 days *in vitro* is the best way to minimize glial growth and ensure long-term neuronal survival.

We have repeated these explantation studies several times and found that not all explantations are successful. Despite consistent explantation and culture conditions, we have only been successful in establishing 4 out of 6 human fetal cultures. In the two failed cultures neurons formed small clusters, but did not begin to elaborate processes on day 4 or 5. After more time in culture, all cells developed a granular appearance consistent with necrosis and were discarded without further study. This variability in neuronal survival does not appear to correlate either with differences in fetal age at time of explantation or with differences in transit time before explantation. Of course, with human tissue obtained after abortion there are many factors that cannot be well controlled. Among these are maternal health, fetal genetics, and the possibility of fetal exposure to viral pathogens or

[9] G. J. Brewer, J. R. Torricelli, E. K. Evege, and P. J. Price, *J. Neurosci. Res.* **35,** 567 (1993).

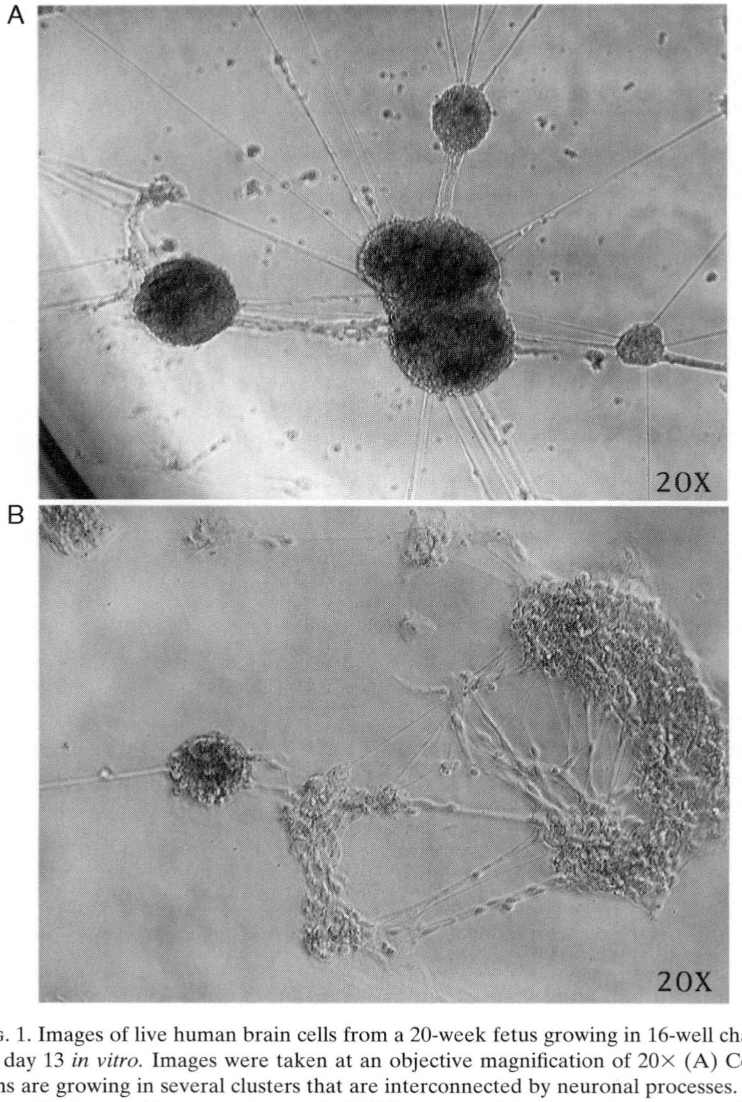

FIG. 1. Images of live human brain cells from a 20-week fetus growing in 16-well chamber slides, day 13 *in vitro.* Images were taken at an objective magnification of 20× (A) Cortical neurons are growing in several clusters that are interconnected by neuronal processes. There are few cells observed in the background. N-B27 medium. (B) Sister cultures maintained in IMEM with 10% fetal bovine serum (see text) have fewer cables. Neuronal clusters are not so tightly packed, and bipolar glial cells are growing beneath the clusters.

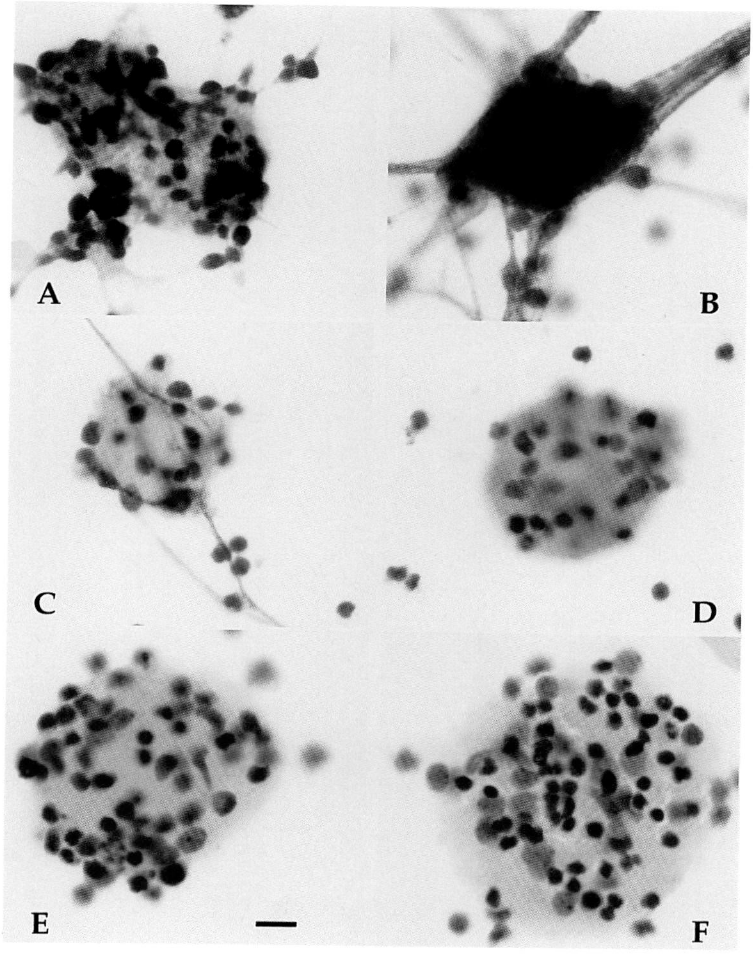

FIG. 2. Immunostains of fixed human fetal brain cells derived from a 22-week fetus and grown in N-B27 medium, day 20 *in vitro*. All staining was done using DAB as the chromogen and hematoxylin as the counterstain. (A) The CXCR2 receptor is present in association with cells within the neuronal clusters. Rabbit polyclonal antibody. (B) Human NCAM is found in great abundance within the clusters and along the neuronal processes. Mouse monoclonal antibody. (C) Antibodies (mouse monoclonal) directed against glial fibrillary acidic protein (GFAP) reveal the presence of a few glial cells within some clusters. Background cells are not GFAP-positive. (D) The macrophage marker CD68 (mouse monoclonal antibody from clone EB11) is not found in the cultures. (E and F) Cells reacted with irrelevant antibodies (rabbit polyclonal or mouse monoclonal, respectively) do not show staining. Bar: 10 μm.

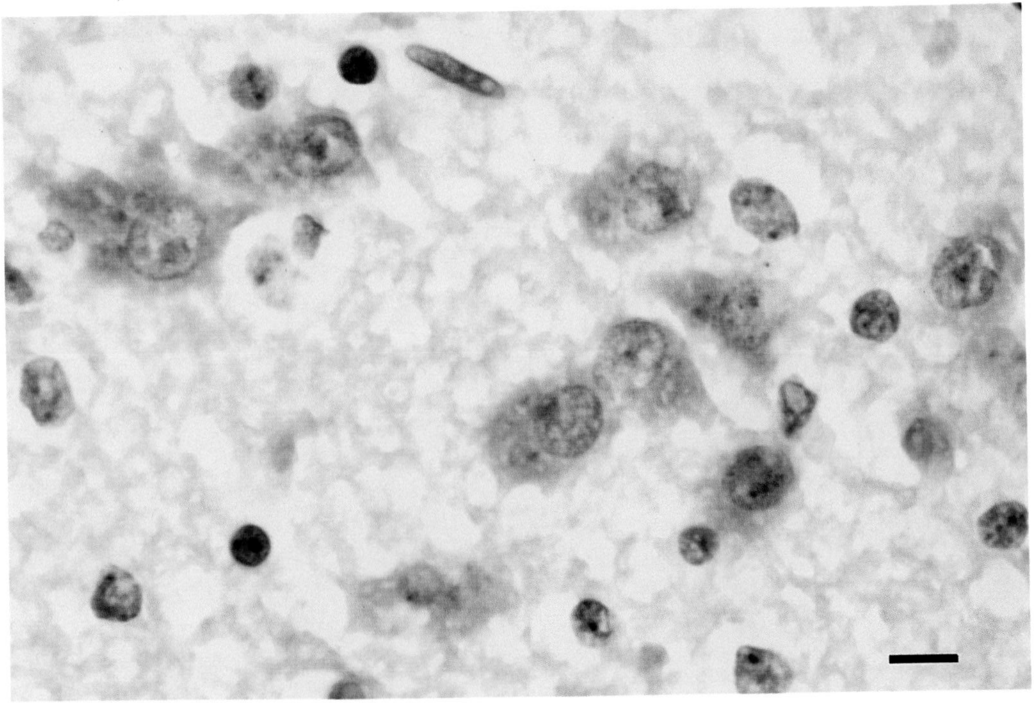

FIG. 3. Formalin-fixed, paraffin-embedded section of brain from a 22-week-old human fetus showing intense cytoplasmic staining of neurons with antibodies directed against CXCR2. Bar: 20 μm.

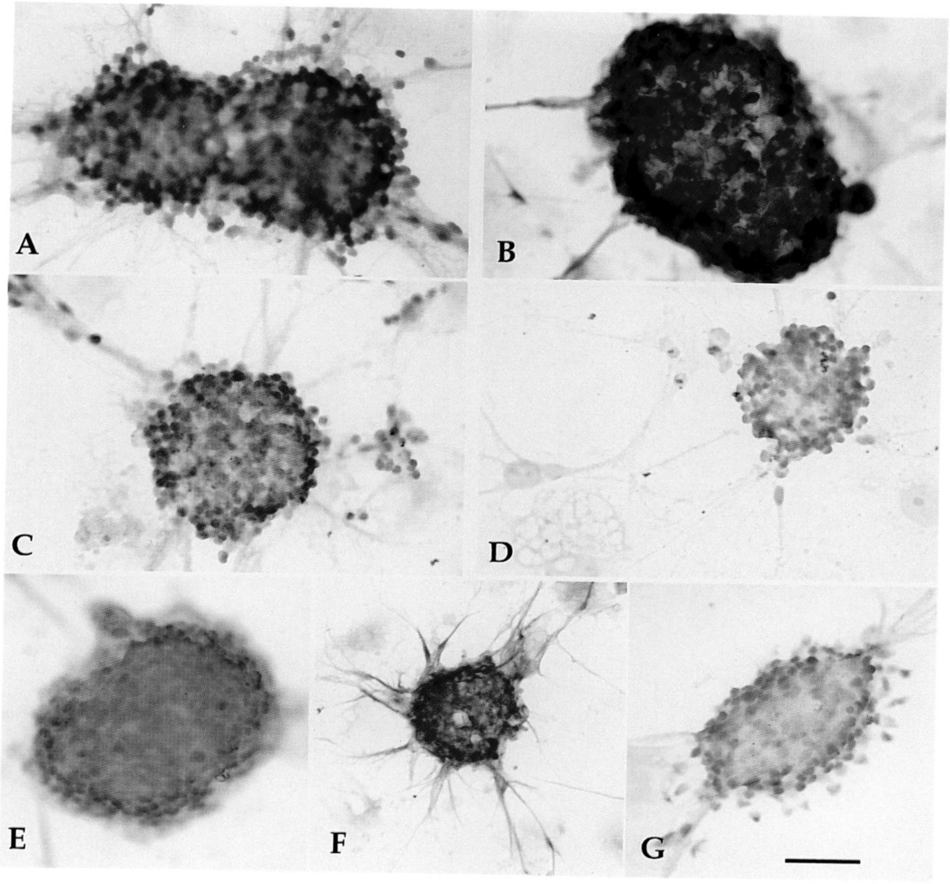

FIG. 4. hNT cells in culture fixed and immunostained by antibodies directed against (A) CCR1, which predominantly stains hNT cell bodies; (B) fusin, which stains some proximal processes as well as cell bodies; (C) CCR5, which also stains cell bodies; (D) CCR5 incubated with peptide antigen, showing that staining is blocked; (E) CD4, which does not react with the cultures; (F) neurofilament, which shows intense staining of cell bodies and processes; and (G) GFAP, which does not react with any cell type in the hNT cultures. Bar: (A–E, G) 50 μm; (F) 100 μm.

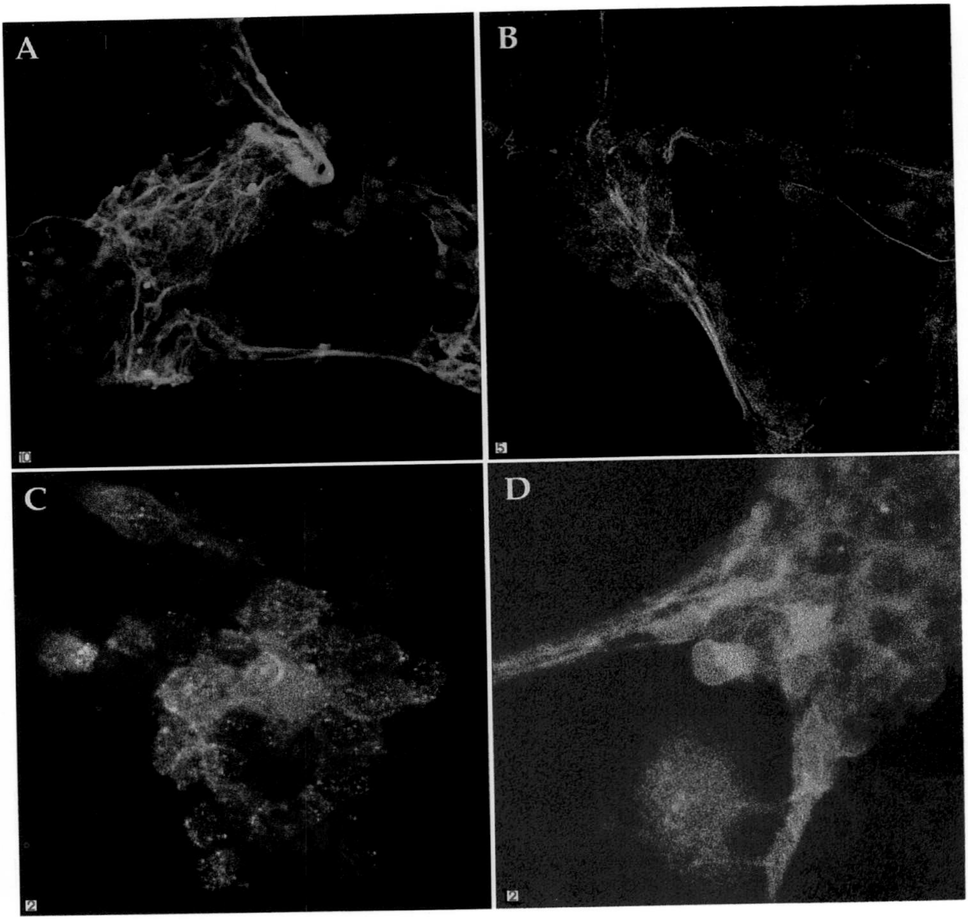

FIG. 5. Confocal images of hNT cells fluorescently labeled with antibodies against (A) neurofilament (green) and CXCR2 (red); (B) neurofilament (green) and CCR5 (red); (C) human NCAM (red) and CCR1 (green); and (D) fusin (pseudocolored, with red and yellow indicating higher density of receptor). Fusin appears to be staining the outlines of neuronal cells and their processes. The hNT cell surface as outlined by the neuronal surface marker NCAM indicates that receptors (green) are present in both cytoplasmic and surface locations, where overlap of red and green signals produces a yellow color. In (A) and (B) receptors do not colocalize with the neurofilament proteins. Small yellow squares define an area of 10 μm per side (A), 5 μm per side (B), and 2 μm per side (C and D).

drugs of abuse. Even the stated gestational age cannot be accepted with absolute certainty, as we have noted that two relatively intact brains, both reported to be 22 weeks of gestational age, were substantially different in size. For these reasons, it is not possible for us to establish the best gestational age for explantation of human brain neocortex. It is, however, reasonable to assume that cortical tissues from brains of about 16 to 17 gestational weeks (although more difficult to obtain because abortions are less commonly performed at these fetal ages) will be the most viable, based on our knowledge of cortical development *in vivo*.

In brief, the human cortex is formed principally from waves of neuroblast migration that begin in the eighth week of gestation.[10] Neuroblasts leave the ventricular zone and, using the radial glia as a scaffolding, ascend to an area just below the marginal zone where they will eventually form layers VI through II of the neocortex. As migration continues, the more recent arrivals pass over the previous cells and form the upper layers of neocortex in an inside-out pattern of migration. The marginal zone becomes layer I. This process is primarily over by gestational week 16. Maturation of the cortical neurons also has an inside-out pattern, with the lower layers of cortex receiving afferent synaptic contacts and elaborating dendrites even before cortical migration is totally complete. With more time cortical neurons in the upper layers begin to mature, elaborate processes, and make appropriate synaptic connections. Thus, cortical tissue from brains at 16 to 17 gestational weeks should have nearly a full complement of neurons, most of which would be immature enough to be explanted successfully (i.e., they would not be encumbered by extensive process development and synapse formation). Cortical neurons from this age of human fetus are roughly equivalent in maturity to cells from rat brain cortex at 15 to 16 days of gestation. Cells from fetuses of this age are routinely used to establish CNS cultures, and, indeed, our human cultures have a morphology similar to that seen in fetal rodent explants.

In summary, we have been able to grow human brain cells *in vitro* successfully for up to 20 days in 4 out of 6 attempts. Results from all the viable cultures are fundamentally identical. Neurons grow in small clusters (about 100 μm in diameter) and begin to extend processes after 4 days *in vitro* if the cultures are healthy. By day 12 *in vitro* there is a complex network of neuronal processes connecting island clusters of neurons, especially in serum-free cultures. When cells are cultured in the presence of serum, astrocytes form a carpet of cells beneath the neurons, which they eventually overgrow. Cells of macrophage morphology are also present in serum-fed cultures and increase in number as the cultures age.

[10] R. L. Sidman and P. Racik, *Brain Res.* **62,** 1 (1973).

hNT Neuron Cultures

Because the human brain cell explants present many problems with respect to consistency of brain samples and reliability of explantation, we decided to use a different human brain cell culture system, the hNT neurons, to continue our studies. hNT neurons are derived from a proliferating cell line, the NTera2 cells, which are obtained from a clonal cell line (Ntera2/clone D1) derived from a human embryonal carcinoma.[8,11] The NTera2 cell line is obtained from Stratagene (La Jolla, CA). Cells are passaged weekly and maintained in Dulbecco's modified Eagle's medium (DMEM) with high glucose (4 m*M*) supplemented with 10% heat-inactivated FBS, 2 m*M* L-glutamine, and penicillin (50 U/ml)/streptomycin (50 μg/ml).

To induce neuronal differentiation, hNT cells are trypsinized and plated at 2×10^7 cells per 225 cm^2 flask in medium with 10 μM retinoic acid (RA) for 5 weeks, with fresh medium changes biweekly. The cells are then replated at 1×10^7 cells either in 225-cm^2 flasks or in 100-mm culture dishes, in medium containing the mitotic inhibitors 1β-D-arabinofuranosylcytosine (1 m*M*), fluorodeoxyuridine (10 m*M*), and uridine (10 m*M*) to inhibit nonneuronal cell division. After 2 weeks neuronal cells (hNT) are separated from precursor cells by differential trypsinization. This is achieved by mild trypsin treatment for 30 sec followed by a gentle knock-off of the differentiated neuroblastlike cells (hNT) from the flask. hNT cells are plated at 5×10^6 cells per 75-cm^2 flask and maintained for up to 1 week in 50:50 (v/v) mixture of fresh medium plus spent cell culture medium (0.22-μm filtered) from the mitotic inhibitor treatment. For immunohistochemical analysis, cells are plated 1×10^6 cells/ml into either 8- or 4-well plastic chamber slides. hNT cells have been shown to be neuronlike on the basis of expression of neurofilament and human neuronal cell adhesion molecule (NCAM) antigens by indirect immunofluorescence labeling.[8]

Human Autopsy Material: Fetal Brain

Fetal brains obtained at autopsy or sections of fetal brain from aborted fetuses are fixed in phosphate-buffered 10% formalin (v/v) for several weeks. Areas of brain are then cut into blocks, embedded in paraffin, and sectioned at 5-μm thickness. Fetal ages range from 17 to 23 weeks of gestation.

Antibody Preparation

CXCR2 antibody is from Santa Cruz Biotechnology (Santa Cruz, CA). The CXCR4 monoclonal antibody 12G5 is from Dr. James Hoxie.[12] The

[11] P. W. Andrews, *Dev. Biol.* **103,** 285 (1984).
[12] M. J. Endres *et al.*, *Cell* (*Cambridge, Mass.*) **87,** 745 (1996).

Fy6 monoclonal antibody to the Duffy blood group antigen (DARC) is as previously described.[13] Polyclonal antisera to CCR1 and CCR5 are raised in New Zealand White rabbits by subcutaneous and intramuscular injection with the corresponding amino-terminal domains for CCR1 conjugated to keyhole limpet hemocyanin (KLH) and for CCR5 conjugated to glutathione *S*-transferase (GST). Following primary immunization and six challenges with peptide, CCR1 antiserum from several pooled bleeds is collected and purified over a 5 ml protein A-Sepharose HiTrap column (Pharmacia, Uppsala, Sweden). Following initial immunization with CCR5-conjugated peptide and three challenges with the antigen, aliquots of anti-CCR5 serum are absorbed with GST-glutathione-Sepharose to remove antibodies to the GST moiety of the fusion protein. The absorbed antiserum is used to probe Western blots containing GST, CCR5-GST, and the CCR5 peptide released from the fusion protein by thrombin cleavage. Purified antibody is analyzed by indirect immunofluorescence fluorescence-activated cell sorting (FACS) for binding to CCR1 and CCR5 transfected human kidney 293 cells. The CD4 antibody MT310, the CD68 antibody KPI, and the neurofilament antibody 2F11 are from DAKO (Carpinteria, CA). Monoclonal antibodies to glial fibrillary acidic protein (GFAP) are obtained from BioGenex (San Ramon, CA), and monoclonal antibodies to human NCAM are from Chemicon (Temecula, CA).

Immunohistochemical Methods

Both immunohistochemistry with diaminobenzidine (DAB) as chromogen and immunofluorescence are used to stain the fetal brain tissues and cultures. For an in-depth review of immunochemical methods, see elsewhere in this volume.[14] Many different variations of immunochemical methods are possible, and the ones we have chosen can be modified to suit different experimental conditions. In particular, we want to emphasize that we used Triton X-100 to permeabilize the cells to facilitate antibody penetration into the small, three-dimensional clusters of neurons. This procedure also allows antibodies and reagents to penetrate the cell somata, revealing both cytoplasmic and surface antigens.

In brief, after 12 to 20 days in culture, cells grown on 4-, 8-, or 16-well plastic (Thermanox) or glass chamber slides are selected for immunohistochemical staining. Live cells are rinsed in phosphate-buffered saline (PBS) and then fixed *in situ* with 2% paraformaldehyde and 1.5% sucrose in PBS,

[13] M. E. Nichols, P. Rubinstein, J. Barnwell, S. R. de Cordoba, and R. E. Rubinstein, *J. Exp. Med.* **166,** 776 (1987).

[14] A. R. Glabinski, M. Tani, V. K. Tuohy, and R. M. Ransohoff, *Methods Enzymol.* **288,** 182 (1997).

pH 7.3. The fixative is removed after 1 hr, and the cells are carefully washed twice in PBS and then stored in PBS at 4° in an airtight container until the time of immunostaining (never store for longer than 2 to 3 weeks unless PBS is replaced because evaporation will alter osmolarity). Before staining the cells are blocked for 30 min in PBS containing 0.005% Triton X-100 (v/v) and 10% (v/v) serum from either goat or donkey (depending on the species of the secondary antibody). The blocking solution is removed, and the primary antibodies (diluted in the same buffer used for blocking) are added without a rinse step. After 1 hr of incubation at room temperature, the primary antibodies are removed and the tissues are rinsed in PBS.

For immunohistochemistry the staining reaction is completed using a peroxidase-linked (BioGenex) avidin–biotin staining kit appropriate to the species of the primary antibody. The reaction is visualized with DAB as chromogen, and the cells are lightly counterstained with hematoxylin to increase nuclear definition. Cells grown on glass slides are dehydrated and coverslipped with Permount (Fisher Scientific, Pittsburgh, PA). Those grown on plastic slides are air-dried and then mounted with glass coverslips using Gelmount (Sigma). Cells are viewed on a Zeiss Axioskop (Zeiss, Thornwood, NY) and photographed with an attached Fuji HC-2000 (Fuji Medical Systems, Stamford, CT), three-chip CCD digital camera. Images are first edited in Adobe Photoshop 3.0 (Adobe Systems, San Jose, CA) to enhance contrast and white balance and then printed on a Fuji Pictrography 3000 digital printer.

For immunofluorescence studies, the slides are reacted with a secondary antibody (IgG) conjugated to fluorescein isothiocyanate (FITC), Texas Red, or rhodamine. After a 60-min incubation in the dark the slides are rinsed in PBS and coverslipped with an aqueous medium (Supermount, BioGenex). Ideally, a mountant containing an antifading agent such as DAKO Fluorescent Mounting Medium should be used to minimize the decay of fluorescence.

Immunohistochemical staining of the primary human brain cell cultures reveals the presence of the IL-8 receptor, CXCR2, in cells of neuronal phenotype (Fig. 2, see color insert). Some staining is also found in background cells, presumably of glial origin. CXCR2 is also found in fetal brain whole sections (Fig. 3, see color insert), consistent with its presence in adult brain.

Studies have been continued in hNT cells, revealing the presence of the chemokine receptor CXCR2 as well as receptors CCR1, CCR5, and fusin (CXCR4) on cultured cells with neuronal morphology (Fig. 4, see color insert). When primary antibodies are incubated with their appropriate receptor peptides before application, staining is blocked. Antibodies against DARC and CD4 fail to produce staining.

Determination of the exact cellular distribution of the antigens under study is difficult by light microscopy because of the three-dimensional aspects of the neuronal cell clusters. Although irrelevant antibody controls and peptide blocking produce negative staining, there is still a question of the DAB reaction product becoming trapped within the clusters and giving a false indication of the actual staining intensity and antigen distribution. It is also difficult to photograph stained cells in the larger clusters because a common focal plane cannot be found at higher magnifications. To address these issues we have turned to confocal microscopy, which allows us to make serial optical sections through the cell clusters.

Confocal Microscopy

Confocal microscopy uses an intense beam of laser light to scan a semitransparent, three-dimensional object having a thickness that is, optimally, less than 200 μm. The coherent laser light allows different focal planes to be imaged separately without diffusion. A photomultiplier tube picks up light that is either reflected or emitted (fluorescence) from the tissue and amplifies it. The image collected from each focal plane is digitized by computer and stored as part of a stack of images that can later be digitally reconstructed, rotated, or viewed as a movie. Examination of these images can give highly specific information about colocalization of antigens and the three-dimensional relationships among two or three different fluorescent probes. We have used this technology to verify that the chemokine receptors are indeed primarily associated with neuronal cells *in vitro* and to confirm that they are present in both cytoplasmic and cell surface locations.

hNT cells are double-labeled for confocal microscopy using FITC for one probe and either Texas Red or rhodamine for the other. This produces a good visual contrast and allows us to compare the distribution of the various chemokine receptors with the different neuronal and glial markers. The stained cells are viewed on an inverted confocal microscope, the MultiProbe 2010 (Molecular Dynamics, Sunnyvale, CA). Digitized images are saved as TIFF files for export to a Macintosh 8100 Power PC. On this platform the images are edited in Photoshop 4.0 (Adobe Systems) and printed on a Pictrography 3000 (Fuji).

Confocal imaging confirms the relationship between chemokine receptors and neuronal cells found in DAB-labeled sections. This technique clearly shows a granular staining pattern for the chemokine receptors CXCR2, CCR1, and CCR5 (Fig. 5, see color insert) and CXCR4 (data not shown). The receptors are strongly associated with cells of neuronal morphology, although background cells (partly differentiated stem cells) also show some light staining, particularly in paranuclear cytoplasmic areas consistent with the Golgi apparatus. In neuronal cells granular staining is found in both cytoplasmic and surface locations.

Acknowledgments

We thank Ms. Virginia Del Vecchio, Berlex Biosciences, for preparation of the slides and the immunostains. Dr. Susan Palmieri, Stanford University Cell Sciences Imaging Facility, kindly allowed us to view our samples with the MultiProbe 2010 confocal microscope (Molecular Dynamics). Dr. David Hanzel, Molecular Dynamics, was also very generous with time and equipment, helping us to download and process the confocal images.

[4] Expression of Chemokine Receptors in Insect Cells Using Baculovirus Vectors

By ZI-XUAN WANG, YING-HUA CEN, HAI-HONG GUO, JIAN-GUO DU, and STEPHEN C. PEIPER

Introduction

The objective of this chapter is to describe the application of a contemporary eukaryotic expression technology that is an alternative to traditional approaches for expression of receptors in mammalian cells. This strategy entails the production of a recombinant insect virus and infection of target cells, which typically result in expression of the gene of interest at high levels. Posttranslational modifications, such as proteolytic processing,[1–3] phosphorylation,[4–7] glycosylation (both N- and O-linked),[8–13] and acyla-

[1] A. M. Lebacq-Verheyden, P. G. Kasprzyk, M. G. Raum, K. Van Wyke Coelingh, J. A. Lebacq, and J. F. Battey, *Mol. Cell. Biol.* **8,** 3129 (1988).

[2] K. Kuroda, C. Hauser, R. Rott, H. D. Klenk, and W. Doerfler, *EMBO J.* **5,** 1359 (1986).

[3] K. Kuroda, A. Gröner, K. Frese, D. Drenckhahn, C. Hauser, R. Rott, W. Doerfler, and H.-D. Klenk, *J. Virol.* **63,** 1677 (1989).

[4] A. Höss, I. Moarefi, K. H. Scheidtmann, L. J. Cisek, J. L. Corden, I. Dornreiter, A. K. Arthur, and E. Fanning, *J. Virol.* **64,** 4799 (1990).

[5] H. Piwnica-Worms, N. G. Williams, S. H. Cheng, and T. M. Roberts, *J. Virol,* **64,** 61 (1990).

[6] R. E. Lanford, *Virology* **167,** 72 (1988).

[7] R. Herrera, D. Lebwohl, A. G. de Herreros, R. G. Kallen, and O. M. Rosen, *J. Biol. Chem.* **263,** 5560 (1988).

[8] K. Kuroda, H. Geyer, R. Geyer, W. Doerfler, and H.-D. Klenk, *Virology* **174,** 418 (1990).

[9] D. L. Jarvis and M. D. Summers, *Mol. Cell. Biol.* **9,** 214 (1989).

[10] M. Svoboda, M. Przybylski, J. Schreurs, A. Miyajima, K. Hogeland, and M. Deinzer, *J. Chromatogr.* **562,** 403 (1991).

[11] D. L. Domingo and I. S. Trowbridge, *J. Biol. Chem.* **263,** 13386 (1988).

[12] D. R. Thomsen, L. E. Post, and A. P. Elhammer, *J. Cell. Biochem.* **43,** 67 (1990).

[13] W. Y. Chen and O. P. Bahl, *J. Biol. Chem.* **266,** 8192 (1991).

0076-6879/97 $25.00

tion,[14–19] that are observed in host insect cell lines are similar to those observed in mammalian cells. In addition to providing an efficacious vehicle for expression of proteins, the phylogenetic distance between mammalian genes and the insect background results in an environment that potentially offers biological neutrality, that is, the possibility to reconstitute signaling pathways by expression of the individual elements. This raises the potential for the analysis of interactions between receptors and signal-transducing molecules in a system that permits coexpression of the proteins of interest in a milieu with minimal interfering activities.[20–23]

Baculoviruses are rod-shaped viruses that occur predominately in insects. Their genome is composed of double-stranded DNA ranging from 80 to 200 kilobase pairs (kbp).[24] The two occluded baculovirus (*Eubaculovirinae*) strains most commonly used as vehicles for expression are *Autographa californica* multiple nuclear polyhedrosis virus (AcMNPV) and *Bombyx mori* nuclear polyhedrosis virus (BmNPV). AcMNPV is the viral system more commonly used for *in vitro* culture applications because the target cell lines have a faster doubling time (18–24 hr versus 4–5 days for BmNPV),[25] a broad spectrum of transfer vectors derived from AcMNPV is available, and an array of promoters is active during infection. In contrast, BmNPV systems are preferable for *in vivo* expression because the infectable larvae are larger (~5 g) than those for AcMNPV (0.3–0.5 g). The genome of AcMNPV is 130 kbp. The ease with which the viral capsid can accept large inserts together with its large genome do not place any size restrictions for insertions into the viral genome. Inserts of approximately 15 kb have been reported. Thus, the major size limitation is imposed by the capacity of the transfer vector for efficient replication.

Insight into the biology of AcMNPV provides an important foundation

[14] R. E. Lanford, V. Luckow, R. C. Kennedy, G. R. Dreesman, L. Notvall, and M. D. Summers, *J. Virol.* **63,** 1549 (1989).

[15] L. Luo, Y. Li, and C. Y. Kang, *Virology* **179,** 874 (1990).

[16] S. Morikawa, T. F. Booth, and D. H. L. Bishop, *Virology* **183,** 288 (1991).

[17] P. N. Lowe, M. J. Page, S. Bradley, S. Rhodes, M. Sydenham, H. Paterson, and R. H. Skinner, *J. Biol. Chem.* **266,** 1672 (1991).

[18] M. Kloc, B. Reddy, S, Crawford, and L. D. Etkin, *J. Biol. Chem.* **266,** 8206 (1991).

[19] J. E. Buss, L. A. Quilliam, K. Kato, P. J. Casey, P. A. Solski, G. Wong, R. Clark, F. McCormick, G. M. Bokoch, and C. J. Der, *Mol. Cell. Biol.* **11,** 1523 (1991).

[20] D. A. Brickey, R. J. Colbran, Y. L. Fong, and T. R. Soderling, *Biochem. Biophys. Res. Commun.* **173,** 578 (1990).

[21] T. A. Vik, L. J. Sweet, and R. L. Erikson, *Proc. Natl. Acad. Sci. U.S.A.* **87,** 2685 (1990).

[22] L. L. Parker, S. Atherton-Fessler, M. S. Lee, S. Ogg, J. L. Falk, K. I. Swenson, and H. Piwnica-Worms, *EMBO J.* **10,** 1255 (1991).

[23] I. Freisewinkel, D. Riethmacher, and S. Stabel, *FEBS Lett.* **280,** 262 (1991).

[24] S. Burgess, *J. Gen. Virol.* **37,** 501 (1977).

[25] S. Maeda, *Annu. Rev. Entomol.* **34,** 351 (1989).

for understanding how this virus can be harnessed as a vehicle for expression of recombinant proteins.[26] Two mechanisms for the development of viral progeny are operative during AcMNPV infection. Viral particles that transverse the plasma membrane are formed by budding early during infection and represent the form primarily involved in dissemination of infection, both *in vivo* and *in vitro*. A second mechanism is the development of occluded virus. This mechanism, which is active in the nucleus in the late phases of infection, results in the formation of viral particles that are surrounded by a crystalline matrix composed of the polyhedron protein. Because viruses embedded in the matrix are resistant to degradation by environmental agents, this mechanism is critical for horizontal viral transmission between insects but is not necessary for lateral infection.

Thus, the polyhedron (*polh*) gene is not essential for replication in cell culture,[27] which enabled initial strategies to be based on replacement of this gene with a recombinant gene. This results in recombinant viruses that can be released by budding. Because they lack the ability to form nuclear occlusion bodies (occ^-) and are not stabilized by a matrix, they are not efficient agents for oral infection of insects and, therefore, do not persist in the environment. During infection with wild-type virus the *polh* promoter is highly active in the very late time course, corresponding to greater than 20 hr postinfection (pi).[28] At this time, when viral infection results in decreased expression of host genes, approximately 20% of mRNA transcripts are derived from the *polh* promoter,[29,30] and polyhedron levels in *Spodoptera frugiperda* (Sf) cell lines infected with wild-type virus accumulate to 0.5–1.0 g/ml, which corresponds to up to 50% of the total cellular protein. In this context, there are examples where expression of the target gene under the transcriptional control of the *polh* promoter in recombinant viruses has been reported to achieve the levels of endogenous *polh* expression during infection, which approaches 25–50% of total cellular protein.[31] However, it is clear that levels of expression are different for each target gene system, and more typically the recombinant protein constitutes 1–5% of total cellular protein. The absence of infectivity for insects in the environment coupled with the finding that the typical baculovirus promoters have minimal activity in mammalian cells[32,33] makes the potential for biohazard from such recombinant viruses negligible.

[26] L. K. Miller, J. E. Jewell, and D. Browne, *J. Virol.* **40,** 305 (1981).
[27] G. E. Smith, M. J. Fraser and M. D. Summers, *J. Virol.* **46,** 584 (1983).
[28] B. G. Ooi, C. Rankin, and L. K. Miller, *J. Mol. Biol.* **210,** 721 (1989).
[29] M. J. Adang and L. K. Miller, *J. Virol.* **44,** 782 (1982).
[30] P. Faulkner and E. B. Carstens, *Curr. Topics Microbiol. Immunol.* **131,** 1 (1986).
[31] Y. Matsuura, R. D. Possee, H. A. Overton, and D. H. L. Bishop., *J. Gen. Virol.* **68,** 1233 (1987).
[32] P. C. Hartig, M. C. Cardon, and C. Y. Kawanishi, *J. Virol. Methods* **31,** 335 (1991).
[33] F. M. Boyce and N. L. Bucher, *Proc. Natl. Acad. Sci. U.S.A.* **93,** 2348 (1996).

Insect expression systems have been used to express a variety of types of receptor proteins at high levels. They have been applied extensively to the expression of members of the serpentine receptor superfamily to perform functional studies and to purify the receptor protein for the production of immunologic reagents, as summarized in Table I. These receptors are predicted to have seven hydrophobic α helices assumed to function as transmembrane spanning domains. Most transmit signals through interaction with heterotrimeric guanine nucleotide binding proteins (G proteins). Members of this superfamily bind a broad spectrum of ligands and are found in organisms from bacteria (bacteriohodopsin, although it does not couple to G proteins) and yeast (pheromone receptors) to mammalian cells. The baculovirus expression system has been used to study functional aspects of serpentine receptor biology, such as ligand binding and consequent signaling events in insect cells (see Table I). Expression in insect cells of the human receptor for the bacterial polypeptide f-Met-Leu-Phe, which functions as a leukocyte chemoattractant, revealed that the receptor was not coupled to G proteins and could not bind the cognate ligand at high affinity in this environment.[34] The potential for a model system to study signaling and the possibility of purifying receptors for immunogens to produce immunological reagents provided a rationale for expressing chemokine receptors in insect cells using baculovirus vectors.

Chemokines bind to receptors that are members of the serpentine receptor superfamily. Few immunologic reagents that are specific for and block the biological activity of these receptors are available. Although the receptors are recognized to be coupled to G proteins, the available insight into mechanisms by which chemokine receptors transduce cytosolic signals following ligand binding has not yet reached the level that has been attained for other members of the serpentine receptor family. In fact, one chemokine receptor, the Duffy chemokine receptor (DARC), does not transmit a calcium flux following association with ligand, and it is unclear whether this receptor signals at all.[35] Finally, many chemokine receptors, CCR2, CCR3, and CCR5 in particular, are expressed at low levels in primary cells, as well as in transfectants. High-level expression of chemokine receptors in insect cells using baculovirus vectors provides an experimental approach that potentially offers a strategy for purification of native receptor protein for the purpose of producing immunologic reagents and a model system for studying signal transduction following interaction with ligands. This chapter describes our approach to the expression of receptors for members of the α- and β-chemokine families. The major aspects of this approach

[34] O. Quehenberger, E. R. Prossnitz, C. G. Cochrane, and R. D. Ye, *J. Biol. Chem.* **267,** 19757 (1992).

[35] K. Neote, J. Y. Mak, L. F. Kolakowski, and T. J. Schall, *Blood* **84,** 44 (1994).

TABLE I
EXPRESSION OF SERPENTINE RECEPTORS IN INSECT CELLS

Receptor	Source	Insect cell line	Vector	Biological feature(s)	Ref.
N-Formyl peptide (f-Met-Leu-Phe)	Human	Sf9	pVL1393	Low-affinity binding, lack of $G_i\alpha$ protein coupling	*a*
Gastrin-releasing peptide	Murine	Sf9	pVL1393	High-level expression, G-protein and phospholipase C coupling preserved, glycosylation	*b*
Y1 neuropeptide Y	Human	Sf21	pVL1392	High-level expression, normal K_D	*c*
Muscarinic acetylcholine (m1–m5)	Human, rat	Sf9	pVL1393	Same substrate specificity	*d*
Muscarinic acetylcholine (m1, m2)	Human	Sf9	pVL941	Phosphorylation, G-protein coupling, desensitization, pertussis toxin inhibition	*e*
Muscarinic acetylcholine (m3)	Rat	Sf9	pVL Mac3, pVLCam3	Receptor stabilized by ligand binding, heterogeneous glycosylation	*f*
5-Hydroxytryptamine (2C)	Rat	Sf9	IpDC-126	High-level expression, inositol phosphate linkage and antagonist function preserved	*g*
5-Hydroxytryptamine (1A)	Human	Sf9	pVL1392, pVL941	Inhibition of cAMP by agonists, Ser/Thr phosphorylation, desensitization	*h*
β-Adrenergic	Turkey	Sf9	pVL1392, pVL941	Purification, activate G_s in phospholipid vesicles	*i*
α-Adrenergic (2A-C10, 2B-C2)	Human	Sf9	pBlueBacll	Coupled to signal transduction, alter cAMP production	*j*
Cholecystokinin B	Human	Sf9	pVLMelMyc (pVL1392)	High-level expression, normal K_D, glycosylation, purification	*k*
Thrombin	Human	Sf9	pVL1392	Normal endogenous ligand activation, Ca flux	*l*
Neurotensin	Rat	Sf9	pVETLZ/*NheI*	Normal K_D, G-protein coupling, inositol phosphate signaling	*m*
Dopamine (D2L)	Human	Sf9	pJVETLZ	Phosphorylation, palmitoylation, ligand binding sensitive to G-protein coupling	*n*
Cannabinoid	Rat	Sf9	pVL1393	High-level expression, ligand binding	*o*
Prostaglandin E_2 (EP3)	Mouse	MG1	p2Bac	High-affinity binding, site-directed mutagenesis	*p*
Prostaglandin E_2 (EP3α)	Mouse	MG1	p2BAC	High-affinity binding, site-directed mutagenesis	*q*
Luteinizing hormone (LH)	Pig	Sf9	PVL941	High-level expression, high-affinity binding of ectodomain truncation	*r*
Adenosine (A1, A2A)	Human	Sf9	pVL1393	Epitope tagging, normal ligand binding	*s*
Melanocortin 1	Human	Sf9	pFASTBAC1	Epitope tagging, normal K_D	*t*
Oxytocin	Human	Sf9	pVLMelMyc (pVL1392)	Epitope tagging, low affinity binding, high affinity binding with cholesterol-cyclodextrin complex	*u*
Diuretic hormone	*M. sexta*	Sf9	pVL1393	High level expression, high-affinity binding, cAMP signaling	*v*

TABLE I (*continued*)

Receptor	Source	Insect cell line	Vector	Biological feature(s)	Ref.
Thyrotropin	Human	Sf9, High Five	pVL1393 pAcMP3	Novel vector with early promoter, expression of ectodomain, glycosylation, Graves disease autoantibody binding, ligand binding	*w*

[a] O. Quehenberger, E. R. Prossnitz, C. G. Cochrane, and R. D. Ye, *J. Biol. Chem.* **267,** 19757 (1992).
[b] T. Kusui, M. R. Hellmich, L. H. Wang, R. L. Evans, R. V. Benya, J. F. Battey, and R. T. Jensen, *Biochemistry* **34,** 8061 (1995).
[c] M. Munoz, M. Sautel, R. Martinez, S. P. Sheikh, and P. Walker, *Mol. Cell. Endocrinol.* **107,** 77 (1995).
[d] G. Z. Dong, K. Kameyama, A. Rinken, and T. Haga, *J. Pharmacol. Exp. Ther.* **274,** 378 (1995).
[e] R. M. Richardson and M. M. Hosey, *J. Biol. Chem.* **267,** 22249 (1992).
[f] S. Vasudevan, E. C. Hulme, M. Bach, W. Haase, J. Pavia, and H. Reilander, *Eur. J. Biochem.* **227,** 466 (1995).
[g] J. Labrecque, A. Fargin, M. Bouvier, P. Chidiac, and M. Dennis, *Mol. Pharmacol,* **48,** 150 (1995).
[h] C. G. Nebigil, M. N. Garnovskaya, S. J. Casanas, J. G. Mulheron, E. M. Parker, T. W. Gettys, and J. R. Raymond, *Biochemistry* **34,** 11954 (1995).
[i] A. Luxembourg, *Hybridoma* **14,** 261 (1995).
[j] C. C. Jansson, M. Karp, C. Oker-Blom, J. Nasman, J. M. Savola, and K. E. Akerman, *Eur. J. Pharmacol.* **290,** 75 (1995).
[k] G. Gimpl, J. Anders, C. Thiele, and F. Fahrenholz, *Eur. J. Biochem.* **237,** 768 (1996).
[l] X. Chen, K. Earley, W. Luo, S. H. Lin, and W. P. Schilling, *Biochem. J.* **314,** 603 (1996).
[m] H. Boudin, J. Labrecque, A. M. Lhiaubet, M. Dennis, W. Rostene, and D. Pelaprat, *Biochem. Pharmacol.* **51,** 1243 (1996).
[n] G. Y. Ng, B. F. O'Dowd, M. Caron, M. Dennis, M. R. Brann, and S. R. George, *J. Neurochem.* **63,** 1589 (1994).
[o] D. A. Pettit, V. M. Showalter, M. E. Abood, and G. A. Cabral, *Biochem. Pharmacol.* **48,** 1231 (1994).
[p] C. Huang and H. H. Tai, *Biochem. J.* **307,** 493 (1995).
[q] C. Huang and H. H. Tai, *Arch. Biochem. Biophys.* **327,** 161 (1996).
[r] E. Pajot-Augy, L. Couture, V. Bozon, J. J. Remy, G. Biache, M. Severini, J. C. Huet, J. C. Pernollet, and R. Salesse, *J. Mol. Endocrinol.* **14,** 51 (1995).
[s] A. S. Robeva, R. Woodard, D. R. Luthin, H. E. Taylor, and J. Linden, *Biochem. Pharmacol.* **51,** 545 (1996).
[t] H. B. Schioth, A. Kuusinen, R. Muceniece, M. Szardenings, K. Keinanen, and J. E. Wikberg, *Biochem. Biophys. Res. Commun.* **221,** 807 (1996).
[u] G. Gimpl, U. Klein, H. Reilander, and F. Fahrenholz, *Biochemistry* **34,** 13794 (1995).
[v] J. D. Reagan, *Insect Biochem. Mol. Biol.* **25,** 535 (1995).
[w] G. D. Chazenbalk and B. Rapoport, *J. Biol. Chem.* **270,** 1543 (1995).

are (1) the preparation of the chemokine receptor transfer vector construct, (2) cotransfection of the transfer vector construct and linear baculovirus DNA to rescue infectious virus, (3) characterization and purification of the recombinant virus, and (4) infection of target cells for the analysis of chemokine receptor expression.

Preparation of Transfer Vector Construct

The open reading frame (ORF) encoding the majority of the chemokine receptors is contained in a single, uninterrupted exon. Thus, it is possible to derive the sequences to direct expression of the chemokine receptors of interest using polymerase chain reaction (PCR) to amplify products from genomic DNA templates. This strategy serves to decrease the risk of introducing sequence errors by PCR amplification following reverse transcription of complementary DNA (cDNA) from mRNA templates. Amplification of the coding sequences by PCR also provides the opportunity to insert appropriate restriction endonuclease cleavage sites to enable subsequent forced, directional cloning of the sequences encoding the chemokine receptor into the transfer vector. This serves as a facile alternative to standard molecular cloning approaches.

A variety of transfer vectors for subcloning the chemokine receptor coding sequences are commercially available. These vectors contain segments of AcMNPV to permit homologous recombination with the replication incompetent linearized variant of the parental virus. The majority use the *polh* promoter and have varying repertoires of cloning sites. We have used several transfer vectors but have had consistent success with pVL1393 (pVL1392 is identical, but the orientation of the polylinker sequences is reversed). This vector contains a small segment of the *polh* open reading frame upstream of the multiple cloning site that serves to enhance the expression level of the target gene, which is located at position +35 of this gene. Although the ATG codon has been mutated to ATT, there is some evidence that this modified codon may be utilized in rare instances. Thus, if problems in expression of the targeted gene are encountered, it is necessary to determine whether the translational frame matches that of *polh,* raising the possibility of utilization of the upstream ATT codon at the +1 position. The pAcC12/13 vectors are closely related to pVL1392/3 but have a different polylinker sequence. We routinely confirm the nucleotide sequence of the construct, particularly if the insert has been created by PCR.

Infectious recombinant baculoviruses are generated by rescue of viral variants through complementation by homologous recombination with the recombinant transfer plasmid. Linearized viral DNAs for cotransfection with transfer vectors using *polh* promoters are commercially available. We have used AcMNPV DNA linearized by cleavage at three genetically engineered *Bsu*361 restriction endonuclease sites, including one in the 3′ region of ORF1629, which results in a deletion of sequences that interfere with expression of a protein essential for viral replication. This minimizes the production of nonrecombinant viruses following cotransfection of the linearized (replication incompetent) viral DNA. Rescue of infectious viruses is dependent on homologous recombination with the transfer plasmid, which contains the sequences deleted from ORF1629 in the linearized viral DNA. Similar products are available from many biotechnology companies (Invitrogen, San Diego, CA; Pharmingen, San Diego, CA; and Clontech, Palo Alto, CA), although we only have experience with reagents from Invitrogen and Clontech.

Cell Culture and Cotransfection for Rescue of Infectious Recombinant Virus

The insect cell lines that are the most common targets for cotransfection for the production of recombinant virus and infection with the virus are the lepidopteran line Sf21, which is derived from the ovary of the fall armyworm (*Spodoptera frugiperda*) pupa, and the subclone of this line, Sf9. The latter cell line assumes a more uniform shape during growth and, thus, forms monolayers that are more regular. Plaques generated in this background are better defined and easier to recognize. High Five cells, derived from ovary of the cabbage looper (*Trichoplusia ni*), are more spindle-shaped and can be used for the amplification of viral supernatants for the production of high-titered stocks. We have used the High Five cell line (Invitrogen) for cotransfection with limited success. These cells are recommended primarily for high-level expression of (secreted) proteins, where they are more efficient than the other two lines. We have been unable to apply the advantage of the purported high level of expression of High Five cells to chemokine receptors because this line showed significant nonspecific binding to mouse myeloma proteins, both monoclonal antibodies as well as control immunologic reagents.

The two Sf cell lines may be grown in suspension or monolayer culture. In the latter setting, the cells are loosely adherent. During culture they form a monolayer, and as saturation density is approached nonadherent cells increase in number. When grown at room temperature these cell types can be passaged approximately once per week, but when grown in an incubator at 27° this interval is decreased to approximately every 4 days. They do not require carbon dioxide for growth. Cells may be removed from the flask by physical disruption, optimally by gentle tapping and pipetting. The yield of viable cells obtained is better when harvesting is performed without excessive traumatic shock. The cells are grown in Grace's medium supplemented with 10% fetal bovine serum, but they can be adapted for growth without fetal calf serum. Serum-free supplements are now available to maintain cells during the production of secreted proteins to facilitate purification.

The culture conditions for the Sf target cell lines are critical for successful cotransfection, infection, and protein expression. Adherent cultures should be maintained in log growth and passaged when the cells reach confluency, the point when the surface of the culture vessel is completely covered, but clusters of cells have not yet formed and cells have not begun to detach. Habitual passage of cells after confluence has been reached results in decreases in viability, attachment, and increased doubling times. Passaging of cells prior to confluency also has similar negative effects on the quality of the culture, including tighter adherence to the culture dish. Cultures should be passaged at approximately five-fold dilutions of the cells that can readily be released from the flask (which constitute approximately 50% of the total adherent cells). Periodic analysis of doubling time and viability should be performed, with favorable results in the range of 18–24 hr and greater than 95%, respectively. Cultures with altered doubling times and/or decreased viability should not be used for infection or expression experiments. When Sf cell cultures have been passaged extensively (≥30 times) they become less suitable for transfection and infection experiments, and fresh cultures of cells should be initiated.

To generate an infectious recombinant virus, the transfer vector containing the ORF encoding the chemokine receptor is cotransfected along with the linearized AcMNPV viral DNA into Sf9 or Sf21 cells using cationic liposomes. The insect cell lines used as targets in these experiments are relatively resistant to the toxic effects of these agents, and thus a conspicuous level of cell death, as is characteristic of mammalian cells, is not seen. As several detailed protocols are available for this procedure, only the ciritical issues are discussed. The first is the condition of the target cells. The Sf cells from cultures with at least 90% viability should be plated at 0.8×10^6 to 1.0×10^6 cells in a 35-mm dish. Optimal cotransfection efficiencies occur when the cells are at 50–70% confluence. The cells should be allowed to attach to the dish for at least 1 hr, and this should be confirmed by examination by phase microscopy. A second major consideration is the quality of the DNA to be transfected. Sf cells are sensitive to contaminants in DNA preparations, and control wells without linearized virus may be analyzed for toxic effects of plasmid DNA preparations in transfection experiments. DNA should be prepared using a resin-based purification method or cesium chloride banding. We typically use plasmid that has been extracted using kits from Qiagen (Chatsworth, CA), precipitated with ethanol, and washed extensively with 70% (v/v) ethanol to further remove contaminants.

Purification and Characterization of Recombinant Virus

Following the cotransfection the cells are grown for approximately 72 hr, the medium is harvested, and fresh medium is added. This medium from the second phase is harvested after growth for an additional 3–4 days. At the end of this second interval of growth it is possible to see some morphological evidence of baculovirus infection, although the virus, if present, is at a low titer. The first two culture supernatants are pooled, and virus is amplified by infection of fresh Sf cells. As the titer increases in the stock supernatant, incubation with fresh Sf cells will result in morphological manifestations of viral infection that are more readily detectable. When this cytopathological effect is evident, it is appropriate to verify and purify the recombinant virus by plaque assay in monolayer cultures overlaid with soft agar. The use of trypan blue or neutral red facilitates the identification and quantitation of viral plaques because the Sf cells lysed by viral infection take up the vital dye. At this point it is possible to pick several single plaques, expand and amplify them by reinfection of Sf cells, confirm the presence of the insert by PCR, and prepare viral stocks from clonal viral isolates derived from a single plaque. However, plaques should not be isolated from cultures containing the vital dyes because of the potential for the introduction of mutations by these compounds. Viral stocks should be amplified from single plaques that have been shown to contain a recombinant virus. Once high-titered viral stocks are obtained and titered using plaque assays, it is possible to infect target cells to determine whether the virus efficiently directs the expression of the gene of interest (i.e., chemokine receptor).

Analysis of Chemokine Receptor Expression

Multiple strategies can be applied to determine the level of expression of chemokine receptor expression. Ligand-binding studies have been used to detect and quantitate the expression of many mammalian homologs of members of the serpentine receptor family. In almost all of these instances, in which high-level expression of receptors with K_D values virtually identical to those observed in the mammalian expression system was obtained, the ligand was a biochemical compound or a short, highly conserved polypeptide, but not a larger protein ligand. Some serpentine receptors may not be capable of binding to their cognate ligands at high affinity when uncoupled from $G_i\alpha$ proteins by pertussis toxin. This agent has been shown to abrogate interleukin-8-induced signaling in neutrophils. The importance of $G_i\alpha$ protein coupling of chemoattractant receptors is further supported by the finding that only low-affinity binding of f-Met-Leu-Phe was observed when the human *N*-formyl peptide receptor was expressed in Sf9 cells using a baculovirus expression system, presumably owing to the lack of appropriate G_i proteins.[34] It has been our experience (Z.-X. Wang, R. Horuk, and S. C. Peiper, personal communication) that ligand binding is not an effective approach to determine the expression of chemokine receptors in insect cell lines, although specific details are beyond the scope of this chapter. Although it is possible to epitope tag the receptors, we have avoided this approach because it introduces a variable that could conceivably alter the conformation of the receptor.

Thus, we have relied on immunologic approaches to demonstrate expression of chemokine receptors, because the flow cytometric analysis of viable cells exclusively detects the presence of proteins on the cell surface. Monoclonal antibodies to DARC, CXCR2, and CCR5 that recognize epi-

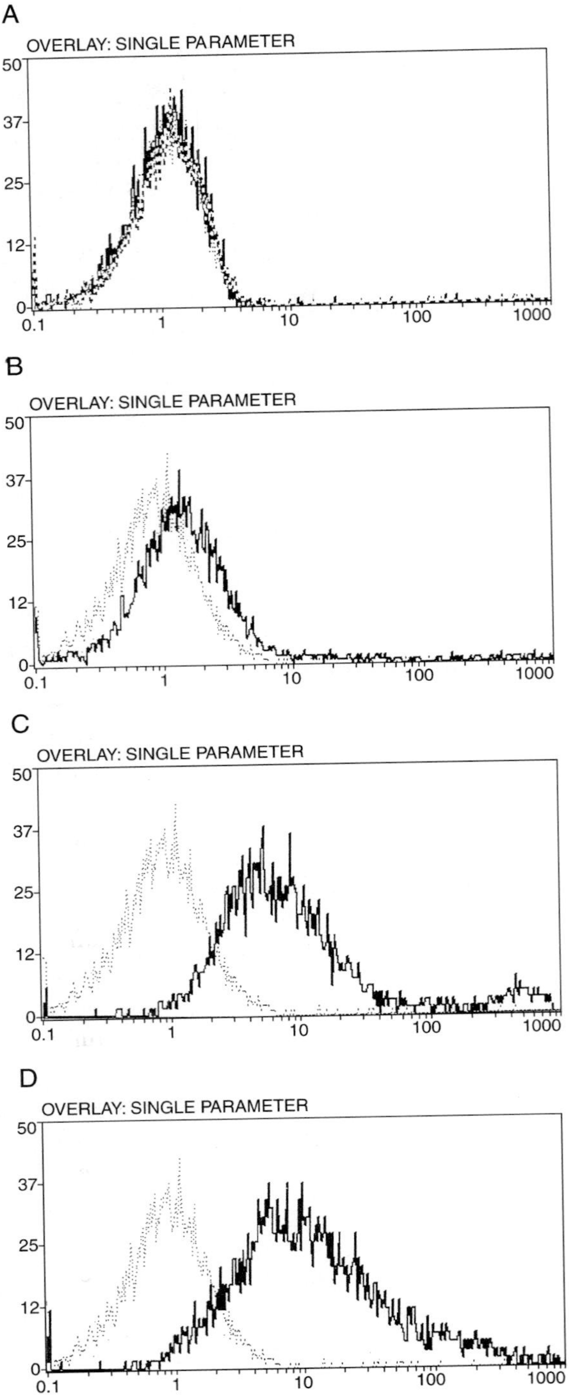
A
OVERLAY: SINGLE PARAMETER
50
37
25
12
0
0.1
1
10
100
1000
B
OVERLAY: SINGLE PARAMETER
50
37
25
12
0
0.1
1
10
100
1000
C
OVERLAY: SINGLE PARAMETER
50
37
25
12
0
0.1
1
10
100
1000
D
OVERLAY: SINGLE PARAMETER
50
37
25
12
0
0.1
1
10
100
1000

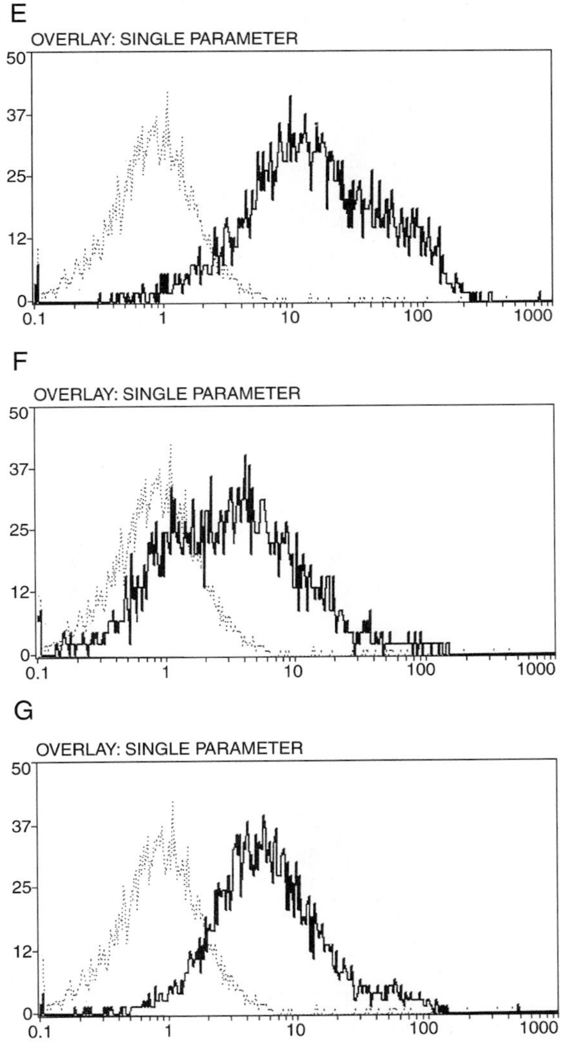

FIG. 1. Flow cytometric analysis of baculovirus-mediated expression of DARC in insect cells. Sf21 cells were analyzed for expression of DARC by immunofluorescent staining and flow cytometry using Fy6, a monoclonal antibody to the amino-terminal ectodomain. Cells were analyzed prior to infection (A) and at varying intervals following infection with a recombinant baculovirus encoding DARC under the transcriptional control of the *polh* promoter (B, 12 hr; C, 18 hr; D, 24 hr; E, 36 hr; F, 48 hr; G, 60 hr). Monoclonal antibodies: anti-Fy6, solid line; control mouse myeloma protein, broken line.

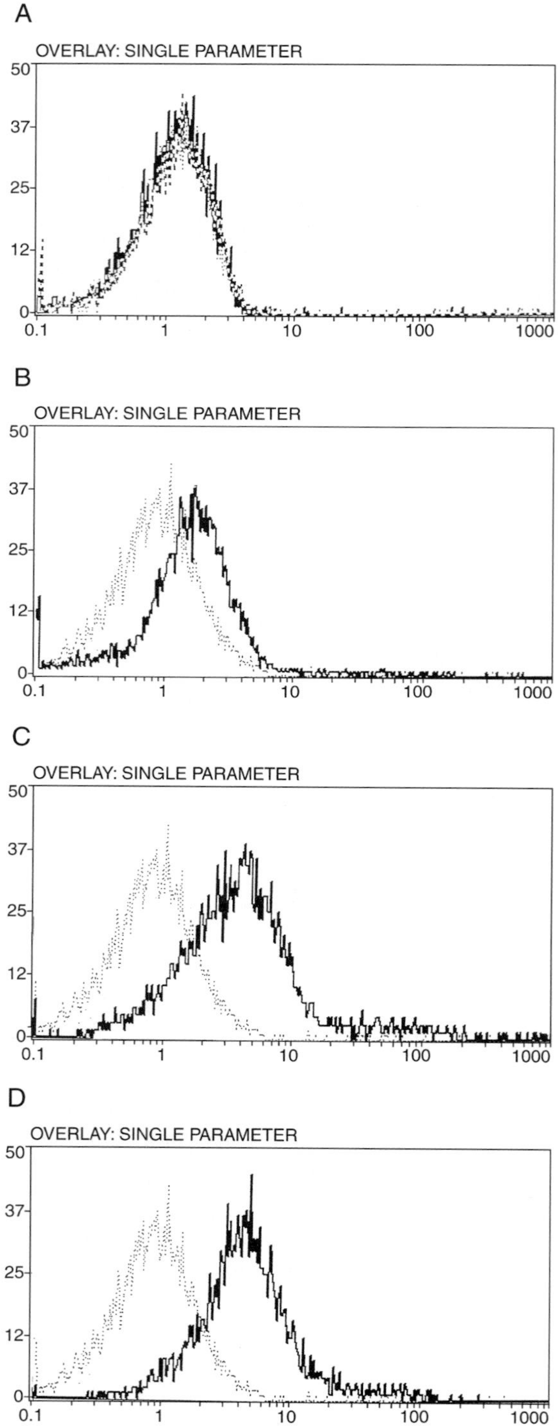
A
OVERLAY: SINGLE PARAMETER
50
37
25
12
0
0.1
1
10
100
1000
B
OVERLAY: SINGLE PARAMETER
50
37
25
12
0
0.1
1
10
100
1000
C
OVERLAY: SINGLE PARAMETER
50
37
25
12
0
0.1
1
10
100
1000
D
OVERLAY: SINGLE PARAMETER
50
37
25
12
0
0.1
1
10
100
1000

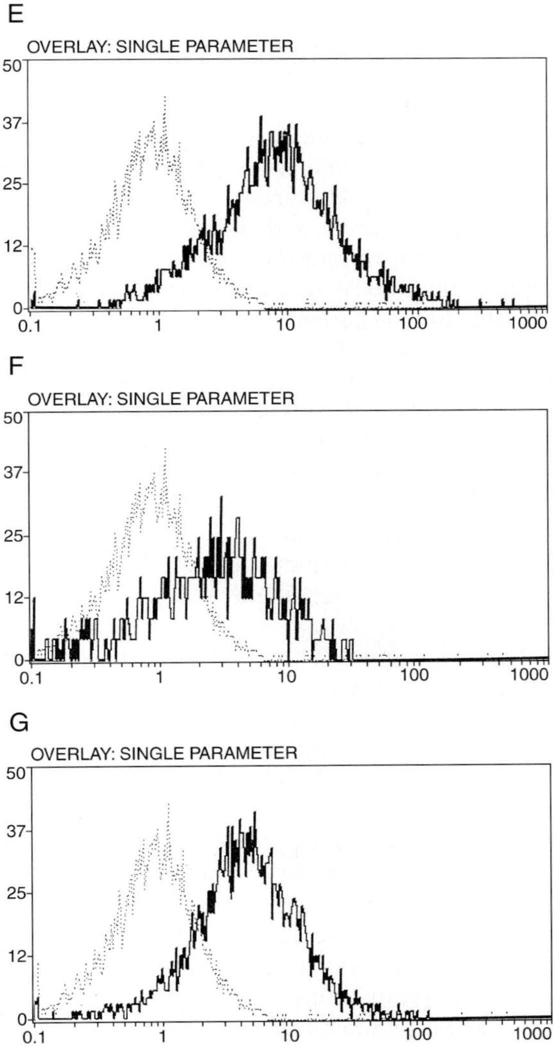

FIG. 2. Parallel analysis of DARC expression shown in Fig. 1 using a monoclonal antibody that recognizes an epitope in the extracellular loops of DARC. Reactivity with control cells is shown in (A) and at various time points postinfection (B)–(G) (B, 12 hr; C, 18 hr; D, 24 hr; E, 36 hr; F, 48 hr; G, 60 hr). Monoclonal antibodies: anti-Fy3, solid line; control mouse myeloma protein, broken line.

topes in predicted extracellular domains have been used for this purpose. In many instances, it has been possible to demonstrate the specificity of the immunologic reagents by blocking their binding with polypeptides that were used as immunogens.

Optimization of several parameters of infection is required to attain optimal expression of the chemokine receptor in the target insect cell line. It is important to utilize Sf cells grown under optimal conditions (i.e., maintained in log phase growth and having a high viability, ≥95%) as targets for infection in expression assays. Cells should be plated to achieve densities of approximately 90% in culture dishes (3×10^6 cells in 5 ml medium per 60-mm dish). The cells should be allowed to adhere for at least 1 hr prior to infection. The multiplicity of infection (MOI) is an important factor in the expression of recombinant proteins because the synchronous infection of Sf cells enhances the efficiency of expression. We try to infect cultures at an MOI of at least 10 and preferably 20. This requires 3.0×10^7 to 6.0×10^7 viral particles (for the 3.0×10^6 cells seeded), as calculated from titers deduced from the plaque assays.

The second major parameter to be optimized is the time course of infection. Typically, we infect replicate cultures at time zero and process the cells for expression of the chemokine receptor by immunofluorescence staining and flow cytometry at the various time points (18, 24, 36, 48, and 60 hr pi). Following indirect immunofluorescence staining with the appropriate immunologic reagents using standard methods, the cells are fixed with paraformaldehyde in phosphate-buffered saline and held at 4° in the dark until the batch of samples can be analyzed by flow cytometry using an Epics Elite flow cytometer (Coulter Corporation, Miami, FL).

We have used the baculovirus system to express CXCR2, DARC, CCR1, and CCR5 on the plasma membrane of Sf21 and Sf9 insect cell lines. These two lines have yielded similar levels of chemokine receptor expression following baculovirus infection, and neither offers a significant advantage for this application. A time course experiment for expression of DARC in Sf21 cells is shown in Figs. 1 and 2. Cells were infected at an MOI of 20 and incubated at room temperature for the period indicated. Expression was analyzed by indirect immunofluorescence using one monoclonal antibody that binds to an epitope in the N-terminal ectodomain (Fy6) (Fig. 1) and a second that binds to an epitope in the extracellular loops of the receptor (Fy3) (Fig. 2). Expression of the receptor is inconspicuous at 12 hr, but clearly evident at 18 hr. The expression detected by flow cytometry reaches maximal levels by 36 hr, after which time there may be some decrease in the mean peak fluorescence among the viable cells, which are identified by electronic bit-mapped gating. Parallel analysis of forward angle and orthogonal light scatter characteristics of the cells reveals increasing changes associated with cell death. Thus, for many applications, it is prefera-

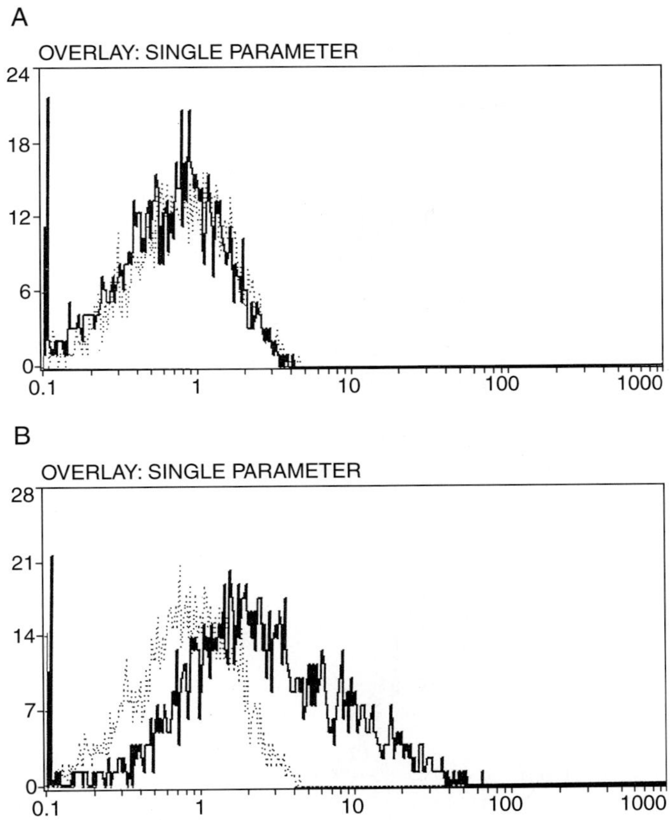

FIG. 3. Expression of CXCR2 using the baculovirus system. Sf21 cells were analyzed prior to (A) and 36 hr following (B) infection with a recombinant baculovirus encoding CXCR2. Cells were stained with a monoclonal antibody to CXCR2 (kindly provided by J. Kim, Genentech, South San Francisco, CA) (solid line) or a control mouse myeloma protein (broken line). Following washing and staining with a fluoresceinated secondary antibody, the cells were washed, fixed, and analyzed by flow cytometry.

ble to analyze the cells at an earlier time after infection to optimize the viability of the infected cells and avoid protein degradation due to cell death.

Figures 3 and 4 demonstrate similar high-level expression of CXCR2 and CCR5 in Sf21 cells at 36 hr pi. It is of note that levels of CCR5 expression are similar to those obtained with the other receptors despite the difficulty in expressing this receptor in stable transfectants. The specificity of the binding of the 12D1 monoclonal antibody to CCR5 is demonstrated in Fig. 4. This monoclonal antibody was produced to a glutathione *S*-transferase (GST) fusion protein that contained the N-terminal extracellular

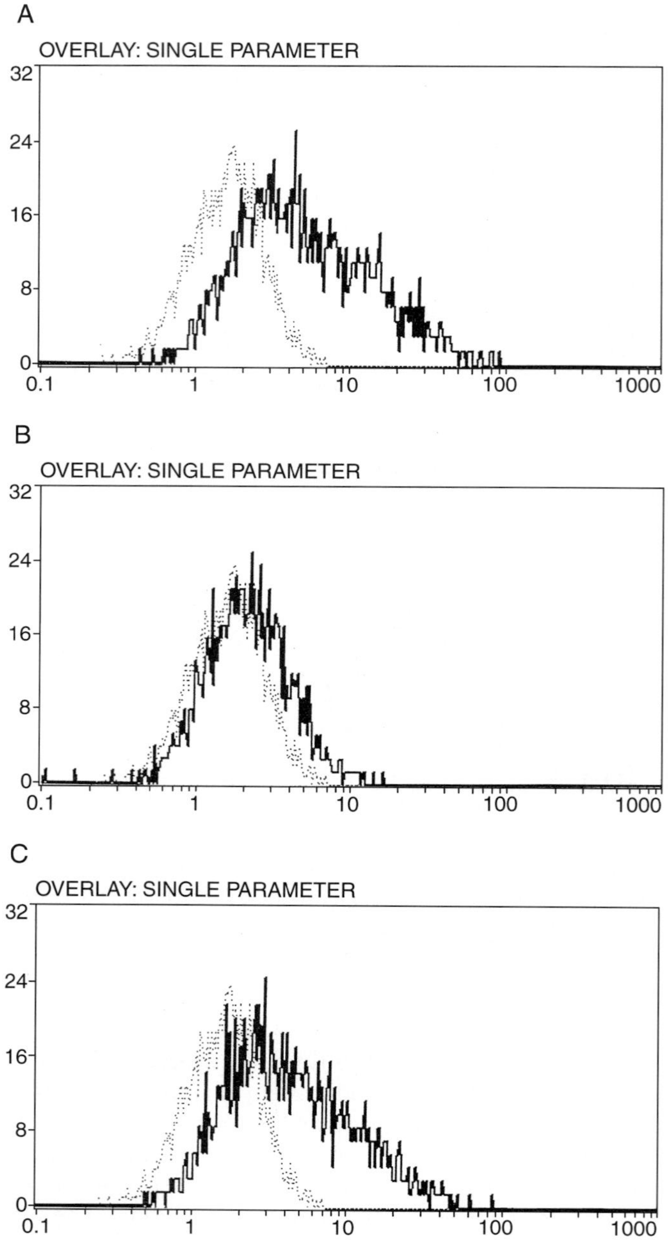
A
OVERLAY: SINGLE PARAMETER
32
24
16
8
0
0.1
1
10
100
1000
B
OVERLAY: SINGLE PARAMETER
32
24
16
8
0
0.1
1
10
100
1000
C
OVERLAY: SINGLE PARAMETER
32
24
16
8
0
0.1
1
10
100
1000

domain of CCR5 (amino acid residues 1–35). This domain of CCR5 was liberated from the fusion protein by cleavage with thrombin and purified by chromatography. Preincubation of the monoclonal antibody with the CCR5 N-terminal polypeptide at concentrations as low as 10 n*M* resulted in inhibition of binding to the transfectants (data not shown).

In summary, the baculovirus system is an effective approach for the expression of chemokine receptors. This will, at minimum, provide a practical, efficient method for the characterization and standardization of immunologic reagents for chemokine receptors. In addition, we have used this system to map the epitopes recognized by mononclonal antibodies to chemokine receptors using receptor chimeras.[36] Judging from the marked loss in ligand-binding activity of the formyl peptide receptor when expressed in Sf9 cells observed by Quehenberger *et al.*,[34] this system may not lend itself to the study of chemokine binding. However, these results raise the possibility that this may be an excellent system for studying the linkage of chemokine receptors to the α subunits of G proteins.

Note Added in Proof

Subsequent to the submission of this chapter, a transposon-based system has become available for preparing recombinant baculoviruses (Life Technologies, Gaithersburg, MD). The target gene is cloned into a transfer vector under the control of the *polh* promoter and flanked by transposable elements. This construct is then transformed into *Escherichia coli* containing a baculovirus shuttle vector. This bacmid contains a transposon attachment site within the *lacZ* gene, as well as the AcNPV genome. Site-specific transposition of the target gene inactivates the *lacZ* gene, which enables selection of white recombinant colonies that contain the expression cassette in the context of the baculovirus shuttle vector. Recombinant bacmids confirmed by mapping or PCR are directly transfected into insect cell lines, and the consequent infectious recombinant baculovirus is harvested from the supernatant. We have used this approach to develop a recombinant virus that directs the expression of human CXCR4 at high levels. This system promises to facilitate and streamline the production of recombinant baculoviruses.

[36] Z.-H. Lu, Z.-X. Wang, R. Horuk, J. Hesselgesser, Y. C. Lou, T. J. Hadley, and S. C. Peiper, *J. Biol. Chem.* **270,** 26239 (1995).

FIG. 4. Staining with anti-CCR5 monoclonal antibody is receptor specific. Sf21 cells infected with a recombinant baculovirus encoding CCR5 for 24 hr. Cells were stained with a monoclonal antibody to the amino-terminal ectodomain of CCR5 (A, B, C). The staining was performed in the presence of excess quantities of a recombinant peptide corresponding to the amino-terminal ectodomain of CCR5 (B) and CXCR4/LESTR (C). Monoclonal antibodies: 12D1, solid line; control myeloma protein, broken line.

[5] Chimeric Chemokine Receptors for Analysis of Structure–Function Relationships

By STEPHEN C. PEIPER, ZHAO-HAI LU, TIAN-YUAN ZHANG, and ZI-XUAN WANG

Introduction

Receptors are prime targets for structure–function studies because they represent the initial mechanism for triggering cellular responses to physiological and pathological stimuli. The armamentarium of approaches for the study of structure–function relationships includes several variations on mutagenesis to generate receptor variants that may provide insight into domains that directly impact biological activity. There are several major families of receptors expressed on the plasma membrane of cells, including those with tyrosine kinase activity, those with serine–threonine kinase activity, members of the cytokine receptor family, and serpentine receptors, which are typically coupled to G proteins. The latter superfamily contains more than 300 members, with a projection of more than 1000 total, and has a significant degree of sequence conservation.[1,2] This level of similarity of primary structure indicates that the members of this superfamily share a common topology, which is characterized by seven hydrophobic α helices that are predicted to function as transmembrane spanning domains.

The oldest member of this family, and perhaps the best understood at the structural level, is bacteriorhodopsin, a bacterial proton pump. High resolution structural analysis has elucidated a topology characterized by an amino-terminal extracellular domain, three extracellular loops, three cytoplasmic loops, and a carboxyl-terminal tail, each of which is physically demarcated by a transmembrane spanning domain.[3] Analysis of the photoreceptor bovine rhodopsin by two-dimensional electron crystallography at 9 Å resolution indicates differences in the conformation of the transmembrane spanning helices from that of bacteriorhodopsin but reveals a similar overall architecture.[4] Although all other types of receptors also have an extracellular and a cytoplasmic component, most of the functionally relevant domains/motifs lack topological boundaries that are physical in nature. In contrast, the members of the serpentine receptor family are, in essence,

[1] J. M. Baldwin, *Curr. Opin. Cell Biol.* **6,** 180 (1994).
[2] T. Gudermann, B. Nürnberg, and G. Schultz, *J. Mol. Med.* **73,** 51 (1995).
[3] J. Hoflack, S. Trumpp Kallmeyer, and M. Hilbert, *Trends Pharmacol. Sci.* **15,** 7 (1994).
[4] G. F. X. Schertler, C. Villa, and R. Henderson, *Nature* (*London*), **362,** 770 (1993).

METHODS IN ENZYMOLOGY, VOL. 288
0076-6879/97 $25.00

composed of eight domains, which could be considered as a type of modular architecture. The redundancy of receptors in this family, that is, the existence of multiple receptors with similar or overlapping repertoires of ligand binding, as well as the limited structural complexity of many of the ligands, makes them good candidates for characterization by exchanging modular domains between similar receptors. The creation of receptor chimeras, hybrid molecules composed of complementary portions of related receptors that have distinguishing biochemical or biological characteristics, is a powerful genetic engineering approach to the identification of domains that impact activity, such as specificity of chemokine[5–10] and human immunodeficiency virus type-1 (HIV-1) envelope glycoprotein (gp120) binding in extracellular domains, signaling activity through coupling with G proteins by sequences in the cytosolic domains, and HIV-1 fusion cofactor activity.[11,12]

The receptor chimera strategy has been successfully applied to members of the serpentine receptor family and, specifically, to chemoattractant receptors. Surprisingly, the findings from the application of this approach have demonstrated that different types of serpentine receptors have biological functions localized to different domains. For example, the sequences that form the binding pocket of receptors for small, biochemical ligands, such as chemoattractants, typically involve the transmembrane helices,[13–16] whereas sequences that confer ligand binding specificity to larger, polypeptide li-

[5] R. B. Gayle III, P. R. Sleath, S. Srinivason, C. W. Birks, K. S. Weerawarna, D. P. Cerretti, C. J. Kozlosky, N. Nelson, T. Vanden Bos, and M. P. Beckman, *J. Biol. Chem.* **268,** 7283 (1993).

[6] G. J. LaRosa, K. M. Thomas, M. E. Kaufman, R. Mark, M. White, L. Taylor, G. Gray, D. Witt, and J. Navarro, *J. Biol. Chem.* **267,** 25412 (1992).

[7] Z.-H. Lu, Z.-X. Wang, R. Horuk, J. Hesselgesser, Y.-C. Lou, T. J. Hadley, and S. C. Peiper, *J. Biol. Chem.* **270,** 26239 (1995).

[8] L. Boring, J. Gosling, F. S. Monteclaro, A. J. Lusis, C. L. Tsou, and I. F. Charo, *J. Biol. Chem.* **271,** 7551 (1996).

[9] F. S. Monteclaro and I. F. Charo, *J. Biol. Chem.* **271,** 19084 (1996).

[10] S. K. Ahuja, J. C. Lee, and P. M. Murphy, *J. Biol. Chem.* **271,** 225 (1996).

[11] J. Rucker, M. Samson, B. J. Doranz, F. Libert, J. F. Berson, Y. Yi, R. G. Collman, C. C. Broder, G. Vassart, R. W. Doms, and M. Parmentier, *Cell (Cambridge, Mass.)* **37,** 457 (1996).

[12] R. E. Atchison, J. Gosling, F. S. Monteclaro, C. Franci, L. Digilio, I. F. Charo, and M. A. Goldsmith, *Science* **274,** 5294 (1996).

[13] J. A. DeMartino, Z. D. Konteatis, S. J. Siciliano, G. Van Riper, D. J. Underwood, P. A. Fisher, and M. S. Springer, *J. Biol. Chem.* **270,** 15966 (1995).

[14] S. J. Siciliano, T. E. Rollins, J. DeMartino, Z. Konteatis, L. Malkowitz, G. Van Riper, S. Boudy, J. Rosen, and M. H. Springer, *Proc. Natl. Acad. Sci. U.S.A.* **91,** 1214 (1994).

[15] H. D. Perez, R. Holmes, L. R. Vilander, R. R. Adams, W. Manzana, D. Jolley, and W. J. Andrews, *J. Biol. Chem.* **268,** 2292 (1993).

[16] O. Quehenberger, E. R. Prossnitz, S. L. Cavanagh, C. G. Cochrane, and R. D. Ye, *J. Biol. Chem.* **268,** 18167 (1993).

gands, including chemokines, tend to be localized to the extracellular domain(s), typically the amino terminus.[5–7,17,18] Similarly, whereas the sequences involved in coupling to G proteins of the adrenergic[19,20] and α-factor[21] receptors have been mapped to the third cytoplasmic loop, analogous sequences in the *N*-formyl peptide (chemoattractant) receptor appear to be contained in the second loop and the cytoplasmic tail.[22]

The goal of this chapter is to describe strategies for making chemokine receptor chimeras, including (1) theoretical considerations for creating hybrids with the proper topological architecture for intracellular trafficking and expression on the cell surface, (2) genetic engineering technical approaches for producing hybrid receptors with the desired composition, and (3) experimental methods for the analysis expression on the cell surface. Experimental approaches for designing and assembling chimeras composed of one domain each from two parental receptors are illustrated. However, the techniques and considerations presented can be adapted to generate more complex chimeric receptors.

Genetic Engineering of Chemokine Receptor Chimeras: Theoretical Considerations

The initial decision in the development of receptor chimeras regards the selection of the receptor pair that will serve as donors of sequences. Clearly, this is largely dependent on the nature of the function to be studied. In most instances, a specific aspect of the biology of a particular receptor will be analyzed; thus, one receptor can often be considered a "host" and the second receptor, which lacks the function under investigation, as a "graft." In this way, modules of the host receptor are replaced by the corresponding modules of the graft, and the resulting chimeric receptors can be tested in functional assays. In general, receptors that are closely related at the level of primary structure but differ in critical aspects of function are selected for the creation of chimeras. In some instances, there is sufficient functional divergence in the same receptor type from different

[17] I. Ji and T. Ji, *J. Biol. Chem.* **266,** 14953 (1991).

[18] S. Kosugi, T. Ban, T. Akamizu, and L. D. Kohn, *Mol. Endocrinol.* **6,** 166 (1992).

[19] M. A. Kjeldsberg, S. Cotecchia, J. Ostrowski, M. G. Caron, and R. J. Lefkowitz, *J. Biol. Chem.* **267,** 1430 (1992).

[20] S. Cotecchia, S. Exum, M. G. Caron, and R. J. Lefkowitz, *Proc. Natl. Acad. Sci. U.S.A.* **87,** 2896 (1990).

[21] C. Boone, N. G. Davis, and G. F. Sprague, Jr., *Proc. Natl. Acad. Sci. U.S.A.* **90,** 9921 (1993).

[22] R. E. Schreiber, E. H. Prossnitz, R. D. Ye, C. G. Cochrane, and G. M. Bokoch, *J. Biol. Chem.* **269,** 326 (1994).

species to enable the creation of hybrids from different homologs. This offers advantages that are discussed in subsequent sections.

Although the technical aspects of producing receptor chimeras requires the application of complex, sophisticated techniques, the true art of preparing these hybrids is designing the junction(s) of the chimera. This is not a complex process in the case of chimeras composed of complementary segments of homologs of the same receptor type from different species, because the sequences can be aligned with a high degree of accuracy. This permits the splicing of complementary segments that match essentially flawlessly. In contrast, the chimera junction is a major consideration in chimeras composed of segments of related, but divergent receptors, as is frequently necessary to discriminate functional activities. If compatible junctions uniting sequences that correspond with relative precision are not selected, a "monster" will be created that is not trafficked to the cell surface.

It is appealing to produce chimeras that do not disrupt topological landmarks, because inclusion of an intact domain facilitates interpretation of the results. The use of molecular modeling approaches to identify transmembrance helices has been used as a guideline to design junctions of chimeras with limited success. The highest degree of success has been achieved by alignment of the predicted amino acid sequences of the parental receptors and the formation of junctions in zones where there is a high degree of conservation. If junctions are desired that make demarcations in regions lacking significant homology, the best results have been obtained by joining the receptors in the middle of a neighboring domain. For example, if chimeras containing varying complements of extracellular loops are desired, junctions could be designed in the central region of a predicted cytosolic loop. In general, in the absence of significant conservation of amino acid sequence, junctions near predicted transmembrane spanning domains, either the extracellular or cytoplasmic aspect, are avoided.

Genetic Engineering of Chemokine Receptor Chimeras: Technical Approaches

Three basic genetic engineering approaches can be applied to the creation of chemokine receptor chimeras: (1) fusion of restriction endonuclease fragments, (2) DNA amplification by polymerase chain reaction (PCR) using overlapping oligonucleotide primers (overlap PCR), and (3) ligation of receptor modules generated by PCR (PCR–ligation–PCR). Although formally possible, the simple fusion of restriction fragments does not merit further discussion at this point, because the occurrence of restriction endonuclease sites that generate compatible cohesive ends at the desired sites is a rare event, even when the chimeras are composed of complementary portions of the same receptor from different species. However, this strategy can be merged with the two DNA amplification strategies, as described in subsequent sections. The incorporation of restriction sites into PCR primers for the amplification of a receptor segment to use sites already existing in the partner receptor offers a powerful tool to overcome the complication of mutations resulting from amplification by DNA polymerases with limited proofreading activity, which can pose a significant problem.

We have successfully applied both overlap PCR and PCR–ligation–PCR strategies for the creation of hybrid receptor proteins. Overlap PCR is depicted in Fig. 1 (see color insert). In this approach, the component segments of the two parental receptors are amplified with a complementary internal receptor that is either a "single" primer designed from sequences that encode a highly homologous region shared by the two parental receptors or, alternatively, a "compound" primer that is composed of sequences encoding the amino-terminal moiety of one parental receptor followed by sequences encoding the segment of the carboxyl-terminal portion from the other. The upstream and downstream primers reflect sequences encoding regions that flank the translational initiation codon of the N-terminal segment of one parental receptor and the termination codon of the C-terminal segment of the other. Typically, the 5′ and 3′ untranslated regions are removed when chimeras are created, which often enhances levels of expression.

Two separate reactions are amplified by PCR, one containing the specific upstream primer, the antisense version of the shared internal primer, and the template corresponding to the N-terminal parental receptor (#1) segment and the other containing the sense version of the shared internal primer, the downstream primer, and the template corresponding to the parental receptor (#2) that constitutes the C-terminal segment. The products of these two reactions, which contain a region of overlap ranging from 25 to 40 base pairs, are separated by electrophoresis in agarose gels and purified from the bands excised from the ethidium bromide-stained gel. The two purified products serve as template(s) in a subsequent PCR amplification containing the upstream and downstream primers from the corresponding parental receptors. The full-length hybrid template is formed in the first cycle following the denaturation phase when some of the segments encoding the N-terminal (#1) and C-terminal (#2) segments anneal through the homologous region resulting from the

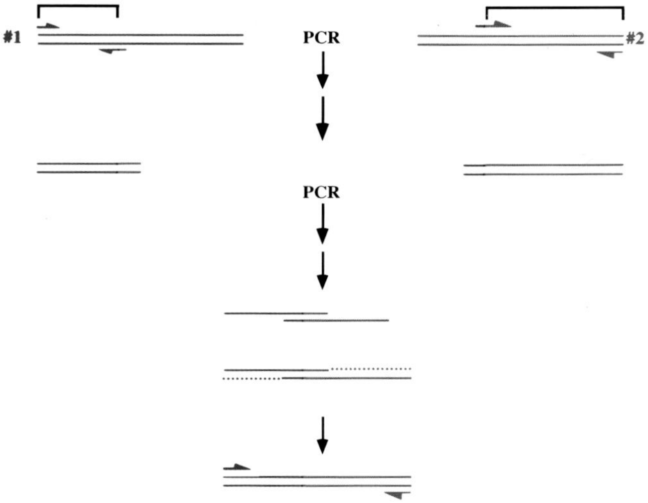

FIG. 1. Schematic representation of the overlap PCR strategy for the production of chimeric receptors.

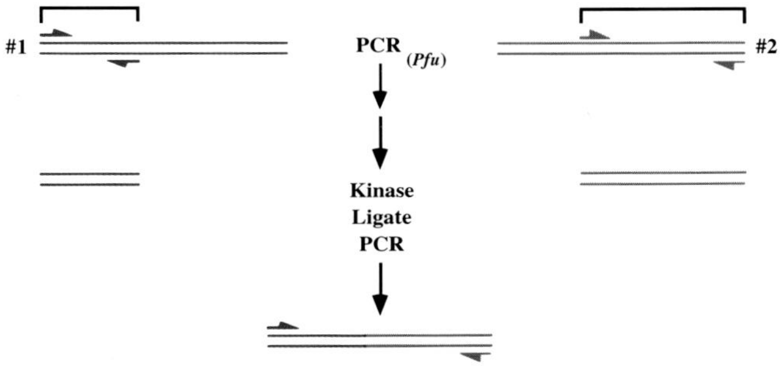

FIG. 2. Schematic representation of the ligation–PCR–ligation strategy for the production of chimeric receptors.

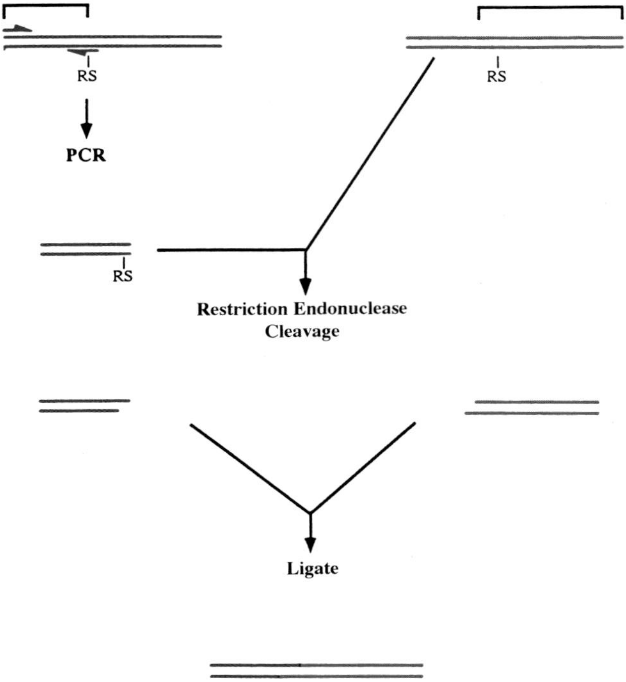

FIG. 3. Schematic representation of the restriction site introduction strategy for the production of chimeric receptors. The presence of a unique restriction site is represented as RS.

overlapping internal primer and form heterologous templates that are extended in the first synthesis phase. Then, after formation of the full-length hybrid template encoding the segments derived from both parental receptors constituting the chimera, the corresponding 5′ and 3′ primers from the two receptor elements serve to amplify the hybrid segment encoding the chimeric receptor.

Several pitfalls complicate this strategy. Perhaps the most limiting complication arises from the design of the "chimeric" complementary internal primers. If a single primer homologous to both receptors is employed, there is always significant mismatch, which can alter the efficiency of amplification of both of the DNA fragments encoding components of the hybrid and result in a low yield of the product. Although DNA amplification represents a powerful tool and the magnitude of the quantity of the product can compensate for many difficulties, the ability to work with the purified component DNA fragments greatly enriches the subsequent amplification of the desired hybrid, which, in turn, facilitates later steps in the molecular cloning and identification of the chimeric product. Alternatively, if a compound internal primer pair is used, it is by necessity much longer than the upstream and downstream primers. This results in a mismatch in the melting point (T_m) of the primers, again resulting in a decrease in the efficiency of the amplification reactions, with the consequences described above. To minimize this effect, it is possible to either ramp the annealing temperature or develop a linked series of cycles in which the annealing temperature is increased as a gradient finally to reach a level intermediate between the higher and lower T_m values.

At first glance, PCR–ligation–PCR, which is illustrated in Fig. 2 (see color insert), has many advantages over the overlap PCR approach. In this strategy, the complementary modules of the parental receptors that compose the desired hybrid are amplified in standard PCR using specific upstream and downstream primers. However, *Pfu* or *Vent* polymerase is used in place of *Taq* polymerase to produce DNA fragments with blunt ends instead of the overhangs that are generated by the latter enzyme. In addition, the former two polymerases have proofreading activity, thereby decreasing the number of mutations that are introduced during amplification. An unfavorable corollary of this activity is that extent of amplification is not as efficient as that catalyzed by *Taq* polymerase. As described above, this limits the purification of the DNA segments that will be used in the second phase of this approach.

Following purification of the DNA segments encoding the two modules of the chimeric receptor, they are phosphorylated using T4 polynucleotide kinase and ligated. The ligation reaction then serves as the source of template in a PCR reaction containing the appropriate upstream and downstream primers and catalyzed by *Taq* polymerase. Although multiple products result from the blunt end ligation of the initial two PCR products, only the orientation encoding the chimeric receptor can be amplified using the correct primer pair corresponding to 5′ sequences encoding the N-terminal portion of parental receptor #1 and sequences complementary to the 3′ region encoding the C-terminal portion of parental receptor #2 in the subsequent amplification reaction.

This strategy avoids imbalance between the T_m values of the primer pairs, as may occur in the overlap PCR approach. It also permits greater flexibility in the placement of the junction in the chimera, since the two DNA segments are amplified independently, obviating the need for a region of homology between the two receptors.

In addition to the pitfalls to PCR–ligation–PCR mentioned above, two other obstacles are common to the two approaches. One such problem relates to the oligonucleotide primers. Typically, it is not ncessary to purify primers for efficient PCR reactions. However, if the oligonucleotides are not purified, some of the products isolated using these strategies may reflect incorporation of primers that are not full length. Therefore, only purified primers are used in the creation of receptor chimeras to avoid this potential difficulty. The second and more formidable problem is the introduction of mutations during DNA amplification resulting from the infidelity of *Taq* polymerase. Somewhat surprisingly, this represents a significant obstacle. In fact, nucleotide sequence analysis of the hybrid DNA segments produced using these approaches reveals that it is infrequent to derive an open reading frame that does not have mutations resulting in changes in amino acid residues, although they are not infrequently conservative. The strategy that we have developed to compensate for this problem is described later.

Following the production of the hybrid DNA segment encoding the chimeric receptor, the DNA is subcloned into a vector adapted to incorporate fragments with overhangs generated by *Taq* polymerase, such as the TA vector. Clones containing inserts of the appropriate size are subjected to nucleotide sequence analysis using standard dideoxy techniques. Typically, the nucelotide sequence analysis at this point is limited to the chimeric junction and the upstream and downstream regions flanking this interface. This is done because it is frequently easier to excise the region surrounding the chimeric junction *en bloc* using restriction endonucleases that cut in the relevant neighboring regions of the parental receptors and then to ligate this segment to upstream and downstream regions of the parental receptors. This "anticipatory repair" may often be more expeditious than cloning and sequencing multiple candidates because the end product obtained using this modification virtually always results in an open reading frame with the correct nucleotide sequence. Clearly, in this approach, the extent of nucleotide sequence analysis is dependent on the location of unique restriction sites in the parental cDNAs.

A third approach, which is shown in Fig. 3 (see color insert), includes DNA amplification and insertion of restriction endonuclease fragments to minimize the introduction of PCR mutations. This strategy can be applied to chimeras that involve relatively short segments that are close to one end of the receptor if there is an appropriate restriction site. In this context, a unique restriction site is identified in the recipient receptor at a position appropriate for the chimera junction. The receptor segment to be added can be amplified using a standard external primer (designed from the receptor nucleotide sequence) and an internal primer (corresponding to the sequences that will form the junction) into which the identical restriction endonuclease site present in the recipient receptor has been introduced to generate an overhang that results in fusion of the two segments in the appropriate translational frame. In this way, after a relatively short DNA amplification product (subcloned into a vector such as TA) having the correct nucleotide sequence has been identified, it can be liberated by restriction endonuclease digestion and ligated to compatible cohesive ends in the recipient chemokine receptor cDNA (and in the vector). Regardless of the strategy employed, the complete nucleotide sequence of the final construct encoding the chimeric receptor is confirmed.

Analysis of Cell Surface Expression of Chemokine Receptor Chimeras

A prerequisite for the proper interpretation of functional assays on chimeric receptors is knowledge that they undergo appropriate cellular trafficking and are effectively expressed on the cell surface. Multiple strategies can be applied to demonstrate chemokine receptor expression on the cell surface, such as ligand binding, G-protein-mediated signaling following ligand stimulation, and immunologic detection. Because chimeric receptors are frequently analyzed to gain insight into domains involved in functions such as the formation of chemokine binding pockets, expression of these hybrid proteins on the plasma membrane can be best confirmed by immunologic approaches. The limited repertoire of immunologic reagents to detect chemokine receptors is complicated by the fact that it is necessary to have definitively mapped the epitope recognized by monoclonal antibodies or to have an array of domain-specific polyvalent (or monoclonal) antibodies.

The majority of the available monoclonal antibodies to chemokine receptors bind epitopes in the amino-terminal extracellular domain, including those that recognize CXCR1,[23] CXCR2, CCR5 (S. C. Peiper, Z.-H. Lu, T.-Y. Zhang, and Z.-X. Wang, unpublished data, 1996), and DARC (Duffy chemokine receptor).[24] A monoclonal antibody that inhibits infection of selected target cells by HIV has been shown to bind to CXCR4 (LESTR/fusin)[25] but to recognize an epitope localized to the extracellular loops. One that binds to the loops of DARC has also been identified.[7] Polyvalent antibodies have been prepared to the N terminus of CCR1 and CXCR4. These reagents appear to be receptor-specific and do not cross-react with other chemokine receptors.

Although detection of chimeric receptor expression using antibodies, either monoclonal or polyvalent, to the parental receptors avoids introduction of exogenous sequences, it is possible to genetically engineer epitope tags into chimeric chemoattractant receptors to monitor cell surface expression. Numerous groups have successfully employed the influenza hemagglutinin tag,[11] and some reports describe insertion of the FLAG sequence[12,15] for this purpose as well. Typically, these amino acids are inserted at the N terminus. Ligand binding studies have shown that the effect on the affinity of chemokine binding is minimal.

Typically, transfectants are stained with monoclonal antibodies at a final concentration in the range of 10 μg/ml. However, it is important to titer the antibodies on transfectants that express the parental receptor prior to the analysis of chimeras. Transfectants that grow as adherent monolayers are removed by incubation in phosphate-buffered saline (PBS) lacking calcium and magnesium ions, and single cell suspensions are prepared by pipetting. The 293 human embryonic kidney cell line is particularly useful for these analyses since it can be transfected at high efficiency (for transient or stable expression), it is removed easily from the tissue culture flask, and it has been shown to have the components required for effective coupling to G proteins. Following washing with PBS, transfectants (adherent or suspension cells) are incubated with the primary antibody for 45 min at room temperature. The cells then are washed with PBS and incubated with a fluorochrome-labeled secondary antibody for 45 min at room temperature in the dark. Fluorescein-conjugated secondary antibodies are commonly

[23] A. Chuntharapai, J. Lee, J. Burnier, W. I. Wood, C. Hébert, and K. J. Kim, *J. Immunol.* **152,** 1783 (1994).

[24] K. Wasniowska, D. Blanchard, D. Janvier, Z.-X. Wang, S. C. Peiper, T. J. Hadley, and E. Lisowska, *Mol. Immunol.* **33,** 917 (1996).

[25] M. J. Endres, P. R. Clapham, M. Marsh, M. Ahuja, J. D. Turner, A. McKnight, J. F. Thomas, B. Stoebenau-Haggarty, S. Choe, P. J. Vance, T. N. C. Wells, C. A. Power, S. S. Sutterwala, R. W. Doms, N. R. Landau, and J. A. Hoxie, *Cell (Cambridge, Mass.)* **87,** 745 (1996).

employed and represent the logical initial choice. Secondary antibodies coupled to phycoerythrin (PE) are preferable for the detection of receptors that are expressed at very low levels because the level of autofluorescence at wavelengths of light used to excite this fluorochrome is lower than with fluorescein, making immunocytochemical staining with PE conjugates somewhat more sensitive. The cells are washed with PBS after this incubation, resuspended in the residual PBS present following decanting the wash fluid from the pellet, and fixed in formaldehyde buffered with PBS. Fluorescence intensity is then measured by flow cytometry.

Either transient or stable expression approaches may be applied to the analysis of chimeric receptors. The pcDNA3 vector and the pREP episomal vectors have been used in our laboratory for transient and stable expression of chemokine receptor chimeras, respectively. Flow cytometric analysis typically shows a significant shift in the mean peak fluorescence in comparison to the control (subtype matched) mouse myeloma protein and to the monoclonal antibody incubated with the target cell line transfected with the control vector.

Creation of Chimeric Receptors Composed of Duffy Chemokine Receptor and CCR1

Reciprocal receptor chimeras composed of complementary portions of DARC and CCR1 have been created to gain insight into domains of DARC and CCR1 involved in RANTES and macrophage inflammatory protein-1α (MIP-1α) binding and domains of DARC involved in interleukin-8 (IL-8) and melanoma growth stimulatory activity (MGSA) binding as well. Because we previously demonstrated that the amino-terminal extracellular domain of DARC is sufficient to confer a promiscuous chemokine binding repertoire when attached to the loops and transmembrane helices of CXCR2, we elected to further analyze this domain of DARC and CCR1.

To this end, two hybrid receptors were created, one in which the amino-terminal extracellular domain of DARC was "grafted" onto the remainder of CCR1 and the second in which this domain from CCR1 was "grafted" onto the remainder of DARC. The overlap PCR strategy was employed for the production of these chimeras. The junction for the chimeras was placed near the first transmembrane spanning domain because 6 of 15 amino acid residues present in this region of CCR1 are the same as in DARC and a conserved proline residue is predicted to occur immediately before the transmembrane helix in both. Thus, the apparent boundary of the first transmembrane spanning domain in both receptors is similar.

The nucleotide and predicted amino acid sequences of the regions of DARC and CCR1 forming the chimera junction and the primers employed to make these chimeras are shown in Fig. 4. Standard upstream and downstream primers (24-mers with T_m values between 52° and 59°) were designed to amplify DARC and CCR1. The internal, shared primer contained 17 nucleotides from CCR1 and 18 nucleotides from DARC in the CCR1 N terminus/DARC chimera ($T_m \approx 90°$) and vice versa in the DARC N terminus/CCR1 chimera ($T_m \approx 87°$). In the initial phases of amplification before significant amounts of the product that has incorporated the compos-

A

DARC-5'	TTCCCAGGAGACTCTTCCGG
DARC-3'	CTTTAATTCAGGTTGACAGGTGGG
CCR1-5'	ATGGAAACTCCAAACACCACAGAG
CCR1-3'	CTCTCTATCCAAGAGGCGGACTCC

B

```
DARC   AACCTGCTGGATGACTCTGCA.................CTG........CCCTTCTTCATCCTCACCAGTGTCCTGGGT
        N  L  L  D  D  S  A           L     P* F  F  I  L  T  S  V  L  G
        N  E  R  A  F  G  A  Q  L  L  P  P* L  Y  S  L  V  F  V  I  G
CCR1   AACGAGAGGGCCTTTGGGGCCCAACTGCTGCCCCCTCTGTACTCCTTGGTATTTGTCATTGGC
```

C

DARC-↓-CCR1

5'-ATGACTCTGCACTGCCCCTGTACTCCTTGGTATTT
3'-TACTGAGACGTGACGGGGACATGAGGAACCATAAA

CCR1-↓-DARC

5'-CCCAACTGCTGCCCCCTTTCTTCATCCTCACCAGT
3'-GGGTTGACGACGGGGGAAAGAAGTAGGAGTGGTCA

FIG. 4. Primers employed for the creation of chimeric receptors derived from DARC and CCR1. The nucleotide sequences of the upsteam and downstream primers used for the amplification of the two parental receptors and segments of the chimeras are shown in (A). The nucleotide sequence and predicted amino acid sequence of the two parental receptors are compared in (B). The proline residue that forms the boundary between the segments of the two receptors is designated with an asterisk. Conserved amino acid residues are shaded. Residues predicted to constitute the first transmembrance α helix are underlined. The nucleotide sequence of the two internal composite primers used to amplify the two segments of the chimeras is shown in (C). The junction of the portions derived from DARC and CCR1 is marked with arrows.

ite primer is formed, the relevant T_m is limited to the segment that anneals to the DARC (~44°) and CCR1 (~43–46°) parental templates. In later phases of amplification, when products formed that have incorporated the composite central primer are more abundant, the operative T_m reflects the 35 bp that include both DARC and CCR1 components. Thus, the optimal annealing temperatures for the early and late cycles of the PCR process are different. Although calculation of T_m values in this range are of limited accuracy, this serves to demonstrate that the T_m values of the internal compound are significantly higher than those of the 5′ and 3′ primers for DARC and CCR1.

The following PCR amplifications were performed: (1) 5′ DARC primer, antisense internal compound primer (5′-DARC/CCR1-3′), DARC template; (2) sense internal compound primer (5′-DARC/CCR1-3′), 3′ CCR1 primer, CCR1 template; (3) 5′ CCR1 primer, antisense internal compound primer (5′-CCR1/DARC-3′), CCR1 template; and (4) sense internal compound primer (5′-CCR1/DARC-3′), 3′ DARC primer, DARC template. The fragments generated in reactions (1) and (2) will be combined to form the DARC N terminus/CCR1 hybrid, and those from reactions (3) and (4) will be combined to form CCR1 N terminus/DARC. Typically, 30 standard cycles of PCR are performed using an annealing temperature intermediate between the two T_m values. We have used *Pfu* polymerase in the first round of amplification to minimize the introduction of mutations. Amplification in this round of PCR is typically not highly efficient.

The reactions were concentrated and then subjected to electrophoresis in an agarose gel. The ethidium bromide-stained bands of the proper size were excised from the gel, and the DNA fragments were extracted using a standard approach, such as resin binding. The DNA fragments encoding complementary portions of the hybrid were then combined in a second amplification reaction using *Taq* polymerase and containing approximately 10 ng of the two overlapping DNA fragments and the appropriate 5′ and 3′ primers. The PCR reactions were subjected to five cycles in which the annealing temperature was increased from approximately 50° to 72° over an interval of 5 min to optimize the formation of the heterologous template, followed by 30 standard PCR cycles with an annealing temperature approximately 2–5° below the calculated T_m values, which are reasonably well matched at this point. Analysis of this amplification product in agarose gels typically yielded multiple bands, but the fragment of the predicted size is usually the predominant band. This band was excised from the gel, and the DNA was extracted and cloned into the TA vector. Candidate recombinant clones were then subjected to nucleotide sequence analysis, and those with the correct sequence were subcloned into an appropriate eukaryotic expression vector.

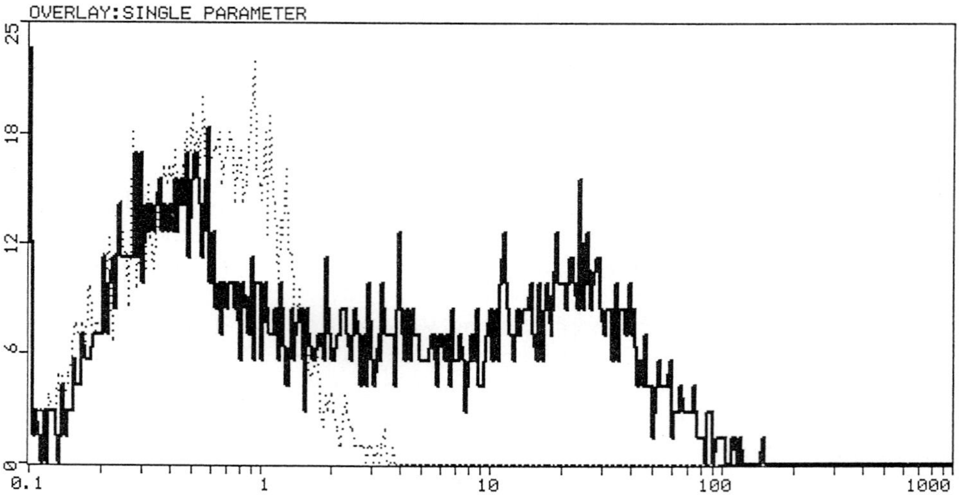

FIG. 5. Cell surface expression of the DARC-E1/CCR1 chimera. Sf21 cells were infected with a recombinant baculovirus encoding the DARC-E1/CCR1 chimeric receptor. The cells were stained approximately 36 hr following infection with Fy6 (solid line), a monoclonal antibody to the amino-terminal ectodomain of DARC, or a control myeloma protein (broken line). The cells were stained with a fluoresceinated secondary antibody after washing and analyzed by flow cytometry. The negative region of the biphasic population probably represents uninfected cells, perhaps due to infection at a low multiplicity of infection (MOI).

We have successfully employed the pREP and pCEP episomal vectors, pcDNA3, and pRC/CMV vectors for expression in human cell lines, such as K562 human erythroleukemia and 293 cells. In addition, we have subcloned the cDNA encoding the chimera into the pVL1393 transfer vector and rescued infectious virus following cotransfection of this construct with linearized AcMNPV DNA. The resulting infectious recombinant virus can then be used to direct the expression of the chimera in an insect cell line, such as Sf21 cells. This was the strategy employed for the DARC–CCR1 hybrid receptors.

Because details for the preparation of recombinant baculoviruses and infection of target cell lines are covered elsewhere in this volume[26] and immunofluorescent staining has already been discussed in this chapter, experimental details are not presented at this point. Briefly, recombinant baculoviruses containing inserts that encode a hybrid receptor composed

[26] Z.-X. Wang, Y.-H. Cen, H.-H. Guo, J.-G. Du, and S. C. Peiper, *Methods Enzymol.* **288,** 38 (1997).

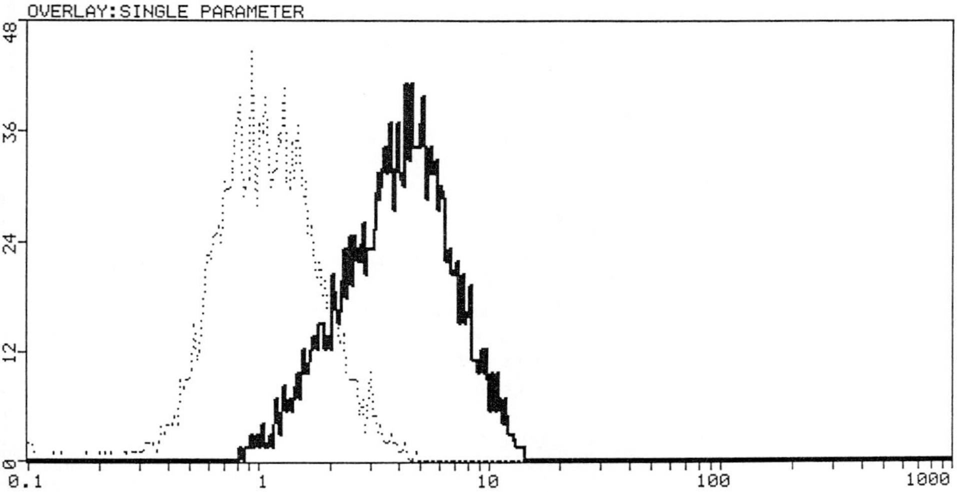

FIG. 6. Cell surface expression of the CCR1-E1/DARC chimera. SF21 cells were infected with a recombinant baculovirus encoding the CCR1-E1/DARC chimeric receptor. The infected cells were stained with anti-Fy3 (solid line), a monoclonal antibody that binds to the loops of DARC, or a control myeloma protein (broken line), as described for Fig. 5.

of the DARC amino-terminal extracellular domain (DARC-E1) fused to the complementary portion of CCR1 (DARC-E1/CCR1) and the reciprocal, CCR1-E1/DARC, were used to infect Sf21 (*Spodoptera fugiperda* fall armyworm ovary) cells. Staining of control cells failed to reveal any difference between the nonimmune control myeloma protein and the Fy6 and Fy3 monoclonal antibodies (data not shown). Staining of Sf21 cells infected with the virus encoding DARC-E1/CCR1 with Fy6, a monoclonal antibody that recognizes an epitope in DARC-E1, showed high level expression of this chimera on the cell surface at 24 hr following infection (Fig. 5). Similarly, immunofluorescent staining of cells infected with a virus encoding CCR1-E1/DARC with Fy3, a monoclonal antibody that binds to an epitope in the extracellular loops of DARC, demonstrates that this hybrid is expressed on the cell surface (Fig. 6).

Conclusion

In summary, chimeric receptors can be powerful tools for the study of structure–function relationships. Although the presentation of such functional studies is beyond the scope of this chapter, theoretical considerations and experimental strategies have been presented for the creation of such hybrids. Pitfalls regarding trafficking to the cell surface have been emphasized, and strategies for the detection of receptors at the cell surface have been presented.

Note Added in Proof

Since the submission of this chapter, we have analyzed the coreceptor activity of a broad panel of chimeric chemokine receptors that contain segments of CCR5 (B. Doranz, Z.-H. Lu, J. Rucker, T.-Y. Zhang, M. Sharron, Y.-H. Cen, Z.-X. Wang, H.-H. Guo, J.-G. Du, M. A. Accavitti, R. W. Doms, and S. C. Peiper, *J. Virol.* in press) and CXCR4 [Z.-H. Lu, J. F. Berson, Y. J. Chen, J. D. Turner, T.-Y. Zhang, M. Sharron, M. H. Jenks, Z.-X. Wang, J. Kim, J. Rucker, J. A. Hoxie, S. C. Peiper, and R. W. Dooms, *Proc. Natl. Acad. Sci. U.S.A.* **94,** 6426 (1997)] in HIV-1 env-mediated fusion assays.

We have employed a novel strategy to develop a large battery of hybrid receptors between human and mouse CCR5 using a random chimeragenesis approach [J. Y. Kim and P. N. Devrotes, *J. Biol. Chem.* **269,** 28724 (1994)]. In this method, the two parental chemokine receptors are cloned in tandem in a plasmid in which they are separated by at least two unique restriction endonuclease cleavage sites. *Escherichia coli* are transformed with the plasmid that has been linearized by digestion with both of these restriction enzymes, which makes the ends incompatible for ligation. Selection of bacteria for expression of the dominant selectable marker encoded by the plasmid results in the recovery of plasmids that have circularized by homologous recombination. This recombination occurs through homologous regions of the two receptors, thus generating chimeras with junctions in regions with a high degree of (nucleotide) sequence similarity.

[6] Molecular Approaches to Identifying Ligand Binding and Signaling Domains of C-C Chemokine Receptors

By FELIPE S. MONTECLARO, HIDENORI ARAI, and ISRAEL F. CHARO

Introduction

Chemokine receptors belong to a family of G-protein-coupled, seven-transmembrane segment receptors, many of which have been cloned. There has been considerable interest in identifying functional domains of these receptors, with respect not only to their chemokine ligands, but also to other ligands such as the human immunodeficiency virus (HIV) envelope glycoproteins. Several strategies employing recombinant DNA techniques have been used to elucidate structure–function relationships. Construction of hybrid or chimeric receptors, whereby a region of one receptor is exchanged with the complementary region of the other receptor, can provide

METHODS IN ENZYMOLOGY, VOL. 288
0076-6879/97 $25.00

a useful starting point for identifying domains involved in ligand binding and signaling. The chemokine receptors are particularly good candidates for this approach because their high degree of similarity favors the formation of stable and functional chimeras while retaining distinguishable functional characteristics, such as ligand binding, signaling, and G protein coupling. Site-directed mutagenesis may be used to further delineate functional domains through substitutions or deletions of specific residues. An important caveat is that loss of function of a particular chimera may be simply due to its lack of expression at the cell surface. However, incorporation of an epitope tag at the extreme amino terminus of the receptor provides a rapid and quantitative means of determining surface expression and allows for assessment of receptor function.

The monocyte chemoattractant protein-1 (MCP-1) receptor (CCR2) and CCR1, the receptor for macrophage inflammatory protein-1α (MIP-1α) and regulated on activation normal T-expressed and secreted (RANTES) protein belong to the C-C subfamily of chemokine receptors. CCR2 consists of two isoforms, type A (CCR2A) and type B (CCR2B), which differ only in the amino acid sequence of the cytoplasmic tail. Here we describe the construction of CCR1 and CCR2B receptor chimeras as an example of the molecular approach for identifying ligand binding and signaling domains. This receptor pair was chosen because of the high degree of sequence identity (~50%), the clear differences in ligand selectivity between the two receptors, and the fact that both couple to $G_i\alpha$ to inhibit adenylyl cyclase (adenylate cyclase). Methods to determine ligand binding and receptor signaling are detailed below, and the construction of additional chimeras to further delineate ligand binding domains is discussed. Finally, the approach to identifying intracellular loops involved in G protein coupling is described.

Selection of Receptor Pairs

Related receptors can be identified through a BLAST search.[1] A number of molecular biology programs are available that perform sequence alignments, such as DNAsis (Hitachi, San Bruno, CA) and PC Gene/Geneworks (Intelligenetics, Campbell, CA) for the Macintosh or IBM-compatible computers, respectively, and GCG program suites (Genetics Computer Group, Madison, WI) for Unix. Sequence alignment of CCR2B with CCR1 (Fig. 1) indicates that these two receptors are 51% identical overall and are most divergent in the extracellular domains. Although

[1] S. F. Altschul, W. Gish, W. Miller, E. W. Myers, and D. J. Lipman, *J. Mol. Biol.* **215,** 403 (1990).

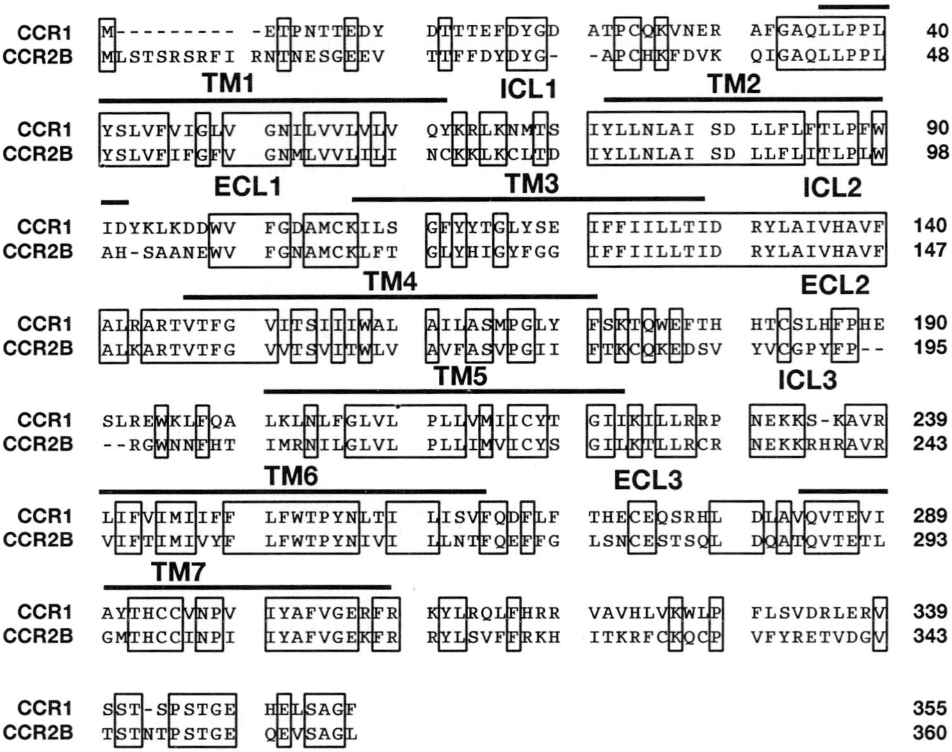

FIG. 1. Sequence alignment between CCR1 and CCR2B generated by the computer program Geneworks. The bars above the sequence denote the putative transmembrane (TM) domains. Boxed residues mark identical residues, and dashes represent gaps introduced in the sequence for optimal alignment. ICL, Intracellular loop; ECL, extracellular loop. Note that a major difference in sequence identity occurs within the extracellular domains.

highly related, CCR2B and CCR1 show distinct ligand specificity.[2] MIP-1α and RANTES are functional ligands for CCR1,[3] whereas MCP-1, MCP-4,[4,5]

[2] S. J. Myers, L. M. Wong, and I. F. Charo, *J. Biol. Chem.* **270,** 5786 (1995).

[3] K. Neote, D. DiGregorio, J. Y. Mak, R. Horuk, and T. J. Schall, *Cell (Cambridge, Mass.)* **72,** 415 (1993).

[4] P. Schulz-Knappe, H.-J. Mägert, B. Dewald, M. Meyer, Y. Cetin, M. Kubbies, J. Tomeczkowski, K. Kirchhoff, M. Raida, K. Adermann, A. Kist, M. Reinecke, R. Sillard, A. Pardigol, M. Uguccioni, M. Baggiolini, and W.-G. Forssmann, *J. Exp. Med.* **183,** 295 (1996).

[5] M. Uguccioni, P. Loetscher, U. Forssmann, B. Dewald, H. Li, S. H. Lima, Y. Li, B. Kreider, G. Garotta, M. Thelen, and M. Baggiolini, *J. Exp. Med.* **183,** 2379 (1996).

and MCP-5[6] are functional ligands for CCR2. MCP-3 is a ligand for both CCR2 and CCR1.[7,8] Both receptors couple to $G_i\alpha$, consistent with the high degree of amino acid sequence conservation in the intracellular loops.

Construction of CCR1/CCR2 Chimeras

Receptor chimeras are constructed from cDNAs of CCR2B[9] and CCR1.[3,10] Each receptor contains the prolactin signal sequence, followed by the Flag epitope sequence[11] fused to the second translation codon.[12] The epitope-tagged receptors are then subcloned into the expression vector pcDNA3 (Invitrogen, San Diego, CA) to yield the tagged wild-type receptors. Addition of the Flag epitope at the extreme amino terminus of each of these chimeras facilitates quantitation of receptor expression and also allows selection of stable cell lines with comparable receptor expression at the cell surface by fluorescence-activated cell sorting. The designations 2222 (CCR2B) and 1111 (CCR1) are used to represent the amino-terminal extension and each of the three extracellular loops of each receptor and chimera. Figure 2 depicts the set of CCR1 and CCR2 chimeras.

The exchange of regions in the construction of chimeric receptors is greatly facilitated by unique restriction sites shared by the receptor pairs at conserved domains. Receptors with the amino-terminal exchange are created by using an *Apa*I site located in transmembrane one of CCR1 and CCR2. Thus, in mutant 1222 amino acids 1–32[9] are from CCR1, whereas in mutant 2111 amino acids 1–40 are from CCR2 (Fig. 2).

Receptors that exchange the amino terminus and the first extracellular loop are created by overlapping polymerase chain reaction (PCR)[13] because of the absence of a unique restriction site in the conserved region of the

[6] M. N. Sarafi, E. A. Garcia-Zepeda, J. MacLean, I. F. Charo, and A. D. Luster, *J. Exp. Med.* **185,** 99 (1997).

[7] C. Franci, L. M. Wong, J. Van Damme, P. Proost, and I. F. Charo, *J. Immunol.* **154,** 6511 (1995).

[8] C. Combadiere, S. K. Ahuja, J. Van Damme, H. L. Tiffany, J.-L. Gao, and P. M. Murphy, *J. Biol. Chem.* **270,** 29671 (1995).

[9] I. F. Charo, S. J. Myers, A. Herman, C. Franci, A. J. Connolly, and S. R. Coughlin, *Proc. Natl. Acad. Sci. U.S.A.* **91,** 2752 (1994).

[10] Nucleotide (nt) numbers correspond to Genbank accession numbers U03905 and L10918, respectively.

[11] K. Ishii, L. Hein, B. Kobilka, and S. R. Coughlin, *J. Biol. Chem.* **268,** 9780 (1993).

[12] Nucleotide number 84 for CCR2B and nucleotide 66 for CCR1.

[13] S. N. Ho, H. D. Hunt, R. M. Horton, J. K. Pullen, and L. R. Pease, *Gene* **77,** 51 (1989).

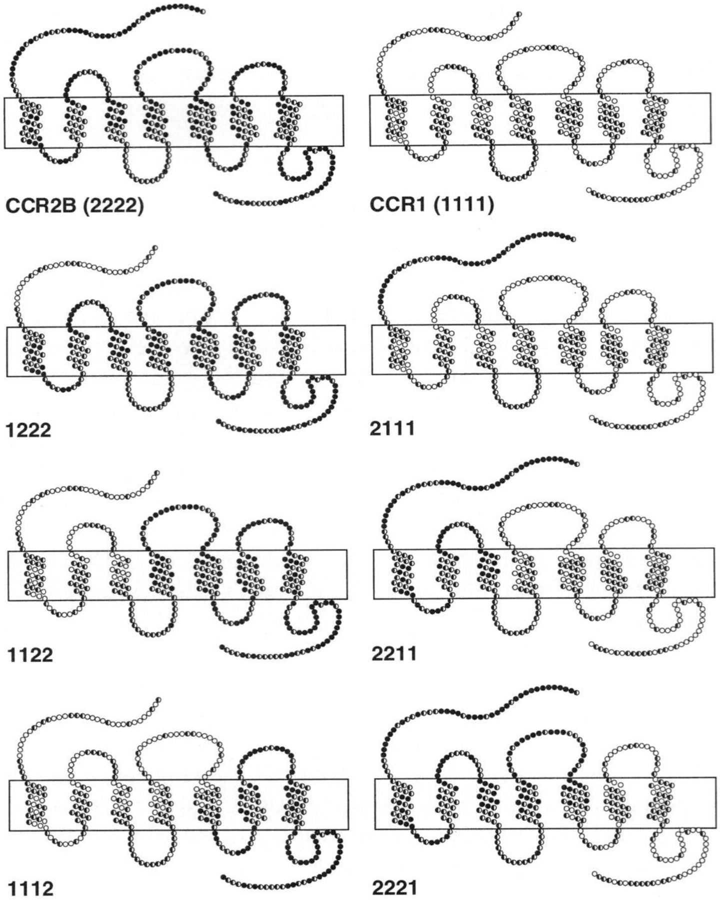

FIG. 2. Schematic diagram of the CCR1, CCR2B, and chimeric receptors. Each dot represents an amino acid residue. Filled circles are CCR2-derived residues, and open circles are CCR1-derived residues. The half-filled circles denote residues that are identical in CCR1 and CCR2. Chimera 2111 depicts the CCR1 receptor with the amino terminus exchanged with the amino terminus of CCR2B. Chimera 1222 depicts the CCR2B receptor with the amino terminus exchanged with the amino terminus of CCR1. In chimeras 2211 and 1122, the region corresponding to the amino terminus through the first extracellular loop was exchanged between CCR1 and CCR2B. In chimeras 2221 and 1112, the region corresponding to the amino terminus through the second extracellular loop was exchanged between CCR1 and CCR2B. The junctions in the chimeric receptors are located within the conserved amino acid sequences of the receptors, as described in the text. (Reproduced from Ref. 15, with permission.)

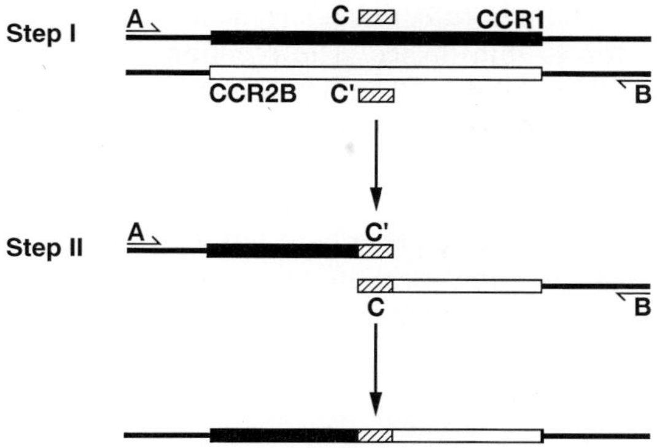

FIG. 3. PCR strategy for generating chimeric receptors. Primers A and B correspond to vector sequences; primers C and C′ are complementary and correspond to sequences common to both receptors. In step 1, specific fragments of the receptors were first amplified, A to C′ for CCR1 and C to B for CCR2B. In step 2, the amplified fragments in step 1 were amplified by PCR using flanking primers A and B to generate CCR1/CCR2B chimeras. The PCR fragments were then subcloned into the expression vector pcDNA3.

receptors. The PCR strategy is shown in Fig. 3. Complementary pairs of primers (~24 nucleotide bases) that correspond to the sequence RYLAIVHA, nucleotides (nt) 492–514 and 453–476 of CCR2 and CCR1, respectively, are synthesized. A few base pair mismatches between complementary primers are usually well tolerated but should be avoided near the 3′ end of the oligonucleotide. The amino-terminal and carboxyl-terminal halves of CCR1 and CCR2 are amplified in one round of PCR. A second round of PCR is performed using vector-derived primers that flank the cDNA insert. Purified fragments of the amino-terminal half of CCR2 and the carboxyl-terminal half of CCR1 are then used as templates to create the mutant 2211. To generate the mutant 1122, the second round of PCR is performed using purified fragments of the amino-terminal half of CCR1 and the carboxyl-terminal half of CCR2 as templates. Amplified fragments are cloned into expression vectors and sequenced. Chimeras 1112 and 2111 are constructed in a similar manner using overlapping PCR.

Ligand Binding

Preparation of Labeled Ligand

MCP-1 (R&D, Minneapolis, MN) is labeled using the Bolton–Hunter reagent (diiodide, NEN, Wilmington, DE). Five micrograms of ligand in

10 μl of 100 mM sodium borate, pH 8.5, is incubated with 1.0 mCi Bolton–Hunter reagent for 15 min on ice. The reaction is terminated by adding 100 μl of 0.5 M ethanolamine, 100 mM sodium borate, 10% glycerol (v/v), and 0.1% xylene cyanol (w/v). Unconjugated iodide is separated from labeled protein by gel filtration with Sephadex G-25 (Pharmacia, Piscataway, NJ) and an Econo column (Bio-Rad, Richmond, CA). Labeled MCP-1 is eluted in the void volume with phosphate-buffered saline (PBS) containing 0.1% gelatin (w/v). The final protein concentration is determined by an enzyme-linked immunosorbent assay (ELISA) kit using the manufacturer's protocol (Quantikine, R&D), and the specific activity is determined by counting gamma emissions.

Binding Assay

Equilibrium binding is performed by adding ^{125}I-labeled ligand, with or without a 100-fold excess of unlabeled ligand, to 0.5×10^6 stably transfected human embryonic kidney (HEK) 293 cells in a 12 × 75 mm propylene tube (Falcon, Becton-Dickinson, Lincoln Park, NJ) in a total volume of 300 μl of binding buffer [50 mM HEPES, pH 7.4, 1.0 mM $CaCl_2$, 5.0 mM $MgCl_2$, and 0.5% bovine serum albumin (w/v)]. After incubation for 90 min at 27° on an orbital shaker set at 150 rpm, the cells are collected with a Skatron cell harvester (Skatron Instruments, Sterling, VA) onto glass-fiber filters presoaked in 0.3% polyethylenimine and 0.2% bovine serum albumin (w/v). Unbound ligand is removed by washing with 4 ml of buffer [10 mM HEPES, pH 7.4, 0.5 M NaCl, and 0.5% bovine serum albumin (w/v)] for 10 sec. After washing, the filters are removed, and bound ligand is quantitated by counting gamma emissions. Ligand binding by competition with unlabeled ligand is determined by incubating 0.5×10^6 transfected cells (as above) with 1.5 nM of radiolabeled ligand and adding increasing concentrations of unlabeled ligand in a final volume of 300 μl. The samples are incubated, collected, washed, and counted as above. The data are analyzed with the curve-fitting program Prism (GraphPad, San Diego, CA) and the iterative nonlinear regression program Ligand.[14]

The binding of labeled MCP-1 to wild-type receptors and chimeric receptors with the amino-terminal extracellular domains exchanged illustrates both gain and loss of function mutations. In a direct-binding assay, wild-type CCR2 (2222) binds MCP-1 with high affinity (K_d = 0.33 nM), whereas CCR1 (1111) binds MCP-1 with very low affinity (K_d = 5.0 nM). Mutant 2111 binds MCP-1 as well as wild-type CCR2 (K_d = 0.27 nM),

[14] P. J. Munson and D. Rodbard, *Anal. Biochem.* **107,** 220 (1980).

whereas 1222 binds MCP-1 poorly (K_d = 3.5 nM) (Fig. 4). Thus, the replacement of the CCR1 amino-terminal extension with the CCR2 amino-terminal extension results in a gain of high-affinity MCP-1 binding by CCR1. In contrast, replacement of the CCR2 amino terminus with the CCR1 amino terminus results in a loss of affinity for MCP-1. High-affinity binding of MCP-1 therefore correlates with the presence of the amino-terminal extension of CCR2. Scatchard analysis of the binding of MCP-1 to CCR2 reveals both high-affinity (K_d = 0.32 nM) and low-affinity (K_d = 15 nM) binding sites (Fig. 5A). As a further test of the role of the CCR2 amino terminus for high-affinity binding, we have fused the 35 amino-terminal residues of CCR2 onto the single transmembrane domain of CD8 to generate mutant M-Term CD8. In equilibrium binding assays, M-Term CD8 binds labeled MCP-1 with high affinity, whereas the CCR1 amino terminus fused to CD8 shows no detectable binding (data not shown). Scatchard analysis reveals a K_d virtually identical to that of the high-affinity binding site of CCR2 (Fig. 5B). Thus, the CD8 fusion construct, M-Term CD8, demonstrates that virtually all of the high-affinity binding of MCP-1 to CCR2 can be attributed to the amino-terminal extension of the receptor.

The amino terminus of CCR2 plays a critical role not only in binding, but also in signaling.[15] The gain in high-affinity binding seen with mutant 2111 increases signaling more than 100-fold, as measured by the inhibition of adenylyl cyclase activity.[2] In contrast, loss of high-affinity binding (mutant 1222) results in a 30-fold decrease in signaling. Although mutant 1222 binds MCP-1 poorly, the receptor is capable of transducing a signal, albeit at higher ligand concentrations. Taken together, these data suggest a two-step model leading to activation of the MCP-1 receptor in which the amino-terminal extension serves to bind MCP-1 with high affinity and noncovalently tethers it for lower affinity interactions with one or more extracellular loops to initiate signaling.[15]

G Protein Coupling

The creation of chimeric receptors has also yielded valuable insights in the identification of intracellular receptor domains that interact with G proteins and induce signaling. During our initial studies of CCR2, we found that the receptor coupled via G_i to mediate agonist-dependent inhibition of adenylyl cyclase and intracellular calcium release in stably transfected

[15] F. S. Monteclaro and I. F. Charo, *J. Biol. Chem.* **271,** 19084 (1996).

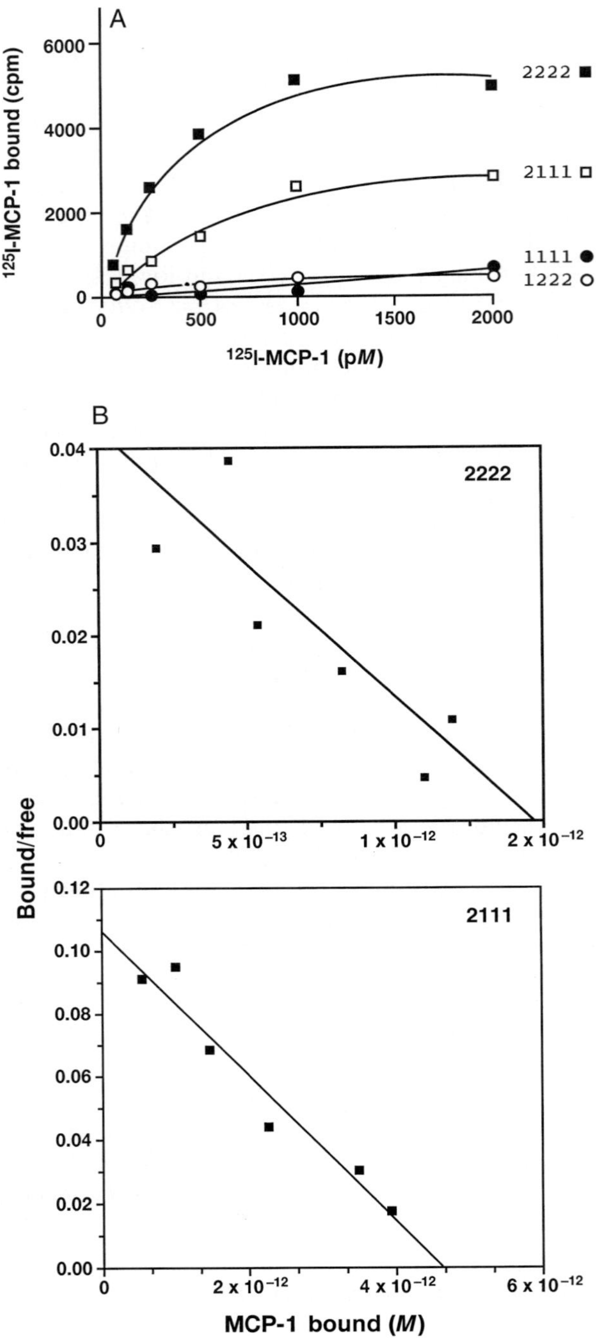
A
6000
4000
2000
0
^{125}I-MCP-1 bound (cpm)
0
500
1000
1500
2000
^{125}I-MCP-1 (pM)
2222
2111
1111
1222
B
0.04
0.03
0.02
0.01
0.00
2222
0
5 x 10^{-13}
1 x 10^{-12}
2 x 10^{-12}
Bound/free
0.12
0.10
0.08
0.06
0.04
0.02
0.00
2111
0
2 x 10^{-12}
4 x 10^{-12}
6 x 10^{-12}
MCP-1 bound (M)

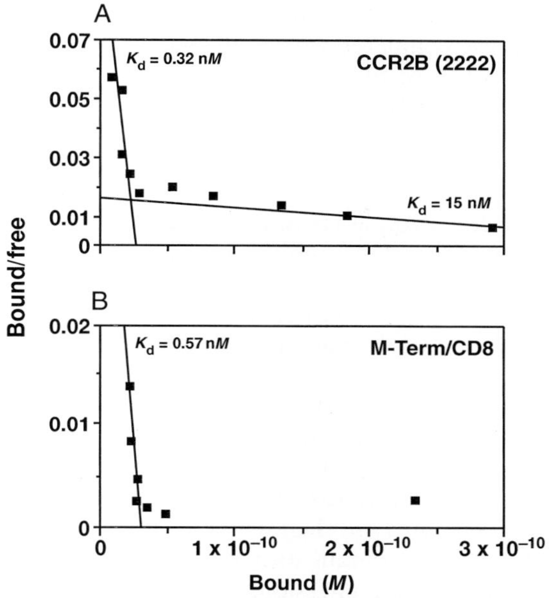

FIG. 5. Scatchard plots obtained from analysis of equilibrium binding assays. (A) The wild-type receptor data (CCR2B) best fit a model that describes two binding sites (K_d values are indicated). (B) The M-Term CD8 data fitted a model with only one binding site. Data were analyzed with the computer program Ligand. Ligand binding by competition was performed as described in the text, using HEK-293 cells stably expressing wild-type and chimeric receptors.

HEK-293 cells.[2] In those studies, signaling was not completely blocked by pretreating the cells with pertussis toxin (PTX), suggesting that the receptors also coupled to additional PTX-resistant G proteins, such as $G\alpha_q$ or $G\alpha_{16}$. To identify G proteins that coupled to C-C chemokine receptors in a PTX-resistant manner, we cotransfected Gα subunits and chemokine receptors in COS-7 cells and measured phosphoinositide (PI) hydrolysis

---

FIG. 4. Binding of ^{125}I-labeled MCP-1 to wild-type and chimeric receptors. Radiolabeled MCP-1 was incubated with HEK-293 cells stably expressing CCR2B and CCR1 receptors, as well as chimeric receptors 1222 and 2111. (A) Binding isotherms. Specific binding (total binding minus nonspecific binding in the presence of a 100-fold excess of unlabeled MCP-1) is shown for the wild-type and chimeric receptors. Nonspecific binding varied between 10 and 20% of total binding. (B) Scatchard plots for MCP-1 binding to CCR2B (2222) and 2111. The K_d values (2222, 0.33 ± 0.12 n*M*; 2111, 0.27 ± 0.10 n*M*) were determined with the computer program Ligand.

as a qualitative measure of G protein coupling and as an index of signal transduction. We also compared G protein coupling in C-X-C [interleukin-8 (IL-8) receptor type A, also known as CXCR1] and C-C chemokine receptors (CCR1 and CCR2).

Inositol Phosphate Formation Assay

Phosphoinositide hydrolysis is assayed in the following manner. Approximately 24 hr after transfection (see below), cells are labeled for 20–24 hr with *myo*-[2-^{3}H]inositol (2 μCi/ml) in inositol-free medium containing 10% dialyzed fetal calf serum (v/v). Labeled cells are washed with inositol-free Dulbecco's modified Eagle's medium (DMEM) containing 10 m*M* LiCl and incubated at 37° for 1 hr with inositol-free DMEM containing 10 m*M* LiCl and the indicated agonist. After incubation with agonists, the medium is aspirated, and cells are lysed by the addition of 0.75 ml of ice-cold 20 m*M* formic acid (30 min). Supernatant fractions are loaded onto AG1-8X Dowex columns (Bio-Rad), followed by immediate addition of 3 ml of 50 m*M* NH_4OH. The columns are then washed with 4 ml of 40 m*M* formate and eluted with 2 *M* ammonium formate. Total inositol phosphates are quantitated by counting beta emissions.

Assessment of Receptor Surface Expression

The surface expression of CCR2, CCR1, and CXCR1 is assessed by ELISA,[11] and typical results are shown in Fig. 6. Briefly, cells are cultured on 24-well plates at 5 × 10^4 cells/well and incubated overnight before transfection as described in the legend to Fig. 6. After 48 hr, the cells are fixed with 4% paraformaldehyde in PBS (w/v) for 15 min. Plates are washed twice with PBS and then incubated with 1 μg/ml of the M1 antibody directed against the Flag epitope (Kodak/IBI, New Haven, CT) in DMEM containing 10 m*M* HEPES and 0.1% bovine serum albumin (w/v) for 1 hr at room temperature. After washing with PBS, plates are incubated with horseradish peroxidase-conjugated second antibodies [Bio-Rad; 1:1000 dilution in DMEM/10 m*M* HEPES/0.1% bovine serum albumin (w/v)] for 30 min at room temperature. After additional washing with PBS, the plates are developed using 2,2′-azinobis(3-ethylbenzthiazoline-6-sulfonic acid) (1 mg/ml) in citrate/phosphate buffer (pH 4.0) with 0.03% hydrogen peroxide (v/v). Absorbance at 450 nm is read after 5–30 min on an ELISA plate reader (V_{MAX}, Molecular Devices, Menlo Park, CA).

Identification of G-Protein-Coupling Domains

To identify the domain(s) of CCR2B that binds to $G\alpha_q$, we replace the second and third intracellular loops of CCR2B with the corresponding

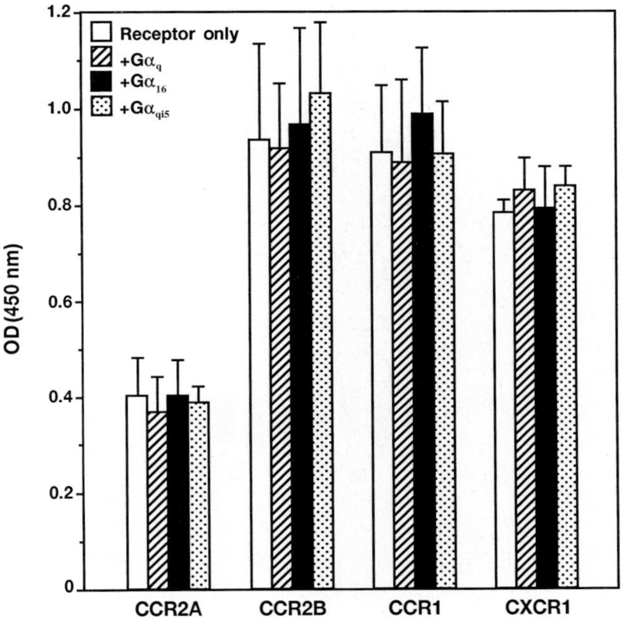

FIG. 6. Cell surface expression of the chemokine receptors. COS-7 cells were seeded into 24-well plates at approximately 0.5×10^5 cells/well. One day later, the cells were transiently transfected by addition of 0.25 μg of plasmid DNA of chemokine receptor and 0.125 μg of G-protein plasmid DNA per well in 0.25 μl Opti-MEM serum-free medium (BRL, Gaithersburg, MD) mixed with 1.5 μl of Lipofectamine (BRL). After incubation at 37° for 5 hr, the transfection medium was replaced with DMEM with 10% fetal bovine serum. (Reproduced from Ref. 16, with permission.)

regions of CXCR1 (Fig. 7). These chimeras are made by overlapping PCR as described above. Replacement of the 23-amino acid third intracellular loop of CCR2B with that of CXCR1 results in a chimera (MM8) that is phenotypically identical to IL-8 receptor type A (CXCR1) in terms of signaling (Fig. 8). Thus, MCP-1-dependent signaling is detected only in the presence of cotransfected $G\alpha_{16}$, and the receptor fails to couple to $G\alpha_q$. Moreover, the complementary construct in which the third intracellular loop of CCR2B is substituted into CXCR1 (chimera 88M) results in a receptor in which signaling in response to IL-8 is indistinguishable from that of CCR2B. Exchange of this loop changes only 14 amino acid residues because the carboxyl ends of the loops are virtually identical in CCR2B and CXCR1 (Fig. 7). In contrast, substitution of the second intracellular loop of CXCR1 into CCR2B (chimera M8M) has no effect on MCP-1-

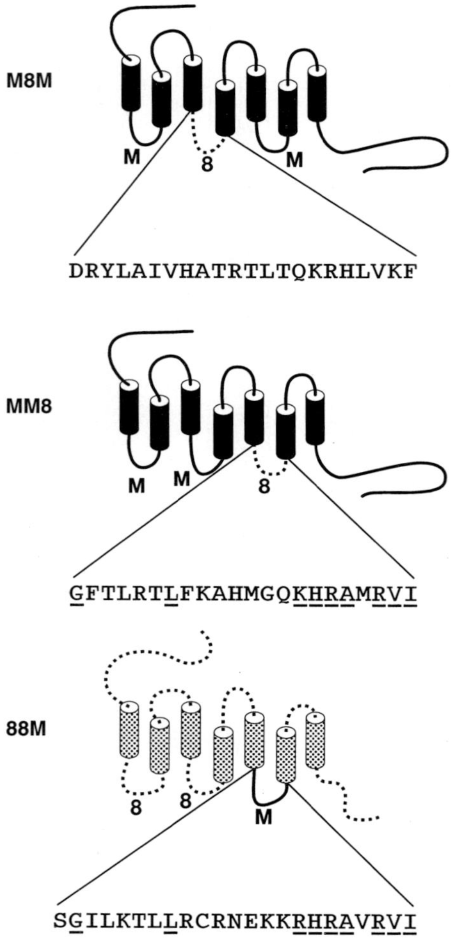

FIG. 7. Exchange of intracellular loops between CCR2B (MCP-IR) and CXCR1 (IL-8RA). The sequences of the intracellular loops are shown. M designates MCP-1 receptor sequences, and 8 indicates IL-8 receptor sequences. The chimera M8M denotes CCR2B into which the second intracellular loop of the CXCR1 has been substituted. MM8 and 88M indicate chimeras in which the third intracellular loops have been interchanged. Amino acids that are identical or conservatively substituted in the third intracellualr loop are underlined. (Reproduced from Ref. 16, with permission.)

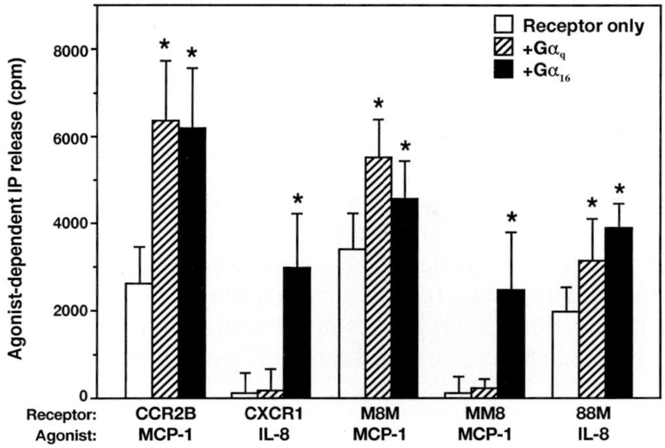

FIG. 8. Coupling of Gα_q to the third intracellular loop of CCR2B. COS-7 cells were transiently transfected with cDNAs encoding the wild-type and chimeric receptors (1 μg/ml) and the indicated G proteins (0.5 μg/ml). The cells were incubated with the indicated agonist (100 nM), and inositol phosphate release was measured as described in the text. Asterisks (*) denote statistically significant results ($p < 0.05$ versus control). (Reproduced from Ref. 16, with permission.)

dependent signaling. Thus, in the context of the receptor, the third intracellular loop of CCR2B is both necessary and sufficient for coupling to Gα_q. The mechanism of Gα_{16} coupling appears to be more complex, in that substitution of the third intracellular loop of CCR1 (which fails to couple to Gα_{16}) into CCR2B does not change coupling to Gα_{16} (data not shown).

To identify PTX-resistant G proteins that couple to CCR2 and related chemokine receptors, we have performed cotransfection experiments in mammalian cells.[16] Signaling by both forms of CCR2 is significantly enhanced by coexpression of Gα_q and Gα_{16}. Phosphoinositide hydrolysis mediated by CCR1 is potentiated by Gα_q but not by Gα_{16}. The chimeric G protein Gα_{qi5} has the carboxyl-terminal five amino acids of G$_i\alpha$, which bind to the receptor, spliced onto Gα_q.[17] Cotransfection of Gα_{qi5} also significantly potentiates signaling by each of the C-C chemokine receptors with coupling to G$_i\alpha$. In contrast, CXCR1 does not induce PI turnover in COS-7 cells unless it is cotransfected with Gα_{16} or Gα_{qi5}. Furthermore,

[16] H. Arai and I. F. Charo, *J. Biol. Chem.* **271,** 21814 (1996).

[17] B. R. Conklin, Z. Farfel, K. D. Lustig, D. Julius, and H. R. Bourne, *Nature* (*London*) **363,** 274 (1993).

CXCR1 signaling is not enhanced by $G\alpha_q$. These results suggest that both forms of CCR2 couple efficiently to $G\alpha_q$, $G\alpha_{16}$, and G_i, whereas the CXCR1 receptor has a preference for $G\alpha_{16}$ and G_i.

Summary

The construction of chimeric receptors provides a useful starting point for the identification of ligand-binding and G-protein-coupling sites on chemokine receptors. Correlation of the binding and signaling properties of a set of complementary receptor chimeras is a powerful approach for probing structure–function relationships. Further molecular resolution can subsequently be achieved by site-directed mutagenesis and/or alanine scanning.[18]

[18] O. Moro, M. S. Shockley, J. Lameh, and W. Sadée, *J. Biol. Chem.* **269,** 6651 (1994).

[7] Characterization of Functional Activity of Chemokine Receptors Using the Cytosensor Microphysiometer

By SIMON PITCHFORD, MARGARET HIRST, H. GARRETT WADA, SAMUEL D. H. CHAN, VÉRONIQUE E. TIMMERMANS, and GILLIAN M. K. HUMPHRIES

Introduction

In this chapter we describe the use of the Cytosensor Microphysiometer System (Molecular Devices, Sunnyvale, CA) to monitor the activation of a variety of primary cells and cell lines that express chemokine receptors. This silicon biosensor-based system monitors the changes in cell metabolism that occur subsequent to receptor activation and has been used extensively to rapidly assess activation of a large number of different receptor types including many members of the seven-transmembrane, G-protein-coupled superfamily of receptors to which the chemokine family belongs.[1]

Chemokines (chemoattractant cytokines) are a family of low molecular weight cytokines that are critically involved in the activation and recruitment of immune cells during the inflammatory response. They are released

[1] For a complete listing of receptors, refer to "Cytosensor Microphysiometer at Work," available from Molecular Devices Corporation, Sunnyvale, CA.

0076-6879/97 $25.00

from many different types of tissues and cells and act through specific chemokine receptors. Because of the ubiquitous role of chemokines in the inflammatory process, intervention in chemokine action may represent a powerful therapeutic strategy in combatting a variety of inflammatory and immune diseases.[2]

Two major families of chemokines have been identified on the basis of the conservation of certain cysteine residues. These have been named the C-C and C-X-C families. A third family, named the C family, has also been described.[3] Members of these families of chemokines interact with various degrees of potency and selectivity with at least seven identified subtypes of chemokine receptors. In addition, there have been reports of many other orphan seven-transmembrane-type receptors, some of which have been proposed to be chemokine receptors.[4–8] Of particular interest is the identification of a novel seven-transmembrane receptor named fusin. Fusin, now renamed CXCR4, has been demonstrated to play a role in the binding of the membrane wall of the human immunodeficiency virus (HIV) to T cells and therefore may be implicated in the ability of HIV to infect immune cells. This fusin protein was demonstrated to have a high degree of sequence homology with known chemokine receptors, suggesting it may be a member of the chemokine receptor family.[9] Subsequently, three separate groups of researchers have identified another chemokine receptor, the CCR5 receptor, as a binding site for HIV.[10–12] These findings further heighten the interest in these ligands and/or receptors as therapeutic targets.

[2] T. J. Schall, *in* "The Cytokines Handbook" (A. Thomson, ed.), p. 419. Academic Press, London, 1994.

[3] G. S. Kelner, J. Kennedy, K. B. Bacon, S. Kleyensteuber, D. A. Largaespada, N. A. Jenkins, N. G. Copeland, J. F. Bazan, K. W. Moore, T. J. Schall, and A. Zlotnik, *Science* **266,** 1395 (1994).

[4] T. Dobner, I. Wolf, T. Emrich, and M. Lipp, *Eur. J. Immunol.* **22,** 2795 (1992).

[5] M. Birkenbach, K. Josefson, R. Yalamanchili, G. Lenoir, and E. Kieff, *J. Virol.* **67,** 2209 (1993).

[6] E. E. Jazin, H. Yoo, A. G. Blomqvist, F. Yee, G. Weng, M. W. Walker, J. Salon, D. Larhammar, and C. Washlestedt, *Regul. Pept.* **47,** 247 (1993).

[7] J. L. Gao and P. M. Murphy, *J. Biol. Chem.* **270,** 17494 (1995).

[8] I. Matsouka, T. Mori, J. Aoki, T. Sato, and K. Kurihara, *Biochem. Biophys. Res. Commun.* **194,** 504 (1993).

[9] Y. Feng, C. C. Broder, P. E. Kennedy, and E. A. Berger, *Science* **272,** 872 (1996).

[10] H. Deng, R. Liu, W. Ellmeier, S. Choe, D. Unutmaz, M. Burkhart, P. Di Marzio, S. Marmon, R. E. Sutton, C. M. Hill, C. B. Davis, S. C. Peiper, T. J. Schall, D. R. Littman, and N. R. Landau, *Nature (London)* **381,** 661 (1996).

[11] T. Dragic, V. Litwin, G. P. Allaway, S. R. Martin, Y. Huang, K. A. Nagashima, C. Cayanan, P. J. Maddon, R. A. Koup, J. P. Moore, and W. A. Paxton, *Nature (London)* **381,** 667 (1996).

[12] G. Alkhatib, C. Combadiere, C. C. Broder, Y. Feng, P. E. Kennedy, P. M. Murphey, and E. A. Berger, *Science* **272,** 1955 (1996).

In general, members of the C-X-C family of chemokines [e.g., interleukin-8 (IL-8)] tend to attract neutrophils, whereas members of the C-C family have been demonstrated to attract monocytes and have varying actions on lymphocytes, basophils, and eosinophils. Lymphotactin, a member of the C family of chemokines, appears to be an attractant for mouse thymocytes.

The chemokine receptors appear to signal through heterotrimeric G proteins belonging to the $G\alpha_{i/o}$ class, as signaling is inhibited by pertussis toxin treatment, suggesting that activation of these receptors will result in an inhibition of adenylate cyclase. One of the first consequences of chemokine receptor activation, however, is a rapid and transient elevation of intracellular calcium that is pertussis toxin sensitive.[13,14] This increase in intracellular calcium appears to be dependent on external calcium, suggesting an influx through plasma membrane calcium channels rather than a release from intracellular calcium stores.[15]

Measurement of intracellular calcium changes has shed some light on the ligand selectivities of some of the chemokine receptors.[16–18] In addition, study of intracellular calcium changes has revealed the potential activation of multiple signaling pathways following receptor stimulation. Bacon *et al.*[19] demonstrated that relatively high concentrations of the chemokine RANTES resulted in two phases of intracellular calcium changes in T-cell clones that could be inhibited by different modulators of signaling pathways. One phase was sensitive to pertussis toxin treatment of the T-cell clones, whereas the other was abolished by treatment with herbimycin A, an inhibitor of protein tyrosine kinase. This suggests either activation of multiple signaling pathways through one receptor or the activation, by RANTES, of multiple receptors coupled to different signaling pathways.

With the discovery of new members of the family of chemokine receptors, there is increased need for rapid, sensitive, and selective assays to measure activity of these receptors and their ligands. Assays for functional

[13] S. R. McColl, M. Hachicha, S. Levasseur, K. Neote, and T. J. Schall, *J. Immunol.* **150,** 4550 (1993).

[14] S. Sozzani, W. Luini, M. Molino, P. Jilek, B. Bottazzi, C. Cerletti, K. Matsushima, and A. Mantovani, *J. Immunol.* **147,** 2215 (1991).

[15] S. Sozzani, M. Molino, M. Locati, W. Luini, C. Cerletti, A. Vecchi, and A. Mantovani, *J. Immunol.* **150,** 1544 (1993).

[16] S. Sozzani, M. Locati, D. Zhou, M. Rieppi, W. Luini, G. Lamorte, G. Bianchi, N. Polentarutti, P. Allavena, and A. Mantovani, *J. Leukoc. Biol.* **57,** 788 (1995).

[17] C. Combadiere, S. K. Ahuja, and P. M. Murphy, *J. Biol. Chem.* **270,** 16491 (1995).

[18] C. Combadiere, S. K. Ahuja, J. Van Damme, H. L. Tiffany, J. L Gao, and P. M. Murphy, *J. Biol. Chem.* **270,** 29671 (1995).

[19] K. B. Bacon, B. A. Premack, P. Gardner, and T. J. Schall, *Science* **269,** 1727 (1995).

chemokine activity have been primarily limited to chemotaxis and the measurement of intracellular calcium. However, certain chemokine-like receptors, such as the Duffy antigen found on red blood cells, have been shown to bind chemokines with high affinity, but binding of the ligand failed to stimulate any transient increase in intracellular calcium, suggesting that not all chemokine receptors may signal through changes in intracellular calcium.[20] This illustrates the need for a method of measuring cell activation in a more comprehensive fashion, the alternative being the need to have the capacity to run assays that will examine all consequences of receptor activation. The Cytosensor Microphysiometer System offers one such assay tool.

Cytosensor Microphysiometer System

The development and utilization of silicon-based sensor systems for biological applications have increased since the late 1980s. One such instrument, the Cytosensor Microphysiometer System, is based on a light-addressable potentiometric sensor (LAPS) that measures changes in solution pH in the microenvironment surrounding the surface of the sensor.[21,22]

Cells in culture metabolize carbon sources (sugars, pyruvate, glutamine, and fatty acids) and generate waste products (e.g., lactic and carbonic acid, Fig. 1). For cells in culture, it is known that glycolysis is by far the preferred catabolic pathway; it also happens to generate the most protons per adenosine triphosphate (ATP) molecule hydrolyzed.[23] Extrusion of the acid metabolites from cells results in the acidification of the extracellular environment. Thus, changes in metabolism occurring in response to alterations in the supply of and/or demand for intracellular ATP are measurable by monitoring extracellular acidification. A large number of perturbations can alter cellular ATP flux. These include, but are not limited to, the stimulation of cell surface receptors and the activation of intracellular signaling events. The production of protons through changes in metabolism, however, cannot completely account for the changes in extracellular acidification that can be measured by the Cytosensor Microphysiometer. The contribution of membrane ion-transport systems (e.g., the Na^+/H^+ ex-

[20] K. Neote, J. Y. Mak, L. F. Kolakowski, Jr., and T. J. Schall, *Blood* **84,** 44 (1994).

[21] D. G. Hafeman, J. W. Parce, and H. M. McConnell, *Science* **240,** 1182 (1988).

[22] H. M. McConnell, J. C. Owicki, J. W. Parce, D. L. Miller, G. T. Baxter, H. G. Wada, and S. Pitchford, *Science* **257,** 1906 (1992).

[23] J. C. Owicki and J. W. Parce, *Biosensors and Bioelectronics* **7,** 255 (1992).

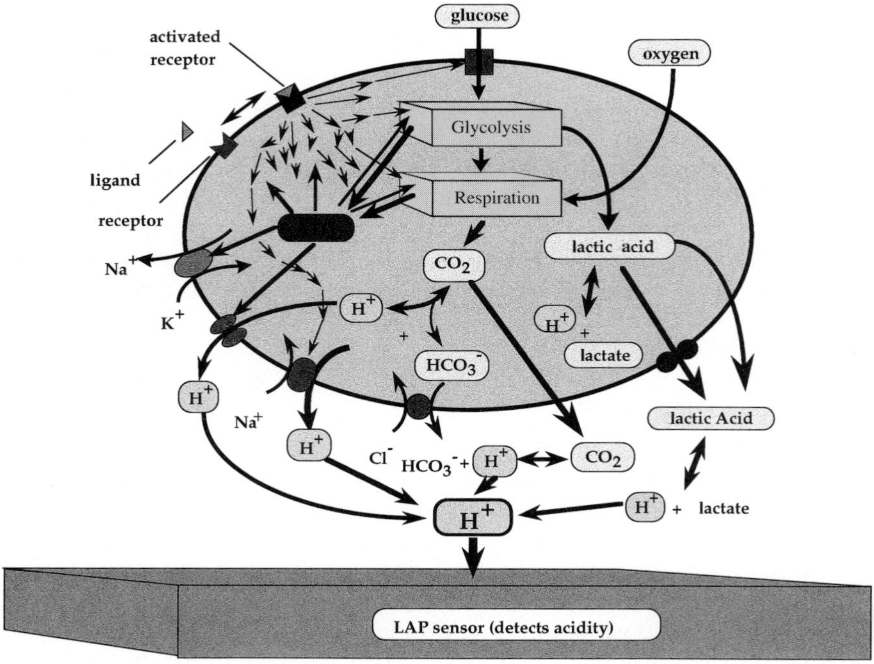

FIG. 1. Cell biology of extracellular acidification. (Reprinted from Ref. 22. Copyright 1992 American Association for the Advancement of Science.)

changer, NHE) are also extremely important in the measured changes in extracellular pH.[24]

Using LAPS technology, the Cytosensor Microphysiometer offers the ability to make rapid, sensitive, and accurate measurements in real time of the rate of extrusion of acid from cells under basal and stimulated conditions, allowing the evaluation of a large number of effector agents. The microphysiometer can make measurements from a wide variety of cells, both eukaryotic and prokaryotic, adherent and nonadherent. Adherent cells are directly seeded onto capsule cups containing a polycarbonate membrane (Fig. 2), whereas nonadherent cells are first immobilized in a low-temperature-melting agarose gel. A second polycarbonate membrane,

[24] J. A. Salon and J. C. Owicki, *in* "Methods in Neuroscience: Receptor Molecular Biology," p. 201. Academic Press, London, 1995.

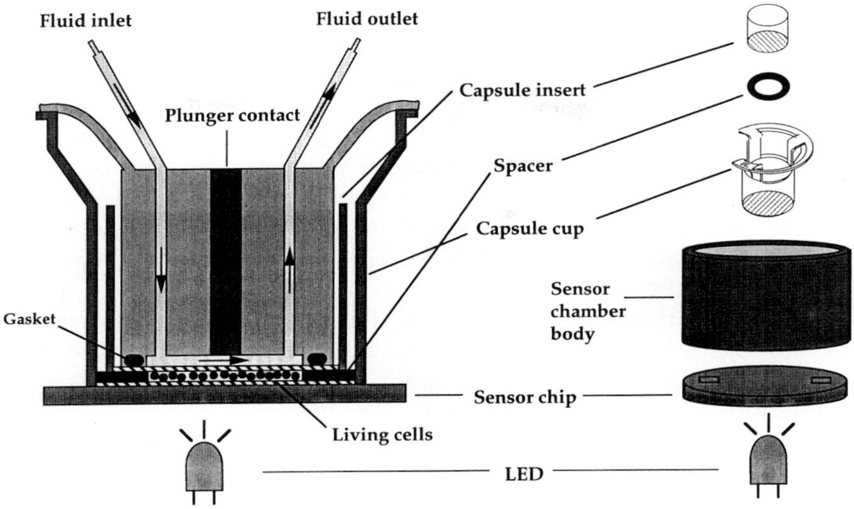

FIG. 2. Schematic representation of the components of the capsule cup and sensor chamber. Cells are held between two track-etched polycarbonate membranes separated by a 50-μm spacer. The lower membrane contacts the surface of the silicon-based LAPS sensor, while low-buffered medium flows across the surface of the top membrane. An LED illuminates the bottom of the LAPS device to activate the sensor.

separated by a 50-μm spacer, is placed on top of the cells. The capsule cup assembly is placed within a sensor chamber with the cells closely apposed to the pH-sensitive surface of the LAPS device. The chamber is perfused by a low-buffered medium, typically a modified form of cell growth medium (the bicarbonate, HEPES, and serum are removed to give a nominal buffering capacity of 1 m*M*). The sensor chamber is placed on a 37° heated pad above the light-emitting diode (LED) that activates the silicon. The Cytosensor system has the capacity to record from eight sensor chambers simultaneously. Fluid for an individual chamber is fed from two separate fluid reservoirs propelled by a peristaltic pump. One of these fluid reservoirs may be replaced by an automated fluid delivery system (the Cytosampler system), which allows unattended operation of the microphysiometer. Medium passes through a debubbler–degasser to a selection valve that controls which of the two fluid streams passes into the chamber. After leaving the chamber, the fluid passes a reference electrode en route to waste. All data acquisition and storage, as well as instrument control, are carried out by a Macintosh computer with dedicated software.

The LAPS device consists of a silicon chip that has a thin, silicon oxynitride insulating layer on the face in contact with the aqueous solution. The silicon surface reversibly binds protons, which results in its surface potential being dependent on the pH of the bathing medium (the response is Nernstian: a voltage excursion of 61 mV is approximately equal to 1 pH unit change at 37°). The microphysiometer makes a voltage measurement that is linearly related to pH once every second. Acid, produced by the cells, alters the pH of the medium and therefore alters the silicon voltage measurement. Further details of the LAPS device and its accompanying electronics can be found elsewhere.[21,22] During periods of flow, the pH of the medium in the chamber is constant (Fig. 3). When the flow stops, the H^+ extruded from the cells acidifies the extracellular medium. The slope of a least-squares fit to this acidification is calculated and plotted as an acidification rate. Typical acidification excursions are less than 0.1 pH unit in magnitude and do not affect cell viability. When flow is resumed, the accumulated acid is flushed out of the chamber, returning cells to the pH of the flowing medium. This cycle of flow-on and flow-off is repeated for all chambers throughout the experiment. Baseline acidification rates are first collected, in the absence of the stimulating factor. Alterations in cellular acidification rates above or below this baseline are then monitored in the presence of an effector agent (Fig. 3). The specificity of the response can be demonstrated by the use of specific antagonists, antibodies, antisense oligonucleotides, or signal transduction inhibitors. In addition, comparison of transfected versus parental, nontransfected cells can be used to demonstrate specificity.

The Cytosensor system has been used successfully to evaluate activation of cell surface receptors representing all of the major receptor classes. This includes receptors that signal through G proteins, tyrosine kinases, and ligand-gated ion channels.[1]

We have used the Cytosensor system to examine the increase in the extracellular acidification rate (ECAR) of a variety of cells in response to the C-C chemokines macrophage inflammatory protein-1α (MIP-1α), RANTES, and macrophage chemotactic protein-1 (MCP-1), and the C-X-C chemokine IL-8. The ECAR response to the chemokines has been examined in both the human monocyte cell line THP-1 and in primary human peripheral $CD4^+$ T cells, monocytes, and neutrophils. In addition, we have used the Cytosensor system to monitor the activity of chemokine receptors that have been transiently transfected into Chinese hamster ovary (CHO) cells. The success of this latter procedure offers the possibility of using the Cytosensor system as a rapid screening assay for novel (orphan) chemokine receptors. Similar procedures have already been used to screen the activity profile of one novel chemokine receptor from the ChemR13 gene, now

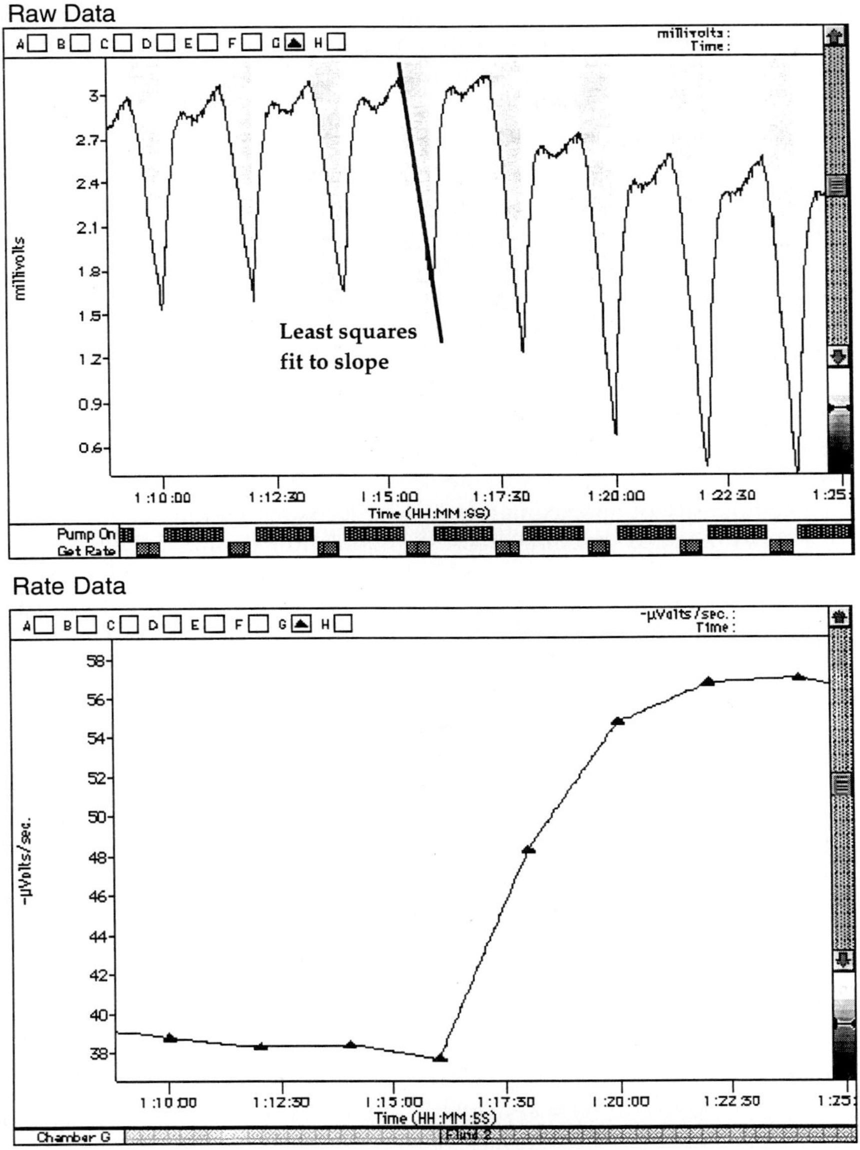

FIG. 3. Data screens taken from Cytosoft, the Cytosensor software. *Top:* Raw data representing the change in pH within the sensor chamber (61 mV $\approx$ 1 pH unit). Fluid flow occurs in constant on–off cycles; during flow-off periods data are fit to a straight line using a least-squares fit. This slope is reported as an acidification rate (*bottom*). Experiments are typically designed to collect acidification rates before, during, and after challenge with a ligand of interest.

identified as the CCR5 receptor.[25] The data presented here indicate that the Cytosensor system can evaluate the chemokine responses in both cell lines and primary cells rapidly and easily.

Cell Preparation: THP-1 Cells and Freshly Prepared Immune Cells

THP-1 Cells. THP-1 cells, obtained from the ATCC (Rockville, MD, TIB 202), are grown and passaged as directed. Generally cells are used within 15 passages of receiving a fresh vial from the ATCC, after which time the acidification response begins to decline. Cells are used either directly from growth medium or following 18 hr of serum starvation. In either case, cells are counted and resuspended at a density of 4×10^6 cells in 300 μl of a low buffered RPMI 1640 medium (Molecular Devices). The cell suspension is then mixed with an aliquot of agarose entrapment medium (Molecular Devices) at a 3:1 (v/v) ratio. A 10-μl aliquot of this cell/agarose suspension is then spotted into the center of a capsule cup, resulting in a final cell density of approximately 1×10^5 cells per capsule cup.

Freshly Isolated Immune Cells. A population of mixed lymphocytes (nonadherent cells) and monocyte/macrophages (adherent population) are prepared from peripheral blood.[26] $CD4^+$ T cells are isolated from the mixed lymphocyte population by positive selection using antibody-linked magnetic particles (Miltenyi Biotech, Auburn, CA). For use in experiments on the Cytosensor system, both the $CD4^+$ T cells and the monocytes are trapped in the agarose entrapment medium. Both monocytes and peripheral blood $CD4^+$ T cells are used at a concentration of $1–2 \times 10^6$ cells/cell capsule.

Freshly prepared human neutrophils are isolated from whole blood by discontinuous density gradient centrifugation using Ficoll–Hypaque, followed by hypotonic lysis of the contaminating red blood cell population.[26] Cells are seeded at a final density of 1×10^5 cells/capsule cup using the agarose entrapment medium (as outlined above).

Transiently Transfected Chemokine Receptors into CHO-K1 Cells

cDNA for the chemokine receptors CCR1, CCR2b, and CCR5 is kindly supplied by Dr. Philip Murphy (National Institutes of Health, Bethesda, MD). These cDNA inserts are cloned into an expression plasmid, pLEC2,

[25] M. Samson, O. Labbe, C. Mollereau, G. Vassart, and M. Parmentier, *Biochemistry* **35,** 3362 (1996).

[26] M. E. Kanoff and P. D. Smith, *in* "Current Protocols in Immunology" (J. E. Coligan, A. M. Kruisbeek, D. H. Margulies, E. M. Shevach, and W. Strober, eds.), Chap. 7. Wiley, New York, 1992.

containing the cytomegalovirus (CMV) promoter upstream of the cloning site and the BGH polyadenylation signal sequence downstream of the cloning site. CHO-K1 cells (ATCC CCL-61) are transiently transfected with individual classes of chemokine plasmids using LipofectAMINE (GIBCO Life Technologies, Grand Island, NY) according to the manufacturer's instructions. Essentially, CHO-K1 cells at 70% confluency in a 10-cm dish are treated for 6 hr with a complex of 6 μg of plasmid DNA and 42 μl of LipofectAMINE reagent in 6 ml of serum-free OptiMEM medium (GIBCO Life Technologies). After 6 hr, the medium is changed to growth medium, Ham's F12 supplemented with 10% fetal bovine serum (FBS), and cultures are incubated overnight in a CO_2 incubator. The cells are then harvested using trypsinization and seeded at 8×10^4 cells per capsule cup in Ham's F12 supplemented with 10% FBS. After incubation overnight to allow adherence to the capsule cup membrane, a capsule spacer and capsule insert are placed in the capsule cup. The assembled cups are placed on the Cytosensor system, and the cells are monitored for acidification rates using the protocol described earlier. After 2 hr, the transfected cells are exposed to chemokines for 6 min.

Running Medium

Because the Cytosensor system measures small changes in excretion of protons from cells, the running medium that is used with the system has to have a low buffering capacity to ensure that the excreted acid causes a measurable pH change. Generally, a medium with a buffering capacity of approximately 1 m*M* is compatible with acidification rate measurements. In the experiments on the THP-1 or the immune cells, cells are bathed with "running buffer," comprising a low-buffer RPMI 1640 medium (Molecular Devices) supplemented with 1 mg/ml human serum albumin (HSA, Bayer, Kankakee, IL). For experiments on the transiently transfected cells, they are bathed in low-buffered, bicarbonate, HEPES and serum-free F12 medium (GIBCO Life Technologies) supplemented with 0.1% HSA.

Flow rates used for these experiments are typically 100 μl/min with a pump cycle of 1.5–2 min (consisting of a flow-on time of 60–80 sec and a flow-off time of 30–40 sec). Acidification rates are measured for 20–30 sec during this flow-off period.

Reagents

All chemokines and neutralizing antibodies are obtained either from R&D Systems (Minneapolis, MN) or Peprotech (Rocky Hill, NJ) and are initially diluted in water to either a 1 mg/ml or 100 μg/ml concentration. Frozen aliquots of these stock chemokines are then dissolved in low-buffered medium to the final concentration required. For the neutralizing antibody experiments, all drugs and antibodies are diluted to their final concentrations in the running medium and then incubated at 37° for 60 min prior to use. Experiments are carried out at 37°.

Effect of Chemokines on THP-1 Cells

THP-1 cells are placed on the Cytosensor system in running medium, and the acidification rates are monitored until a stable baseline rate is obtained for each chamber. After equilibration, individual populations of cells are exposed to 25 n*M* MCP-1 for different times (Fig. 4). This allows us to estimate the optimal exposure time for additional chemokine studies. For all exposure times, on addition of the chemokine there is a rapid increase in ECAR, which reaches a peak by the first rate data point after commencement of the treatment. ECAR responses then begin to decline even in the continued presence of the chemokine. The ECAR

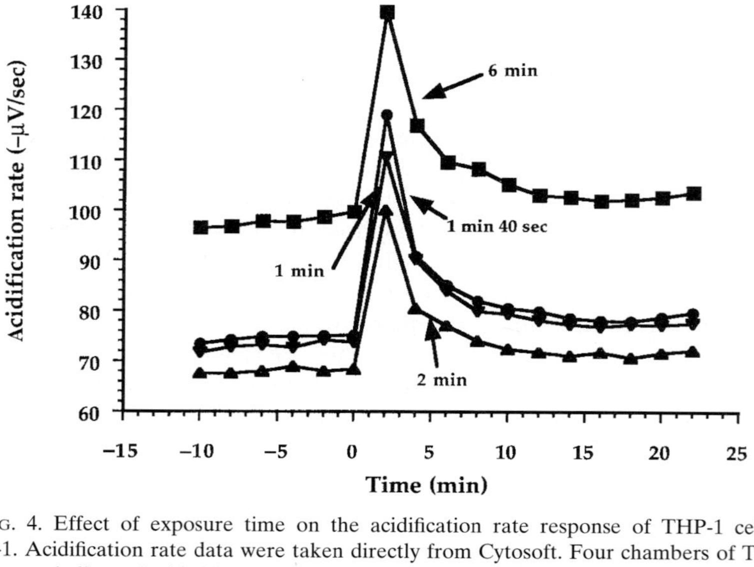

FIG. 4. Effect of exposure time on the acidification rate response of THP-1 cells to MCP-1. Acidification rate data were taken directly from Cytosoft. Four chambers of THP-1 cells were challenged with 25 n*M* MCP-1 for different times. On addition of chemokine the acidification rates of the cells increased rapidly and then declined during the continued exposure to chemokine.

responses generally return to baseline within about 14 min of the initiation of treatment. Further work using shorter exposure times resulted in the determination of an optimal incubation time of 100 sec, with the chemokine being added 60 sec before the start of an acidification rate measurement.

Using this exposure time, separate populations of THP-1 cells are treated with different concentrations of MIP-1α, RANTES, MCP-1, or IL-8. In these experiments, each of the eight separate chambers of THP-1 cells receives a different concentration of chemokine (an example for MIP-1α is shown in Fig. 5). We observe that there is a concentration-dependent increase in extracellular acidification rates to these chemokines. Concentration–response curves are generated by plotting the concentration of each dose of chemokine versus the corresponding peak acidification rate (Fig. 6). The peak responses to both MIP-1α and MCP-1 are very similar (~30% above baseline), whereas the response to the maximal concentration of RANTES resulted in a lower maximal acidification rate (~20%). There is a detectable response to IL-8 in the THP-1 cells, but the response is much less than that observed for the C-C chemokines (peak response <10% of baseline).

The EC_{50} values calculated from these experiments (mean $\pm$ SE; n = 5) are as follows: MCP-1, 1.2 $\pm$ 0.4 nM; MIP-1α, 2.4 $\pm$ 0.9 nM; RANTES,

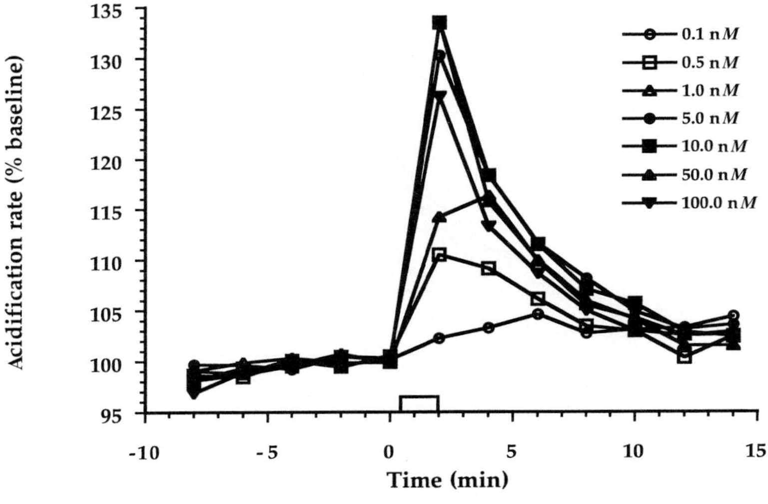

FIG. 5. MIP-1α treatment results in a concentration-dependent increase in acidification rates. Each individual chamber of THP-1 cells was exposed for 100 sec to a different concentration of MIP-1α.

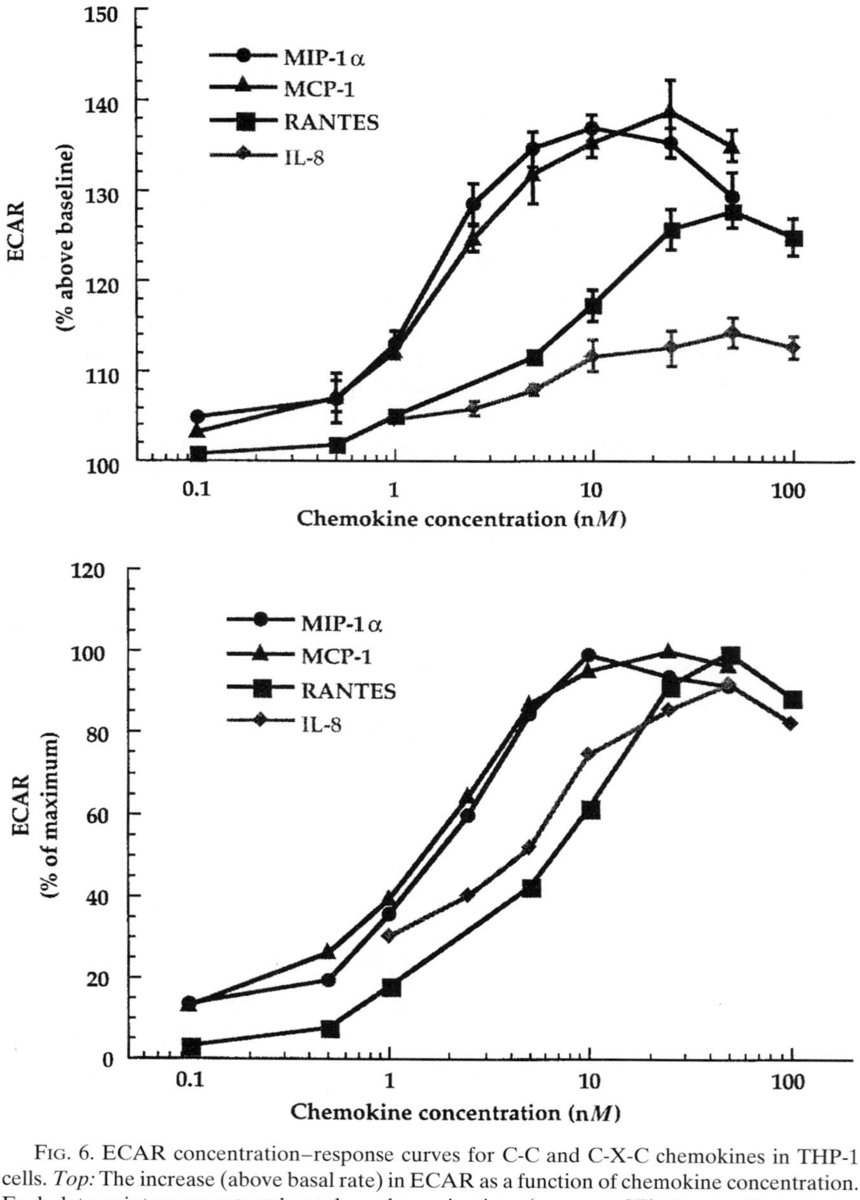

FIG. 6. ECAR concentration–response curves for C-C and C-X-C chemokines in THP-1 cells. *Top:* The increase (above basal rate) in ECAR as a function of chemokine concentration. Each data point represents at least three determinations (means ± SE). *Bottom:* Data shown as percent maximum response.

7.3 ± 0.8 n*M*; and IL-8, 5.7 ± 1.5 n*M*. These values correspond very closely to data collected on monocyte cell lines using increases in intracellular calcium as a measure of chemokine response and to data reported for chemotaxis in human monocytes.[27–29]

With other G-protein-coupled receptors, it is possible to expose a single chamber of cells to a series of ligand concentrations, having previously determined the appropriate incubation time for the ligand and recovery time between successive exposures. When this is attempted with the chemokine ligands in the THP-1 cells, there is a significant shift in the concentration–response curve to the right and a reduction in the maximal acidification response achieved at high concentrations of chemokine (data not shown). Therefore, in all of the experiments carried out to determine the concentration–response profile of the chemokines we use parallel dosing across the eight chambers of the Cytosensor system. This effect is attributed to the desensitization characteristics of the chemokine family of receptors.

Desensitization of THP-1 Cells to C-C Chemokines

One of the common characteristics of the chemokine receptors is the marked homologous and, in many cases, heterologous desensitization following an initial treatment with chemokine.[16,27] For example, in the monocyte cell line Monomac 6, Charo *et al.*,[27] using measurement of intracellular calcium changes demonstrated that treatment of cells with MCP-1 would abolish subsequent responses to this same chemokine and dramatically reduce the response to subsequent treatment with RANTES. Conversely, treatment with 100 n*M* RANTES resulted in an abolition of a response to subsequent RANTES treatment but left the response to MCP-1 unchanged. We have tested the effect of combinations of chemokine treatments on THP-1 cells to see whether similar desensitization characteristics could be observed using the Cytosensor system. After a steady baseline rate is established, the cells are exposed to 10 n*M* MIP-1α, 10 n*M* MCP-1, or 50 n*M* RANTES for 100 sec. After a recovery period of about 20 min, the cells are reexposed for 100 sec to MIP-1α, MCP-1, or RANTES.

[27] I. F. Charo, S. J. Myers, A. Herman, C. Franci, A. J. Connelly, and S. R. Coughlin, *Proc. Natl. Acad. Sci. U.S.A.* **91,** 2752 (1994).
[28] C. Franci, L. M. Wong, J. Van Damme, P. Proost, and I. F. Charo, *J. Immunol.* **154,** 6511 (1995).
[29] M. D. Uguccioni, M. Apuzzo, and M. Loetscher, *Eur. J. Immunol.* **25,** 64 (1995).

The data from this experiment (Fig. 7) indicate that MIP/MIP or MCP/MCP exposure causes substantial (>50%) homologous desensitization. Heterologous exposures of MIP/MCP, MCP/MIP, or RAN/MIP resulted in partial (<50%) desensitization of the response to the second chemokine. The heterologous combination of MIP/RAN shows desensitization similar to the homologous exposures.

Effect of Methylisobutylamiloride on Extracellular Acidification Rate Response to C-C Chemokines

Figure 8 shows Cytosensor system data from an experiment in which THP-1 cells were stimulated with 50 n*M* MIP-1α or 50 n*M* RANTES in the presence or absence of methylisobutylamiloride (MIA), an NHE inhibitor. The cells are pretreated with vehicle (ethanol) or 10 μM MIA for 15 min, then exposed for 100 sec to 50 n*M* MIP-1α or RANTES in the continued presence or absence of MIA.

In the absence of MIA, the cells responded with an approximately 30% increase in ECAR when exposed to MIP-1α and about a 25% increase in ECAR when exposed to RANTES. In the presence of MIA these responses are substantially reduced, suggesting that the sodium/hydrogen exchanger is critically involved in the ECAR response to chemokines. This is a characteristic of the ECAR response to many of the G-protein-coupled receptors, particularly those that are believed to couple primarily through the $G\alpha_i$ family.[30,31]

Differentiating Extracellular Acidification Rate Responses to C-C and C-X-C Chemokines

After equilibration, individual populations of cells are exposed for 100 sec to one of the following: MIP-1α alone, IL-8 alone, IL-8 neutralizing antibody alone, MIP-1α + IL-8, MIP-1α + IL-8 neutralizing antibody, IL-8 + IL-8 neutralizing antibody, and MIP-1α + IL-8 + IL-8 neutralizing antibody. All exposures are for 100 sec. The data shown in Fig. 9 are the means ± the range of two determinations performed in two separate experiments.

Exposing THP-1 cells to MIP-1α or IL-8 alone caused increases in the ECAR of approximately 36 and 10%, respectively. Exposing the cells to

[30] C. L. Chio, M. E. Lajiness, and R. M. Huff, *Mol. Pharmacol.* **45,** 51 (1993).

[31] C. L. Chio, R. F. Drong, D. T. Riley, G. S. Gill, J. L. Slightom, and R. M. Huff, *J. Biol. Chem.* **269,** 11813 (1994).

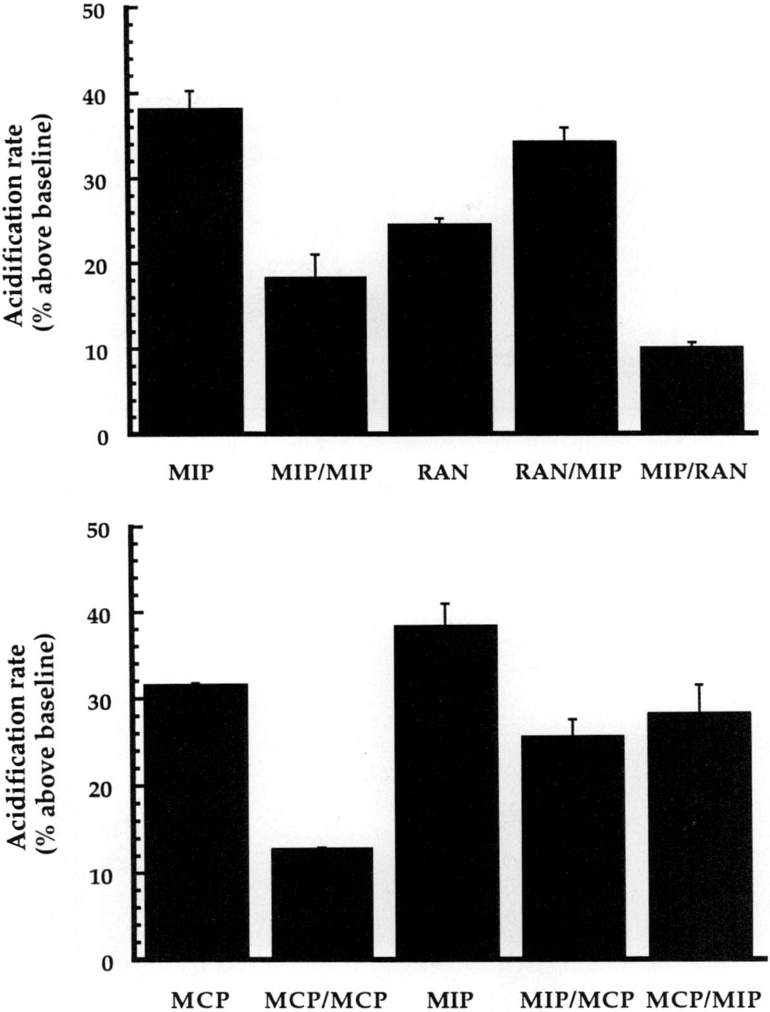

FIG. 7. Desensitization of THP-1 cells to C-C chemokines. The percent increase in acidification rate refers to the ECAR at the peak of the response compared to the baseline value. *Top:* Desensitization to MIP-1α and RANTES. MIP, 10 n*M* MIP-1α alone; MIP/MIP, 10 n*M* MIP-1α followed by 10 n*M* MIP-1α; RAN, 50 n*M* RANTES alone; RAN/MIP, 50 n*M* RANTES followed by 10 n*M* MIP-1α; MIP/RAN, 10 n*M* MIP-1α followed by 50 n*M* RANTES. *Bottom:* Desensitization to MCP-1 and MIP-1α. MCP, 10 n*M* MCP-1 alone; MCP/MCP, 10 n*M* MCP-1 followed by 10 n*M* MCP-1; MIP, 10 n*M* MIP-1α alone; MIP/MCP, 10 n*M* MIP-1α followed by 10 n*M* MCP-1; MCP/MIP, 10 n*M* MCP followed by 10 n*M* MIP-1α.

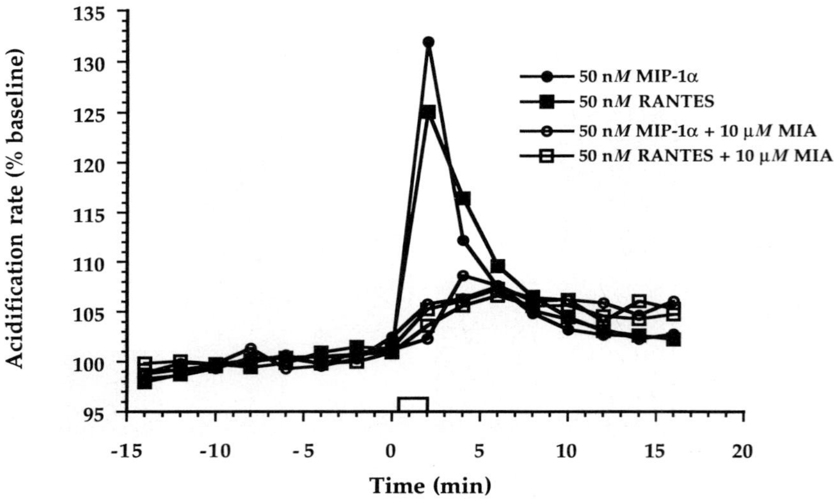

FIG. 8. The effect of methylisobutylamiloride (MIA) on the ECAR response to the C-C chemokines MIP-1α and RANTES.

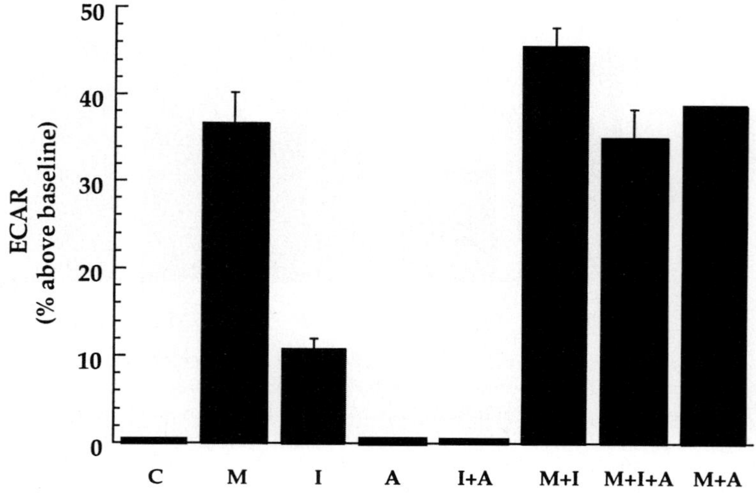

FIG. 9. Differentiating the C-C and C-X-C chemokine ECAR responses. C, Control, no treatment; M, 10 n*M* MIP-1α; I, 50 n*M* IL-8; A, IL-8 neutralizing antibody; I+A, IL-8 plus 10 μg/ml IL-8 neutralizing antibody; M+I, 10 n*M* MIP-1α plus 50 n*M* IL-8; M+I+A, MIP-1α plus IL-8 plus IL-8 neutralizing antibody; M+A, MIP-1α plus IL-8 neutralizing antibody.

MIP-1α and IL-8 together gave an additive response of approximately 45%. The IL-8 response is blocked by the presence of an IL-8 neutralizing antibody (I+A, M+I+A), but the response to MIP-1α is not affected by the same antibody (M+A). These data are consistent with the responses to C-C and C-X-C chemokines being triggered by separate receptors, which can be differentiated using the Cytosensor system.

Extracellular Acidification Rate Response of Primary Human CD4⁺ T Cells and Monocytes to Chemokines

Primary $CD4^+$ cells and monocytes are isolated from normal human peripheral blood and their ECARs are monitored following treatment with various chemokines (Fig. 10). As with the THP-1 cells we observe a rapid increase in ECARs following treatment with the chemokines MIP-1α, MCP-1, and IL-8, suggesting the presence of both C-C and C-X-C chemokine receptors on these cells.

Response of Human Neutrophils to IL-8

Although THP-1 cells and $CD4^+$ T cells give fairly weak responses to the C-X-C chemokine IL-8, freshly prepared human neutrophils respond very strongly to this ligand. Human neutrophils treated with 50 n*M* IL-8 for 100 sec respond with an extremely rapid and large increase in acidification rates, reaching a peak 120% above the original baseline of the cells. The neutrophils also respond to a similar treatment with 50 n*M* MCP-1, though the response is much smaller, only reaching about 50% above the baseline (data not shown). Because of the rapid nature of the response to IL-8, a second test is carried out in which the incubation of chemokine commences 10 sec prior to the collection of a rate data point (total incubation time is 50 sec). To this challenge the cells respond with an increase in ECARs that is 7 to 8-fold above the original baseline. Pretreatment of the IL-8 with 1 μg/ml of a neutralizing antibody results in a complete abolition of the ECAR response to IL-8 treatment (Fig. 11).

Transient Expression of Chemokine Receptors in CHO-K1 Cells

When the cDNA for CCR-1, -2b, and -5 is transfected into CHO cells, the cells respond to the chemokine ligands MIP-1α, MCP-1, and RANTES, respectively (Fig. 12). Similarly, the cDNA for the C-X-C receptors CXCR1 and CXCR2 imparts responsiveness to IL-8 in the CHO cells (Fig. 13). The kinetics of the extracellular acidification response to the chemokines

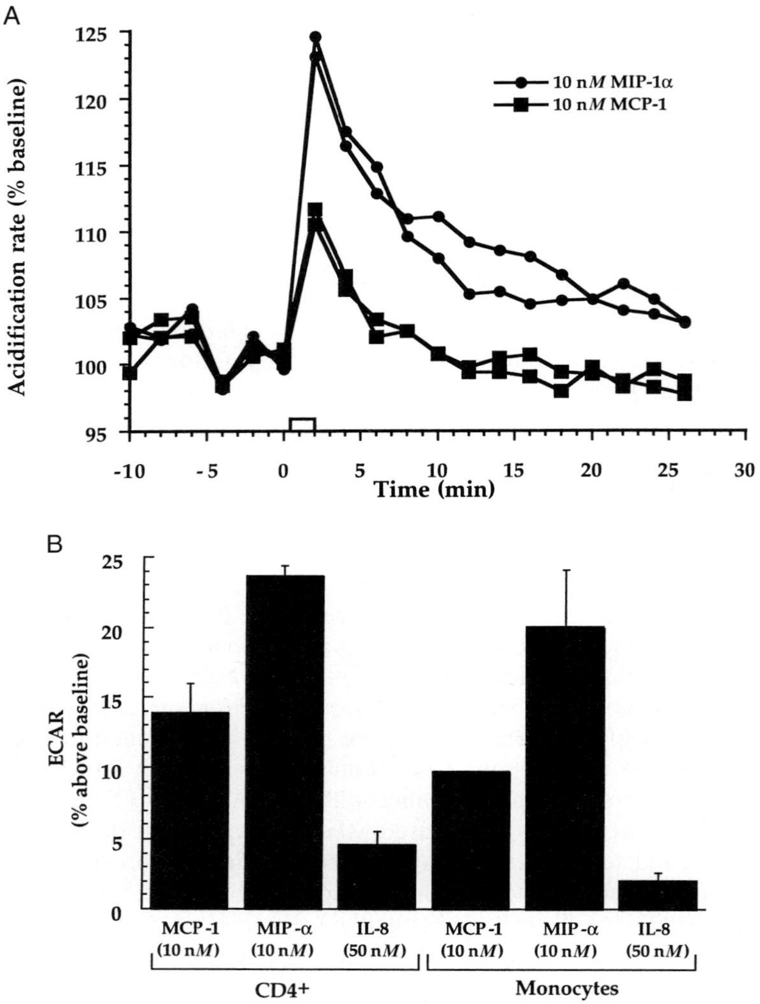

FIG. 10. (A) Acidification rate data showing the response of duplicate chambers of CD4$^+$ cells to 10 n*M* MCP-1 and MIP-1α. (B) Averaged data (means ± SE) of three experiments on freshly prepared CD4$^+$ cells and monocytes showing the responses to the chemokines MCP-1 (10 n*M*), MIP-1α (10 n*M*), and IL-8 (50 n*M*). All exposure times were 100 sec.

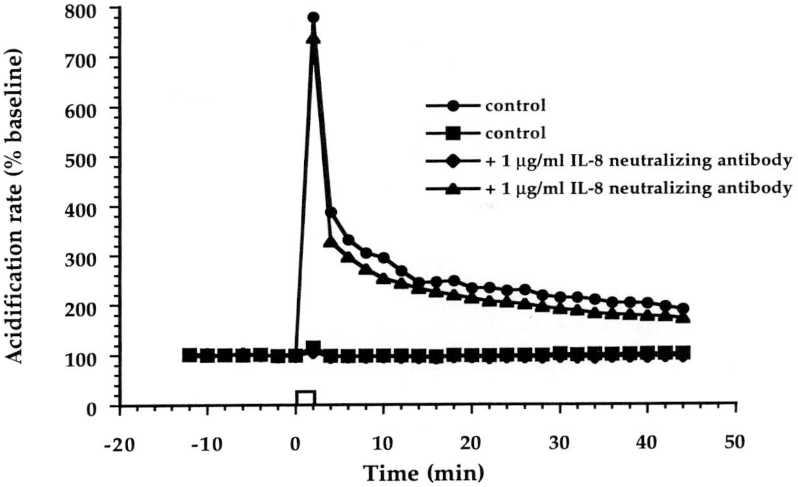

FIG. 11. Acidification rate data showing the response of duplicate chambers of freshly prepared human neutrophils to 50 n*M* IL-8 in the absence and presence of pretreatment with 1 μg/ml of an IL-8 neutralizing antibody.

in transfected CHO cells are similar to that observed in THP-1 cells in that the increase is rapid, occurring within 2–4 min, and there is a rapid return to basal acidification rates after the ligands are removed. We also observe, for all of the transfected chemokine receptors tested, that a second treatment with chemokine stimulates an increase in rate that is reduced by approximately 50% from the primary stimulation, suggesting receptor desensitization (see Fig. 13A). With regard to the G-protein coupling of the transfected receptors, stimulation of CCR1 transfected CHO cells by MIP-1α is abolished by an overnight (18-hr) treatment with 100 ng/ml pertussis toxin (Fig. 14). This suggests that the CCR1 receptor is coupled through $G\alpha_i$ to metabolic pathways.

Discussion

The data collected by ourselves and other researchers demonstrate that the measurement of changes in extracellular acidification using the Cytosensor system may be used to rapidly evaluate the effects of the family of chemokines and their receptors in a variety of cell types, including preparations where the receptors are transiently expressed in a host cell line. Treating THP-1 cells with C-C and C-X-C chemokines resulted in

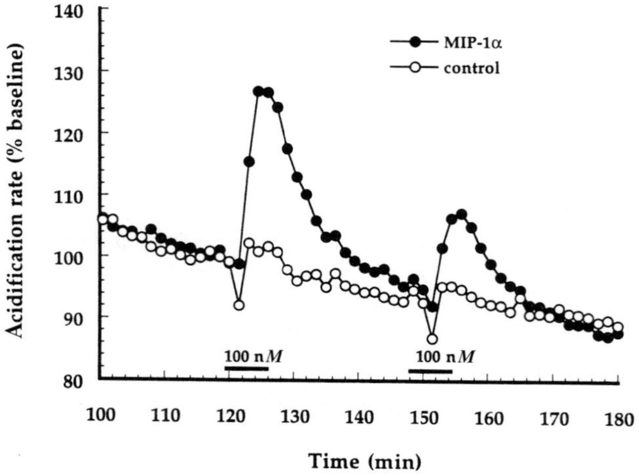
MIP-1α
control
Acidification rate (% baseline)
100 nM
100 nM
Time (min)
140
130
120
110
100
90
80
100
110
120
130
140
150
160
170
180

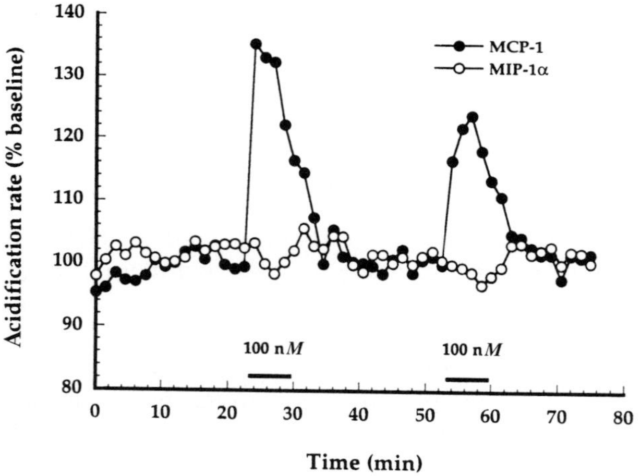
MCP-1
MIP-1α
Acidification rate (% baseline)
100 nM
100 nM
Time (min)
140
130
120
110
100
90
80
0
10
20
30
40
50
60
70
80

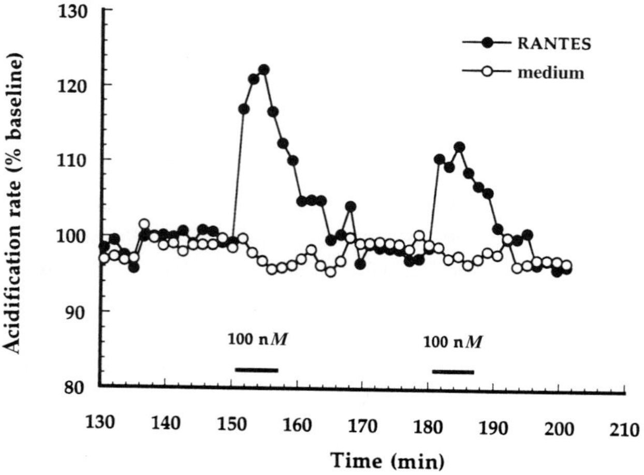
RANTES
medium
Acidification rate (% baseline)
100 nM
100 nM
Time (min)
130
120
110
100
90
80
130
140
150
160
170
180
190
200
210

an increase in the extracellular acidification rate that was concentration dependent. The EC_{50} values ranged from 1 to 7 n*M*, and the order of potency for C-C chemokines in THP-1 cells was MCP-1 $\geq$ MIP-1α $>$ RANTES. The response to the C-X-C chemokine IL-8 was comparable to RANTES in potency, but IL-8 gave an acidification response that was much lower in magnitude. In freshly prepared neutrophils, however, the response to IL-8 was extremely marked. Pleass *et al.*[32] have further characterized the acidification response of THP-1 cells to the chemokines MCP-1 and MCP-3 and have demonstrated that the EC_{50} values obtained through microphysiometry were equivalent to those measured from calcium signals in this cell line.[27,28]

An increasing number of examples have appeared demonstrating the use of the Cytosensor system in detecting cellular responses following activation of chemokine receptors in a number of different cell types. In 1994, Vaddi and Newton reported the use of the Cytosensor system to detect responses to the chemokines MCP-1, MIP-1α, and RANTES in human monocytes.[33] They demonstrated that, not only could one easily detect acidification rate changes by treatment of cells with these three chemokines, but also the apparent signaling pathway leading to ECAR changes differed from that involved in the alteration of intracellular calcium. The kinetics and magnitude of the ECAR changes in the monocytes were similar to those illustrated above in the THP-1 cells. Importantly, however, the ECAR response observed in the monocytes was abolished by pretreatment of the cells with pertussis toxin, which uncouples the receptor from the G protein and has been shown to inhibit chemotactic responses to chemokine activation. Interestingly, the chemokine-induced calcium signal was not abolished by a similar cell manipulation in this preparation. We have also found that pertussis toxin treatment completely blocked the MIP-1α signaling, leading to acidification rate increases in the CCR1 transfected CHO cells, demonstrating the involvement of the receptor coupling to $G\alpha_i$ in the signaling pathway.

There are an increasing number of reports of the use of the Cytosensor system to evaluate chemokine receptors exogenously expressed in cells

[32] R. D. Pleass, U. M. Moore, A. G. Roach, and R. J. Williams, *B.P.S. Meeting,* P110 (1995).
[33] K. Vaddi and R. C. Newton, *FASEB J.* **8,** A502 (1994).

FIG. 12. CHO cells transiently transfected with either the CCR1 (*top*), CCR2b (*middle*), or CCR5 (*bottom*) receptors respond to the chemokines MIP-1α, MCP-1, and RANTES, respectively.

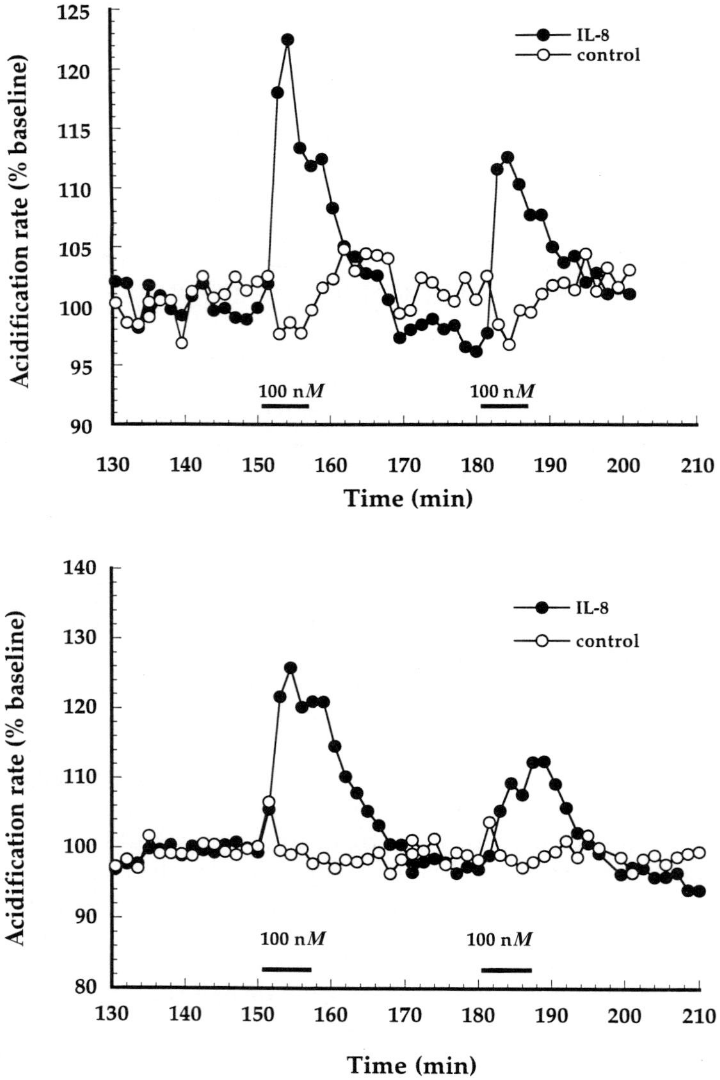

FIG. 13. CHO cells transiently transfected with either the CXCR1 (*top*) or CXCR2 (*bottom*) receptors respond to the chemokine IL-8.

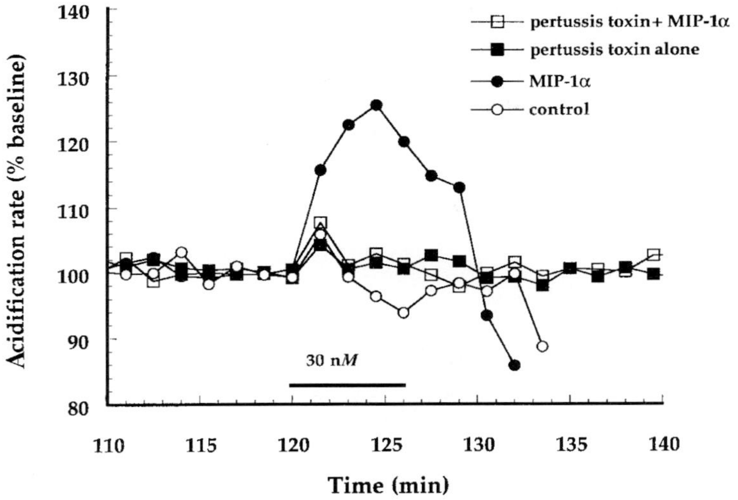

FIG. 14. Pretreatment of CHO cells transiently transfected with the CCR1 receptor with pertussis toxin abolishes the acidification response to MIP-1α treatment.

either stably or transiently. An important example was reported by Samson *et al.* (1996) on the cloning of the CCR5 receptor.[25] They transfected this receptor into CHO cells and used microphysiometry as the sole method to demonstrate functional activity, reporting significant responses to the chemokines MIP-1α, MIP-1β (the first receptor demonstrated to respond physiologically to this factor), and RANTES. In this study, when the researchers attempted to assess their transfection efficiently in binding studies using iodinated MIP-1α, they were unable to detect any specific binding activity even at high levels of ligand. In addition, they reported that CHO cells that had been transfected with the CXCR1 receptor exhibited a marked increase in acidification rates (~60% above basal rates) when exposed to IL-8. The latter studies, along with the demonstration of activity of chemokine receptors transiently transfected into CHO cells, illustrate the usefulness of the Cytosensor system in the identification and assessment of new chemokine receptors.

In conclusion, we have been able to detect activity of both native and exogenously expressed chemokine receptors using the Cytosensor Microphysiometer system. The measurement of acidification rates allowed a rapid differentiation between the responses of a particular receptor to the different C-C and C-X-C chemokines, and it allowed us to quickly assess their relative desensitization properties and signal transduction pathways. As new gene products are being identified that may encode new chemokine receptors and with the discovery of an expanding role for these agents in immune function, use of microphysiometry along with other techniques will provide the necessary information to assess the function and activation profiles of these new receptors.

[8] Calcium Flux Assay of Chemokine Receptor Expression in *Xenopus* Oocytes

By PHILIP M. MURPHY

Introduction

Since the pioneering studies of John Gurdon and colleagues in 1971, demonstrating that unfertilized oocytes from the African clawed toad *Xenopus laevis* synthesize rabbit globin after microinjection of rabbit reticulocyte mRNA,[1] these cells have become established as a versatile and valuable tool in molecular and cell biology for studying the properties of a wide variety of foreign proteins.[2,3] Their use for the study of chemokine receptors began in 1991[4] as a logical extension of earlier work in oocytes on other G-protein-coupled receptors,[5] including receptors for the nonchemokine "classic" leukocyte chemoattractants fMet-Leu-Phe, C5a, and platelet-activating factor (PAF).[6–9]

Although oocytes can be used for expression cloning of cDNAs, as they were for the guinea pig lung PAF receptor,[9] and for detailed study of

[1] J. B. Gurdon, C. D. Lane, H. R. Woodland, and G. Marbaix, *Nature* (*London*) **233,** 177 (1971).

[2] A. Colman, *in* "Transcription and Translation—A Practical Approach" (B. D. Hames and S. J. Higgins, eds.), p. 271. IRL Press, Oxford, 1984.

[3] A. Soreq and B. Seidman, *Methods Enzymol.* **207,** 225 (1992).

[4] P. M. Murphy and H. L. Tiffany, *Science* **253,** 1280 (1991).

[5] T. Kubo, K. Fukuda, A. Mikami, A. Maeda, H. Takahashi, M. Mishina, T. Haga, K. Haga, A. Ichiyama, K. Kangawa, M. Kojima, H. Matsuo, T. Hirose, and S. Numa, *Nature* (*London*) **323,** 411 (1986).

[6] P. M. Murphy, E. K. Gallin, H. L. Tiffany, and H. L. Malech, *FEBS Lett.* **261,** 353 (1990).

[7] P. M. Murphy, E. K. Gallin, and H. L. Tiffany, *J. Immunol.* **145,** 2227 (1990).

[8] P. M. Murphy and D. McDermott, *J. Biol. Chem.* **266,** 12560 (1991).

[9] Z. Honda, M. Nakamura, I. Miki, M. Minami, T. Watanabe, Y. Seyama, H. Okado, H. Toh, K. Ito, T. Miyamoto, and T. Shimizu, *Nature* (*London*) **349,** 614 (1991).

0076-6879/97 $25.00

protein processing and function,[2] their application to chemokine receptors has so far been restricted to determining the chemokine specificity of orphan open reading frames (ORFs) identified by another means, for example, by cross-hybridization with DNA probes made from known G-protein-coupled receptors or by finding interesting sequence similarity between a known chemokine receptor and an orphan sequence present in the gene databases.[10]

The power of coupling "orphan cloning" to heterologous expression analysis is illustrated by the fact that 10 of the 12 known chemokine receptor subtypes have been identified this way, in most cases even before the corresponding native receptor had been characterized pharmacologically in primary leukocytes. Five of the 12 subtypes were first characterized by heterologous expression of cloned ORFs in oocytes, including the first C-X-C and C-C chemokine receptors, CXCR2 and CCR1[4,11]; the first virally encoded chemokine receptor, ECRF3 of *Herpesvirus saimiri*[12]; two splice variants of CCR2[13]; and CCR4.[14] The other 7 were first characterized by heterologous expression of cloned ORFs in a variety of mammalian somatic cell lines.[15]

A detailed review of strategies for cloning chemokine receptor cDNAs and genes can be found in Ref. 10. This chapter focuses instead on methods for expressing cloned chemokine receptors in oocytes, and it includes a simple calcium flux assay for detecting expression.

Oocyte Anatomy and Physiology

The ovaries fill the abdominal cavity of female *Xenopus*. Each ovary is made up of multiple connected lobes, each of which is formed of multiple grapelike clusters of individual oocytes in different stages of maturation. Oocyte maturation has been classified into six stages by DuMont based on size and appearance.[16] For microinjection of mRNA and expression of foreign proteins, stage V and VI oocytes are required. In wild-type animals, these are approximately 1 mm in diameter or greater and have a well-

[10] P. M. Murphy, *Methods* (*San Diego*) **10,** 104 (1996).

[11] J.-L. Gao, D. B. Kuhns, H. L. Tiffany, D. McDermott, X. Li, U. Francke, and P. M. Murphy, *J. Exp. Med.* **177,** 1421 (1993).

[12] S. K. Ahuja and P. M. Murphy, *J. Biol. Chem.* **268,** 20691 (1993).

[13] I. F. Charo, S. J. Myers, A. Herman, C. Franci, A. J. Connolly, and S. R. Coughlin, *Proc. Natl. Acad. Sci. U.S.A.* **91,** 2752 (1994).

[14] C. A. Power, A. Meyer, K. Nemeth, K. B. Bacon, A. J. Hoogewerf, A. E. I. Proudfoot, and T. N. C. Wells, *J. Biol. Chem.* **270,** 19495 (1995).

[15] P. M. Murphy, *Cytokine Growth Factor Rev.* **12,** 593 (1994).

[16] J. N. Dumont, *J. Morphol.* **136,** 153 (1972).

demarcated equator separating a dark brown hemisphere or "animal pole," which harbors the nucleus, and a pale yellow hemisphere or "vegetal pole," which contains the cytoplasm and yolk platelets. Many laboratories also use oocytes from albino *Xenopus,* in which case size is the chief characteristic used to identify desirable oocytes.

The oocyte can be viewed as part of a miniorgan, existing in intimate communication with a contiguous single cell layer of follicular cells that is itself surrounded by a second connective tissue or thecal cell layer with capillaries and epithelial attachments to the ovarian wall. The oocyte plasma membrane is covered by an acellular fibrous barrier termed the vitelline membrane. For microinjection experiments, the stromal and follicular cells must be removed from the oocyte by either enzymatic digestion or mechanical methods, but the vitelline membrane can remain.

Compared to bacteria, yeast, and cultured somatic cells, oocytes are large (~1 μl total volume) and relatively quiescent with respect to endogenous gene transcription and protein translation. Nevertheless, they contain large amounts of tRNA and ribosomes for the postfertilization translational burst, which can be used instead for translating foreign mRNAs of interest introduced by microinjection. High levels of protein production can be achieved, and the protein products are directed to the appropriate subcellular or extracellular location. Importantly, oocytes do not constitutively express chemokine receptors, but they do express native G_i-type heterotrimeric G proteins capable of interacting with chemokine receptors, coupling them to endogenous downstream oocyte effectors.[17]

Microinjection of *Xenopus* Oocytes

Materials

Adult female *Xenopus laevis*
0.15% (w/v) Ethyl 3-aminobenzoate (also known as tricaine) in water
OR-2 solution: 82.5 m*M* NaCl, 1 m*M* $MgCl_2$, 2.4 m*M* KCl, 5 m*M* HEPES, pH 7.4
ND96 solution: 96 m*M* NaCl, 2 m*M* KCl, 1.8 m*M* $CaCl_2$, 1 m*M* $MgCl_2$, 2.4 m*M* sodium pyruvate, 100 U/ml penicillin, 100 μg/ml streptomycin, 5 m*M* HEPES, pH 7.4
Collagenase type II, ~130 U/mg (Worthington, Freehold, NJ)
Dissecting microscope
Micromanipulator and microinjector (available, e.g., from Narishige, Greenvale, NY)

[17] J. Olate, S. Martinez, P. Purcell, H. Jorquera, and J. Allende, *FEBS Lett.* **30,** 2687 (1990).

Pipette puller (manufactured, e.g., by Kopf, Tujunga, CA)
Borosilicate glass capillary tubes, 1 mm outer diameter
Two pairs of watchmaker forceps
3-O suture materials (absorbable or nonabsorbable)
Scalpel or razor blade
Needle holder
Pair of scissors
Blunt forceps
Parafilm

Methods

ANIMAL HUSBANDRY. Laboratory-bred *Xenopus* can be purchased from commercial suppliers (e.g., Nasco, Fort Atkinson, WI; Xenopus Ltd., South Nuffield, UK). We generally purchase 2-year-old virgin females. When maintained properly, they can survive for more than a decade in the laboratory. Because only a fraction of the complement of approximately 40,000 oocytes/animal is needed for a typical experiment, the same animal can be used repeatedly. We have had good success with oocytes from animals operated as frequently as six times per year. The quality of oocytes can vary considerably in different animals, and multiple animals may need to be screened before one with suitable oocytes is found (see below for definition of suitable oocytes). Moreover, some seasonal variation in oocyte quality can occur in the same animal.

Xenopus is a fully acquatic animal. They are maintained in the laboratory in tanks containing at least 1 gallon water/animal at 19–23° that has been dechlorinated either by aging at room temperature for at least 24 hr or by adding a chelating agent (e.g., ResQ from Nasco). The water can be filtered continuously, or, if this is not possible, it can be changed manually at least once weekly. *Xenopus* thrives on Purina trout chow, given as about 5 pellets/animal, once weekly.

OOCYTE REMOVAL AND PREPARATION

1. To anesthetize the animal, place it in an ice bucket containing 500 ml of 0.15% aqueous solution of tricaine at room temperature, until it is limp (generally 5–10 min). Because tricaine is a potential carcinogen, gloves should be worn.

2. Transfer the animal ventral side up to a 4 $\times$ 4 gauze pad saturated with tricaine solution on the laboratory bench.

3. To expose the ovary, make a 1- to 2-cm lateral abdominal incision. Using blunt forceps and scissors, remove the desired number of lobes, taking care to avoid cutting the gut, which can sometimes be retracted along with the ovary. Place the tissue in a 60-mm dish containing OR-2

solution, and close the gut wall and skin with two or three interrupted sutures using 3-O suture material. Place the animal in 100 ml water in an ice bucket until it regains consciousness, taking care to keep the nostrils above the water line to avoid drowning. When the animal is fully alert, place it back in the original water tank. Sterile technique is not required for this procedure, and infection of the operative site rarely occurs.

4. To defolliculate the oocytes, tease apart the ovarian lobes with two pair of forceps and then incubate in OR-2 solution containing 2 mg/ml collagenase for approximately 2 hr with gentle rocking at room temperature. Single oocytes, oocytes loosely associated with each other, and stromal tissue will be present. Mechanically disrupt the remaining associations by pipetting the oocytes up and down using a 2-ml pipette fitted with a thumb-controlled pipette device. Transfer stage V and VI oocytes to ND96 solution, and inspect microscopically. Cells that have not been completely defolliculated can be identified by the presence of surface capillaries, and with practice two pair of watchmaker forceps can be used to remove the remaining follicular jackets. Healthy oocytes can be maintained for many days after harvesting. Reject cells that are small, deformed, or discolored. The half-life of an oocyte population varies greatly among different animals. In our experience, oocytes from the best animals die at a rate of approximately 5% per day.

OOCYTE MICROINJECTION. Borosilicate glass capillary tubes with an outer diameter of 1 mm are pulled to form a fine-tipped pipette, and the tip is trimmed against a size standard with watchmaker forceps using a stereo microscope to achieve an internal tip diameter of about 10 μm. Tips with larger diameters can damage the oocyte, whereas tips with smaller diameters are susceptible to occlusion by oocyte material. The pipette is mounted on a micromanipulator and connected to a microinjector. Integrated oocyte injection systems are now commercially available; alternatively, a system can be constructed from individual components. If the tip is properly trimmed, 50 nl of water should be ejected from the pipette using an injection time of 15 msec at 150 kilopascals. Pressure and injection time can be inversely varied to achieve the same ejection volume. An inexpensive injection chamber can be made by fixing a 3×3 cm plastic mesh containing 1 mm^2 units to the bottom of a 60-mm polystyrene dish, using a small amount of chloroform to melt the polystyrene, which then acts as the fixative. During microinjection, the oocytes will sit immobilized in the square units. With practice, more than 200 oocytes can be injected per hour. The number of different RNA solutions tested and the quality of the pipette tip are much more important factors than the total number of oocytes injected in determining the duration of the microinjection session. A stage for preparing RNA dilutions and for filling the pipette can be made from Parafilm draped taut over a 100-mm petri dish. The solutions are simply deposited as a drop on the stage, and the pipette is then lowered into it under microscopic control. We have found that using a swan neck fiber optic light source works best under these conditions.

Individual oocytes in ND96 solution are injected at the equator with approximately 50 nl total volume. This is about 5% of the oocyte total volume, and it causes a visual swelling of the cell, which ascertains injection. After removing the pipette from the oocyte, a small amount of yolk platelets will normally stream out through the wound before the membrane reseals. After injection, the oocytes can be maintained in ND96 solution on the laboratory bench at room temperature (between 19° and 23°). The medium is changed daily.

Preparation of Receptor RNA

The RNA encoding chemokine receptors can be natural or synthetic. Standard molecular biology methods can be used to isolate natural RNA from cell types known to respond to a chemokine of interest, and the RNA can be enriched for mRNA by oligo(dT) affinity chromatography and for chemokine receptor mRNA by sucrose gradient size selection, using standard methods.[18,19] Alternatively, candidate or known chemokine receptor ORFs can be cloned into any transcription-competent vector, and defined cRNA can be synthesized. RNA transcription kits are commercially available (e.g., from Stratagene, La Jolla, CA; Promega, Madison, WI), all of which use the same basic methodology, and so only the general method is given here. First, the plasmid is linearized by restriction enzyme digestion of a unique site in the polylinker region at the 3′ end of the ORF. Second, a bacteriophage RNA polymerase (e.g., T3, T7, or Sp6) that recognizes a specific promoter site present in the plasmid 5′ to the ORF is used to synthesize sense mRNA. One microgram of linearized plasmid DNA will yield several micrograms of synthetic RNA. Many vectors have two distinct RNA polymerase recognition sites, one 5′ and the other 3′ to the multiple cloning site, which allows for synthesis of both sense and antisense RNA from the same plasmid.

Capping the polynucleotide with diguanosine triphosphate and polyadenylation improve RNA stability in the oocyte, but we have observed that neither is essential. There are examples of proteins that can be expressed in oocytes using ORF sequence alone, and other examples exist where

[18] D. Julius, A. B. McDermott, R. Axel, and T. M. Jessell, *Science* **241,** 558 (1988).

[19] T. Maniatis, E. F. Fritsch, and J. Sambrook, "Molecular Cloning: A Laboratory Manual," 2nd Ed. Cold Spring Harbor Laboratory, Cold Spring Harbor, New York, 1989.

the 3′-untranslated region (3′-UTR) sequence has important effects on expression, acting at the level of mRNA stability.[20]

Calcium Flux Assay of Chemokine Receptor Expression

Chemokine receptors typically signal through the G_i–phospholipase C pathway, which leads to transient increases in intracellular calcium. This can be detected in oocytes in several ways. Electrophysiological monitoring of a calcium-dependent chloride conductance in voltage-clamped oocytes has been most commonly used for G-protein-coupled receptors of this type, and it has also been used for the chemokine receptor CCR4.[14] A simpler technique, however, requiring no sophisticated instrumentation, is the $^{45}Ca^{2+}$ release assay first described by Williams *et al.*[21]

Materials

OR-2 and ND96 solutions (see above)
$^{45}CaCl_2$ 25 mCi/mg (New England Nuclear, Boston, MA)
96-Well flat-bottomed polystyrene plates
Multichannel pipettor

Methods. Oocytes are injected with an RNA of interest and then incubated as a group in 500 μl of OR-2 solution containing 50 μCi $^{45}Ca^{2+}$ per milliliter for 3 hr at room temperature. After this time the cells are "loaded" typically with 1000–5000 cpm of $^{45}Ca^{2+}$, which remains inside the cell unless a calcium-mobilizing receptor is activated. The oocytes are washed at least five times with 1-ml exchanges of ND96 solution, then transferred to 6-mm-diameter flat-bottomed polystyrene wells in a 96-well plate format, with one oocyte and 100 μl ND96 solution per well. Washing is then continued using a multichannel pipettor and five 100-μl exchanges of ND96. To minimize the risk of oocyte damage during pipetting, oocytes are positioned at the 6 o'clock position of the well by tilting the plate at a 45° angle, and the pipette tip is positioned at 12 o'clock.

Oocytes are incubated for 20 min in the final 100-μl wash, which is then harvested using the multichannel pipettor to replicate 96-well plates. Scintillation fluid is added to each sample, and β emissions are determined. This is the prestimulated release of $^{45}Ca^{2+}$, symbolized as x. Then 100 μl ND96 containing the desired final concentration of chemokine is added to

[20] V. Kruys, M. Wathelet, P. Poupart, R. Contreras, W. Fiers, J. Content, and G. Huez, *Proc. Natl. Acad. Sci. U.S.A.* **84,** 6030 (1987).

[21] J. A. Williams, D. J. McChesney, M. C. Calayag, V. R. Lingappa, and C. D. Logsdon, *Proc. Natl. Acad. Sci. U.S.A.* **85,** 4939 (1988).

the oocyte, plates are incubated for 20 min at room temperature, and wash fluid is collected in the same way for counting. This value is designated y. The oocyte is then lysed in 1% sodium dodecyl sulfate (SDS) and counted, and the value designated as z. The percent release is calculated as $(y - x/x + y + z) \times 100$.

After the washing protocol, the oocyte releases very little $^{45}Ca^{2+}$ into the medium spontaneously; after the agonist is added, $^{45}Ca^{2+}$ is released rapidly into the medium, and the rate of release declines rapidly after 5 min (Fig. 1). A positive oocyte response on application of agonist implies that the mRNA injected encodes a functional receptor. With healthy oocytes, this assay has extremely low background, essentially equivalent to environmental background, so that very high signal-to-noise (S/N) ratios can be observed. Oocytes begin to respond to receptor agonists within 2 days after RNA is injected. In good experiments, they continue to express receptors and to respond beyond 7 days after RNA injection (Fig. 2).

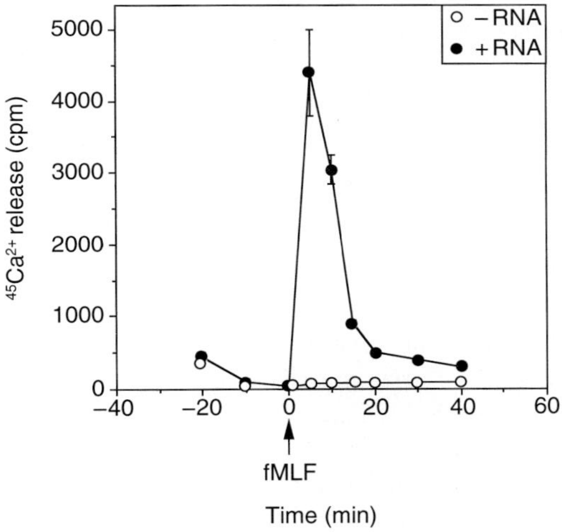

FIG. 1. Kinetics of $^{45}Ca^{2+}$ release from oocytes expressing chemoattractant receptors. Oocytes injected with 50 ng of HL60 granulocyte RNA were loaded with $^{45}Ca^{2+}$. Each time point corresponds to the counts per minute present in a 100-μl wash incubated for the length of the interval shown. Arrow indicates the time of addition of 100 nM fMet-Leu-Phe (fMLF), which activates the N-formyl peptide receptor encoded by transcripts in this natural RNA sample. Data are means ± SEM of six oocytes per condition. Oocytes contained approximately 10,000 cpm at time 0. Control oocytes injected with water (open circles) did not respond. For most purposes, the assay can be simplified by measuring counts per minute in the bath at time 0 and +20 min and by measuring the remaining counts per minute in the oocyte as described in the text. Similar results are obtained with chemokine receptors.

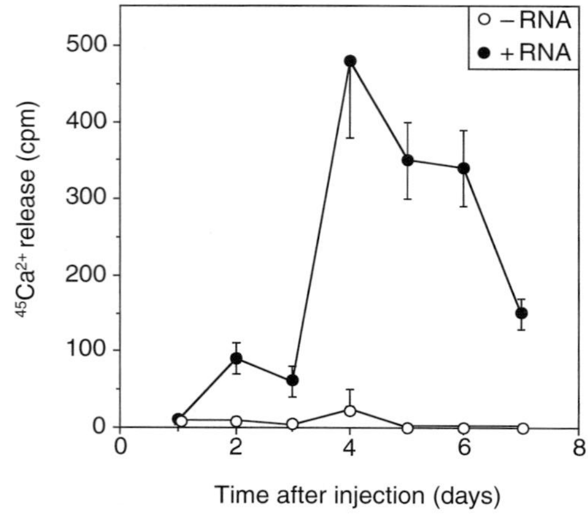

FIG. 2. Time course for functional expression of chemoattractant receptors. Oocytes were injected with 50 ng of HL60 granulocyte RNA. The $^{45}Ca^{2+}$ release assay was performed on each day after injection using 100 n*M* fMet-Leu-Phe as the agonist. Data are means ± SEM of six oocytes per condition. Oocytes were loaded with approximately 1000 cpm on each day. Control oocytes injected with water (open circles) did not respond. For most purposes, the oocytes can be assayed on day 3 or 4, when activity peaks for most chemoattractant receptors we have studied, including the chemokine receptors.

Advantages and Disadvantages

Advantages

The study of chemokine receptor expression in *Xenopus* oocytes has several advantages. (1) Large matrices of RNAs versus agonists can be tested in parallel in a single cell assay, with large numbers of replicates per condition. This is a major advantage over calcium fluorimetry performed on mammalian cells expressing chemokine receptors, and over electrophysiological assay of receptor expression in oocytes, which are carried out one sample at a time in series. (2) The assay measures receptor function. (3) Because of low background and signal amplification, the assay is very sensitive.

Disadvantages

Disadvantages of the expression system are as follows. (1) Microinjection requires good eye–hand coordination and training. (2) Many *Xenopus* do not produce oocytes suitable for injection, and considerable effort may be required to identify and maintain reliable stocks. (3) Microinjection requires the use of laboratory animals. (4) The assay does not examine ligand–receptor interactions directly.

Comparison to Results Obtained Using Mammalian Cells

When examined in parallel, chemokine specificities for cloned chemokine receptors defined in oocytes have sometimes differed from the specificities for the same receptor expressed in mammalian cell types. The following are the known examples of this. (1) CCR4 was reported to activate transmembrane currents in oocytes in response to macrophage inflammatory protein-1α (MIP-1α), RANTES, and macrophage chemotactic protein-1 (MCP-1); however MCP-1 did not bind specifically to CCR4 expressed in HL60 cells.[14,22] (2) The apparent agonist rank order differs for CXCR2 expressed in oocytes versus mammalian HEK 293 cells.[4,12,23] (3) fMet-Leu-Phe was reported to be a ligand and agonist for rabbit CXCR2 expressed in oocytes, whereas only IL-8 could bind to the same molecule expressed in COS cells.[24,25]

Supplementary and Alternative Assays of Chemokine Receptors in Oocytes

A nonradioactive alternative to the radioactive calcium flux assay has been developed using oocytes from albino *Xenopus* loaded with the protein aequorin, which fluoresces in a calcium-dependent manner.[26] Intact oocytes and oocyte membranes can also be tested using radiolabeled chemokine ligands in direct binding assays.[4,11] Finally, measurement of chemokine-induced transmembrane currents in voltage-clamped oocytes is another powerful technique, but it requires substantial training and nonroutine technical expertise. Readers interested in learning more about these techniques are directed to Refs. 2 and 3.

[22] A. J. Hoogewerf, D. Black, A. E. I. Proudfoot, T. N. C. Wells, and C. A. Power, *Biochem. Biophys. Res. Commun.* **218,** 337 (1996).

[23] S. K. Ahuja and P. M. Murphy, *J. Biol. Chem.* **271,** 20545 (1996).

[24] K. M. Thomas, H. Y. Pyun, and J. Navarro, *J. Biol. Chem.* **265,** 20061 (1990).

[25] K. M. Thomas, L. Taylor, and J. Navarro, *J. Biol. Chem.* **266,** 14839 (1991).

[26] E. Giladi and E. R. Spindel, *Biotechniques* **10,** 744 (1991).

[9] Cell–Cell Fusion Assay to Study Role of Chemokine Receptors in Human Immunodeficiency Virus Type 1 Entry

By JOSEPH RUCKER, BENJAMIN J. DORANZ, AIMEE L. EDINGER, DEBORAH LONG, JOANNE F. BERSON, and ROBERT W. DOMS

Introduction

Human immunodeficiency virus type 1 (HIV-1), the causative agent of acquired immunodeficiency syndrome (AIDS), encodes an integral membrane glycoprotein termed envelope (env).[1] Entry of HIV-1 into a cell requires binding of the viral env protein to receptors on the cell surface followed by fusion of the viral envelope with the cellular membrane. Following membrane fusion, the viral genome enters the host cell cytoplasm and directs the host cell to produce viral proteins including env. Newly synthesized env protein is expressed on the surface of the cell where it can assemble with other viral proteins to form new virion particles. Surface expression of env can also lead to cell–cell fusion and the formation of multinucleated giant cells (syncytia).

Much interest has been generated by the discovery that chemokine receptors function as necessary cofactors for HIV-1 entry.[2–8] It has long been known that HIV-1 env initially binds the virus to the cell surface via a high affinity interaction with CD4. Although the interaction between env and CD4 has been well characterized, the presence of CD4 alone is not sufficient to allow viral entry. For example, expression of human CD4 in

[1] J. A. Levy, "HIV and the Pathogenesis of AIDS." ASM Press, Washington, D.C., 1994.

[2] Y. Feng, C. C. Broder, P. E. Kennedy, and E. A. Berger, *Science* **272,** 872 (1996).

[3] G. Alkhatib, C. Combadiere, C. C. Broder, Y. Feng, P. E. Kennedy, P. M. Murphy, and E. A. Berger, *Science* **272,** 1955 (1996).

[4] T. Dragic, V. Litwin, G. P. Allaway, S. R. Martin, Y. Huang, K. A. Nagashima, C. Cayanan, P. J. Maddon, R. A. Koup, J. P. Moore, and W. A. Paxton, *Nature* (*London*) **381,** 667 (1996).

[5] H. Deng, R. Liu, W. Ellmeier, S. Choe, D. Unutmaz, M. Burkhart, P. D. Marzio, S. Marmon, R. E. Sutton, C. M. Hill, C. B. Davis, S. C. Peiper, T. J. Schall, D. R. Littman, and N. R. Landau, *Nature* (*London*) **381,** 661 (1996).

[6] B. J. Doranz, J. Rucker, Y. Yi, R. J. Smyth, M. Samson, S. C. Peiper, M. Parmentier, R. G. Collman, and R. W. Doms, *Cell* **85,** 1149 (1996).

[7] H. Choe, M. Farzan, Y. Sun, N. Sullivan, B. Rollins, P. D. Ponath, L. Wu, C. R. Mackay, G. LaRosa, W. Newman, N. Gerard, C. Gerard, and J. Sodroski, *Cell* **85,** 1135 (1996).

[8] J. F. Berson, D. Long, B. J. Doranz, J. Rucker, F. R. Jirik, and R. W. Doms, *J. Virol.* **70,** 6288 (1996).

0076-6879/97 $25.00

nonhuman cells generally does not allow for virus infection or env-mediated syncytia formation.[9,10] In addition, there are a number of examples of human cell lines where expression of CD4 does not allow for HIV-1 entry.[11] Finally, HIV-1 strains often show marked tropism for either T cells (T-tropic) or macrophages (M-tropic), both of which are CD4-positive.[12] The discovery of the link between chemokine receptors and HIV-1 entry has resolved many of these issues. The expression of CD4 in conjunction with the appropriate chemokine receptor renders otherwise nonpermissive cells susceptible to HIV-1 entry and to env-mediated syncytia formation. T-tropic viruses have been shown to primarily use the C-X-C chemokine receptor CXCR4 (also called LESTR/fusin) in conjunction with CD4,[2] whereas M-tropic viruses have been shown to primarily use the C-C chemokine receptor CCR5.[3–7] Finally, other chemokine receptors such as CCR2b and CCR3 have been shown to be used by a small subset of HIV-1 strains.[6,7]

There are two general approaches for delineating the role of chemokine receptors in HIV entry. One method is to infect cells that express CD4 and the chemokine receptor of interest with HIV-1 and to assess viral entry [PCR (polymerase chain reaction) or gene reporter assays] or replication [p24 structural protein quantitation or reverse transcriptase (RT) assays].[13] Virus infection is straightforward but has a number of significant drawbacks. Not all cells are efficiently infected by HIV-1. A number of nonhuman cells have postentry blocks at various stages of the virus life cycle; these blocks can dramatically lower the sensitivity of the various detection methods. In addition, HIV-1 infection assays require specialized containment facilities.

Another approach is to monitor cell–cell fusion between cells expressing env and cells expressing CD4 and the appropriate chemokine receptor. Syncytia can be monitored visually by counting multinucleated cells in a microscopic field. Fusion can also be evaluated by fluorescent dye redistribution/dequenching assays that monitor membrane lipid mixing and cytosolic content mixing.[14–16] The former technique, though useful, suffers

[9] P. J. Maddon, A. G. Dalgleish, J. S. McDougal, P. R. Clapham, R. A. Weiss, and R. Axel, *Cell* **47,** 333 (1986).

[10] P. A. Ashorn, E. A. Berger, and B. Moss, *J. Virol.* **64,** 2149 (1990).

[11] B. Chesebro, R. Buller, J. Portis, and K. Wehrly, *J. Virol.* **64,** 215 (1990).

[12] C. C. Broder, and E. A. Berger, *Proc. Natl. Acad. Sci. U.S.A.* **92,** 9004 (1995).

[13] A. Aldovini, and B. Walker, eds., "Techniques in HIV Research." Stockton Press, New York, 1990.

[14] D. S. Dimitrov, H. Golding, and R. Blumenthal, *AIDS Res. Hum. Retroviruses* **7,** 799 (1991).

[15] S. J. Morris, J. Zimmerberg, D. P. Sarkar, and R. Blumenthal, *in* "Membrane Fusion Techniques: Part B" (N. Düzgünes, ed.), p. 42. Academic Press, New York, 1993.

[16] V. Litwin, K. A. Nagashima, A. Ryder, C.-H. Chang, J. M. Carver, W. C. Olson, M. Alizon, K. W. Hasel, P. J. Maddon, and G. P. Allaway, *J. Virol.* **70,** 6437 (1996).

from an inability to be accurately quantitated as well as difficulty in distinguishing low levels of fusion. The latter technique requires the use of equipment not often found in biologically oriented laboratories and thus is not accessible to most researchers. As an alternative, we have found that a gene reporter system using firefly luciferase as a reporter to measure cytosolic mixing is quite versatile. This assay was originally developed by Nussbaum *et al.*,[17] using β-galactosidase as a gene reporter, and was the assay used to identify CXCR4 as a fusion cofactor. This assay has been extensively modified by our laboratory for the study of chemokine receptors.[6,17a] This chapter gives a detailed description of this reporter assay and discusses its variations and limitations.

General Scheme of Gene Reporter Assay

The general scheme of the reporter assay is shown in Fig. 1. Throughout this chapter, cells expressing env protein will be designated as effectors and cells expressing CD4 and the chemokine receptor of interest will be designated as targets. Effector cells are coinfected with vaccinia viruses that express HIV-1 env and T7 RNA polymerase. Target cells are cotransfected with plasmids expressing CD4, the chemokine receptor of interest, and the firefly luciferase gene under the control of the T7 promoter. Effector and target cells are incubated overnight to allow protein expression before mixing the following morning. If fusion occurs, T7 polymerase from the effector cells activates the luciferase gene in the target cells and causes production of luciferase. After approximately 8 hr, the cells are lysed and luciferase activity is assayed.

Effector Cell Requirements

There are a number of requirements for both effector and target cells. Effector cells express the env glycoprotein as well as T7 RNA polymerase that drives production of the gene reporter. We typically use a vaccinia-based system to express these proteins at high levels, although other approaches can also be used. A detailed description of vaccinia virus as a protein expression system is given in *Current Protocols in Molecular Biology*.[18] Many of these vaccinia constructs are available from the National Institutes of Health AIDS Research and Reference Reagent Program (NIH,

[17] O. Nussbaum, C. C. Broder, and E. A. Berger, *J. Virol.* **68,** 5411 (1994).

[17a] J. Rucker, M. Samson, B. J. Doranz, F. Libert, J. F. Berson, Y. Yi, R. J. Smyth, R. G. Collman, C. C. Broder, G. Vassart, R. W. Doms, and M. Parmentier, *Cell* **87,** 437 (1996).

[18] P. Earl, and B. Moss, *in* "Current Protocols in Molecular Biology" (F.M. Ausubel *et al.*, eds.), p. 16.15.1. Wiley-Interscience, New York, 1991.

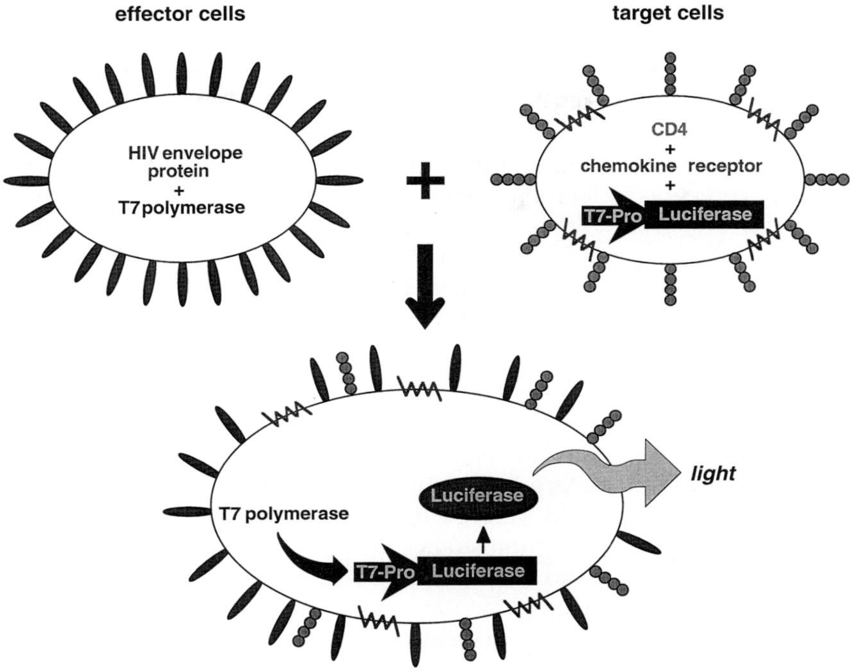

FIG. 1. Schematic of the cell–cell fusion assay. See text for details.

Bethesda, MD) or the American Type Culture Collection (ATCC, Rockville, MD). Vaccinia virus vectors result in high levels of protein expression beginning approximately 4 hr postinfection, and they can be used to infect most mammalian and avian cells [with the notable exception of Chinese hamster ovary (CHO) cells]. One requirement for effector cells is that they withstand the cytopathic effects (CPE) of vaccinia virus infection. We typically use HeLa cells, which are quite resistant to vaccinia CPE, although both B-SC-1 and NIH 3T3 cells have been used by others with good success.[2] Another requirement for effector cells is that they must properly process the env glycoprotein.[19–21] Such processing includes glycosylation, cleavage of the gp160 precursor to SU (gp120) and TM (gp41) subunits, and transport

[19] J. M. McCune, L. B. Rabin, M. B. Feinberg, M. Lieberman, J. C. Kosek, G. R. Reyes, and I. L. Weissman, *Cell* **53,** 55 (1988).

[20] R. L. Willey, J. S. Bonifacino, B. J. Potts, M. A. Martin, and R. D. Klausner, *Proc. Natl. Acad. Sci. U.S.A.* **85,** 9580 (1988).

[21] K. Kozarsky, M. Penman, L. Basiripour, W. Haseltine, J. Sodroski, and M. Krieger, *J. Acquir. Immune Defic. Syndr.* **2,** 163 (1989).

to the cell surface. Cleavage of gp160 is particularly important because this posttranslational modification is required to activate fusion activity in env. We have found that HeLa cells can be successfully used to express fusion-active forms of numerous divergent env proteins.

It is important to note that vaccinia virus is infectious for humans and requires class 2 biological safety precautions.[18] An attenuated vaccinia protein expression system has been developed that is replication-deficient in many cell types and is thus much safer for laboratory usage.[22] This system may also alleviate many issues of CPE, although perhaps at the cost of lowered protein expression. We have not tested this system in the fusion assay, but it is potentially a worthwhile modification to examine.

Target Cell Requirements

The requirements for target cells are somewhat more stringent. The most important requirement is that they be nonpermissive for fusion when expressing CD4 alone. As a result, most human cells should be regarded with caution. For example, HeLa cells express CXCR4 and thus are not amenable to structure–function studies of that receptor.[17] However, most nonhuman cell lines fail to support env-mediated fusion when expressing human CD4 alone, indicating that they lack a functional homolog of either CCR5 or CXCR4.[10] This issue becomes even more important when dealing with env proteins that can use multiple receptors such as YU2, which can utilize CCR5 and CCR3, and 89.6, which can use CCR5, CXCR4, CCR3, and CCR2b.[6,7] One might expect these proteins to be able to more readily use nonhuman homologs of chemokine receptors. For example, cells expressing 89.6 env are able to fuse, in a CD4-dependent fashion, with CCCS+L cells (a feline kidney cell line) without the addition of a human chemokine receptor, indicating the presence of a functional feline homolog of at least one of four possible receptors.[22a] Target cells must also express human CD4. Human CD4 can be introduced by transient transfection, by infection using vaccinia virus, or by introduction of a selectable stable gene. A number of nonhuman cell lines that stably express CD4 are available.

Although use of cells stably expressing cofactors is probably the most reproducible method for assessing cofactor usage, such stable cell lines do not provide the flexibility to examine large numbers of cofactor constructs. Likewise, vaccinia virus-based expression systems can be impractical for large numbers of receptor constructs. In light of this, we use a transfection-based protocol to allow transient expression of CD4 and the chemokine

[22] L. S. Wyatt, B. Moss, and S. Rozenblatt, *Virology* **210,** 202 (1995).

[22a] B. J. Willett, M. J. Hosie, J. C. Neil, J. D. Turner, and J. A. Hoxie, *Nature* (*London*) **385,** 587 (1997).

receptor of interest and introduction of a gene reporter under the control of the T7 promoter. Therefore, target cells should be easily transfectable to ensure a high percentage of transfected cells and thus a high signal. Quail QT6 and murine PA317 cells are easily transfectable using the calcium phosphate method, which is probably the most economical and simplest method of transfection. For less transfectable cells, liposome-mediated transfection can be used with good results.

Whatever transfection method is used, sufficient amounts of CD4 and chemokine receptor need to be produced by the transfected cells. We have found that many promoters give ample expression, including the widely used cytomegalovirus (CMV) promoter and Rous sarcoma virus (RSV) promoter as well as a myeloproliferative sarcoma virus promoter. In all cases, CD4 and chemokine receptor production is sufficient to allow cells to undergo env-mediated fusion.

Gene Reporters

Two gene reporters have been employed in these cell–cell fusion assays. The original assay as developed by Nussbaum *et al.* employed β-galactosidase as a gene reporter.[17] β-Galactosidase is useful because it can be used both to stain syncytia *in situ* and as a colorimetric assay of fusion using cell lysates. However, many recombinant vaccinia virus constructs express β-galactosidase (which is used as a selection marker) and thus cannot be used with a β-galactosidase gene reporter.[18] For this reason, we have employed luciferase as a gene reporter. The luciferase gene reporter is more sensitive than β-galactosidase and has enabled us to detect extremely low levels of fusion. Some experiments have given signal-to-noise (*S/N*) ratios in excess of 1000 : 1. Other gene reporters that have potential use in these assays include chloramphenicol acetyltransferase (CAT), secreted alkaline phosphatase (which may be useful for kinetic studies), and green fluorescent protein (which can be used as an *in situ* fluorescent probe). Whatever gene reporter chosen, it must be under the control of the T7 promoter; any gene reporter plasmid that is used for a T7-based *in vitro* transcription/translation system will be useful in this assay system.

Standard Fusion Assay

Transfection of QT6 Cells by Calcium Phosphate Method

In our standard fusion assay, quail QT6 cells, which can be transfected with efficiencies of 50 to 100%, are used as targets and grown in 24-well plates. The recommended growth medium for quail cells is Medium 199

(M199, GIBCO-BRL, Gaithersburg, MD) supplemented with 1% chicken serum (GIBCO-BRL), 5% fetal bovine serum (FBS, Hyclone, Logan, UT), 2 m*M* glutamine, 2 m*M* penicillin/streptomycin, and 10% tryptose phosphate broth (29.5 g/liter, GIBCO-BRL). One day before transfection, QT6 cells should be split into 24-well plates so that they are approximately 50% confluent on the day of transfection and thus actively growing. Several hours prior to transfection (optimally), the medium on the cells is changed to Dulbecco's modified Eagle's medium (DMEM) supplemented with 10% fetal bovine serum, 2 m*M* glutamine, and 2 m*M* penicillin–streptomycin (DMEM-10) because tryptose phosphate broth present in the quail medium may inhibit transfection by calcium phosphate precipitation. We use 0.5 ml of medium per well in a 24-well plate.

Two micrograms of each plasmid encoding the desired coreceptor, CD4, and T7–luciferase are used to transfect each well. DNA is resuspended in a solution composed of 0.25 *M* $CaCl_2$, 10 m*M* Tris (pH 7.4), 2.5 m*M* EDTA (pH 8.0), and 150 m*M* NaCl in a total volume of 25 μl per well (of a 24-well plate) being transfected. Separately, a 2× stock of transfection buffer composed of 50 m*M* HEPES (pH 7.1), 180 m*M* NaCl, and 2 m*M* Na_2HPO_4 is freshly prepared. The pH of the HEPES solution is critical for calcium phosphate transfections and needs to be adjusted to exactly pH 7.1. The DNA cocktail is added dropwise to an equal volume of transfection buffer with constant vortexing at low to medium speed in a microcentrifuge tube. Fifty microliters of this transfection mixture is then added dropwise to the 0.5 ml of medium in each well of the 24-well plate with gentle agitation of the plate to ensure mixing. Cells are incubated at 37° and approximately 4–6 hr later are gently washed once with PBS and fresh medium added. Allowing the transfectant to remain on the cells for longer times can result in higher transfection efficiency but can also reduce cell viability depending on cell type. QT6 cells are particularly sensitive to calcium phosphate precipitation. Fusion assays can be performed anywhere between 12 and 48 hr posttransfection, with optimal results obtained within 24 hr. The calcium phosphate transfection procedure given in *Current Protocols in Molecular Biology* can be substituted with equivalent results.[23]

Infection of HeLa Cells by env-Encoding Vaccinia Virus

Vaccinia virus infection protocols are discussed in *Current Protocols in Molecular Biology*.[18] HeLa cells are grown to 70–80% confluence in a 10-cm plate using DMEM-10. Vaccinia viruses at a multiplicity of infection (MOI) of 10, encoding both env protein and T7 polymerase (vTF1.1), are

[23] R. E. Kingston, C. A. Chen, and H. Okoyama, *in* "Current Procotols in Molecular Biology" (F. M. Ausubel *et al.*, eds.) p. 9.1.1. Wiley-Interscience, New York, 1990.

incubated with an equal amount of bovine pancreatic trypsin (0.25 mg/ml; Sigma, St. Louis, MO) for 30 min at 37° with occasional vortexing and then resuspended in 4 ml DMEM-10. The medium is aspirated off the HeLa cells, and the 4 ml of medium containing virus is added. The infection is allowed to proceed at 37° for 1.5–3 hr. The cells are then washed once with phosphate-buffered saline (PBS) and gently lifted off the plate using trypsin–EDTA (GIBCO-BRL). The trypsinized cell are resuspended in 10 ml DMEM-10 in a 50-ml conical tube, centrifuged at 1000 rpm for 5 min, washed with 10 ml PBS, centrifuged, and resuspended in 10 ml of DMEM-10 supplemented with rifampicin (100 μg/ml, Sigma). Rifampicin inhibits vaccinia virus assembly, minimizes CPE, and decreases the amount of exogenous virus particles that could lead to an increase in gene reporter background.[24] Rifampicin should be prepared as a 1000× solution (100 mg/ml) in sterile dimethyl sulfoxide (DMSO), as it is extremely insoluble in aqueous solutions, and stored at −20°. Rifampicin should be solubilized in medium prior to adding to cells; it may be necessary to warm the medium at 37° to allow complete solubilization. The HeLa cells are incubated overnight at 32° in the 50-ml conical tubes. The tubes should be tilted at about 30° from horizontal and loosely capped to allow for CO_2 exchange. Incubation at 32° reduces vaccinia CPE. High levels of env and T7 polymerase expression should begin within 4–6 hr postinfection and continue to increase up to approximately 24 hr.

Assessing Cell–Cell Fusion

After overnight expression, wash effector cells twice with cold PBS centrifuging at 1000 rpm for 5 min between washes, and resuspend at 2×10^6 cells/ml in DMEM-10 supplemented with rifampicin (100 μg/ml, Sigma) and araC (cytosine β-D-arabinofuranoside, 10 μM, Sigma). Target cell medium should be replaced with 0.5 ml per well of DMEM-10 supplemented with rifampicin and araC. AraC is an inhibitor of vaccinia virus DNA replication and late-gene expression and thus will inhibit production of new T7 polymerase (which is under the sole control of the vaccinia late promoter in the vaccinia virus vTF1.1, previously known as vP11gene1).[25,26] VTF7.3, another T7 polymerase-encoding virus, carries T7 polymerase under control of the vaccinia synthetic early/late (SEL) promoter, and araC will attenuate but not completely block production of new T7 polymerase

[24] P. M. Grimley, E. N. Rosenblum, S. J. Mims, and B. Moss, *J. Virol.* **6,** 519 (1970).

[25] W. A. Alexander, B. Moss, and T. R. Fuerst, *J. Virol.* **66,** 2934 (1992).

[26] B. Moss, *in* "Recombinant Poxviruses" (M. M. Binns and G. L. Smith, eds.), p. 45. CRC Press, Boca Raton, FL, 1992.

by this virus.[27] The combination of viral inhibitors, overnight incubation at 32°, and extensive washing helps to reduce background luciferase activity in the assay due to infection of target cells by residual or newly synthesized T7 polymerase vaccinia virus.

To initiate the fusion reaction, approximately 1×10^5 effector cells are added to each well of a 24-well plate containing target cells. Cells can be added on ice if a synchronized start is desired (useful for kinetic studies). Cells are allowed to fuse for 6–10 hr with incubation at 37°. Fusion can occur at lower temperatures but proceeds more slowly and gives no advantage with regard to S/N. To measure the amount of fusion, medium should be aspirated and the cells lysed with 150 μl of either 0.5% (v/v) Nonidet P-40 (NP-40) in PBS or Luciferase Reporter Lysis buffer (Promega, Madison, WI).

The luciferase activity of cell lysates can be assayed using either a luminometer or a scintillation counter with luminescence capabilities. We find that detectors that read a 96-well microtiter plate format are extremely convenient. We have used a Wallac (Turku, Finland) 1450 Microbeta Plus liquid scintillation counter with good results. This detector requires the use of solid (usually white) 96-well microtiter plates. To assay luciferase activity, 20 μl of each cell lysate is transferred to a solid white 96-well microtiter plate, and 50 μl of luciferase assay reagent (Promega) is added to each well. The luciferase reagent is light-sensitive and should be added to the wells under dim light. Plates should be counted for 1 to 2 sec per well in a luminometer or scintillation counter. Cell lysates should be assayed for luciferase activity soon after lysis. Lysates can be frozen but lose approximately 50% activity on reassay.

Luminescence detectors have maximum detection limits. Exceeding the linear range of the detector can drastically alter the quantitative and qualitative interpretation of data; it is important that the assay signal be adjusted to within the linear range. The amount of cell lysate can be changed to bring signals into the appropriate range. Signals can also be brought into the linear range by use of a red filter (luminescence foil, Wallac), which also eliminates cross-talk of high signals between wells. Finally, it should be noted that the amount of luciferase assay reagent generally used (50 μl per 96-well) is well above enzyme saturation and adjusting the amount of assay reagent does not usually change output signal.

Figure 2 shows the quality of the data from a typical experiment comparing the abilities of CCR5 and CD4, CXCR4 and CD4, and CD4 alone to support fusion by three HIV-1 env proteins: JR-FL (M-tropic), BH8 (T-tropic), and 89.6 (dual tropic). This graph clearly shows use of CXCR4 by

[27] T. R. Fuerst, E. G. Niles, F. W. Studier, and B. Moss, *Proc. Natl. Acad. Sci. U.S.A.* **83,** 8122 (1986).

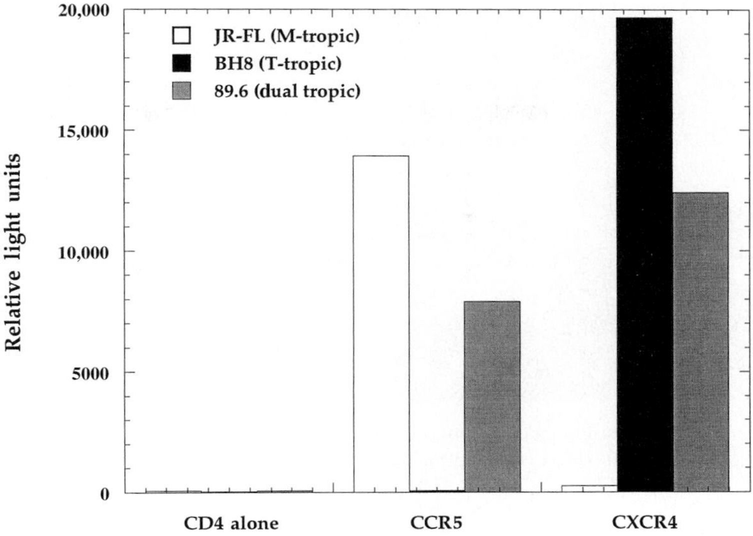

FIG. 2. Cell–cell fusion of transfected QT6 cells using the standard fusion assay. QT6 cells expressing either CCR5 and CD4, CXCR4 and CD4, or CD4 alone were mixed with HeLa cells expressing either JR-FL (M-tropic), BH8 (T-tropic), or 89.6 (dual tropic) env proteins. Cells were allowed to fuse for 8 hr at 37° before lysis and assay for luciferase activity.

a T-tropic strain, CCR5 utilization by an M-tropic strain, and utilization of both receptors by a dual-tropic strain. The S/N ratio in this experiment is on the order of 1000 to 1. Signal-to-noise ratios of 20 to 1 to 1000 to 1 can be routinely obtained in this assay.

Fusion can also be assessed visually if desired, as shown in Fig. 3. After, incubation cells are fixed with 1% (v/v) formaldehyde/0.2% (v/v) glutaraldehyde in PBS for 5 min. The cells should then be washed twice with PBS and stained with a solution of 0.5% (w/v) methylene blue and 0.17% (w/v) pararosaniline in methanol, or any other cell stain. Syncytia can often be seen without staining if a substantial amount of fusion has occurred. This is most often seen with the primary cofactors CXCR4 and CCR5. Not all cells form visible syncytia even if cytoplasmic mixing occurs. Figure 3 shows QT6 target cells expressing CCR5 and CD4 after mixing with Hela effector cells expressing either BH8 (Fig. 3A) or JR-FL (Fig. 3B) env for 8 hr. Syncytia can clearly be seen with JR-FL, whereas cells remain unfused with BH8.

Other Methods of Expressing env

As described above, the standard fusion assay requires env to be expressed by vaccinia virus. Although making recombinant vaccinia virus is

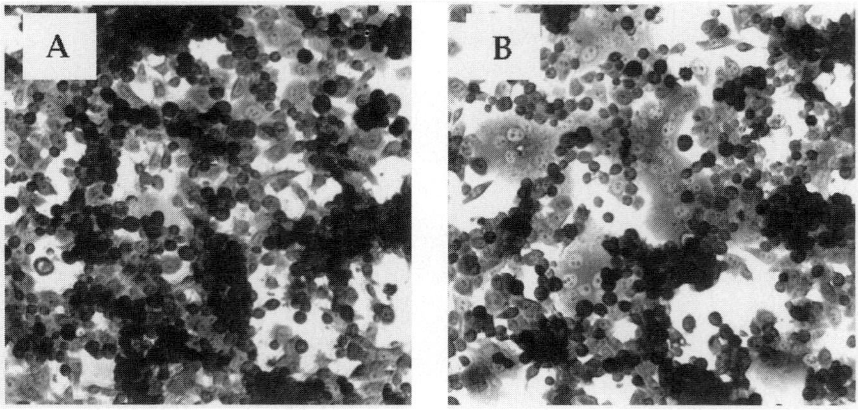

FIG. 3. Cell–cell fusion evaluated by syncytia formation. QT6 cells expressing CCR5 and CD4 were mixed with HeLa cells expressing either BH8 (panel A) or JR-FL (panel B) and allowed to fuse for 8 hr at 37°. Cells were fixed and stained as per the text discussion, and syncytia formation was evaluated microscopically.

straightforward, it is time-consuming and thus impractical for looking at large numbers of env constructs. Thus it is often necessary to use a plasmid-based method of env expression. We have found that an infection–transfection protocol works well; vaccinia-encoded T7 polymerase is used to overexpress a plasmid containing the env of interest under the control of the T7 promoter.

To prepare effector cells in this fashion, cells are infected with vaccinia virus encoding T7 polymerase (vTF1.1) for 30 min to 2 hr. Following infection, medium is aspirated, cells are gently washed once with PBS, and the medium is replaced with DMEM-10 supplemented with rifampicin. The cells are then transfected using the calcium phosphate procedure. Transfectant is normally left on for only 2 to 4 hr before new medium (with rifampicin) is added. After transfection, cells are shifted to 32° for overnight incubation. The effector cells can be incubated overnight as adherents, or they can be trypsinized and resuspended in medium in conical tubes for overnight incubation as in the basic protocol. If trypsinizing, cells can be briefly washed with EDTA (5 m*M* in PBS) before adding trypsin, as the calcium phosphate transfection precipitate can interfere with the action of the trypsin–EDTA. After overnight incubation, adherent effector cells should be lifted by EDTA treatment (not trypsin–EDTA, which would digest surface-expressed env) and resuspended by vigorous pipetting. At this point, the basic protocol should be followed, with effector cells being washed twice with PBS before being resuspended in medium supplemented with araC and rifampicin and added to target cells.

Analysis of Cell-Surface Expression

Cell-surface expression is assumed for chemokine receptor constructs that are functional. However, it is necessary to look for cell-surface expression to distinguish between transport defects and actual functional defects for constructs that are nonfunctional. We have found immunofluorescence microscopy to be a useful tool for these studies. This technique requires that the constructs be N-terminally labeled with an epitope tag or that antibodies to the chemokine receptor of interest be available. N-terminally tagged CCR5 constructs have been shown to be fully functional.[28,29] In addition, a number of CCR5 and CXCR4 antibodies have been developed to allow questions such as these to be answered.[30,31]

We use an immunofluorescence protocol to look at both total chemokine receptor expression and surface expression. QT6 cells are grown and transfected with the chemokine receptor of interest on glass coverslips. QT6 cells are very sensitive and will slough off coverslips if treated roughly. This problem can be avoided by washing and fixing coverslips in staining trays rather than in wells.

Surface expression is distinguished from total expression by staining with primary antibody either before or after cells are fixed and permeabilized with a solution of 4% (w/v) paraformaldehyde/0.15% (v/v) Triton X-100 in PBS with calcium (133 mg/liter) and magnesium (100 mg/liter). Cells are blocked with 5% FBS in PBS (with calcium and magnesium) before incubating with primary antibody. Ethidium bromide (10 μg/ml, Sigma), a fluorescent DNA stain, is added to cells along with the primary antibody. After extensive washing, fluorescein isothiocyanate (FITC)-conjugated secondary antibody is incubated with cells along with DAPI (4,6-diamidino-2-phenylindole, Sigma, 0.5 μg/ml), another DNA staining fluorescent dye. After a final wash, coverslips are mounted on slides.

In this procedure, green (FITC) fluorescence indicates the presence of chemokine receptor, either on the surface or throughout the cell. The DNA stains ethidium bromide and DAPI are designed to assess the health of cells at the time of staining. Ethidium bromide (red) will enter cells only

[28] R. Liu, W. A. Paxton, S. Choe, D. Ceradini, S. R. Martin, R. Horuk, M. E. MacDonald, H. Stuhlmann, R. A. Koup, and N. R. Landau, *Cell* **86,** 367 (1996).

[29] J. Rucker, M. Samson, B. J. Doranz, F. Libert, J. F. Berson, Y. Yi, R. J. Smyth, R. G. Collman, C. C. Broder, G. Vassart, R. W. Doms, and M. Parmentier, *Cell* **87,** 437 (1996).

[30] M. Endres, P. Clapham, M. Marsh, M. Ehuja, J. D. Turner, A. McKnight, J. F. Thomas, B. Stoebenau-Haggarty, S. Choe, P. J. Vance, T. N. C. Wells, C. A. Power, S. S. Sutterwala, R. W. Doms, N. R. Landau, and J. A. Hoxie, *Cell* **87,** 745 (1996).

[31] S. Rana, G. Besson, D. G. Cook, J. Rucker, R. J. Smyth, Y. Yi, J. Turner, H.-h. Guo, J.-g. Du, S. C. Peiper, E. Lavi, M. Samson, F. Libert, C. Liesnard, G. Vassart, R. W. Doms, M. Parmentier, and R. G. Collman, *J. Virol.* **71,** 3219 (1997).

if they have been permeabilized before incubation with primary antibody (total expression). An ethidium bromide stain in cells that were fixed after addition of primary antibody (surface expression) indicates that the particular cells are dead and/or unhealthy; the presence of FITC staining in such cells should not be considered as evidence of chemokine receptor surface expression. Staining of all cells by DAPI is also used to assess the health of cells as well as estimate cell density. The addition of DNA stains enables one to avoid false positives. Such controls are extremely important in cases where a chemokine receptor mutant may not be expressed on the cell surface, such as the truncated CCR5 polymorphism present in a large percentage (16%) of the Caucasian population.[28,32]

Other techniques can also be used to evaluate cell surface expression. Flow cytometry can be used in a fashion similar to immunofluorescence microscopy. Endoglycosidase H sensitivity of glycosylated chemokine receptors, such as CXCR4, can be used as a probe of proper posttranslational processing of chemokine receptor proteins.[29] Finally, probes of chemokine receptor function such as direct binding of chemokines or calcium flux/inositol trisphosphate activity in response to chemokine addition can be used to determine if functional receptor is present on the cell surface.[28] A description of these techniques is beyond the scope of this review.

Blocking Studies

One the earliest links between the chemokine and HIV fields was the discovery that the β-chemokines RANTES (regulated on activation, normal T cell expressed and secreted), MIP-1α (macrophage inflammatory protein-1α), and MIP-1β could block infection of M-tropic strains of HIV-1.[33] It was this observation, coupled with the discovery that CXCR4 was the T-tropic virus cofactor, that lead to the identification of CCR5, which uses these three chemokines, as the M-tropic HIV coreceptor. This has lead to research exploring the possible use of chemokines, or chemokine analogs, as anti-HIV therapeutics.[34]

[32] M. Samson, F. Libert, B. J. Doranz, J. Rucker, C. Liesnard, C.-M. Farber, S. Saragosti, C. Lapouméroulie, J. Cognaux, C. Forceille, G. Muyldermans, C. Verhofstede, G. Burtonboy, M. Georges, T. Imai, S. Rana, Y. Yi, R. J. Smyth, R. G. Collman, R. W. Doms, G. Vassart, and M. Parmentier, *Nature (London)* **382,** 722 (1996).

[33] F. Cocchi, A. L. DeVico, A. Garzino-Demo, S. K. Arya, R. C. Gallo, and P. Lusso, *Science* **270,** 1811 (1995).

[34] F. Arenzana-Seisdedos, J.-L. Virelizier, D. Rousset, I. Clark-Lewis, P. Loetscher, B. Moser, and M. Baggiolini, *Nature (London)* **383,** 400 (1996).

As new pharmacological agents are developed that can bind chemokine receptors, a rapid method for screening their ability to block HIV-1 entry will be needed. Antibodies against CD4, antibodies against env, and peptides designed against env protein have all readily inhibited cell–cell fusion in our hands (J. Rucker, B. J. Doranz, A. L. Edinger, D. Long, J. F. Berson, and R. W. Doms, unpublished results, 1996). Agents against the chemokine receptors, such as chemokines and antibodies to the chemokine receptors, are not as active in preventing cell–cell fusion (J. Rucker, B. J. Doranz, A. L. Edinger, D. Long, J. F. Berson, and R. W. Doms, unpublished results, 1996). Previous studies that have blocked cell fusion at the chemokine receptor level have used much larger quantities of chemokines than are required to inhibit viral entry.[3] Other groups have also found that the level of chemokine receptor expression can dramatically influence the ability to block cell–cell fusion.[4] Nevertheless, by altering conditions of fusion, such as shortening the time of fusion to 2 to 3 hr and using cells with low to moderate amounts of receptor, adequate blocking results may be obtained.

Kinetics

Kinetic studies are an extremely useful technique for probing the mechanism of membrane fusion. An example of fusion kinetics is given in Fig. 4. This assay looked at the fusion of HeLa effector cells expressing JR-FL

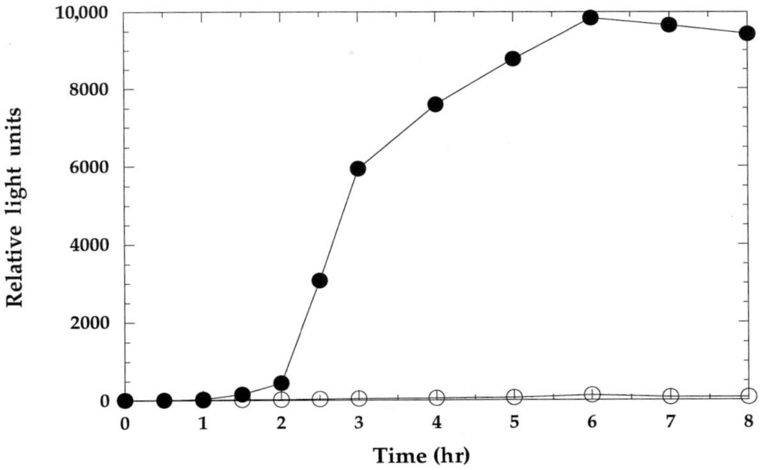

FIG. 4. Kinetic analysis of cell–cell fusion. NIH 3T3 cells stably expressing either CCR5 and CD4 (filled circles) or CXCR4 and CD4 (open circles) were mixed with HeLa cells expressing JR-FL. The time course was generated by successive lysis of wells during 8 hr. Fusion is clearly seen with cells expressing CCR5 and CD4, whereas cells expressing CXCR4 and CD4 remain negative for fusion during the assay.

env with NIH 3T3 cells stably expressing CCR5 and CD4. The time course was generated by the successive lysis of wells over the course of 8 hr. Fusion, as measured by this assay, seems to have a lag time of approximately 2 hr. This is followed by a rapid increase in fusion over the course of 2 to 3 hr, reaching a maximum at about 6 hr postmixing. Such sigmoidal behavior is often seen in cell–cell fusion and is representative of a highly cooperative process.[35,36] Changes in this sigmoidal behavior are potentially a more sensitive method of examining differences between env proteins or chemokine receptors containing subtle changes such as point mutations.

The successive lysis of wells is a relatively inaccurate method of following kinetics because variability in cell number between wells can introduce large uncertainties in signal output. Accurate quantitation of fusion kinetics would be aided by a secreted gene reporter that can be assayed without lysis or fixation of cells and thus allow one to follow the fusion of a single sample over time. A secreted form of alkaline phosphatase is available.[37] Aliquots of medium can be sampled over time, and the aliquots assayed for alkaline phosphatase activity. We are presently involved in integrating such a secreted gene reporter into the fusion assay.

Assay Problems

Two problems commonly arise in the fusion assay: high background and low signal or no signal. High luciferase activity in the negative controls indicates activation of the luciferase gene reporter in the absence of cell–cell fusion and cytosolic mixing. High background may be caused by direct infection of target cells by residual or nascent T7 polymerase vaccinia virus. Residual virus can be the result of incomplete washing of effector cells. The effect of residual T7 polymerase virus infection of target cells is minimized, but not eliminated, by the presence of araC during cell mixing. New vaccinia virus production should be almost completely inhibited by the presence of rifampicin and araC. However, even in the presence of these compounds, an increase in background is sometimes seen at $\geq$10 hr, perhaps due to a low level of vaccinia-induced cell fusion or a low level of virus production and infection. For this reason, fusion assays should not be run for much longer than 8 hr.

The S/N ratio will also be decreased if the signal is beyond the linear range of the detector. This can be corrected by bringing the signal into the linear range of the detector as previously discussed. Finally, the health of

[35] T. Danieli, S. L. Pelletier, Y. I. Henis, and J. M. White, *J. Cell Biol.* **133,** 559 (1996).

[36] R. Blumenthal, D. Sarkar, S. Durell, D. E. Howard, and S. J. Morris, *J. Cell Biol.* **135,** 63 (1996).

[37] J. Berger, J. Hauber, R. Hauber, R. Geiger, and B. R. Cullen, *Gene* **66,** 1 (1988).

the cells is also an important determinant of S/N ratios. The treatment of target cells during transfection is particularly critical. For this reason, use of DMSO or chloroquine shocks to improve transfection efficiency should be avoided.

Low signal or no signal is due to inefficient transfection of target cells or infection of effector cells. Vaccinia infection is generally quite straightforward and is only a problem when dealing either with cell types that are inefficiently infected by vaccinia or with degraded virus stocks. A more likely cause of poor signal is transfection failure. As previously mentioned, the efficiency of calcium phosphate transfections is highly pH-dependent and can be severely reduced by shifts away from pH 7.1. For many cell lines, transfection efficiency can be improved by use of a liposome-mediated transfection protocol. In addition, plasmid quality is an important factor, and plasmids should be prepared either by cesium chloride banding, polyethylene glycol (PEG) precipitation, or by use of Qiagen (Chatsworth, CA) plasmid purification kits. Plasmids can degrade over time and should be checked either spectrophotometrically or by agarose gel electrophoresis if low signal is an endemic problem. All of these factors should be considered when troubleshooting the fusion assay.

An interesting conceptual issue is whether HIV-1 env-mediated cell–cell fusion is mechanistically equivalent to env-mediated virus–cell fusion. One might suspect that the structural and energetic requirements for fusing two cell membranes would be substantially different than fusing a cell membrane with a viral membrane, yet experimentally this does not appear to be the case for HIV. The ability of a virus to enter cells seems to be well correlated with the ability of the env protein of the virus to mediate cell–cell fusion in the presence of the proper chemokine receptor and CD4. However, as the issues of virus entry are explored in greater detail, it is certain that some differences will arise. This is already suggested by the difficulty in using chemokines to block cell–cell fusion versus the ability of chemokines to block virus–cell fusion. The cell–cell fusion assay should only be considered a useful model of virus entry. Given this caveat, it is our hope that this assay will open up the study of HIV-1 entry to a broader range of researchers than would otherwise be possible.

Acknowledgments

We thank Chris Broder for discussions and assistance in developing this assay.

[10] Iodination of Chemokines for Use in Receptor Binding Analysis

By GREGORY L. BENNETT and RICHARD HORUK

Introduction

An understanding of the interaction between cell surface receptors and extracellular factors such as polypeptide hormones is critical to gain insight into the mechanisms by which these molecules can regulate intracellular processes. The rapid advances made in understanding cellular signaling can in part be attributed to the development of reliable receptor binding assays. Crucial to this rapid progress in receptor research has been the ability to effectively radiolabel receptor ligands with full retention of their biological activity. Indeed, successful radiolabeling of a receptor ligand requires that the radiolabeled ligand will behave in the same manner as the native ligand. In general the isotope most usually selected to radiolabel receptor ligands that are proteins is iodine-125. The dual emission of iodine-125, both γ radiation and X rays, facilitates its detection both by γ counting to analyze receptor binding experiments in a rapid manner and by autoradiography, where it has found use as a radiolabel to visualize receptors cross-linked to their iodinated ligands. In this chapter, we review the most common methods for radiolabeling proteins and provide detailed protocols and critiques of each method. In addition we discuss methods of assessing the purity and purification of the radioligands, stability, and specific activity as well as methods to validate the biological properties of the radiolabeled material. We illustrate these methods with reference to specific chemokines.

In general the procedures for radioiodinating proteins can be grouped into two types. One set of procedures that include chloramine-T, Iodogen, and lactoperoxidase are characterized by the oxidation of iodide to iodine and its subsequent incorporation into aromatic rings of tyrosine residues.[1–4] In contrast, the other major labeling procedure, the Bolton–Hunter method, is distinguished by the incorporation of the previously labeled reagent into lysine and terminal amine residues via amide bonds.[5] The choice of which

[1] W. M. Hunter and F. C. Greenwood, *Nature* (*London*) **194,** 495 (1962).
[2] M. Morrison and G. S. Bayse, *Biochemistry* **9,** 2995 (1970).
[3] P. J. Fraker and J. C. Speck, Jr., *Biochem. Biophys. Res. Commun.* **80,** 849 (1978).
[4] A. E. Bolton, "Radioiodination Techniques," Review 18. Amersham International, Amersham, UK, 1984.
[5] A. E. Bolton and W. M. Hunter, *Biochem. J.* **133,** 529 (1973).

METHODS IN ENZYMOLOGY, VOL. 288 0076-6879/97 $25.00

type of iodination procedure to use depends on a variety of factors. For example, the Bolton–Hunter procedure would be chosen for labeling proteins that do not have tyrosine residues such as the chemokine melanoma growth stimulating activity (MGSA).[6] The choice could also be determined empirically, by demonstrating that one type of iodination results in tracers with superior characteristics. It is important to note that care should be taken not to generalize or extrapolate from the results of a single set of iodination conditions. Claims of one method being superior to another may be true for one chemokine but not for another. Also, what is commonly overlooked is the importance of optimizing the procedures being evaluated before drawing conclusions.

A method is deemed suitable when one is able to add a label to the protein in such a way as to retain normal biological function or at least to minimize deleterious effects. In receptor binding, for example, it is essential for the labeled and unlabeled forms to have similar receptor binding affinities. Because the process of iodinating is one of adding a foreign group or groups, the protein is no longer truly "native." The presence of such a group could even induce a conformational change, thereby altering receptor binding. In many cases, however, the presence of iodine is not detrimental. The presence of an iodine, although not causing a conformational shift, may in fact be detrimental for steric reasons. The modification of a residue in the "active" site of a chemokine could easily block receptor binding.

Iodination reactions are difficult to control in the sense that it is difficult to know just where the iodines will end up. Only the overall efficiency of incorporation of iodine into the protein as a whole can be predicted. The resulting radiolabeled compound is most likely a complex mix of various forms. Heterogeneity in the protein itself is one important variable. Another is the number of iodinatable residues. The location of these residues in the protein is significant in that, because of steric issues, some are more likely to be labeled than others. The fact that each tyrosine can be singly or doubly labeled adds more complexity. By maximizing labeling efficiency, labeling heterogeneity can be minimized, but in the process the prospect of having a biologically active tracer is minimized. One limitation of the oxidative methods is the lack of control and specificity of the oxidizing reaction itself, whether it be chemical or enzymatic. Of concern is the oxidation of the protein itself and the effect that would have on the function of the protein. The residues susceptible to oxidation, cysteine and methionine, may be in critical regions, where receptor binding could be altered because of steric or conformational reasons. One approach to deal with this possibility is to deliberately label to a low specific activity. This is

[6] R. Horuk, *Immunol. Today* **15,** 169 (1994).

often done with the lactoperoxidase methods. The hope is to minimize the deleterious effects of oxidation by diluting out the effect. This is accomplished by iodinating in the presence of a large excess of the protein, such that only a small percentage of the molecules will be labeled. Presumably, oxidation of the protein is also minimized. Another approach that we have successfully employed in our laboratory is to titer out the lactoperoxidase to the minimum amount required. We have successfully labeled various chemokines to moderately high specific activities while minimizing protein oxidation.

One of the characteristics of chemokines that allow for detailed characterizations of iodinations are their relatively small sizes. This has allowed us to take advantage of mass spectrometry to evaluate tracers prepared with "cold" (i.e., nonradioactive) iodine. The degree of resolution is adequate to indicate the addition of iodine or oxygen atoms to the protein. Chemokines are also ammenable to the use of reversed-phase high-performance liquid chromatography (RP-HPLC). This procedure can serve multiple purposes. First and foremost, it is an effective method for desalting free iodine, which does not bind to the column. By evaluating relative peak areas, the chromatogram can also be a picture of iodination efficiency (i.e., the percentage of iodine incorporated). Multiple peaks indicate multiple tracer forms. As discussed previously, these forms may include different locations of labeling, mono- versus diiodo forms, and oxidized variants. Heterogeneity of the protein could also be present. Although not evident in a radioactive compound, it is possible to separate the noniodinated from the iodinated forms.

Determination of Optimal Level of Iodination

To determine the desirable level of iodination of a typical protein several factors have to be considered. These can best be illustrated by reference to the chemokine MIP-1α (macrophage inflammatory protein-1α). The maximum theoretical specific radioactivity for one radioactive atom of iodine per MIP-1α molecule is 2200 Ci per mmol. Thus carrier-free ^{125}I-MIP-1α would have a specific activity of 275 μCi/μg protein. To minimize the production of diiodotyrosines, which reduce the biological activity of the protein, we typically label the molecule at around one-half of its theoretical specific activity. For MIP-1α with two tyrosine residues the maximum desirable level of incorporation of iodine is 1.0 per molecule. Knowing the specific activity of the Na^{125}I, it is easy to calculate the desirable incorporation in a given iodination. For example, if the specific activity of the Na^{125}I is 500 μCi/μl and we want to label MIP-1α at a final specific activity of 140 μCi/μg of protein; then we would add 4 μl of Na^{125}I (2000 μCi) to 10 μg of protein and aim to incorporate 140 × 10/2000 = 70% of the radiolabeled iodine. Depending on the protocol of choice (see below) we design the reaction conditions and time of iodination to achieve incorporations close to these values.

The rate of incorporation is checked most easily by carrying out trichloroacetic acid (TCA) precipitations at various incubation times of the protein as described below. After a reasonable interval of iodination a 5-μl aliquot of the radiolabeled protein is made to 1 ml with 0.5% (w/v) bovine serum albumin (BSA), and 50 μl of this is made to 500 μl with 5% (w/v) BSA (2000-fold dilution). Then 500 μl of 10% (w/v) TCA is added, and the tube is well mixed and centrifuged at 5000 *g* for 2 mins. The supernatant and precipitated BSA are counted, and the level of incorporation of the radiolabeled iodine can be readily determined.

Methods of Iodination

A number of methods exist to radioactively label a protein. These can be conveniently grouped into two types, direct and indirect. The direct method uses the introduction of a radioactive iodine into the protein of interest, usually on a tryrosine or a histidine residue. Direct methods of labeling include chloramine-T, lactoperoxidase, and Iodogen.[1–4] In contrast, indirect labeling introduces a radioactive iodine into a protein by attachment to a conjugating molecule; the best example of this method of labeling is the Bolton–Hunter region.[5] Both direct and indirect methods of labeling are discussed below, together with their attendant advantages and disadvantages. In addition we offer some typical protocols that we have found useful for labeling chemokines.

Chloramine-T

Chloramine-T iodinations may be performed with chloramine-T in solution[1] or bound to a solid phase (Pierce, Rockford, IL, Iodobeads).[7] Chloramine-T chemically oxidizes iodide (I^-) to iodine (I^-I^+) followed by electrophilic substitution (I^+) usually into the aromatic ring of a tyrosine residue. The reaction rate is determined by the concentrations of the reactants, and the extent of iodination depends on the time of reaction. Typically, 10 μg is added to 10 μg protein and 1 mCi iodine and allowed to incubate for 1 to 2 min. The reaction is usually carried out in phosphate buffer at neutral pH and is generally highly efficient with regard to the incorporation of iodine into the tyrosine residues of the protein or peptide being labeled.

[7] M. A. K. Markwell, *Anal. Biochem.* **125,** 427 (1982).

Radioactive proteins with high specific activities are usually desirable for assay sensitivity in radioimmunoassays and competitive receptor binding assays. For a given number of radioactive disintegrations, translated into counts per minute (cpm), less mass of tracer needs to be inhibited from binding in comparison to a lower specific activity tracer, where more mass must be inhibited. However, the side effects of such efficient incorporation of iodine can lead to inactivation of the biological activity of a protein and is always a compromise.

Because of the speed of the reaction it is possible to overlabel the tyrosines and to oxidize susceptible residues. One way to control the duration of the reaction is to stop it at a given time by the addition of the reducing reagent such as sodium metabisulfite. The amount added is equivalent to the amount of chloramine-T added. This is an effective way of preventing the iodination from continuing, by converting the unincorporated iodine back to iodide with its negative charge. Unfortunately, however, this procedure carries with it the risk of reducing disulfide bonds and thus disrupting the secondary protein structure. This could induce a conformational change in the protein that would render the protein unrecognizable to specific antibodies or receptors.

Another method of controlling the rapid chloramine-T reaction is by quenching the reaction. The most common method is to add an excess amount of nonradioactive, or cold, potassium or sodium iodide. With this added, it becomes statistically unlikely that any additional radioactive iodine will be incorporated into the protein. The problem with this approach is that iodine itself continues to be incorporated into available tyrosines.

At this point, it is important to consider the effect of iodination on a protein. One must obviously be concerned about the effect of γ radiation on the protein and the damage it could cause. One of the important functions of carrier proteins added to proteins following iodination is not only to prevent nonspecific binding of tracers to surfaces, but also to absorb much of the γ radiation and thus limit the amount that the tracer protein itself absorbs.

The other effect of iodination on a protein is simply the physical presence of the iodine itself. The presence of iodine in a particular location could create steric hindrances and thus partially or completely block the binding of the tracer to antibodies or receptors. In this regard, the addition of cold iodine to quench the chloramine-T reaction may have undesirable consequences.

Another method of quenching a chloramine-T reaction is to add an excess of free tyrosines, in the form of *N*-acetyltyrosine. Although the oxidation reaction may continue, iodine will only be incorporated in readily available free tyrosines.

Because the conditions of chloramine-T reactions are often considered to be too harsh, modifications have been made to minimize the damaging effects. One method is to add much smaller doses of chloramine-T in pulsatile fashion. Three doses of 2 μg are added followed by incubations of 2, 1.5, and 1 min. Because the amount of chloramine-T added is significantly lower with this method, the use of a reducing agent is not necessary.

Reactions using Iodobeads (Pierce), with solid-phase chloramine-T,[7] have the advantage of allowing for improved control. The rate of reaction is determined by the number of beads chosen and the volume used for the reaction. A minimum volume of 100 μl for one bead is required. The reaction can be stopped by separating the solution from the beads. No reducing agent is needed. Iodobeads are compatible with many buffers, including Tris and HEPES, and with detergents such as sodium dodecyl sulfate (SDS), Triton X-100, and Nonidet P-40 (NP-40), and they remain active in the presence of denaturants such as urea. However, these beads are not compatible with reducing agents such as dithiothreitol (DTT) or mecaptoethanol.

The following protocol gives an example of iodination of MIP-1α with Iodobeads.

1. To 10 μl of MIP-1α (1 mg per ml stock) add 2 mCi of $Na^{125}I$ and make up to 100 μl with phosphate-buffered saline (PBS). Add one Iodobead (Pierce) and vortex vigorously to mix. Incubate until the desired specific incorporation of iodine is achieved, usually 5 to 10 min at room temperature.
2. At the end of the incubation add 400 μl PBS in 0.1% BSA and transfer the 500 μl to a Sephadex G-25 column (Pharmacia, Piscataway, NJ, PD-10) equilibrated in PBS/0.1% BSA. Let the level of the buffer fall to the glass sinter that marks the top of the gel bed. Wash the column with 1.5 ml of the same buffer and discard. Then wash the column with 0.5-ml aliquots and collect the next 10 fractions.
3. Count 2-μl aliquots of each fraction in a γ counter and pool the fractions containing the first peak of radioactive counts, to yield the radiolabeled MIP-1α.

Iodogen

The second direct method of radioiodination of proteins involves the use of the oxidant 1,3,4,6-tetrachloro-3α,6α-diphenylglycoluril,[3] sold by Pierce under the trade name Iodogen. Iodogen is insoluble in aqueous solution and is typically dissolved in organic solvents such as chloroform. The solvent is then evaporated with a gentle stream of nitrogen, and the Iodogen is coated on a small area on the bottom of the tube. Care must be taken to create an even layer that attaches well to the surface of the tube. This minimizes the dislodging of Iodogen from the surface into solution. When thoroughly dry, tubes coated with Iodogen can be stored for several months.

Protein and iodine are then added to the Iodogen tube and allowed to incubate for several minutes; the reaction is stopped by removing the solution from the tube. A variation of this is called the indirect Iodogen method and involves only the addition of iodine to the Iodogen tube. The iodide that has been oxidized by Iodogen is then transferred to another tube containing the protein. This method has been used successfully with vascular endothelial gowth factor (VEGF) and other proteins. The beauty of this method is that the protein to be iodinated is never exposed to oxidizing conditions that can result in oxidation of protein residues. Care must be taken in transferring activated iodine between tubes, because it is in its most volatile state. The following protocol gives an example of iodination with Iodogen.

1. Dissolve Iodogen (Pierce) in chloroform to a final concentration of 200 μg/ml.
2. Pipette 50-μl aliquots of the Iodogen into a series of tubes and evaporate the chloroform by blowing through a gentle stream of nitrogen. This creates a thin layer of Iodogen over the bottom of the tubes.
3. For iodination take one of the prepared coated Iodogen tubes and add 20 μl of 5 *M* K_2HPO_4, pH 7.4, and 1 mCi $Na^{125}I$.
4. Cover the tube and incubate for 10 min at room temperature to activate the ^{125}I.
5. With great care, transfer the entire contents of the tube containing the activated ^{125}I to a 1.5-ml microcentrifuge tube containing 0.25 to 0.5 nmol protein.
6. Cover the tube and incubate for 10 min on ice.
7. Quench the reaction by adding PBS/0.05% Tween 20 (for PD-10) or distilled water (for HPLC) to a final volume of 0.3 ml.
8. Remove 5 μl for specific activity determination by TCA precipitation.

Either desalt the radiolabeled protein on a PD-10 column or chromatograph by RP-HPLC using a C_4 analytical column and an acetonitrile/trifluoroacetic acid (TFA)/water gradient as described above.

Lactoperoxidase

The final example of direct radioiodination of proteins is based on the enzyme lactoperoxidase, which oxidizes iodide to iodine.[2] It is considered a gentler method of iodination than chloramine-T and is used commonly in the preparation of radioactive ligands for receptor binding assays. As with chloramine-T, lactoperoxidase can be used in solution or immobilized on solid phase (Enzymobeads, Bio-Rad, Richmond, CA). Reactant concentrations and reaction times determine the rate and extent of incorporation. However, pH also is a crucial variable. The optimum pH is 5.6 and reaction rates diminish at pH values approaching neutrality. Standard methods of lactoperoxidase iodinations use high amounts of the enzyme to achieve higher specific activities. In solution the enzyme is added to the protein and iodine mixture, and hydrogen peroxide is added to initiate the reaction as an electron acceptor. When the enzyme is used in the solid phase as Enzymobeads, a second enzyme, glucose oxidase, is added along with D-glucose as an electron acceptor.

The method commonly used to preserve the biological function of the radioligand is to deliberately iodinate the protein to a very low specific activity by using an excess of protein in relation to the other reactants. Presumably, the deleterious side effects of the iodination will more likely affect the noniodinated molecules. Another version of the lactoperoxidase method uses significantly lower levels of the enzyme in addition to stoichiometric amounts of protein, iodine, and hydrogen peroxidase. In several examples, including recombinant interleukin-8 (IL-8) and RANTES, nearly equimolar ratios of iodine to protein have been achieved, while maintaining the biological function of the protein. An example of the iodination protocol for IL-8 is given below

Lactoperoxidase Iodination of Interleukin-8

1. Add 5 μg of human IL-8 in a volume of 10 μl or less to 1 mCi $Na^{125}I$ and 50 μl of 0.4 *M* sodium acetate, pH 5.6.
2. To these reactants are added 10 mU of lactoperoxidase (Calbiochem, La Jolla, CA) in 0.1 *M* sodium acetate, pH 5.6, and 10 μl of 0.003% H_2O_2. These reactants are incubated for 5 min with periodic vortexing.
3. At the end of the incubation distilled water is added to a final volume of 500 μl, and 5 μl is removed for the determination of the specific activity by TCA precipitation.
4. The remainder of the material is loaded onto a C_{18} reversed-phase HPLC column and developed with a gradient of 35–40% (w/v) acetonitrile in 20 min at a flow rate of 1 ml/min. Solvent A is distilled water/0.1% (w/v) TFA; solvent B is acetonitrile/0.1% (w/v) TFA.

A typical profile from the HPLC column separation of lactoperoxidase-labeled material is shown in Fig. 1.

Bolton–Hunter Method

The Bolton–Hunter method is distinguished from the methods discussed above not only because it is an indirect method of radioiodination but

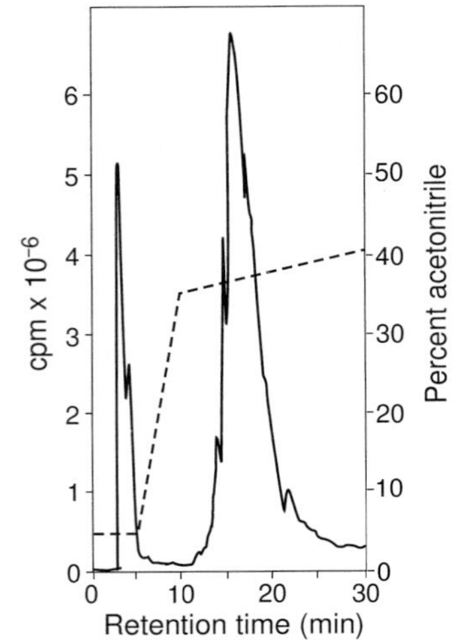

FIG. 1. Separation by HPLC of lactoperoxidase-labeled human IL-8.

also because it is used to label lysines and terminal amino groups. The Bolton–Hunter reagent is an *N*-hydroxysuccinimide ester of iodinated *p*-hydroxyphenylpropionic acid and was first described for the radioiodination of a number of hormones including the chemokine platelet factor 4.[5] Iodinated Bolton–Hunter reagent is routinely supplied as a 1 mCi solution dissolved in benzene and is rapidly unstable in aqueous solutions. The preparation of iodinated Bolton–Hunter reagent involves a chloramine-T iodination followed by a series of solvent extractions. To avoid such a tiresome procedure, most investigators choose to purchase the Bolton–Hunter reagent already iodinated.

The process of iodination begins with evaporating the benzene from the Bolton–Hunter reagent. A gentle stream of nitrogen is introduced through the septum with a needle attached to a tube connected to a house system or tank. The nitrogen exits the vial through a second needle connected to a trap containing activated charcoal, where free iodine is collected. The drying process continues until all the benzene is evaporated and Bolton–Hunter reagent is coated on the surface at the bottom of the vial. It is important to evaporate the benzene gently to avoid splashing and to allow for the reagent to collect at the very bottom of the tube. The evaporation and iodination should be carried out in a properly ventilated iodination hood. Special care should also be taken to avoid exposing skin to vapors during the process. Some free iodine may be released, and it is much more volatile when carried in the benzene.

The actual iodination takes place on addition of the protein to be labeled. The optimum pH range is pH 8–8.5. Suitable buffers include borate and carbonate. Tris must be avoided because it serves as a source of free amino groups that would compete for reaction with the Bolton–Hunter reagent. Because rabbit IL-8 has no tyrosine residues, the Bolton–Hunter method is the method of choice. In other cases, such as with glial-derived neurotropic factor (GDNF), Bolton–Hunter labeling is the method of choice because it results in a radiolabeled protein that is able to bind to its receptor with high affinity. A water-soluble version of the Bolton–Hunter reagent (Pierce) is fully compatible with aqueous solutions and does not require the evaporation of organic solvents prior to use. A protocol for Bolton–Hunter labeling of rabbit IL-8 is given below.

Bolton–Hunter Iodination of Rabbit Interleukin-8

1. The Bolton–Hunter reagent is supplied in a sealed vial in benzene at approximately 1 mCi per 100 μl of [^{125}I]diiodo-labeled Bolton–Hunter reagent (NEN, Boston, MA; Amersham, Arlington Heights, IL). The benzene is evaporated by flushing the vial with a gentle stream of nitrogen using an activated charcoal trap.
2. Once the benzene is removed, 10 μg of IL-8 in 10 μl sodium carbonate, pH 8.5, is added, and the solution is gently mixed and allowed to stand for 1 hr at 4°. At the end of the incubation 475 μl of 0.4 *M* glycine in 0.2 *M* sodium carbonate, pH 8.5, is added to quench the reaction, and 5 μl is removed for the determination of the specific activity by TCA precipitation.
3. The remainder of the material is loaded onto a C_{18} reversed-phase HPLC column and developed with a gradient of 35–40% acetonitrile in 20 min at a flow rate of 1 ml/min. Solvent A is distilled water/0.1% TFA; solvent B is acetonitrile/0.1% TFA.

A typical profile from the HPLC column separation of Bolton–Hunter-labeled material is shown in Fig. 2.

Purity and Purification or Radioligand

Trichloroacetic Acid Precipitation

Trichloroacetic acid precipitation is most often used to determine the specific activities of iodinated tracers. During and after iodination procedures, samples can be taken, diluted, and precipitated in 10% TCA. After 30 min on ice, the precipitate is pelleted by centrifugation for 5 min at 10,000 *g*. An aliquot of the supernatant is counted, and this is compared

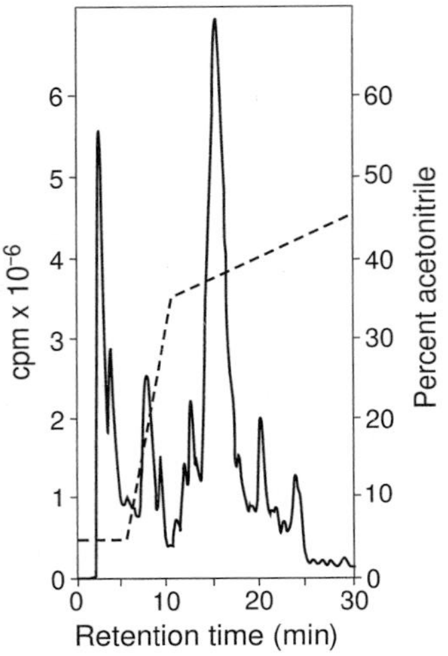

FIG. 2. Separation by HPLC of Bolton–Hunter-labeled rabbit IL-8.

to counts taken prior to precipitation to determine the efficiency of incorporation of iodine in the reaction. As an example an actual iodination of MIP-1β is given below.

Ten micrograms of MIP-1β is iodinated with Iodogen (as described above) in a final volume of 225 μl. At the end of the iodination 5 μl is removed and diluted to 1 ml with water (1:200 dilution precolumn pool). The labeled protein is separated from free iodine by gel filtration on a PD-10 column (as described above), and 0.5-ml fractions are collected. The radiolabeled protein peak is identified by counting aliquots from each of the fractions, after which the fractions are pooled (postcolumn pool) and 5 μl is removed and diluted to 1 ml with water (1:200 dilution postcolumn pool). Fifty microliters of 5% BSA and 50 μl of 60% TCA are added to 200 μl of the 1:200 dilutions of the precolumn and postcolumn pools, respectively. The tubes are mixed, left on ice for 30 min, and centrifuged as described above, and 50 μl of the supernatants are counted.

Fraction	cpm
1:200 dilution precolumn pool	127,697
1:200 dilution postcolumn pool	6,820
TCA 1:200 dilution start	62,415
TCA 1:200 dilution pool	1,238

Because 200 μl of the 1:200 dilution precolumn pool is used for TCA precipitation, the total counts added are 127,697 × 40 = 5,107,880. Total counts left in the TCA supernatant from the 1:200 dilution precolumn pool are 62,415 × 6 cpm per 300 μl = 374,490 cpm. Therefore the TCA precipitability of the 1:200 dilution precolumn pool is 5,107,880 − 374,490 ÷ 5,107,880 × 100 equals 92.7%.

Similarly, on the basis of the data above the TCA precipitability of the 1:200 dilution postcolumn pool is 97%. These data are then used to calculate the specific activity of the radiolabel (see Stability and Measure of Specific Activity).

Electrophoresis

Sodium dodecyl sulfate–polyacrylamide gel electrophoresis (SDS–PAGE) is commonly used to evaluate the radiochemical purity of the radiolabeled protein after iodination. Very little radioactivity needs to be run on the gel. After the gel is run and dry, it can be exposed to film to make an autoradiogram. Ideally, other tracers of known and varied molecular weights can be run as standards. Depending on the amount of tracer added to each lane, the exposure time varies from a few minutes to several days.

Bindability

The ability to adequately radioiodinate a protein while maintaining normal receptor binding affinity and specificity is the ultimate test of labeling techniques. One has to balance the need for an adequate specific activity for statistically acceptable counting efficiency with the danger of overlabeling the protein and loss of bindability. The location of iodination may be critical in whether bindability is maintained. The selection of an alternative iodination technique, such as the Bolton–Hunter method, is required when the labeling of a tyrosine residue in the receptor binding region of the ligand diminishes its bindability. We routinely measure the receptor reactivity or bindability of the proteins that we radioactively label. For example, the bindability of MGSA is measured using a method described by Kermode.[8] Briefly, a trace concentration of radiolabeled MGSA is incubated with increasing concentrations of cells (1×10^6 to 1×10^8 cells/ml) and the specific binding determined. The data are plotted in a double-reciprocal plot and the bindability

[8] J. C. Kermode, *Biochem. J.* **252,** 521 (1988).

determined at the intercept with the *y* axis. The bindability of typical lots of MGSA (labeled using the Bolton–Hunter method) exceeds 90%.

Gel-Filtration Protocols

Gel-filtration chromatography is by far the simplest and most common method of separating free iodine from a radiolabeled protein. Prepoured PD-10 columns, (Pharmacia), containing Sephadex G-25, are convenient to rapidly desalt proteins. For many applications, PBS/0.05% Tween 20 works well as a column buffer. Carrier protein-containing buffers, such as PBS/0.5% (w/v) BSA or PBS/0.1% (w/v) gelatin can also be used. The advantage of using carrier proteins is not only to reduce nonspecific binding of the tracer, but they can also serve a protective function in absorbing radiation that could cause damage to the labeled protein. The advantage of not using carrier proteins is that tracers can be quantified by protein assays, such as the one using fluorescamine, a fluorogenic reagent for assay of primary amines.

For more extensive separation of monomeric forms of the radiolabeled protein from dimers or higher molecular weight aggregates, fractionation by gel filtration on resins containing larger pore sizes can be used.

Chromatography Protocols

A reversed-phase HPLC chromatogram of a radioiodinated chemokine can be a most revealing snapshot of the labeling procedure. The degree of heterogeneity in the resulting tracer can be determined and various forms isolated. With respect to total iodination efficiency, non- versus mono- versus diiodo forms can be identified. It could also be possible to identify different forms in which single tyrosines are multilabeled or in which different tyrosines are labeled.

Reversed-phase HPLC can also be used to separate iodinated protein from the unlabeled protein. In general adding an iodine makes the molecule more hydrophobic, and the labeled protein usually elutes at a retention time greater than that of the unlabeled molecule. A perfect example of this is the separation of labeled from unlabeled MIP-1β (Fig. 3).

Oxidation artifacts can also be by-products of oxidative iodinations. The presence of iodine on a protein makes it more hydrophobic, causing it to elute with longer retention times than the native protein. Oxidation has the opposite effect, creating a less hydrophobic form that would elute

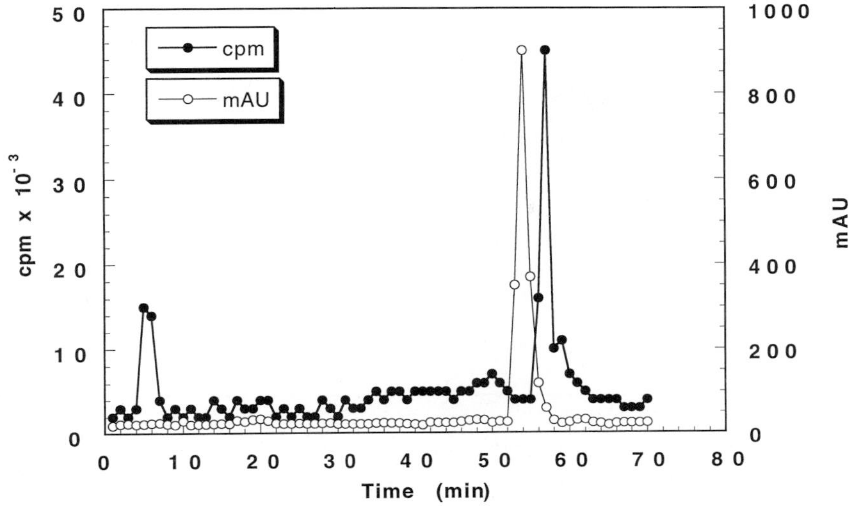

FIG. 3. Separation by HPLC of ^{125}I-labeled MIP-1β from unlabeled MIP-1β.

with a shorter than normal retention time. It is possible to have difficulties in isolating an iodinated and oxidized form from native protein, because of opposing hydrophobic and hydrophilic effects that tend to balance each other.

A valuable tool in tracer characterizations is to use mass spectrometry in conjunction with reversed-phase chromatography of a labeled protein or peptide. With peptides in particular, the ability to achieve atomic resolution allows characterization of a tracer down to the presence of a single oxygen atom on a methionine residue or an additional iodine on tyrosine. With potential multiple sites for iodination or oxidation, especially on larger proteins, one could identify specific locations of modification by trypsin digestion, comparing tryptic maps before and after nonradioactive iodinations by using reversed-phase HPLC and mass spectrometry. This information could be useful in understanding changes in the bindability of the labeled ligand to receptor.

Stability and Measure of Specific Activity

To calculate the specific activity of a radiolabeled protein, we use the example of MIP-1β from the section on TCA precipitation above.

TCA PRECIPITATION DATA SHEET

Specific activity:	127,697 cpm in 5 μl of 1:200 dilution of start × 200 (dilution factor) = original 5 μl × total precolumn volume 215/5 = 1,098 × 10^6 cpm/ml × precolumn TCA = 92.7/100 × column pool TCA = 97/100/mass iodinated (10 μg) = 98.7 × 10^6 cpm/μg/1.54 × 10^6 cpm/μCi (at 70% γ-counter efficiency) = 64.1 μCi/μg.
Concentration:	6,820 cpm per 5 μl of 1:200 dilution of postcolumn pool × 200 (dilution factor) = original 5 μl × 200 per original 1 ml = 272.8 × 10^6 cpm/ml × column pool TCA = 97/100/98.7 × 10^6 cpm/μg = 2.68 μg/ml (MW = 7,819) = 342 n*M*.
Molar ratio:	Specific activity of iodinated protein = 64.1 μCi/μg × 0.5 nmol/1,000 μCi (specific activity of ^{125}I on calibration date)/^{125}I decay factor × molecular weight of protein (7.819 μg/nmol) = 0.25 mol ^{125}I/mol protein.

The only problems with this calculation of the specific activity of the labeled protein is that it assumes effective separation of the labeled protein from any unlabeled protein. Methods to achieve this via RP-HPLC are discussed above.

[11] Expression of Chemokine Receptors by Endothelial Cells: Detection by Intravital Microscopy Using Chemokine-Coated Fluorescent Microspheres

By VICTOR H. FINGAR, HAI-HONG GUO, ZHAO-HAI LU, and STEPHEN C. PEIPER

Introduction

Chemoattractant cytokines, known as chemokines, are multifunctional polypeptides that play a salient role in normal and pathological inflammatory processes.[1] Because chemokines may be secreted by many types of cells, they represent a common mechanism for eliciting inflammation. These agents are small, basic, robust polypeptides that are composed of approximately 75 amino acid residues and are resistant to thermal and biochemical stresses. Their biological activity is sensitive to reduction of the two intrachain disulfide bonds formed by the four canonically conserved cysteine residues.

[1] T. J. Schall, *in* "The Chemokine Handbook" (A. Thomson, ed.), p. 419. Academic Press, London, 1994.

0076-6879/97 $25.00

The primary biological role of chemokines is to summon leukocytes to sites of inflammation. They are potent chemoattractants that induce directed migration of leukocytes along a concentration gradient. Typically, they are secreted by cells residing within tissues, and their final targets, including lymphocytes, monocytes, and granulocytes, are intravascular. Thus, the chemokine must navigate through the interstitium and traverse the endothelial boundary before reaching the zone that contains the relevant target cells. Target cells express receptors for chemokines that are members of the serpentine receptor family, which have seven hydrophobic helices that function as transmembrane spanning domains and transduce signals through coupling to guanine nucleotide binding (G) proteins.[2]

The biological and molecular biological features of two branches of the chemokine family have been elucidated in detail. The structural corollary of these two branches is the configuration of the two amino-proximal cysteine residues. In one family, designated C-X-C or α, these cysteines are separated by a single amino acid residue, whereas they are juxtaposed in the other, which is designated C-C or β. In general, C-X-C chemokines bind to receptors expressed on neutrophils and, thus, play a role in the development of acute inflammation. In contrast, receptors for C-C chemokines are expressed on lymphocytes and monocytes, and, consequently, are inducers of more chronic inflammatory processes. The ligand binding repertoires of the known receptors for C-X-C chemokines, now six in number, and C-C chemokines, now eight in number, overlap within the same branch. The single recognized exception to this rule is the Duffy chemokine receptor (DARC), which binds to members of both the C-X-C and C-C branches.

The primary target cells that express receptors for chemokines, either C-X-C or C-C, are leukocytes. The interaction of chemokines with their cognate receptors results in translocation of integrins to the cell surface, activating cytoadhesion to endothelial cells, which is manifested by rolling, flattening, and, finally, transmigration through the endothelial cell interface. The primary site for this process is at the level of postcapillary venules, which measure 10–15 μm in diameter. Chemokines have been found to bind to subsets of endothelial cells using *in situ* assays,[3] but the available cell lines derived from endothelial cells have not been shown to express chemokine receptors.

The armamentarium of immunological reagents for immunohistochemical analysis of chemokine receptor expression is quite limited. Monoclonal

[2] R. Horuk and S. C. Peiper, *Exp. Opin. Ther. Patents* **5,** 1185 (1995).

[3] A. Rot, *Immunol. Today* **13,** 291 (1992).

antibodies have been produced to CXCR1, CXCR2,[4] CXCR4,[5] CCR3, CCR5, and DARC.[6] However, because there is cross species conservation of chemokine receptors, no immunological reagents have been produced to nonhuman homologs that could be used in animal model systems. Immunohistochemical analysis of human tissues with monoclonal antibodies to receptors that bind interleukin-8 (IL-8) has revealed that DARC, but not other receptors (CXCR1 and CXCR2), is expressed by the subset of endothelial cells that line postcapillary venules.[7] Because several ligands for DARC have been shown to be secreted by endothelial cells, including IL-8,[8] monocyte chemoattractant protein-1 (MCP-1),[8] RANTES,[9] and melanoma growth stimulatory activity (MGSA),[10] and the subset of endothelial cells that are the site of leukocyte diapedesis express DARC, a promiscuous chemokine receptor, it is possible that the biological features of endothelial cells may be modulated by chemokines to regulate their interaction with leukocytes in some fashion.

To study this system in an animal model, we have devised a novel approach to the demonstration of chemokine receptors in endothelial cells in the rat using chemokines coupled to fluorescent microspheres to bind to chemokine receptors and intravital microscopy to localize the FluoSpheres (Molecular Probes, Eugene, OR). This *in vivo* approach allows for the characterization of vessels lined by endothelial cells that express the receptor(s).

The use of video microscopy of blood vessels in living animals, termed intravital microscopy, has facilitated research in a wide array of studies. Direct visualization of blood vessels in a variety of organ sites, including muscle, kidney, lung, and liver, has allowed measurement of microvascular

[4] A. Chuntharapai, J. Lee, J. Burnier, W. I. Wood, C. Hébert, and K. J. Kim, *J. Immunol.* **152,** 1783 (1994).

[5] M. J. Endres, P. R. Clapham, M. Marsh, M. Ahuja, J. Davis Turner, A. McKnight, J. F. Thomas, B. Stoebenau-Haggarty, S. Choe, P. J. Vance, T. N. C. Wells, C. A. Power, S. S. Sutterwala, R. W. Doms, N. R. Landau, and J. A. Hoxie, *Cell (Cambridge, Mass.)* **87,** 745 (1996).

[6] M. E. Nichols, P. Rubinstein, J. Barnwell, S. Rodriguez de Cordoba, and R. E. Rosenfield, *J. Exp. Med.* **166,** 776 (1987).

[7] T. J. Hadley, Z. H. Lu, K. Wasniowska, A. W. Martin, S. C. Peiper, J. Hesselgesser, and R. Horuk, *J. Clin. Invest.* **94,** 985 (1994).

[8] Z. Brown, M. E. Gerritsen, W. W. Carley, R. M. Strieter, S. L. Kunkel, and J. Westwick, *Am. J. Pathol.* **145,** 913 (1994).

[9] D. Schwartz, A. Andalibi, L. Chaverri-Almada, J. A. Berliner, T. Kirchgessner, Z. T. Fang, P. Tekamp-Olson, A. J. Lusis, C. Gallegos, and A. M. Fogelman, *J. Clin. Invest.* **94,** 1968 (1994).

[10] A. Marfaing-Koka, O. Devergne, G. Gorgone, A. Portier, T. J. Schall, P. Galanaud, and D. Emilie, *J. Immunol.* **154,** 1870 (1995).

responses to physiological stimuli and under different pathological states.[11] Typically, data are limited to measurements of vasomotion, counts of leukocyte and platelet adhesion, and blood flow velocities. The use of epifluorescence microscopy permits the visualization of injected dyes and allows the measurement of vessel permeability and vascular leakage.[12] Computer image analysis of fluorescence gray levels provides a means to track dye movement across different tissue compartments.

Various techniques including immunohistochemistry, *in situ* hybridization, and vital staining have been used to detail the presence or absence of various proteins and genetic material. Although these techniques have proven utility for the staining of histological sections, cell culture preparations, or in flow cytometry, they have not found application for intravital microscopy. As such, it has been impossible to monitor endothelial cell expression of various receptor proteins and ligands while these cells are in their native state. The major difficulty is that the systems for intravital microscopy do not have the necessary resolution or sensitivity to image the location of binding events. This is a function of problems with tissue motion and the resulting limitation in useful magnification. The use of fluorescein- or other fluorochrome-labeled antibodies does not produce sufficient signal for visualization in these systems. The introduction of intravital models of confocal microscopy and the availability of better imaging cameras have only begun to solve these problems.

A solution to the difficulty in imaging the binding of antibodies and other probes to the luminal surface of endothelial cells within blood vessels from different organs is to link the probe to small fluorescent microspheres that can be easily observed by intravital microscopy. Radioactively labeled microspheres and, more recently, fluorescent latex microspheres have been used widely to measure blood flow to tissues and are available in a wide ranges of sizes and excitation/emission wavelengths.[13,14] Microspheres sized at 0.1 or 0.3 μm are ideally suited for intravital microscopy because they are small enough to pass through small vessels and capillaries and do not cause measurable disturbances in blood flow. They are large enough to allow observation and resolution of individual microspheres, thereby allowing the location of binding along the tissue microcirculation to be precisely determined. Studies in progress by our group are evaluating the expression of ICAM-1 (intercellular adhesion molecule-1), CD13, and chemokine re-

[11] J. Barker, G. Anderson, and M. Menger, eds., "Clinically Applied Microcirculation Research," pp. 139–148. CRC Press, Boca Raton, Florida, 1995.

[12] L. S. Heuser and F. N. Miller, *Cancer* **57,** 461 (1986).

[13] P. Kowallik, R. Schulz, B. D. Guth, and A. Schade, *Circulation* **83,** 974 (1991).

[14] R. W. Glenny, S. Bernard, and M. Brinkley, *J. Appl. Physiol.* **74,** 2585 (1993).

ceptor location along microvasculature in both lung and muscle using probes tagged with microspheres.

Analysis of the location and relative degree of expresison of chemokine receptors along the vascular endothelium represents an interesting area of study using intravital microscopy as it was hypothesized that endothelial cells from different vascular segments (i.e., arterioles, capillaries, venules) express different amounts of receptors. This suggested that endothelial cell function could be modulated by the physiological conditions in that local environment. In this chapter, we discuss the methods for determining the locations of chemokine receptor binding by intravital microscopy. The facets of this approach are (1) expresison of IL-8 in *Escherichia coli,* (2) covalent linkage of recombinant IL-8 to fluorescent microspheres, and (3) analysis of chemokine receptor expression using intravital microscopy to measure the binding of chemokine-coated fluorescent microspheres to endothelial cells.

Expression of Recombinant Interleukin-8 in *Escherichia coli*

Because significant amounts of IL-8 are required for coupling to fluorescent microspheres, it is necessary to express IL-8 in the laboratory to have sufficient chemokine to make the experiments feasible. This chemokine is expressed in *E. coli* as a recombinant fusion protein to streamline purification of the final protein. A cDNA encoding IL-8 is derived by reverse transcription of polyadenylated RNA from human bone marrow, which, in turn, serves as the source of templates in DNA amplification reactions using primers designed to flank the translational initiation and termination codons. Following confirmation of the nucleotide sequence, a second round of PCR (polymerase chain reaction) is performed using primers to direct the amplification of sequences encoding the mature protein, beginning after the signal peptide and ending with the termination codon. The upstream primer is engineered to introduce a restriction endonuclease cleavage site that will generate an overhang to permit ligation to the pGEXT2 vector in proper translational frame with the gene encoding glutathione *S*-transferase (GST). The nucleotide sequence of the junctions and the segment encoding IL-8 is confirmed.

The expression of the IL-8–GST fusion protein is studied by Coomassie Blue staining of gels following sodium dodecyl sulfate–polyacrylamide gel electrophoresis (SDS–PAGE). Bacteria containing the pGEXT2(IL-8–GST) construct are grown to near saturation density, and the expression of the fusion protein is induced by activation of the *lacZ* operon with isopropylthiogalactoside (IPTG) for 2 hr. Aliquots of bacteria are pelleted in a microcentrifuge and lysed directly in sample buffer containing 2% SDS and 2-mercaptoethanol. Examination of the stained gels reveals the absence of an inducible protein in control *E. coli* and the induction of expression of a protein of 34 kDa, the predicted size of GST plus the IL-8 polypeptide, in bacteria transducing the pGEXT2 (IL-8–GST) construct. Parallel analysis of wild-type GST confirmed a molecular mass of approximately 26 kDa.

The IL-8–GST fusion protein is purified using standard approaches. Analysis of lysates of the induced bacteria reveals that the recombinant protein is in the soluble fraction and not sequestered in inclusion bodies. This finding greatly facilitates the purification process. *Escherichia coli* carrying the pGEXT2(IL-8–GST) construct are fermented on a large scale and induced with IPTG for 2 hr. The cells are pelleted and disrupted by repetitive cycles of freezing and thawing. Following centrifugation, the soluble fractions are applied to a glutathione-Sepharose column, and the fusion protein is isolated by affinity chromatography through the GST moiety. The column is washed extensively, until the absorbance of the buffer flowing through reaches background levels. In pilot experiments, the fusion protein is eluted and characterized by SDS–PAGE with and without digestion with thrombin, which cleaves a site engineered into the carboxyl terminal of the GST polypeptide, thereby permitting liberation of IL-8 from the fusion protein. Characterization by SDS–PAGE confirms that the mature form of IL-8 having the predicted electrophoretic mobility is released from the recombinant GST fusion protein following digestion with trypsin. In the preparative experiments, the IL-8–GST fusion protein is cleaved with thrombin *in situ*, prior to elution from the glutathione-Sepharose with free glutathione, thereby releasing the IL-8 moiety.

Because of its low molecular mass, the purified IL-8 is concentrated by desiccation. The dried pellet is reconstituted in a small volume of water and purified by exclusion chromatography using Sephadex G-25 equilibrated with phosphate-buffered saline (PBS) to remove excess salt. The protein concentration of this material, which yields a single band of the appropriate molecular mass on SDS–PAGE, is determined by Bradford assay. Analysis of the affinity of binding to the cognate receptor using CXCR2 transfectants shows a K_D of less than 10 nM (R. Horuk, personal communication, 1997).

Covalent Linkage of Recombinant Interleukin-8 to Fluorescent Microspheres

Recombinant human IL-8 is coupled to 0.2-μm-diameter carboxylate-modified, red fluorescent latex microspheres (FluoSpheres, Molecular Probes) using a bifunctional, water-soluble carbodiimide. Approximately

250 μg of IL-8 in 250 μl water is added to 250 μl MES buffer, pH 6.0, to which 500 μl of a 1% solution of microspheres is added. Covalent coupling is initiated by the addition of 0.8 mg of EDAC [1-ethyl-3-(3-dimethylamino-propyl)carbodiimide (Molecular Probes)], a water-soluble carbodiimide. Following mixing by sonication, 13 μl of 1.0 *N* NaOH is added. Following mixing by gentle sonication, the reaction mixture is incubated for 2 hr at room temperature. Open sites are blocked by the addition of 15 mg glycine, which is incubated with agitation for 30 min. The coupled microspheres are separated from the reaction mixture by centrifugation at 10,000 rpm for 20 min in a microcentrifuge at 5°. The protein concentration in the supernatant is estimated by Bradford assay using the supernatant from a parallel tube containing microspheres and all of the reactants, including glycine, but lacking IL-8, as a control. Using this approach to calculate the concentration of unbound IL-8, the efficiency of coupling IL-8 to the microspheres is typically greater than 90%. The conjugated microspheres are then washed thrice with PBS and resuspended in 500 μl of PBS. Azide is not added as a preservative because of toxicity that can be encountered when injected into the rat vasculature.

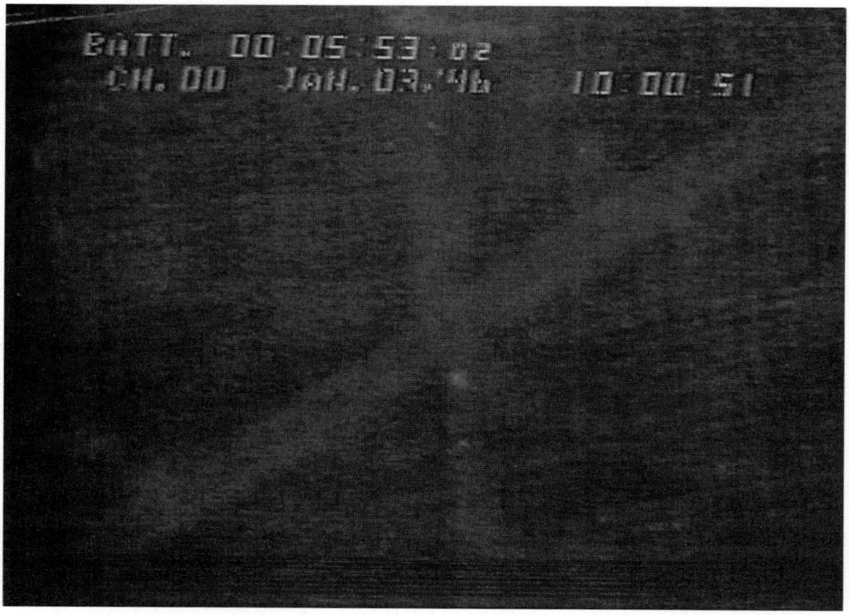

FIG. 1. Fluorescence microscopy image of rat cremaster muscle showing the absence of unlabeled microsphere binding to the microvasculature.

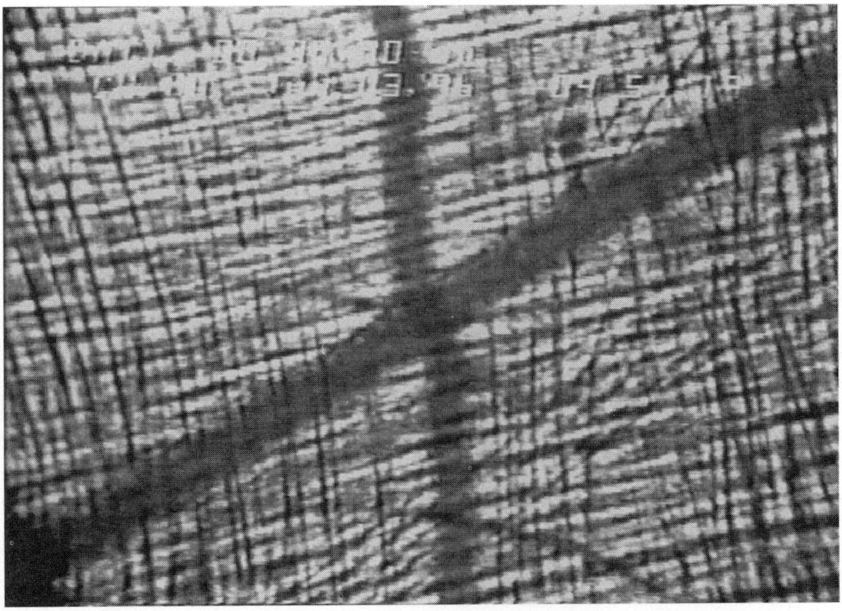

FIG. 2. Transmitted light microscopy image of cremaster muscle corresponding to Fig. 1.

Analysis of Chemokine–FluoSphere Binding by Intravital Microscopy

Before injection, microsphere clumping and aggregation are reduced or eliminated by sonication. A monodispersed suspension of microspheres is prepared by exposing the solution to a sonic dismembrator with a microtip (Model 150, Fisher, Pittsburgh, PA) at a setting of 60% for seven pulses of 30-sec duration under ice with a 30-sec resting period between pulses. The microsphere suspension is then diluted with bacteriostatic saline to a final concentration of 1×10^9 microspheres/ml and given two additional 30-sec pulses under ice with the sonicator. No clumping or aggregation of microspheres is observed by fluorescence.

Intravital Microscopy Model

For these studies, intravital microscopy of the rat cremaster muscle is performed. This muscle is very thin and has an abundant vascular supply. The model has been used extensively for study of vascular physiology.[15]

[15] V. H. Fingar, T. J. Wieman, S. A. Wiehle, and P. B. Cerrito, *Cancer Res.* **52,** 4914 (1992).

Sprague–Dawley rats (100–150 g) are anesthetized with sodium pentobarbital (55 mg/kg, intraperitoneally) and placed on their backs on a temperature-controlled heating pad. Rectal temperature is maintained at 37° and back temperature is monitored with a thermocouple to avoid local overheating of the skin. The right cremaster muscle is prepared for microvascular observations as follows. The muscle is slit on the ventral midline and spread with sutures over a cover glass that is positioned at the top of a temperature-controlled platform and maintained at 35°. The left carotid artery is cannulated for the measurement of mean arterial blood pressure and heart rate, and for the infusion of labeled microspheres.

Microsphere Labeling Protocol

Labeled microspheres are injected intraarterially into animals through a left carotid artery catheter at a dose of 0.5 mg/kg. Microspheres are allowed to circulate for 10 min before fluorescence microscopy.

Brief periods (<5 sec) of epi-illumination using green light (560 nm) are used for fluorescence microscopy of injected red-orange microspheres.

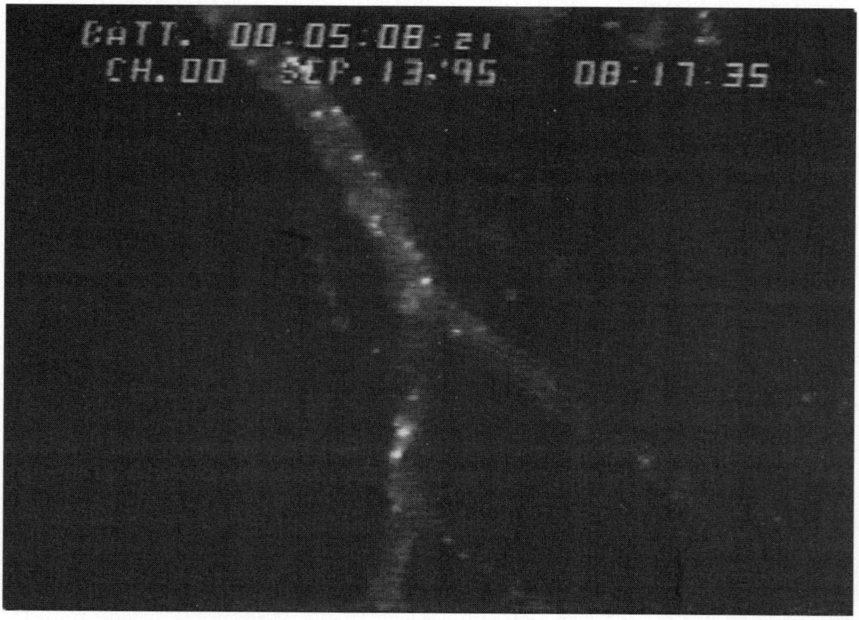

FIG. 3. Fluorescence microscopy image of rat cremaster muscle after injection of IL-8-labeled microspheres. Specific binding of fluorescent microspheres was noted within postcapillary venules with no binding to arterioles or capillaries.

Short exposures are used to minimize photobleaching of the microspheres and to preclude any photochemical effects on the microvasculature. The images are recorded on videotape with a closed-circuit television system. A Cohu SIT (silicon intensifying target) television camera (Cohu Electronics, San Diego, CA) is used to work with very low fluorescent light intensities. A 1-hr equilibration period precedes each experiment.

Control Studies

A number of control experiments must be done to evaluate the extent of nonspecific binding of microspheres to chemokine receptors. Specificity of binding is addressed by injecting a series of control animals with unlabeled or albumin-coated microspheres. Saturation of chemokine receptors by labeled microspheres is determined by injecting different ratios of labeled versus unlabeled microspheres in test animals.

Dependency of binding on blood flow velocities is evaluated by mapping the measurements of microsphere binding in individual vessels with relative blood flow velocities in those vessels. Blocking experiments using free chemokine are done to demonstrate the specificity of binding of chemokine-

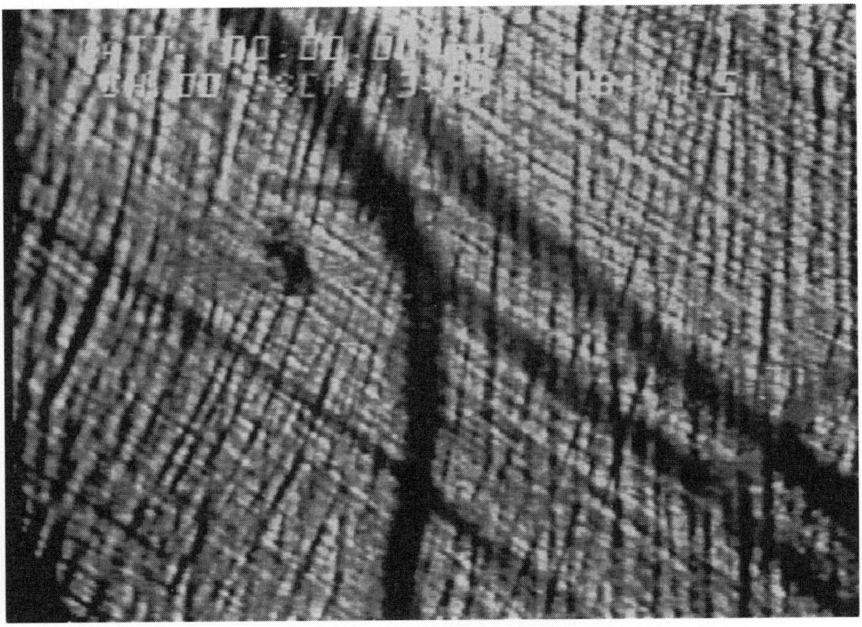

FIG. 4. Transmitted light microscopy image of tissue corresponding to Fig. 3.

coated microspheres. In these experiments, free chemokine is injected 20 min before introduction of microspheres.

Analysis of Videotape

The numbers of adherent microspheres within the mirovasculature are manually counted from videotape images. Image capture hardware and digital image processing have also been useful for counting the number of labeled sites in vessels. Information concerning the histological type of blood vessel under study is done by comparing images from transmitted light microscopy with the fluorescence images. Arterioles are identified by the presence of a smooth muscle wall, rapid blood flow, and blood flow in the direction of large vessels to small vessels. Venules have no muscle wall, exhibit slower blood flow, and have flow in the direction of small vessels to larger vessels.

An example of microsphere binding is shown in Figs. 1–4 (pages 154–157). The fluorescent image showing the absence of nonspecific binding of unlabeled microspheres is shown in Fig. 1. The corresponding light microscopy image outlining the location of arterioles and venules is shown in Fig. 2. Injection of IL-8-coated microspheres showed preferential binding to postcapillary venules with no binding within arterioles or capillaries (Figs. 3 and 4). Experiments where free IL-8 was given before coated microspheres showed no increased binding compared to controls (data not shown), indicating the high specificity of binding.

Conclusion

In summary, intravital microscopic analysis of chemokine-coated fluorescent microspheres is a direct approach to the detection of chemokine receptor expression by endothelial cells. This approach demonstrates that a chemokine receptor that binds IL-8 is expressed in endothelial cells that line venules of approximately 10–15 μm diameter in rodents. These findings directly parallel the immunohistochemical analysis of human tissues, which reveal that DARC is expressed by endothelial cells in a similar distribution. However, these findings could not be verified in rodents by immunohistochemistry because of the lack of immunological reagents for rodent homologs of chemokine receptors. In addition to facilitating the analysis of chemokine receptor expression by endothelial cells in animals, the current approach opens the possibility of observing the dynamics of chemokine receptors in physiological and pathological settings *in vivo*.

Section II

Chemokines in Disease

[12] Neutralization of Interleukin-8 in *in Vivo* Models of Lung and Pleural Injury

By V. COURTNEY BROADDUS and CAROLINE A. HÉBERT

General Issues Concerning Animal Models in Testing Roles of Inflammatory Cytokines

The major test of the function of a cytokine *in vivo* comes from animal studies in which the cytokine can be inhibited. Whereas *in vitro* studies establish the potential functions of a cytokine, for example, whether it is chemotactic, it takes animal studies to determine whether those functions are important *in vivo*. In comparison to *in vitro* studies, the animal model introduces important complexities that will affect the specific roles of the cytokine. Some of these peculiarly *in vivo* features are the presence of other cytokines that may inhibit or act synergistically with the test cytokine or may act in parallel redundant pathways, multiple cell types that may react differently to the cytokine, and the spatial and temporal issues of exactly where and when the cytokine acts *in vivo*. In addition, *in vivo* models establish whether blocking a cytokine could be beneficial to the organism, whether harmful side effects could be expected, and whether blocking the cytokine would be feasible clinically both in space (would the blocking antibody or peptide reach the site of cytokine action) and in time (would inhibition be therapeutic if given minutes to hours after an illness begins as would be necessary in most cases in clinical medicine).

Animal models in which a cytokine is neutralized are more relevant for testing the role of cytokines and the feasibility of clinical therapy than are genetic knockout models in which the cytokine has been absent for the life of the animal. The continuous absence of the cytokine may have led to alterations in animal development in which other cytokines have replaced the function of the lost cytokine. The more useful knockout model for these studies will be the conditional knockout in which specific repressors could be given to the animal to turn off the expression of the cytokine in the adult animal. Although these models have the ultimate ability to eliminate a single cytokine, other animal studies would still be necessary to show feasibility of inhibitory drug delivery in a clinically relevant way and at clinically relevant doses.

Animal models can be judged on several criteria relating to their relevance to human disease.

0076-6879/97 $25.00

1. *How well does the experimental model reflect a human disease?* A close model may be difficult to find, especially if the human disease is complex and poorly understood (see Table I). In the case of acute sepsis, for example, many animal models have proliferated mostly because no one appears to be an exact match of the human condition. Another problem arises when animals do not respond to certain agents as people do. Different species have different responses to gram-negative endotoxin, for example. Another concern is that some human diseases are peculiarly human and cannot be found or mimicked easily in animals, such as idiopathic autoimmune disease. In addition, the model should ideally follow the same time course as the human disease, whether acute or chronic. Conclusions drawn from an animal study lasting hours have limited value when applied to a disease, such as rheumatoid arthritis, lasting years.

2. *What are the outcome variables measured?* The outcome variables measured should first be relevant to the studied action of the cytokine. For example, measuring extravascular infiltration of neutrophils and chemotactic activity of biological fluids is appropriate to determine the chemotactic activity of a cytokine. However, using these variables, the investigators will not be able to state whether the cytokine leads to vascular injury, because the presence of neutrophils does not equate with vascular injury.[1,2] Ideally, multiple variables measuring the same effect should be used. For example, to determine whether neutralization of a cytokine will reduce lung injury, one can measure permeability of the lung vasculature (using vascular tracers, protein concentration in lung edema liquid), function of the lung (e.g., oxygenation), and the consequences of lung injury (extravascular lung water, e.g., edema). The best outcomes to establish the clinical importance of a cytokine are those that have undeniable importance for human disease. In acute experiments, these could include blood pressure or oxygenation and, in chronic experiments, chronic damage of joints or fibrosis of the lung. One undeniably important variable is mortality. Whatever outcome variable is chosen, it is incumbent on the investigators to justify why outcome variables are appropriate, useful, and important measures of the direct or indirect action of the cytokine.

3. *What is the timing of the outcome studied?* Here one wants to judge whether the duration of the experiment was long enough to ensure that the effect was inhibited and not merely postponed. Therefore, if a reduction in neutrophil number is found for 1–2 hr, it would be useful to confirm that this reduction is sustained over a longer period of study.

[1] T. R. Martin, B. P. Pistorese, E. Y. Chi, R. B. Goodman, and M. A. Matthay, *J. Clin. Invest.* **84,** 1609 (1989).

[2] J. P. Wiener-Kronish, K. H. Albertine, and M. A. Matthay, *J. Clin. Invest.* **88,** 864 (1991).

TABLE I
ANIMAL MODELS OF LUNG INJURY

Model	Ref.
Clinically relevant precipitating factors	
Aspiration	
Acid aspiration[a]	*b*
Sepsis	
Escherichia coli intravenous	*c*
Pseudomonas aeruginosa intravenous[a]	*d*
Endotoxin intravenous	*e*
Bacterial/viral pneumonia	
Influenza virus	*f*
Endotoxin intratracheal[a]	*g*
Endotoxin nebulization/aerosolization	*h*
Perforated viscus	
Endotoxin intraperitoneal	*i*
Fecal peritonitis/cecal ligation and puncture	*j*
Mechanical ventilation-related	
High peak inspiratory pressure (PIP)	*k*
Near drowning	
Surfactant depletion by repeated lung lavage	*l*
Fat embolism	
Fat emulsion damage	*m*
Hemorrhagic shock/trauma	
Hemorrhagic shock and aortic clamping	*n*
Pulmonary contusion	*o*
Other precipitating factors	
Immune complex deposition[a]	*p*
Oleic acid intravenous	*q*
Zymosan-activated plasma or intraperitoneal zymosan	*r*
Phospholipase A_2 intratracheal	*s*
Phorbol myristate acetate (PMA) intravenous	*t*
Deoxycholate pancreatitis	*u*

[a] Models in which IL-8 was neutralized.

[b] H. G. Folkesson, M. A. Matthay, C. A. Hébert, and V. C. Broaddus, *J. Clin. Invest.* **96,** 107 (1995); T. Nagase, E. Ohga, E. Sudo, H. Katayama, Y. Uejima, T. Matsuse, and Y. Fukuchi, *Am. J. Respir. Crit. Care Med.* **154,** 504 (1996).

[c] K. M. Schutzer, A. Larsson, B. Risberg, and A. Falk, *Eur. Respir. J.* **7,** 1131 (1994); J. Villar, S. P. Ribeiro, J. B. Mullen, M. Kuliszewski, M. Post, and A. S. Slutsky, *Crit. Care Med.* **22,** 914 (1994); T. Miyata, M. Torisu, H. Toh, and T. Goya, *Circ. Shock* **39,** 44 (1993); D. C. Brockmann, J. H. Stevens, P. O'Hanley, J. Shapiro, C. Walker, F. G. Mihm, J. A. Collins, and T. A. Raffin, *Am. Rev. Respir. Dis.* **134,** 885 (1986); A. K. Sabharwal, S. P. Bajaj, A. Ameri, S. M. Tricomi, T. M. Hyers, T. E. Dahms, F. B. Taylor, Jr., and M. S. Bajaj, *Am. J. Respir. Crit. Care Med.* **151,** 758 (1995); I. C. Dormehl, J. G. Kilian,

(*continued*)

TABLE I (*continued*)

M. Maree, and L. Jacobs, *Am. J. Physiol. Imaging* **5,** 75 (1990); D. H. Hangen, R. J. Bloom, J. H. Stevens, P. O'Hanley, M. Ranchod, J. Collins, and T. A. Raffin, *Am. J. Pathol.* **126,** 396 (1987).

[d] K. Byrne, T. D. Sielaff, B. Michna, P. D. Carey, C. R. Blocher, A. Vasquez, and H. J. Sugerman, *Crit. Care Med.* **18,** 303 (1990); R. A. Mustard, J. Fisher, S. Hayman, A. Matlow, J. B. Mullen, J. Odumeru, M. W. Roomi, B. D. Schouten, and H. T. Swanson, *Lab. Anim. Sci.* **39,** 37 (1989); M. Kadletz, R. J. Dignan, P. G. Mullen, A. C. J. Windsor, H. J. Sugerman, and A. S. Wechsler, *J. Surg. Res.* **60,** 186 (1996).

[e] K. M. Schutzer, A. Larsson, B. Risberg, and A. Falk, *Eur. Respir. J.* **7,** 1131 (1994); A. Castiello, J. F. Paterson, S. A. Shelley, E. M. Haller, and J. U. Balis, *Shock* **2,** 427 (1994); L. H. Pheng, C. Francoeur, and M. Denis, *Inflammation* **19,** 599 (1995); C. R. Turner, M. N. Lackey, M. F. Quinlan, L. W. Schwartz, and E. B. Wheeldon, *Circ. Shock* **34,** 270 (1991); R. K. Simons, R. V. Maier, and E. Y. Chi, *Circ. Shock* **33,** 233 (1991); S. M. Cohn, K. L. Kruithoff, H. R. Rothschild, H. Wang, J. B. Antonsson, and M. P. Fink, *Surg. Forun* **40,** 105 (1989); J. Modig, T. Samuelsson, and R. Sandin, *Acta Chir. Scand.* **153,** 165 (1987); R. F. Jacobs, D. P. Kiel, and R. A. Balk, *Am. Rev. Respir. Dis.* **134,** 745 (1986); J. Modig and T. Borg, *Acta Chir. Scand. Suppl.* **526,** 94 (1985); M. Eriksson, K. Lundkvist, P. Drott, T. Saldeen, and O. Eriksson, *Acta Anaesthesiol. Scand.* **40,** 538 (1996); G. F. Nieman, L. A. Gatto, A. M. Paskanik, B. Yang, R. Fluck, and A. Picone, *Crit. Care Med.* **24,** 1025 (1996); M. P. Fink, B. P. O'Sullivan, M. J. Menconi, P. S. Wollert, H. Wang, M. E. Youssef, and J. H. Fleisch, *Crit. Care Med.* **21,** 1825 (1993); P. E. Forsgren, J. A. Modig, C. M. Dahlback, and B. I. Axelsson, *Acta Chir. Scand.* **156,** 423 (1990); O. Forsgren, S. Jakobson, and J. Modig, *Acta Anaesthesiol. Scand.* **33,** 621 (1989); M. Fuortes, T. W. Pollock, M. J. Holman, M. A. McMillen, B. M. Jaffe, and T. M. Scalea, *I. Trauma* **28,** 1455 (1988); J. Modig and R. Sandin, *Acta Chir. Scand.* **154,** 169 (1988); R. K. Simons, R. V. Maier, and E. S. Lennard, *Arch Surg.* **122,** 197 (1987); J. Modig, T. Samuelsson, and R. Sandin, *Acta Chir. Scand.* **152,** 569 (1986); P. Forsgren, G. Wegenius, and J. Modig, *Acta Anaethesiol. Scand.* **30,** 463 (1986); G. Wegenius, P. Forsgren, and J. Modig, *Acta Radiol.* [*Diagn.*] **27,** 249 (1986); T. Borg, B. Gerdin, and J. Modig, *Acta Anaesthesiol. Scand.* **30,** 47 (1986); P. Banna, M. F. Marcello, R. Murabito, A. Saggio, M. Riggi, C. Cima, and S. Latteri, *Respiration* **47,** 177 (1985); T. Borg and J. Modig, *Acta Chir. Scand.* **151,** 501 (1985); T. Borg, B. Gerdin, and J. Modig, *Acta Anaesthesiol. Scand.* **29,** 831 (1985); T. Borg, A. Alvfors, B. Gerdin, and J. Modig, *Acta Anaesthesiol. Scand.* **29,** 814 (1985); J. E. Rinaldo, J. H. Dauber, J. Christman, and R. M. Rogers, *Am. Rev. Respir. Dis.* **130,** 1065 (1984); N. C. Olson, T. T. Brown, Jr., and D. L. Anderson, *J. Appl. Physiol.* **58,** 274 (1985); F. Sakamaki, A. Ishizaka, T. Urano, K. Sayama, H. Nakamura, T. Terashima, Y. Waki, S. Tasaka, N. Hasegawa, K. Sato, N. Nakagawa, T. Obata, and M. Kanazawa, *Am. J. Respir. Crit. Care Med.* 153 (1996); M. J. Murray, M. Kumar, T. J. Gregory, P. L. Banks, H. D. Tazelaar, and S. J. DeMichele, *Am. J. Physiol. Heart Circ. Physiol.* (1995).

TABLE I (*continued*)

[f] B. Lachmann, *Eur. Respir. J. Suppl.* **3,** 98s (1989).
[g] R. C. Hoch, I. U. Schraufstätter, and C. G. Cochrane; T. P. Shanley, D. Schrier, V. Kapur, M. Kehoe, J. M. Musser, and P. A. Ward, *Infect. Immun.* **64,** 870 (1996); K. Tashiro, K. Yamada, W. Z. Li, Y. Matsumoto, and T. Kobayashi, *Crit. Care Med.* **24,** 488 (1996).
[h] C. R. Turner, M. F. Quinlan, L. W. Schwartz, and E. B. Wheeldon, *Circ. Shock* **32,** 231 (1990).
[i] A. Castiello, J. F. Paterson, S. A. Shelley, E. M. Haller, and J. U. Balis, *Shock* **2,** 427 (1994).
[j] A. R. Webb, R. F. Moss, D. Tighe, M. G. Mythen, N. Al-Saady, A. E. Joseph, and E. D. Bennett, *Intensive Care Med.* **18,** 348 (1992); M. Abe, T. Goya, T. Mitsuyama, M. Morisu, and T. Furukawa, *Prostaglandins Leukotrienes Essential Fatty Acids* **54,** 123 (1996).
[k] K. G. Hickling, *Intensive Care Med.* **16,** 219 (1990).
[l] B. Lachmann, *Eur. Respir. J. Suppl.* **3,** 98s (1989); B. Robertson, *Acta Anaesthesiol. Scand. Suppl.* **95,** 22 (1991); I. M. Cheifetz, D. M. Craig, F. H. Kern, D. R. Black, N. D. Hillman, W. J. Greeley, R. M. Ungerleider, P. K. Smith, and J. N. Meliones, *Crit. Care Med.* **24,** 1554 (1996); D. Hafner, R. Beume, U. Kilian, G. Krasznai, and B. Lachmann, *Br. J. Pharmacol.* **115,** 451 (1995); D. Hafner, P. G. Germann, and D. Hauschke, *Pulm. Pharmacol.* **7,** 319 (1994). M. Lichtwarck-Aschoff, J. B. Nielson, U. H. Sjostrand, and E. L. Edgren, *Intensive Care Med.* **18,** 339 (1992); F. J. Alvarez, L. F. Alfonso, E. Gastiasoro, J. Lopez-Heredia, A. Arnaiz, and A. Valls-i-Soler, *Acta Anaesthesiol. Scand.* **39,** 970 (1995); S. L. Sood, V. Balaraman, K. C. Finn, B. Britton, C. F. Uyehara, and D. Easa, *Am. J. Respir. Crit. Care Med.* **153,** 820 (1996).
[m] A. Jolin, R. Myklebust, R. Olsen, and L. J. Bjertnaes, *Acta Anaesthesiol. Scand.* **38,** 75 (1994).
[n] T. F. Lindsay, P. M. Walker, and A. Romaschin, *J. Vasc. Surg.* **22,** 1 (1995).
[o] S. M. Cohn and P. M. Zieg, *J. Trauma* **41,** 565 (1996).
[p] B. Lachmann, *Eur. Respir. J. Suppl.* **3,** 98s (1989); M. S. Mulligan, M. L. Jones, M. A. Bolanowski, M. P. Baganoff, C. L. Deppeler, D. M. Meyers, U. S. Ryan, and P. A. Ward, *J. Immunol.* **150,** 5585 (1993).
[q] M. Fuortes, T. W. Pollock, M. J. Holman, M. A. McMillen, B. M. Jaffe, and T. M. Scalea, *J. Trauma* **28,** 1455 (1988); M. J. Murray, M. Kumar, T. J. Gregory, P. L. Banks, H. D. Tazelaar, and S. J. DeMichele, *Am. J. Physiol. Heart Circ. Physiol.* (1995); M. C. Papo, P. R. Paczan, B. P. Fuhrman, D. M. Steinhorn, L. J. Hernan, C. L. Leach, B. A. Holm, J. E. Fisher, and B. A. Kahn, *Crit. Care Med.* **24,** 466 (1996); D. P. Schuster, *Am. J. Respir. Crit. Care Med.* **149,** 245 (1994); M. Leeman, *Intensive Care Med.* **17,** 254 (1991); C. Metz and W. J. Sibbald, *Chest* **100,** 1110 (1991); H. P. Grotjohnan, R. M. van der Heijde, J. R. Jansen, C. A. Wagenvoort, and A. Versprille, *Intensive Care Med.* **22,** 336 (1996); S. D. Thies, R. S. Corbin, C. D. Goff, O. A. Binns, S. A. Buchanan, K. S. Shockey, H. J. Frierson, J. S. Young, C. G. Tribble, and I. L. Kron, *Ann. Thoracic Surg.* **61,** 1453 (1996); D. P. Schuster, *Pediatr. Pulmonol.*

(*continued*)

TABLE I (*continued*)

Suppl. **11,** 104 (1995); A. Nahum, R. S. Shapiro, S. A. Ravenscraft, A. B. Adams, and J. J. Marini, *Am. J. Respir. Crit. Care Med.* **152,** 489 (1995); N. S. Shah, D. K. Nakayama, T. D. Jacob, I. Nishio, T. Imai, T. R. Billiar, R. Exler, S. A. Yousem, E. K. Motoyama, and A. B. Peitzman, *Arch. Surg.* **129,** 158 (1994); L. Tachmes, H. Adler, T. T. Woloszyn, M. S. Coons, P. Damiani, C. P. Marini, and J. Horovitz, *Am. Surg.* **57,** 171 (1991); S. T. Sum-Ping, T. Symreng, P. Jebson, and G. D. Kamal, *Crit. Care Med.* **19,** 405 (1991); B. Zwissler, H. Forst, K. Ishii, and K. Messmer, *Res. Exp. Med.* **189,** 427 (1989); B. P. Griffith, R. G. Carroll, R. L. Hardesty, R. L. Peel, and H. S. Borovetz, *J. Appl. Physiol.* **47,** 706 (1979); S. Idell, K. K. James, and J. J. Coalson, *Crit. Care Med.* **20,** 1431 (1992); R. B. Hirschl, R. Tooley, A. Parent, K. Johnson, and R. H. Bartlett, *Crit. Care Med.* **24,** 1001 (1996); S. D. Thies, R. S. Corbin, C. D. Goff, O. A. R. Binns, S. A. Buchanan, K. S. Shockey, H. F. Frierson, Jr, J. S. Young, C. G. Tribble, and I. L. Kron, *Ann. Thoracic Surg.* **61,** 1453 (1996); H. P. Grotjohan, R. M. J. L. Van der Heijde, J. R. C. Jansen, C. A. Wagenvoort, and A. Versprille, *Intensive Care Med.* **22,** 336 (1996); S. Syrbu, R. S. Thrall, and H. M. Smilowitz, *Exp. Lung Res.* **22,** 33 (1996); H. Moriuchi, I. Arai, and T. Yuizono, *Intensive Care Med.* **21,** 1003 (1995).

[r] J. R. Shayevitz, J. L. Rodriguez, L. Gilligan, K. J. Johnson, and A. R. Tait, *Shock* **4,** 61 (1995).

[s] J. Villar, J. D. Edelson, M. Post, J. B. Mullen, and A. S. Slutsky, *Am. Rev. Respir. Dis.* **147,** 177 (1993).

[t] M. Miniati, F. Cocci, S. Monti, E. Filippi, R. Sarnelli, M. Ferdeghini, V. Gattai, and M. Pistolesi, *Eur. Respir. J.* **9,** 758 (1996). R. C. St. John, L. A. Mizer, S. E. Weisbrode, and P. M. Dorinsky, *Am. Rev. Respir. Dis.* **144,** 1171 (1991); I. U. Schraufstatter, S. D. Revak, and C. G. Cochrane, *J. Clin. Invest.* **73,** 1175 (1984); J. B. Waugh, T. B. Op't Holt, J. E. Gadek, and T. L. Clanton, *J. Crit. Care* **11,** 129 (1996); A. Mikulaschek, S. Z. Trooskin, J. Winfield, A. Norin, D. A. Spain, and C. J. Carrico, *J. Trauma Injury, Infect. Crit. Care* **39,** 59 (1995).

[u] H. Murakami, A. Nakao, W. Kishimoto, M. Nakano, and H. Takagi, *Surgery* **118,** 547 (1995); I. A. Goulbourne and G. C. Davies, *J. Surg. Res* **41,** 600 (1986); N. Tanaka, A. Murata, K. I. Uda, H. Toda, T. Kato, H. Hayashida, N. Matsuura, and T. Mori, *Crit. Care Med.* **23,** 901 (1995); R. Milan, Jr., P. M. Pereira, M. Dolhnikoff, P. H. N. Saldiva, and M. A. Martins, *Crit. Care Med.* **23,** 1882 (1995).

4. *For clinically relevant studies, were the route of administration, dose of inhibitor, and timing of administration clinically feasible?* The route of administration that is most practical in clinical medicine is intravenous (or peroral). Therefore, it is most useful if a model can achieve successful cytokine neutralization via an intravenous route. Direct administration, such as into the tissue space, may be necessary to establish the necessary

inhibitor concentrations to neutralize a cytokine, but it would be generally impractical for clinical use. From a pharmacoeconomic point of view, doses greater than 10 mg/kg may not be realistic for treatment when using an expensive recombinant protein. The more important issue, especially for acute studies, is the timing of administration. Neutralization studies in animals are more likely to be successful if the inhibitor is given before the injury is introduced. Probably because of extensive cytokine networks and the amplification of cytokine production, inhibition can be ineffective once the "horse is out of the barn." For most clinical applications, however, a prophylactic approach to treatment is not possible. It is not possible to predict, for example, which patients will develop sepsis or acute lung injury or trauma. Therefore, a therapeutic approach in animal models, one in which the neutralization of the cytokine is tested for effectiveness when given after the illness begins, would be most useful.

5. *Was the cytokine shown to be neutralized?* In the complex *in vivo* environment, it is important to consider whether the inhibitor reached the site of inflammation in sufficient concentration to neutralize the target cytokine. Issues of specificity and affinity of the inhibitor for its target cytokine can be established in *in vitro* studies, but the effectiveness of the *in vivo* delivery is unknown unless directly tested. Some tests of effectiveness we have used are showing either a reduction in chemotactic activity that cannot be further reduced by addition of more inhibitor or a reduction in measurable free cytokine (i.e., unbound by antibody, see below).

We have tested an anti-interleukin-8 (IL-8) monoclonal antibody (MAb) in two anesthetized rabbit models to show an important role for the cytokine IL-8 in inflammation and injury. The pleurisy model, in which an indwelling pleural catheter enabled intrapleural instillation of endotoxin and collection of extravascular liquids and cells, was used in acute 6- to 8-hr studies to demonstrate *in vivo* the unambiguous role of IL-8 in extravascular neutrophil recruitment. The acid-instillation acute lung injury model, in which acid is instilled into the lung, was used in 6- to 24-hr studies to show that IL-8 played an important role in neutrophil-dependent lung injury. However, before going into the details of these protocols we believe it is important to share general comments on the preparation of antibody reagents for efficacy testing *in vivo* and, in particular, methods for testing the quality of the reagent with respect to activity, potency, specificity, and purity.

Preparation of Monoclonal Antibodies for *in Vivo* Experiments

Activity. The anti-IL-8 monoclonal antibody (ARIL8.2) used in our experiments has been developed by immunizing mice with recombinant rabbit IL-8. Procedures for generating neutralizing MAbs to chemokines are described in Chapter 2 of this book. The MAb is selected by virtue of its ability to recognize rabbit IL-8, inhibit binding of rabbit ^{125}I-labeled IL-8 to its receptors, block rabbit IL-8-induced signal transduction in neutrophils (calcium flux assay), and inhibit rabbit IL-8-induced chemotactic activity for rabbit neutrophils.

Potency. The MAb we use has a high affinity for rabbit IL-8 ($K_d = 0.4$ nM). Before *in vivo* administration the MAb preparation is assayed for concentration using amino acid analysis.

Specificity. The MAb used is shown to cross-react with human IL-8 but not with closely related cytokines [human melanoma growth stimulating activity (hMGSA), platelet factor 4, β-thromboglobulin], other human cytokines [IL-1β, tumor necrosis factor (TNF)], or other chemotactic factors [formylmethionylleucylphenylalanine (fMLP) and C5a].

Purity. Before *in vivo* administration the MAbs are tested for endotoxin and checked for purity on a silver-stained sodium dodecyl sulfate (SDS) gel. It is important to verify that the compounds used in models of inflammation are devoid of endotoxin, as very small amounts of endotoxin can induce inflammation and could mask beneficial effects of the compound. The endotoxin concentration of all MAbs used *in vivo* is checked in the *Limulus* amebocyte lysate assay (Whittaker Bioproducts, Walkersville, MD) and determined to be <0.1 endotoxin unit/mg.

Endotoxin Pleurisy Model

In the pleurisy rabbit model, our aim is to determine whether IL-8, a novel cytokine shown in many *in vitro* assays to have neutrophil chemotactic activity, is chemotactic *in vivo.* Despite the potency *in vitro,* there are many doubts that blockade of a single cytokine will be effective at blocking extravascular neutrophil influx. In this rabbit model, we use a monoclonal antibody that is species-specific (developed against rabbit IL-8) and shown to be highly potent by virtue of its ability to block the interaction of rabbit IL-8 and the rabbit IL-8 receptor and to block the activation of that receptor (see above). We chose gram-negative endotoxin as an extremely potent inflammatory stimulus for the rabbit. The pleurisy model is developed with an indwelling catheter in the pleural space (Fig. 1) to introduce the inflammatory stimulus and to measure pleural liquid hourly. Our end points are mainly that of the timing and number of neutrophils entering the pleural space. The antibody is given in excess by both the intravenous and direct intrapleural route to ensure adequate concentrations. We confirm, by Western blot analysis, that IL-8 is generated. We also confirm that endogenous IL-8 is bound by the anti-rabbit IL-8 antibody ARIL8.2 (which we know

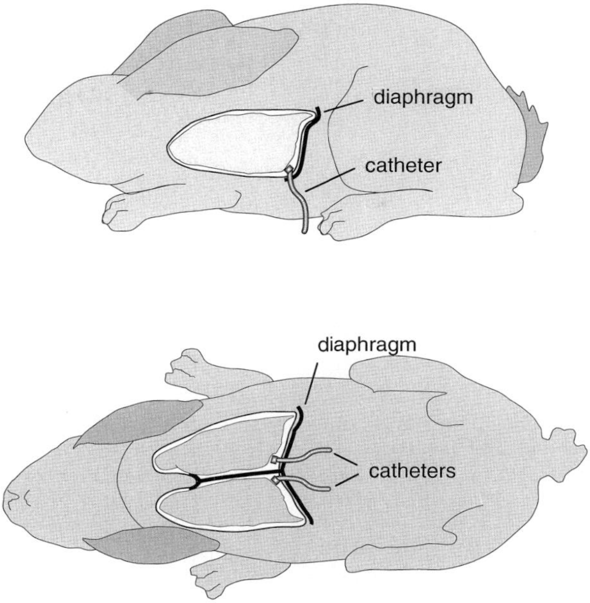

FIG. 1. Pleurisy model. Schematic diagram shows the minimally invasive bilateral pleural catheters in an anesthetized rabbit. The flared tip of the catheter lies in the pleural space where it provides continuous access to the pleural space without causing injury for at least 24 hr. [Reproduced with permission from *Am. J. Physiol.* **267,** L137 (1994).]

is sufficient for neutralization of activity) by using an enzyme-linked immunosorbent assay (ELISA) for free, unbound IL-8 in which ARIL8.2 is used as the capture antibody. In this model, anti-IL-8 blocks 77% of endotoxin-induced neutrophil influx and demonstrates a clear and important role for IL-8 in inflammation *in vivo* (Fig. 2).[3]

Placement of Pleural Catheters. Indwelling pleural catheters are placed as previously described (Fig. 1).[4,5] Briefly, the rabbits are anesthetized with halothane and ventilated via tracheotomy with 60% oxygen/1% halothane at an inspiratory pressure of 18 cm H_2O. The jugular vein and carotid artery are catheterized. For placement of the pleural catheters, the upper abdomen is opened and the ventral diaphragmatic muscle is dissected 1 cm lateral to the midline on each side to expose the diaphragmatic pleura and underly-

[3] V. C. Broaddus, A. M. Boylan, J. M. Hoeffel, K. J. Kim, M. Sadick, A. Chuntharapai, and C. A. Hébert, *J. Immunol.* **152,** 2960 (1994).

[4] A. M. Boylan, C. Rüegg, K. J. Kim, C. A. Hébert, J. M. Hoeffel, R. Pytela, D. Sheppard, I. M. Goldstein, and V. C. Broaddus, *J. Clin. Invest.* **89,** 1257 (1992).

[5] V. C. Broaddus and M. Araya, *J. Appl. Physiol.* **72,** 851 (1992).

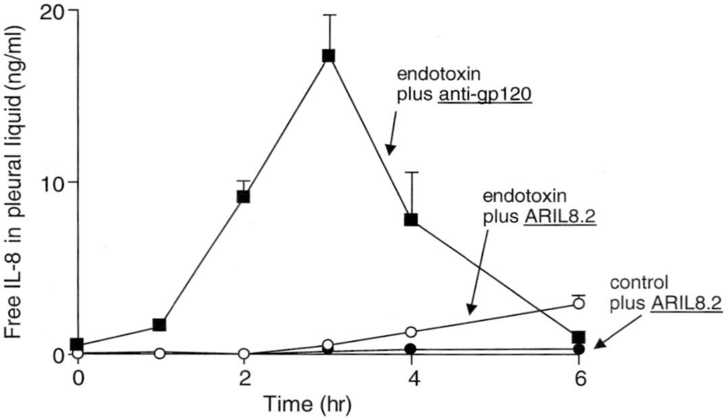

FIG. 2. Concentration of free IL-8 (not bound to neutralizing MAb) in the pleural liquid of rabbits given intrapleural endotoxin together with either the anti-rabbit IL-8 MAb (ARIl8.2) or an irrelevant isotype control MAb (anti-HIV gp120) [1 mg/kg intravenously (i.v.) plus 20 μg/ml intrapleurally]. In the rabbits given the irrelevant MAb, IL-8 concentrations peak at 3 hr. At this time, in the rabbits given ARIL8.2, there is almost no free IL-8 in the pleural liquid. At 6 hr, in the rabbits given ARIL8.2, a small amount of free IL-8 can be detected. Data are means ±SE for three rabbit experiments in each group. [Reproduced from V. C. Broaddus *et al., J. Immunol.,* **152,** 2960 (1994). Copyright 1994, The American Association of Immunologists.]

ing lung. Through a small hole in the pleura, the flared tip of the catheter (PE-160, Clay-Adams, Parsippany, NJ) is inserted into the pleural space, secured in place by circumferential ties, and pulled back flush with the diaphragm (Fig. 1). After the two catheters are placed and the abdomen closed, the rabbit is placed prone with the catheters projecting through a hole in a supporting board. Prepared in this way, the catheter is noninjurious to the pleural space for up to 24 hr and provides continuous, leakproof access to the pleural liquid.[5]

Preparation of Instillates. Escherichia coli endotoxin (serotype O127:B8; Difco, Detroit, MI) instillates are prepared at a final concentration of 20 ng/ml in RPMI 1640 cell culture medium supplemented with calcium gluconate (3 mEq/liter) and human serum albumin (0.1 g/dl). Control instillates are prepared identically, but without endotoxin. Before intrapleural injection, the instillates are warmed to 38° and corrected to pH 7.4.

Delivery of Monoclonal Antibodies in Vivo. To maximize tissue penetration, the MAb is given by two routes: an intravenous dose is given 1 hr before endotoxin instillation, and an intrapleural dose is given at the time of endotoxin instillation. The initial doses of MAb (intravenous, 10 mg/

kg; intrapleural, 200 μg/ml) are chosen to achieve maximal neutralization of generated IL-8.

Experimental Protocol. At 1 hr before endotoxin instillation, MAb or saline is administered intravenously. During the following 1-hr baseline period, arterial and venous blood pressures are recorded every 15 min, and arterial blood is obtained for measuring blood gases. Then, the instillates and MAb are injected into each pleural space via the pleural catheters. Pleural liquid samples (0.8 ml) are withdrawn through the catheters at 0, 1, 2, 3, 4, and 6 hr. Vascular pressures are recorded every 15 min, and blood samples are obtained hourly. After 6 hr, the rabbit is exsanguinated, and the chest is opened. A drop of liquid is cultured aerobically (typically, none of the cultures yield growth after 48 hr of incubation). All available pleural liquid is aspirated and the space lavaged twice with one 10-ml aliquot of sterile saline.

Confirmation of Endotoxin-Induced Generation of Interleukin-8. Western blotting is performed to confirm that rabbit IL-8 has indeed been generated in the anti-rabbit IL-8 MAb group in response to endotoxin. Recombinant rabbit IL-8 (30 μl; 0.2 mg/ml) and pleural liquid (30 μl) from rabbits that received either no endotoxin or intrapleural endotoxin together with anti-HIV (human immunodeficiency virus) coat protein gp120 MAb or anti-rabbit IL-8 MAb are boiled for 2 min in the presence of reducing buffer, separated on a 12.5% SDS–polyacrylamide gel, and then transferred electrophoretically to a polyvinylidene difluoride membrane (Immobilon PVDF, Millipore, Bedford, MA). Following blockade of nonspecific binding sites with 5% dry milk for 1 hr at 4° and three washes with 0.05% Tween 20 in Tris-buffered saline, the blot is incubated for 1 hr at room temperature with an anti-rabbit IL-8 MAb (8C8) generated as described above and selected for its high affinity for denatured rabbit IL-8, as well as for its lack of cross-reactivity with other cytokines from the IL-8 family. After three washes, the blot is incubated first with horseradish peroxidase-conjugated goat anti-mouse immunoglobulin (IgC) in 0.1% Tween 20 in Tris-buffered saline for 1 hr and then with chemiluminescent immunoassay signal reagent (Amersham, Arlington Heights, IL) prior to exposure to autoradiography film.

Chemokine and Cytokine Concentrations. Total chemokine concentration can be measured using commercially available ELISA. Free chemokine (unbound to MAb) concentrations can be measured by an ELISA specifically designed to quantify the amount of chemokine, if any, that has not been neutralized by treatment with the neutralizing antibody in a given fluid [bronchoalveolar lavage (BAL), pleural liquid, plasma, etc.] (Fig. 2). We have developed an ELISA to measure free rabbit IL-8. In this assay, the MAb used to coat the ELISA plate and capture the IL-8 is the same as that used for *in vivo* treatment. Molecules of IL-8 already bound by the MAb *in vivo* are unable to bind the capture MAb because the epitope that could be recognized by that particular clone is made unavailable by the bound neutralizing MAb. All molecules of IL-8 that have been neutralized *in vivo* do not bind the primary, capture MAb and are therefore washed away in the first washing step of the ELISA.

Microtiter plates (96-well; Immuno Plate MaxiSorp, Nunc, Naperville, IL) are coated with neutralizing anti-rabbit IL-8 MAb (6G4; 10 μg/ml), and then blotted dry and blocked with PBS containing 0.5% bovine serum albumin (BSA; Sigma, St. Louis, MO) for 1 hr. Standards of rabbit IL-8 mixed with anti-rabbit IL-8 MAb (MAb:IL-8, 0.5:1 to 10:1, v/v) and pleural liquid samples are diluted (1:10 to 1:80) and added to the wells for a 1-hr incubation. After washing, the secondary antibody (8C1.1.6), conjugated to long-arm biotin (biotin-S-NHS, Research Organics, Cleveland, OH), is added for 2 hr followed by horseradish peroxidase-conjugated streptavidin (1:5000; Zymed Laboratories, South San Francisco, CA) for 1 hr. Tetramethylbenzidine (TMB; two-component system, Kirkegaard & Perry, Gaithersburg, MD) is then added and color allowed to develop at room temperature for 10 min. Optical density is measured with an ELISA plate reader at a wavelength of 450 nm. Sample values are determined by interpolation using a four-parameter program (Genentech, South San Francisco, CA) from a standard curve generated over a range of 2000 to 31 pg/ml.

For the nonneutralized samples, the results for the dilutions are averaged over their linear range. However, when detecting antigen in the presence of a soluble antibody identical to the capture antibody, the ELISA can be nonlinear at increasing dilutions, perhaps because antigen dissociates from the soluble antibody and is subsequently bound by the capture antibody. Therefore, for the neutralized samples, we choose the lowest dilution (1:10, v/v) for quantifying free IL-8, knowing that this may still be an overestimate of the free IL-8 present. When testing the ELISA with standards of rabbit IL-8 mixed with anti-rabbit IL-8 MAb, we find that, as anti-rabbit IL-8 MAb concentrations increase, the free IL-8 detected decreases until, at a molar ratio of 5:1 and higher (MAb:IL-8), no IL-8 can be detected.

Total White Blood Cells and Neutrophils. Total cell counts are calculated as cells per milliliter of pleural liquid and multiplied by the sample volume. The differential cell count can be analyzed on cytospun cells stained with eosin and hematoxylin. An interesting feature of this model is the presence of a noninjurious indwelling catheter which allows for repeated sampling of the pleural milieu during the course of the experiment and quantitation of the neutrophil influx at various times. Furthermore, the ability to sample

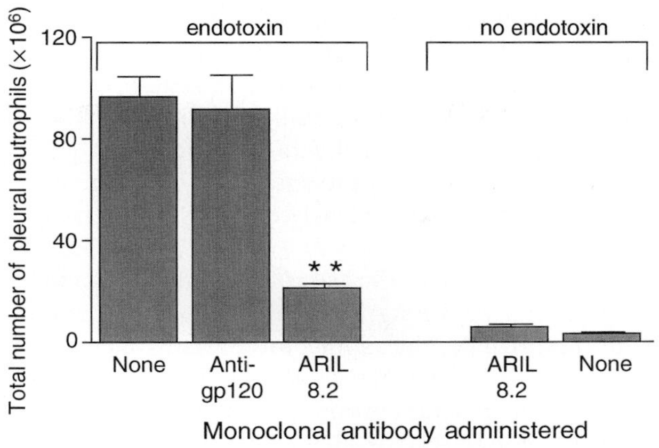

FIG. 3. Total number of neutrophils in the rabbit pleural space 6 hr after the instillation of intrapleural endotoxin (200 ng bilaterally) or control instillates with either MAb to rabbit IL-8 (ARIL8.2), irrelevant isotype control MAb (anti-HIV gp120), or no MAb. In response to endotoxin, there was a large neutrophil influx, which was unaffected by administration of irrelevant anti-gp 120 MAb (1 mg/kg i.v. plus 20 μg/ml intrapleurally). However, there was a 77% decrease in endotoxin-mediated neutrophil influx with the administration of anti-rabbit IL-8 MAb (ARIL8.2; 1 mg/kg i.v. plus 20 μg/ml intrapleurally). Data are means ±SE for three rabbit experiments in each group. Asterisks (**) signify a significant difference from the endotoxin alone and endotoxin plus gp120 MAb, with $p < 0.0001$. [Reproduced from V. C. Broaddus *et al.*, *J. Immunol.*, **152,** 2960 (1994). Copyright 1994, The American Association of Immunologists.]

the entire pleural space allows quantitation of the total number of neutrophils (Fig. 3).

Acid-Aspiration Lung Injury Model

In the acid-aspiration rabbit model, our aims are more ambitious: to show that IL-8 is responsible not only for neutrophil chemotaxis but also for neutrophil-dependent lung injury, that IL-8 can be neutralized by clinically relevant doses and intravenous administration, and that neutralization of IL-8 1 hr after acid injury can be protective.[6] We choose the acid aspiration model because it is clinically relevant; acid aspiration in humans is the second most common cause of the adult respiratory distress syndrome. Our measured outcomes include measures of lung vascular permeability, lung

[6] H. G. Folkesson, M. A. Matthay, C. A. Hébert, and V. C. Broaddus, *J. Clin. Invest.* **96,** 107 (1995).

edema (extravascular lung water), and oxygenation. We also prolong the period of observation to 24 hr which enables us to include the end point of mortality in the experiments; untreated rabbits given the acid die at approximately 12 hr as a result of unremitting hypoxemia, whereas treated rabbits survive for the duration of the 24-hr experiment. Finally, we alter the time of delivery of the anti-IL-8 to show that blockage is effective even when given 1 hr after the acid, a clearly exciting prospect for its eventual use in patients.

Anesthesia and Placement of Catheters. Male New Zealand White rabbits (n = 34, weighing 2.5 to 3.5 kg; Nitabell, Hayward, CA) are surgically prepared as follows. Briefly, the rabbits are initially anesthetized using 4% halothane in 100% O_2; the anesthesia is then maintained with 0.8% halothane in 100% O_2. Pancuronium bromide (0.3 mg/kg/hr; Pavulon, Organon Diagnostics, West Orange, NJ) is given intravenously for neuromuscular blockade. A 22-gauge Angiocath (Deseret Medical, Becton Dickinson, Sandy, UT) is inserted in the marginal ear vein for administering fluids and drugs. A PE-90 catheter (Clay Adams) is inserted in the right carotid artery to monitor systemic blood pressure and obtain blood samples. A 4.0-mm inside diameter endotracheal tube is inserted through a tracheotomy. The rabbits are maintained in a prone position and ventilated as described above.

Preparation of Instillates and Experimental Protocol. A solution of 100 mOsm/kg of NaCl (1/3 normal saline) is prepared with isotonic 0.9% saline and distilled water. The osmolality and pH are chosen to match those of gastric aspirates. Then, HCl is added to the solution and titrated to pH 1.5. In the negative control group, 1/3 normal saline is used as the instillate. Evans blue dye (1 mg; Aldrich, Milwaukee, WI) is added to all instillates to confirm at postmortem examination that the fluid is distributed equally to both lungs. In all experiments, after the surgical preparation, a 1-hr baseline of stable heart rate, systemic blood pressure, and arterial blood gases is required before intratracheal instillation.

For the instillation, a tubing (5 Fr., Accumark Premarked Feeding Catheter; Concord/Portex, Keene, NH) is gently passed through the tracheal tube until it is placed about 1 cm above the carina. Then HCl or 1/3 normal saline (4 ml/kg) is instilled into both lungs over a 3-min period. After the instillation is complete, the tubing is withdrawn. Blood samples are collected as described above. At the end of the experiment (t = 6 or 24 hr), the abdomen is opened, the rabbit exsanguinated by transection of the abdominal aorta, and the lungs removed through a median sternotomy. Lungs, alveolar fluid, and BAL samples are collected as described (*vide supra* and *vide infra*).

Delivery of Monoclonal Antibodies in Vivo. The experimental groups are as follows. The positive control group receives intratracheal HCl at $t = 0$ and intravenous saline (2 ml/kg; or, better, isotype control MAb at 2 mg/kg) at $t = -5$ min or $t = 1$ hr. The negative control group receives intratracheal 1/3 normal saline at $t = 0$ and intravenous saline (2 ml/kg) at $t = -5$ min or $t = 1$ hr. The pretreatment group receives intratracheal HCl at $t = 0$ and intravenous anti-rabbit IL-8 MAb (2 mg/kg) at $t = -5$ min. The treatment group receives intratracheal HCl at $t = 0$ and intravenous anti-rabbit IL-8 MAb (2 mg/kg) at $t = 1$ hr.

Mechanical Ventilation and Blood Gases Analysis. The rabbits are maintained in a prone position during the experiments and ventilated with a constant volume piston pump (Harvard Apparatus, Natick, NJ) with an inspired oxygen fraction of 1.0 and with a peak airway pressure of 15–18 cm H_2O during the baseline period, and supplemented with positive end-respiratory pressure of 4 cm H_2O. During the baseline period, the respiratory rate is adjusted to maintain arterial P_{CO_2} between 35 and 40 mm Hg. Thereafter, the ventilator settings are kept constant throughout the experiment.

The arterial blood can be sampled every 30 to 60 min to measure pH and blood gases. The blood gas readout is used to calculate the alveolar–arterial oxygen tension difference (expressed in mm Hg, Fig. 4), the Pa_{O_2}/PA_{O_2} ratio (pressure of arterial oxygen over pressure of alveolar oxygen), or the Pa_{O_2}/F_{iO_2} ratio (pressure of arterial oxygen over fraction of inspired oxygen). These last two ratios are commonly used in the intensive care unit to monitor the lung function of patients.

Permeability of Lung Vasculature. Vascular permeability is an important physiological end point in models of acute lung injury because it measures the extent of the vascular injury directly. This is quantified by measuring how much plasma leaks into the alveolar compartment. One can measure protein content in the bronchoalveolar lavage, but a more accurate way is to measure how much radiolabeled albumin or human albumin, previously injected intravenously, has accumulated into the lung.

For measurement of lung endothelial permeability to protein, the clearance of the vascular lung protein, ^{131}I-labeled albumin, across the endothelium into extravascular compartments of the lung is measured. In our experiments, 3 μCi of human ^{131}I-labeled serum albumin (Frosst Laboratories, Montreal, Canada) is injected intravenously 15 min into the baseline period. The total extravascular ^{131}I-labeled albumin accumulation in the lung is calculated by taking total lung ^{131}I-labeled albumin (in lung homogenate and in the BAL) and subtracting the vascular space ^{131}I-labeled albumin. The ^{131}I-labeled albumin in the vascular space is calculated by multi-

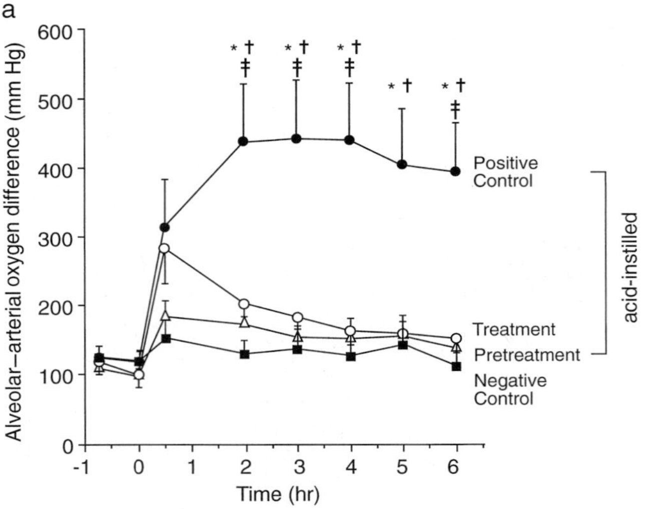

FIG. 4. Alveolar–arterial oxygen tension difference in the positive control, pretreatment, treatment, and negative control groups over 6 (a) and 24 hr (b). In the 6-hr experiments, the alveolar–arterial oxygen tension difference in the pretreatment and treatment groups was significantly less than in the positive control group from 2 hr onward and was no different from that in the negative control group. In the 24-hr experiments, the alveolar–arterial oxygen tension difference was significantly less in the treatment group than in the positive control group by 2 hr and remained low for 24 hr. All the rabbits in the positive control group died at 12–14 hr. Data are means ±SEM. Statistically significant differences are as follows: *, $p < 0.05$ versus the negative control group (a) or the treatment group (b); †, $p < 0.05$ versus the pretreatment group (a); ‡, $p < 0.05$ versus the treatment group (a). (Reproduced from the *Journal of Clinical Investigation,* 1995, Vol. 96, pp. 107–116, by copyright permission of The American Society for Clinical Investigation.)

plying the counts in the final plasma samples by the calculated plasma volume in the lungs.[2,7] The extravascular plasma accumulation of ^{131}I-labeled albumin in the lung is expressed as plasma equivalents (Fig. 5) or the milliliters of plasma that would account for the radioactivity of the lung. A trichloroacetic acid (TCA) precipitation of instillates and selected samples from each experiment confirms that the vascular tracer ^{131}I remains bound to albumin. Note that the same method can be used with human albumin and an ELISA specific for human albumin rather than ^{131}I-labeled albumin.[8]

[7] Y. Berthiaume, N. C. Staub, and M. A. Matthay, *J. Clin. Invest.* **79,** 335 (1987).

[8] T. Yamamoto, O. Kajikawa, T. R. Martin, S. R. Sharar, J. M. Harlan, and R. K. Winn, *J. Immunol.* submitted (1997).

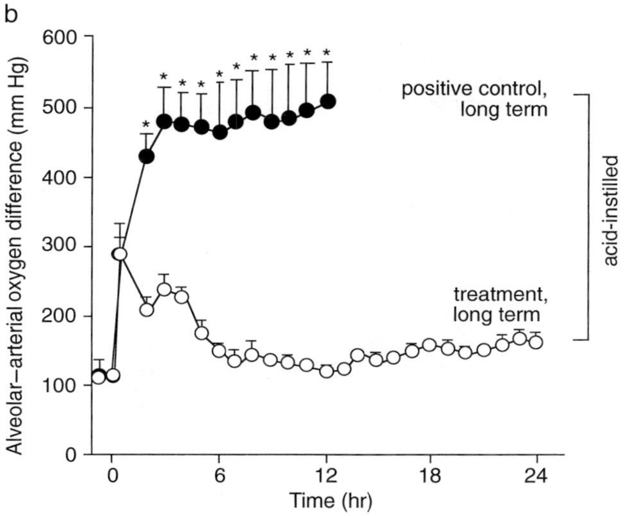

FIG. 4. (*continued*)

Extravascular Lung Water. An alternative method measures the actual water in the lung tissue separate from the water in residual pulmonary blood. The extravascular lung water is determined by measuring the extravascular water-to-dry weight ratio and multiplying that by the dry lung weight. The lung is homogenized and weighed. The homogenate is then dried in an oven and weighed again. The difference of those two weights represents the amount of water that was associated with the lung tissue. This amount of water is represented as grams water/gram dry lung (Fig. 6). By measuring the hemoglobin in the supernatant of the lung homogenate and the wet to dry ratio of the blood itself, the contribution of the blood to the total lung water can be calculated and subtracted. This approach has been well documented.[2,7] If the right lung is lavaged for cell counts (see above), then data for extravascular lung water are obtained for the left lung only.

Sample Collection for Edema Fluid. The alveolar samples are collected by aspirating fluid via a sampling catheter gently passed through the trachea to a wedged position in a distal airway. Of note, analysis of local chemokine and cytokine concentrations is perhaps best achieved when measured in edema fluid. Indeed, in BAL, the area of the lung that is lavaged is variable, and there are no correction factors to compensate for a varying degree of dilution from animal to animal. One drawback of using alveolar fluid collection was that no sample can be collected in negative control animals because their lung is not injured.

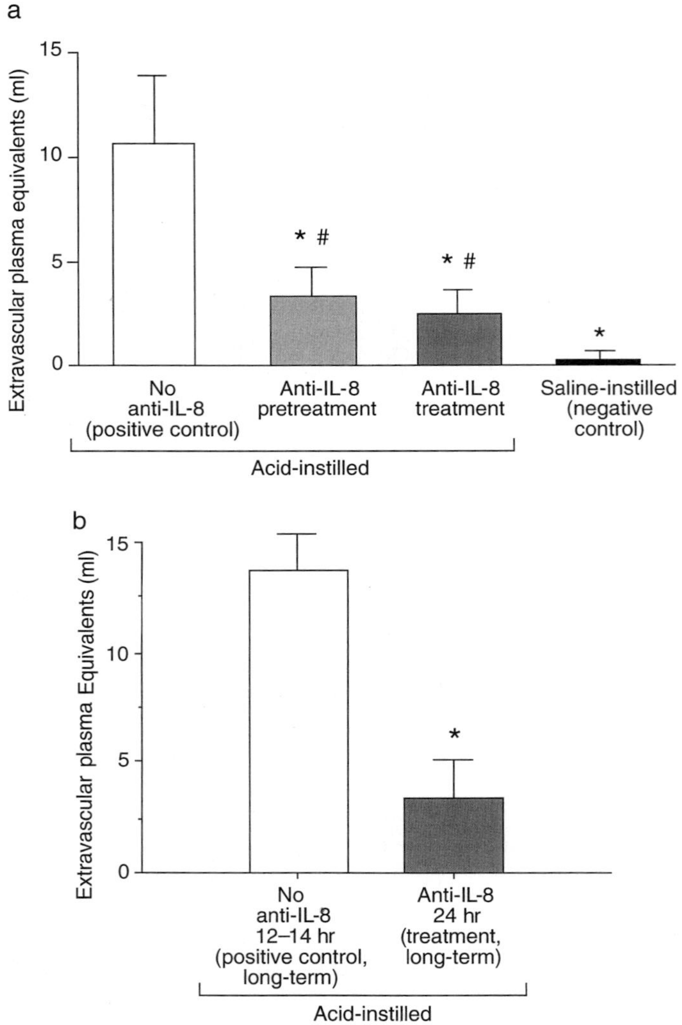

FIG. 5. Endothelial permeability in the lung measured as the accumulation of the vascular protein tracer, ^{131}I-labeled albumin, in the extravascular spaces of the lung and expressed as extravascular plasma equivalents in the positive control, pretreatment, treatment, and negative control groups at 6 hr (a) and in the positive control group at 12–14 hr and treatment group at 24 hr (b). In the 6-hr studies, the extravascular plasma equivalents were decreased by 70% in the pretreatment and treatment groups compared to the positive control group (a). The same reduction in extravascular plasma equivalents was observed in the treatment group at 24 hr compared to the positive control group at 12–14 hr (b). Data are means ± SD. *, $p < 0.05$ versus the positive control group (a and b); #, $p < 0.05$ versus the negative control group (a). (Reproduced from the *Journal of Clinical Investigation,* 1995, Vol. 96, pp. 107–116, by copyright permission of The American Society for Clinical Investigation.)

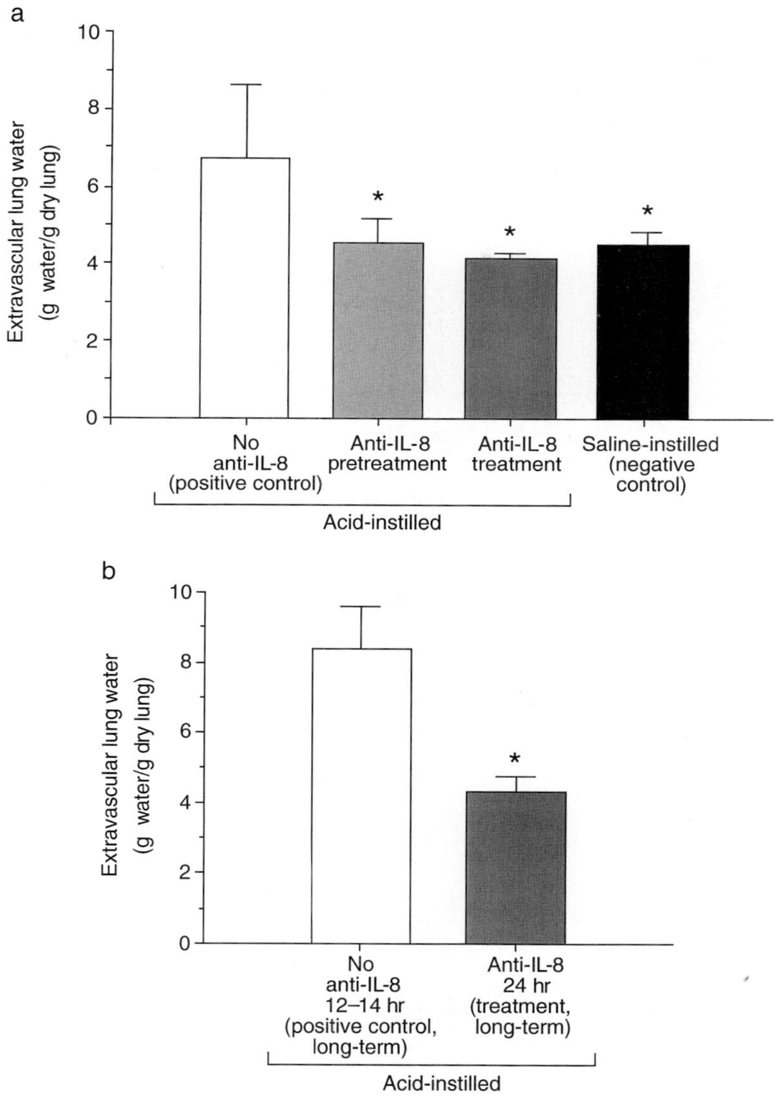

FIG. 6. Extravascular lung water in the positive control, pretreatment, treatment, and negative control groups at 6 hr (a) and in the positive control group at 12–14 hr and treatment group at 24 hr (b). In the 6-hr studies, the extravascular lung water in the pretreatment and treatment groups was 35% lower than in the positive control groups and no different from that in the negative control group (a). The extravascular lung water of a normal uninstilled rabbit lung is 3.2 g water/gram dry lung. In the 24-hr studies, the extravascular lung water was 100% lower in the treatment group than in the positive control group (b). Data are means ±SD. *, $p < 0.05$ versus the positive control group (a and b). (Reproduced from the *Journal of Clinical Investigation,* 1995, Vol. 96, pp. 107–116, by copyright permission of The American Society for Clinical Investigation.)

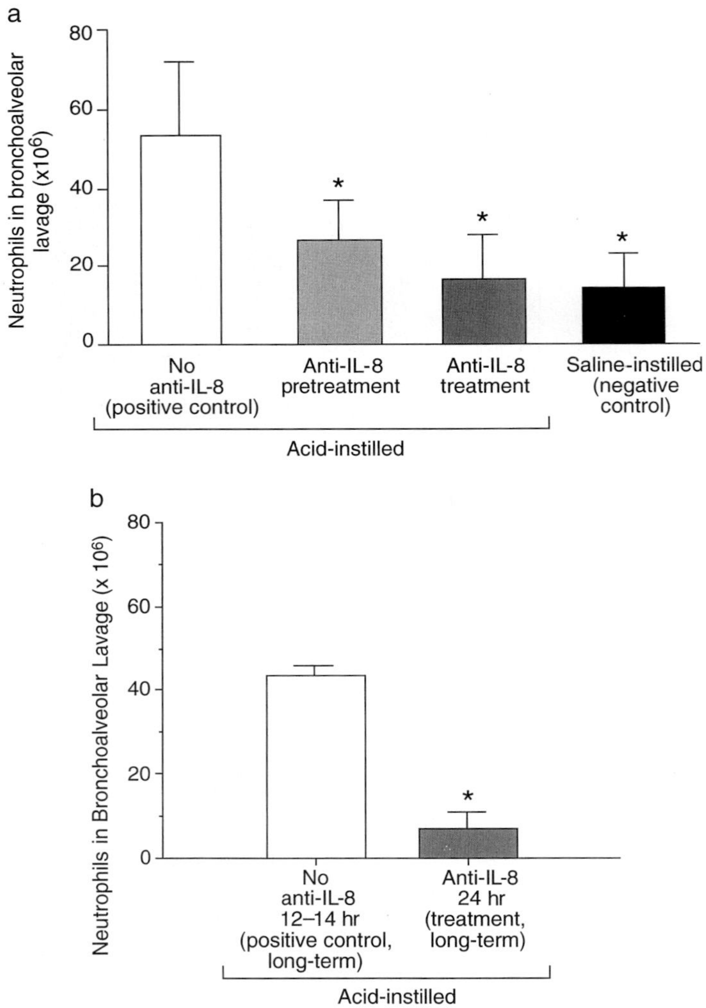

FIG. 7. Number of neutrophils lavaged from the air spaces of rabbits in the positive control, pretreatment, treatment, and negative control groups at 6 hr (a) and in the positive control group at 12–14 hr and treatment group at 24 hr (b). In the 6-hr studies, the number of neutrophils was 50% lower in the pretreatment and treatment groups than in the positive control group (a). In the 24-hr studies, the number of neutrophils was 75% lower in the treatment groups than in the positive control group (b). Data are means ±SD, *, $p < 0.05$ versus the positive control group (a and b). (Reproduced from the *Journal of Clinical Investigation,* 1995, Vol. 96, pp. 107–116, by copyright permission of The American Society for Clinical Investigation.)

Sample Collection for Bronchoalveolar Lavage. Bronchoalveolar lavage samples are collected by lavaging the resected lungs twice using 6 ml of isosmolar 0.9% (w/v) NaCl containing 12 m*M* lidocaine (Sigma) each time. The number of neutrophils present in the BAL is indicative of the severity of the lung injury (Fig. 7).

Statistical Analysis. One-way ANOVA with repeated measurements analysis is used to compare samples obtained at several time points from the same animal. One way ANOVA (factorial) is used when comparing other single groups. Student–Newman–Keuls test is used as a post hoc statistical test.

Conclusion

In conclusion, we have discussed specifically the methods used in two experimental models of inflammation and injury in the context of a general discussion of appropriate models. The pleurisy and acid-aspiration models were developed with different goals and thus had many differences in methods and design. Their common strengths were the use of a potent, species-specific blocking reagent and the use of a ELISA for free, unbound IL-8 to confirm local neutralization of the endogenous IL-8. In addition, in the acid-aspiration model, clinical relevance was established by an intravenous administration of the anti-IL-8, the addition of a mortality end point, and the demonstration of the effectiveness of a delayed delivery of anti-IL-8. There are also limitations, of course, in that no animal model fully replicates the situation in humans. It is hoped, however, that investigators can consider the general points made here when designing their own models or evaluating those of others.

Acknowledgments

The authors thank Drs. A. M. Boylan, H. G. Folkesson, and M. A. Matthay of the University of California, San Francisco, and Dr. M. Sadick of Genentech for active participation in the development of many of the methods presented here, as well as I. Adams and D. Wood of Genentech for manuscript formatting and graphics, respectively.

[13] Murine Experimental Autoimmune Encephalomyelitis: A Model of Immune-Mediated Inflammation and Multiple Sclerosis

By ANDRZEJ R. GLABINSKI, MARIE TANI, VINCENT K. TUOHY, and RICHARD M. RANSOHOFF

Overview: Animal Models of Demyelination

The classic description of multiple sclerosis (MS) pathology by Charcot placed major emphasis on the fact that myelin was destroyed in central nervous system (CNS) tissues of affected individuals, whereas axons were relatively spared.[1] Multiple sclerosis was thus categorized as a disease of primary demyelination. The history of research into MS has marched hand in hand with research of animal models of demyelination, which are grouped in three major categories, according to etiology, as immune-mediated, virus-induced, and toxic demyelinative disorders. The animal models have been used to investigate genetic, biochemical, immunologic, developmental, and metabolic characteristics of demyelination and remyelination.

Immune-Mediated Demyelination

Historically, the leading model of MS has been experimental autoimmune encephalomyelitis (EAE), which was originally developed in the context of rare but tragic reactions to early preparations of rabies vaccine. In particular, rabies vaccine was occasionally and unpredictably observed to cause a devastating syndrome of fever, alteration in mental state and paralysis, often culminating in death.[2] Brains of stricken individuals showed perivenular demyelination and acute inflammation. These pathological alterations were reminiscent of fulminant MS and did not suggest infection by partially inactivated rabies virus.[3] Investigators proposed that the syndrome, termed "neuroparalytic accident," could be caused by allergic reaction to avian brain tissue in which the virus was cultured.[4] This hypothesis was tested and validated in nonhuman primates.[5] The human syndromes of acute disseminated encephalomyelitis (ADEM) and postvaccine encepha-

[1] J. Charcot, *Gaz d hop.* **41,** 554 (1868).
[2] G. Stuart and K. Krikorian, *Lancet* **218,** 1123 (1930).
[3] R. Thompson, *Arch. Pathol.* **12,** 601 (1931).
[4] E. Hurst, *J. Hyg.* **32,** 33 (1932).
[5] T. Rivers and F. Schwenkter, *J. Exp. Med.* **61,** 689 (1935).

METHODS IN ENZYMOLOGY, VOL. 288
0076-6879/97 $25.00

lomyelitis persist and are precise correlates of acute EAE. Subsequent decades of work have led to high refinement in varied animal models of demyelination. The characteristics and uses of such models is discussed in this chapter.

Virus-Induced Central Nervous System Demyelination

It has long been suspected that MS has an environmental trigger that precipitates disease in genetically susceptible individuals.[6] The strongest contemporary support for this notion comes from the high but incomplete concordance for MS in monozygotic twins, and epidemiological considerations suggest that viral infection may provide such a trigger for MS.[7] Naturally occurring viruses cause demyelinating syndromes in humans, mice, and dogs, and virus-induced demyelination has been studied for clues to the etiology and pathogenesis of MS.

Two demyelinating syndromes in small animals have been extensively developed. One such model uses Theiler's murine encephalomyelitis virus (TMEV). This picornavirus is a natural murine pathogen that causes demyelination in susceptible strains of mice.[8] Two aspect of TMEV-induced demyelination have been exploited with particularly productive results. One regards the characteristics of the virus, some strains of which cause poliomyelitis that is rapidly fatal, whereas other strains cause mild gray-matter disease that is followed by demyelinating encephalomyelitis. Determinants of these varying patterns of neurovirulence have been characterized in great detail. A second productive avenue for research concerns the genetics and immunology of the TMEV-related delayed encephalomyelitis.[9] TMEV-induced demyelination is a complex phenomenon that is contingent on precise interactions between the infecting virus, background and immune response genes of the host mouse strain, and the integrity of the immune system. The model has been highly defined in many regards and can be useful for exploring a large variety of issues pertinent to the pathogenesis and treatment of MS.

A second viral model of demyelination is represented by infection of susceptible rats with JHM, a murine hepatitis virus.[10] The characteristics and advantages of these two models have been reviewed.[11]

[6] A. D. Sadovnick and G. C. Ebers, *Can. J. Neurol. Sci.* **20,** 17 (1993).
[7] G. C. Ebers, D. E. Bulmen, and A. D. Sadovnick, *N. Engl. J. Med.* **315,** 1638 (1986).
[8] M. Theiler, *J. Exp. Med.* **65,** 705 (1937).
[9] M. Dal Canto, R. Melvold, B. Kim, and S. Miller, *Microsc. Res. Tech.* **32,** 215 (1995).
[10] F. Cheever, J. Daniels, A. Pappenheimer, and O. Bailey, *J. Exp. Med.* **90,** 181 (1949).
[11] M. Bradl and C. Linington, *Brain Pathol.* **6,** 303 (1996).

Toxic Demyelination

In some circumstances, it is advantageous to obtain precisely timed, focal demyelinative lesions. This characteristic is particularly useful for studying the process of remyelination and for developing imaging protocols for small animals. Models of focal toxic demyelination include intracerebral or intraspinal injection with lysolecithin and cuprizone.[12]

Experimental Autoimmune Encephalomyelitis

Experimental autoimmune encephalomyelitis has been an important model for studying organ-specific autoimmunity since the 1930s. By the early 1950s, it was appreciated that demyelination after immunization with brain tissue was contingent on the inflammatory response. Investigators at that early stage proposed several tasks to their scientific progeny: first, identify the inciting antigen; second, define the mechanism (T cell, antibody, macrophage) of tissue injury.[13] These tasks have been completed since the 1950s, with impressive results. EAE has become a potent model for MS, so much so that putative treatments for MS are routinely subjected to pilot studies in the EAE model.

Selection of Animal Model

Several animals will support the process of EAE, after appropriate immunization. Each of these models has advantages and disadvantages, and preferences should be dictated by the scientific question under examination. Mice have been used most commonly for these studies. Obvious advantages include cost of purchase and maintenance. Additional advantages of work in the murine system include the availability of numerous reagents for characterizing and manipulating the immune system. Further, the immunogenetics of the mouse have been highly defined; the genetics of susceptibility and resistance to EAE have been characterized most recently by whole-genome screens and were reviewed in great detail.[14]

Several strains of transgenic mice can also be exploited for the analysis of specific questions regarding EAE and its pathogenesis. Particularly interesting have been studies of mice with transgenes encoding rearranged encephalitogenic T-cell receptors, with specificity toward myelin antigens.[15]

[12] W. F. Blakemore, *Neuropathol. Appl. Neurobiol.* **4,** 47 (1978).

[13] A. Wolf, *in* "Proceedings of the First International Congress of Neuropathology." Tipgrafia G. Donnini, Perugia, Italy, 1952.

[14] J. Encinas, H. Weiner, and V. Kuchroo, *J. Neurosci. Res.* **45,** 655 (1996).

[15] J. Goverman, A. Woods, L. Larson, L. Weiner, L. Hood, and D. Zaller, *Cell (Cambridge, Mass.)* **72,** 551 (1993).

Mice with targeted deletion in genes that govern the immune system ("knockout" mice) can also be useful for such studies.[16] A particular caveat regarding the use of transgenic and knockout mice regards the host strain. Occurrence of EAE is highly dependent on the genetic background, as noted above. Both H2 and non-H2 genes determine susceptibility toward EAE. Many transgenics are constructed in nonsusceptible strains, for technical reasons related to mouse generation and breeding. Virtually all knockout mice are constructed in the C129 or C57/B16 strains, neither of which are readily susceptible to EAE. It is possible to backcross such mice to a susceptible strain. However, full conversion of a transgenic or knockout strain into a different inbred strain requires approximately 10 generations of backcross, which is prohibitive in terms of both time and money. As an alternative, EAE can be studied in partial backcross mouse strains. However, the resultant mouse strain will be genetically unique so that results must be interpreted with some caution: essentially, such mice can be directly compared only to their littermates. Accordingly, EAE experiments that take advantage of transgenic or knockout mouse strains must be planned and executed with full recognition of the practical barriers.

Experimental autoimmune encephalomyelitis has been extensively studied in rats. The Lewis rat is highly susceptible to EAE and other autoimmune processes, and the disease has been well characterized in the Lewis rat. Seveal advantages pertain to this model. These include the relative ease of surgical or intravenous interventions, based on the larger size of rats compared with mice. Furthermore, rats are more robust than typical laboratory strains of mice so that daily or twice-daily injections during intervention studies are better tolerated. Disadvantages of rats include the relative lack of immunologic reagents and the restricted availability of transgenic animals.

Experimental autoimmune encephalomyelitis has been well characterized since the 1960s in susceptible strains of guinea pigs. One particular advantage of the guinea pig system is that the disease is naturally chronic and relapsing, with clinical similarity to certain forms of human MS. In comparison with mice, rats and guinea pigs are relatively easy subjects for imaging studies, including magnetic resonance imaging (MRI) and magnetic resonance spectroscopy. Clinical/MRI/pathological correlations are of clear and increasing importance for studying and understanding MS.

Experimental autoimmune encephalomyelitis was first studied in nonhuman primates. This line of research has come full circle with a model of

[16] D. Koh, W. Fung-Leung, A. Ho, D. Gray, H. Acah-Orbea, and T. Mak, *Science* **256,** 1210 (1992).

EAE in the *Callithrix jacchus* marmoset.[17] This nonhuman primate offers several advantages including the ease of imaging studies, similarity of the human and marmoset nervous and immune systems, and the unique characteristic that the animals are natural intrauterine chimeras, permitting adoptive transfer studies.

Clinical Form of Experimental Autoimmune Encephalomyelitis

Differing immunization protocols in various animal strains produce varied clinical patterns of EAE that recapitulate the varied forms of clinical MS. The hallmarks of EAE clinically are weight loss and paralysis (as discussed below). EAE studies in strains of mice including PL/J and B10.PL and in Lewis rats produced a disease that was acute and monophasic. The syndrome precisely mirrored human ADEM. However, concerns about the applicability of this model to the chronic, relapsing, multiphasic human disease (MS) persisted. Furthermore, genuine demyelinating pathology was difficult to obtain in several models of acute and passive-transfer EAE. Models have now been developed that incorporate multiphasic chronicity and demyelination. Contemporary investigators have at their disposal several different clinical disease patterns, each of which can be exploited for studying specific aspects of disease.

Acute Experimental Autoimmune Encephalomyelitis following Active Immunization. Acute EAE is perhaps best typified by immunization of several susceptible mouse or rats strains with myelin basic protein (MBP). The immunization protocols include Freund's complete adjuvant (CFA). Frequently, the initial immunization is accompanied by a subsequent injection with inactivated *Bordetella pertussis,* which is believed to disrupt the blood–brain barrier (BBB). As noted above, the disease is acute and monophasic. Frequently, animals that have recovered are resistant to reinduction of disease. This resistant state has been intensively studied and highly characterized. Acute EAE, an inflammatory process with varying degrees of demyelination, is useful for studying CNS inflammation and for examining the afferent limb of immune recognition of myelin. However, the lack of chronicity, the absence of demyelination, and the monophasic nature of the process render it an incomplete model of human MS.

Passive Transfer. Early investigators of EAE were confronted with the question of whether the disease was caused by immune cells, by antibody, or by both components. This question was addressed by classic experiments that showed that EAE could be transferred to a naive recipient with T cells but not with serum. The antigen specificity of encephalitogenic T cells

[17] L. Massacesi, C. P. Genain, D. Lee-Parritz, N. L. Letvin, D. Canfield, and S. L. Hauser, *Ann. Neurol.* **37,** 519 (1995).

has been defined for many inbred mouse strains and for at least three myelin antigens. Using T-cell clones to effect passive-transfer EAE greatly simplifies and refines the approach to certain issues of pathogenesis and has made possible a long series of landmark studies.[18] One particularly useful extension has been introduction of pathogenic T-cell lines into lethally irradiated, bone marrow chimeras, to help define the antigen-presenting cells in the CNS compartment.[19]

The basic design of a passive-transfer experiment is that primary immunization is carried out as for active EAE, but lymph node or spleen cells are harvested from the immunized animal at the peak of primary response (typically 7 days) and cultured *in vitro* with growth factors and antigen. T-cell lines or T-cell clones can be developed from such primary cultures. Passive transfer of EAE is brought about by injection of such cells, after activating cultures with mitogens and antigens. Adoptive transfer protocols typically specify the intravenous injection of large numbers (on the order of 5×10^5 to 3×10^7) of activated, antigen-specific T cells into a syngeneic naive recipient.

Advantages of passive transfer are many. First, the system is highly defined, extremely predictable, and reproducible. In a given strain of recipient, disease will reliably occur within the first week following transfer. Passive immunization is a useful way to study cell recruitment to the CNS. The afferent limb of immune reactivity is sacrificed in this system, but the efferent limb can be thereby subjected to focused investigation. Injecting large numbers of activated T cells in synchronized fashion does not effectively mimic the physiological process of accumulation of antigen-specific T cells in CNS tissues, however.

Chronic-Relapsing Experimental Autoimmune Encephalomyelitis. Human MS typically affects people in the third or fourth decade of life and persists thereafter. Multiple sclerosis therefore has a course of 30 to 50 years. Clinical and immunologic studies indicate significant change in the process over time.[20] This chronic disease appears to be effectively modeled by several currently available forms of chronic-relapsing (C-R) EAE. Chronicity and progression are the hallmarks of MS. These attributes of EAE are a function of the host strain of animal, the antigen employed, and the immunizing regimen. For reasons that are not clear, myelin proteolipid protein (PLP) more commonly causes chronic disease than does MBP.[21] It is also important to note that the immunization protocols, virtually all

[18] A. Ben-Nun, H. Wekerle, and I. R. Cohen, *Eur. J. Immunol.* **11,** 195 (1981).
[19] W. F. Hickey and H. Kimura, *Science* **239,** 290 (1988).
[20] V. Tuohy, R. Fritz, and A. Ben-Nun, *Curr. Opin. Immunol.* **6,** 887 (1994).
[21] V. Tuohy, *J. Neurochem.* **19,** 935 (1994).

of which employ CFA and *Bordetella,* introduce some degree of artificiality into the experimental system. Nevertheless, the generation of a chronic, multiphasic, and demyelinating disease with a single immunization is a considerable achievement and has distinctly advanced MS research. One protocol for generating C-R EAE in mice is described below.

Role of Myelin Antibodies. T-cell function is essential for the occurrence of EAE and, presumably, MS. T-cell transfer studies establish this principle, whose validity remains unchallenged. However, the importance of myelin-specific antibodies has become a focus of renewed interest. This line of investigation began with the discovery that antibodies to a minor myelin component, myelin oligodendroglial glycoprotein (MOG), could synergize with anti-myelin T cells to mediate widespread demyelination in Lewis rats.[22] Subsequent work has confirmed and extended these observations. Human immune responses to MOG are under investigation in the setting of MS and other related disorders. In some settings, it will be important to examine both the T cell and the humoral limbs of the immune response to myelin in evaluating potential therapeutic interventions.[23]

Induction Protocol for Chronic-Relapsing Experimental Autoimmune Encephalomyelitis

Care and Breeding of Mice Preinduction. SWR females and SJL/J males are obtained from Jackson Laboratories (Bar Harbor, ME), and are maintained for at least 2 weeks in the Cleveland Clinic Foundation (CCF) animal facility without food restriction, with maximum of four animals per cage. Mice are maintained in isolated cages that are opened only in a laminar hood. A breeding colony at the CCF is maintained for generation of (SWR×SJL)F_1 mice (SWXJ mice). Females age 7 to 10 weeks, weighing between 18 and 25 g, have been used for all experiments.

Immunization. Immunization of SJL/J females with a PLP encephalitogenic peptide representing amino acid residues 139–151 results in reproducible acute EAE with high incidence.[24] The same peptide produces C-R EAE in SWXJ mice.[25] PLP peptide 139–151 has been synthesized manually using solid-phase method and purified by high-performance liquid chromatography (HPLC).[26]

[22] H. Lassman, C. Brunner, M. Bradl, and C. Linington, *Acta Neurol.* **75,** 566 (1988).

[23] C. P. Genain, M. H. Nguyen, N. L. Letvin, R. Pearl, R. L. Davis, M. Adelman, M. B. Lees, C. Linington, and S. L. Hauser, *J. Clin. Invest.* **96,** 2966 (1995).

[24] R. Sobel, V. Tuohy, and M. Lees, *J. Immunol.* **146,** 543 (1991).

[25] M. Yu, A. Nishiyama, B. Trapp, and V. Tuohy, *J. Neuroimmunol.* **64,** 91 (1996).

[26] V. K. Tuohy, Z. Lu, R. A. Sobel, R. A. Laursen, and M. B. Lees, *J. Immunol.* **142,** 1523 (1989).

The immunization procedure has been previously described.[25] On day zero, mice are injected intravenously (tail vein) with 0.6×10^{10} *B. pertussis* bacilli (Michigan Dept. of Public Health). Two hours later mice are immunized by subcutaneous injections in the abdominal flanks with 200 μl of an emulsion of equal volumes of water and IFA (Freund's incomplete adjuvant) containing 100 nmol of PLP peptide 139–151 (154 μg) and 400 μg of *Mycobacterium tuberculosis* H37RA (Difco, Detroit, MI). Both the immunizing emulsion and *B. pertussis* are sterilized by irradiation with 2000 rads before injection. On day 3, *B. pertussis* (0.6×10^{10} bacilli) is injected intravenously. Response to intravenous injection of *B. pertussis* differs among mouse strains and from lot to lot. The maximum dose of *B. pertussis* that does not cause death should be used.

Monitoring Chronic-Relapsing Experimental Autoimmune Encephalomyelitis

Beginning with day zero mice are ear-marked; weight and clinical score are checked every day at the same time in the morning. The clinical scoring system has been published.[27] Onset of the first attack of C-R EAE typically occurs between days 14 and 28 postimmunization (PI). Usually mice lose about 10% of body weight within 1 day of development of neurological signs. Attacks last from 2 to 4 days, during which mice exhibit worsening, followed by stable neurological impairment. The onset of remission is heralded by weight gain, associated with rather abrupt improvement in neurological function. Following third or fourth relapses of C-R EAE, mice may stabilize at neurological severity scores of 3–4 (hind-limb paraplegia or quadriparesis). During this phase of disease, dehydration may occur and can produce complexity in clinical monitoring. It is therefore difficult at times to define relapses beyond this point in disease.

Sacrifice of mice with EAE and harvesting of CNS tissues has been discussed elsewhere. Histological severity of EAE is an important outcome parameter. There are many published protocols for defining histological EAE severity. We perform histological analysis on sagittal sections of brain and spinal cord sections that are fixed in 10% (w/v) phosphate-buffered formalin and stained with hematoxylin/eosin and Luxol-fast blue. Inflammation is graded on a semiquantitative five-step scale as follows. (1) Perivascular and submeningeal inflammatory infiltrates in spinal cord and brain are counted and scored: 0, absence of infiltrates; 1+, 1–5 infiltrates; 2+,

[27] M. Yu, J. Johnson, and V. Tuohy, *J. Exp. Med.* **183,** 1777 (1996).

>5 infiltrates. (2) Meningeal inflammation is graded as 0, absent, or 1+, present. (3) Scores at all three sites (brain, spinal cord, and meninges) are summed. The maximal score is five: brain (2+), spinal cord (2+), and meninges (1+).

[14] *In Vitro* and *in Vivo* Systems to Assess Role of C-X-C Chemokines in Regulation of Angiogenesis

By DOUGLAS A. ARENBERG, PETER J. POLVERINI, STEVEN L. KUNKEL, ARMEN SHANAFELT, and ROBERT M. STRIETER

Introduction

Angiogenesis is one of the most pervasive and essential biological events encountered in vertebrate animals.[1–5] A number of physiological and pathological processes, such as embryonic development, the formation of inflammatory granulation tissue during wound healing, chronic inflammation, and the growth of malignant solid tumors, are strictly dependent on neovascularization. Normally, physiological angiogenesis occurs infrequently, yet can be rapidly induced in response to a number of diverse physiological stimuli. Among the most extensively studied of these angiogenesis-dependent physiological processes is wound healing.[6] An important feature of wound-associated angiogenesis is that it is locally controlled and transient. The rate of normal capillary endothelial cell turnover in adults is typically measured in months or years.[7,8] However, when quiescent endothelial cells are stimulated, they will degrade basement membrane and proximal extracellular matrix, migrate directionally, divide, and organize into new functioning capillaries invested by a new basal lamina, all within a matter of days. This dramatic amplification of the microvasculature is nevertheless temporary. As rapidly as they are formed, they virtually disappear, re-

[1] J. Folkman and R. Cotran, *Int. Rev. Exp. Pathol.* **16,** 207 (1978).
[2] R. Auerbach, "Angiogenesis-Inducing Factors: A Review. Lymphokines 69." Academic Press, New York, 1981.
[3] P. J. Polverini, "Cytokines," p. 54. Karger, Basel, 1989.
[4] J. Folkman, *Adv. Cancer. Res.* **43,** 175 (1985).
[5] J. Folkman and M. Klagsbrun, *Science* **235,** 442 (1987).
[6] S. J. Leibovich and D. M. Weisman, *Clin. Biol. Res.* **266,** 131 (1988).
[7] R. L. Engerman, D. Pfaffenenbach, and M. D. Davis, *Lab. Invest.* **17,** 738 (1967).
[8] I. F. Tannock and S. Hayashi, *Cancer Res.* **32,** 77 (1972).

0076-6879/97 $25.00

turning the tissue vasculature to a homeostatic environment. This demonstrates the essential aspects of the angiogenic response during wound repair; the formation of new microvasculature is rapid and transient, and it is characterized by regression to a physiological steady-state level.

The abrupt termination of angiogenesis that accompanies the resolution of the wound response suggests two possible mechanisms of control, neither of which are mutually exclusive. First, under circumstances not well understood, there is probably a marked reduction in the synthesis and/or elaboration of angiogenic mediators. Second, a simultaneous increase occurs in the levels of substances that inhibit new vessel growth.[9] In contrast to the strict regulation of angiogenesis that accompanies wound repair, dysregulation of angiogenesis can lead to an imbalance in angiogenic and angiostatic factors that contributes to the pathogenesis of either chronic inflammatory diseases or solid tumor growth. Thus, the complement of positive and negative regulators of angiogenesis may vary among different physiological and pathological settings, and the recognition of this dual mechanism of control is necessary to gain a more thorough understanding of this complex process and its significance in regulating net angiogenesis.

Phases of Angiogenesis

The postcapillary venule appears to be the target vessel for the initiation of angiogenesis in the adult organism. Within these vessels are endothelial cells that are in juxtaposition to basement membrane and are surrounded by an interrupted layer of pericytes and smooth muscle cells invested by extracellular matrix. After injury, the first phase of the angiogenic response is associated with the alteration of adhesive interactions between adjacent endothelial cells, pericytes, and smooth muscle cells.[10–15] Activated endothelial cells undergo reorganization of cytoskeletal elements and express cell surface adhesion molecules (e.g., integrins, selectins) followed by the generation of extracellular matrix components.[16–22] These endothelial cells

[9] N. Bouck, *Cancer Cells* **2,** 179 (1990).
[10] T. H. Bar and J. R. Wolff, *Z. Zellforsch.* **133,** 231 (1972).
[11] D. H. Ausprunk, D. R. Knighton, and J. Folkman, *Dev. Biol.* **38,** 237 (1974).
[12] D. H. Ausprunk and J. Folkman, *Microvasc. Res.* **14,** 53 (1977).
[13] D. H. Ausprunk, K. Falterman, and J. Folkman, *Lab. Invest.* **38,** 284 (1978).
[14] G. D. Philips, R. A. Whitehead, and D. R. Knighton, *Am. J. Anat.* **192,** 257 (1991).
[15] D. E. Sims, *Tissue Cell* **18,** 153 (1986).
[16] J. Madri, B. Pratt, and A. Tucker, *J. Cell Biol.* **106,** 1375 (1988).
[17] D. E. Ingber and J. Folkman, *J. Cell Biol.* **109,** 317 (1989).
[18] D. E. Ingber, *J. Cell. Biochem.* **47,** 236 (1991).

subsequently produce a variety of proteases, which degrade the basement membrane, allowing the cells to migrate into the surrounding extracellular matrix.[23,24] This process initiates the formation of capillary buds, releases sequestered growth factors in the basal lamina and adjacent extracellular matrix, and induces endothelial cell and other mesenchymal cell-derived growth factor expression that facilitates the subsequent phases of angiogenesis. This initial phase can proceed in the absence of endothelial proliferation.[25]

Subsequent phases of angiogensis leading to the formation of microvessels requires continual exposure to angiogenic mediators.[26–30] These molecules function in an autocrine and paracrine manner to control endothelial cell proliferation, elongation, orientation, and differentiation, leading to reestablishment of the basement membrane, lumen formation, and anastomosis with other neovessels or preexisting vessels.[19,31,32] The neovasculature may persist as capillaries, differentiate into mature venules or arterioles, or undergo regression. The signals responsible for this latter event are now beginning to be identified and may either function to initiate apoptosis or induce cell cycle arrest of endothelial cells.[22,33,34]

---

[19] J. R. Gamble, U. Matthias, and G. Meyer, *J. Cell Biol.* **121,** 921 (1993).

[20] M. Nguyen, N. A. Strubel, and J. Bischoff, *Nature* (*London*) **365,** 267 (1993).

[21] P. C. Brooks, R. A. Clark, and D. A. Cheresch, *Science* **264,** 569 (1994).

[22] P. C. Brooks, A. M. P. Montgomery, M. Rosenfeld, R. A. Reisfeld, T. Hu, G. Klier, and D. A. Cherish, *Cell* (*Cambridge, Mass.*) **79,** 1157 (1994).

[23] D. B. Rifkin, J. L. Gross, D. Moscatelli, and E. Jaffe, *in* "Pathobiology of the Endothelial Cell" (H. Nossel and H. J. Vogel, eds.), p. 191. Academic Press, New York, 1982.

[24] M. S. Pepper, J. D. VassaUi, L. Orci, and R. Montesano, *in* "Angiogenesis: Key Principles" (R. Steiner, P. B. Weisz, and R. Langer, eds.), p. 137. Birkhauser, Basel, 1992.

[25] M. M. Sholly, G. P. Fergusen, H. R. Seibel, J. L. Montour, and J. D. Wilson, *Lab. Invest.* **51,** 624 (1984).

[26] A. Baird and N. Ling, *Biochem. Biophys. Res. Commun.* **126,** 358 (1985).

[27] I. Vlodavski, J. Folkinan, R. Sullivan, R. Fridman, R. Ishai-Michaeli, J. Sasse, and M. Klagsbrun, *Proc. Natl. Acad. Sci. U.S.A.* **84,** 2292 (1987).

[28] V. Sarma, F. W. Wolf, R. M. Marks, T. B. Shows, and V. M. Dixit, *J. Immunol.* **148,** 3302 (1992).

[29] M. E. Gerritsen and C. M. Bloor, *FASEB J.* **7,** 523 (1993).

[30] L. Liaw and S. M. Schwartz, *Arterioscler. Thromb.* **13,** 985 (1993).

[31] N. Paweletz and M. Knierim, *Crit. Rev. Oncol. Hematol.* **9,** 197 (1989).

[32] M. A. Konerding, C. VanAckern, F. Steinberg, and C. Streffer, *in* "Angiogenesis: Key Principles" (R. Steiner, P. B. Weisz, and R. Langer, eds.), p. 40. Birkhauser, Basel.

[33] F. Re, A. Zanetti, M. Sironi, N. Polentarutti, L. Lanfrancone, E. Dejana, and F. Colotta, *J. Cell Biol.* **127,** 537 (1994).

[34] A. D. Luster, S. M. Greenberg, and P. Leder, *J. Exp. Med.* **182,** 219 (1995).

Regulation of Angiogenesis by Both Angiogenic and Angiostatic Factors

It has become increasingly apparent that net angiogenesis is determined by a dual, yet opposing system of angiogenic and angiostatic factors[3,5,35–38] (Table I). The majority of angiogenic factors are polypeptides that induce endothelial cell migration, proliferation, and differentiation into tubular structures. These proangiogenic molecules are produced by an array of cells and function as ligands in an autocrine and paracrine manner to facilitate endothelial cell activation. These molecules can stimulate angiogenesis either by directly interacting with specific receptors on endothelial cells or by indirectly attracting and activating accessory cells, such as macrophages, that in turn produce additional angiogenic factors.[3,39,40] In addition, various molecules can function as cofactors for the promotion of angiogenesis. For example, proteases can lead to the release of an active angiogenic factor (e.g., transforming growth factor-β; TGF-β) from a latent molecule.[41] Moreover, heparin, a glycosaminogylcan sequestered in the extracellular matrix, can play a key role in stabilizing and/or enhancing the function of angiogenic molecules; heparin facilitates basic fibroblast growth factor (bFGF) interaction with high affinity receptors on endothelial cells.[5,27]

A role for inhibitors in the control of angiogenesis was first suggested by Eisenstein and colleagues,[42] and Sorgente and associates,[43] who observed that hyaline cartilage was particularly resistant to vascular invasion. They reported that a heat-labile guanidinium chloride extract prepared from cartilage contained an inhibitor of neovascularization. Subsequently other investigators showed that a similar or identical extract from either rodent neonatal or shark cartilage was able to effectively block neovascularization and growth of tumors *in vivo*.[44,45] Similar angiostatic factors have been

[35] M. Klagsbrun and P. A. D'Amore, *Annu. Rev. Physiol.* **53,** 217 (1991).
[36] M. A. Moses and R. Langer, *Biotechnology* **9,** 630 (1991).
[37] S. L. Leibovich, *in* "Human Cytokines," p. 539–555. Blackwell Scientific, Cambridge, Mass.
[38] L. A. DiPietro and Polverini, *Behring Inst. Mitt.* **92,** 238 (1993).
[39] P. J. Polverini, R. S. Cotran, M. A. Gimbrone, Jr., and E. R. Unanue, *Nature (London)* **269,** 804 (1977).
[40] P. J. Polverini and S. J. Leibovich, *Lab. Invest.* **51,** 635 (1984).
[41] A. B. Roberts and M. B. Spom, *Am. Rev. Respir. Dis.* **140,** 1126 (1989).
[42] R. Eisenstein, K. E. Kuettner, C. Neopolitan, L. W. Sobel, and N. Sorgente, *Am. J. Pathol.* **81,** 337 (1975).
[43] N. Sorgente, K. E. Kuettner, L. W. Soble, and R. Eisenstein, *Lab. Invest.* **32,** 217 (1975).
[44] H. Brem and J. Folkman, *J. Exp. Med.* **141,** 427 (1975).
[45] A. Lee and R. Langer, *Science* **221,** 1185 (1983).

TABLE I
REPRESENTATIVE EXAMPLES OF ANGIOGENIC AND ANGIOSTATIC FACTORS

Proangiogenic mediators of angiogenesis	Endogenous inhibitors of angiogenesis
Growth factors	Proteins and peptides
Acidic fibroblast growth factor (aFGF)	Angiostatin
Basic fibroblast growth factor (bFGF)	Eosinophilic major basic protein
Epidermal growth factor (EGF)	Angiostatin
Interleukin-1 (IL-1)	Eosinophilic major basic protein
Interleukin-2 (IL-2)	High molecular weight hyaluronan
Scatter factor/hepatocyte growth factor (SF/HGF)	Interferon-α
	Interferon-β
Transforming growth factor-α (TGF-α)	Interferon-γ
Transforming growth factor-β (TGF-β)	Non-ELR C-X-C chemokines
Tumor necrosis factor-α (TNF-α)	Interleukin-1
Vascular endothelial growth factor (VEGF)	Interleukin-4
Carbohydrates and lipids	Interleukin-12
12(*R*)-Hydroxyeicosatrienoic acid (compound D)	Laminin and fibronectin peptides
	Placental RNase (angiogenin) inhibitor
Hyaluronan fragments	Somatostatin
Lactic acid	Substance P
Monobutyrin	Thrombospondin 1
Prostaglandins E_1 and E_2	Tissue inhibitor of metalloproteinases (TIMPs)
Other proteins and peptides	
Angiogenin	Lipids
Angiotensin II	Angiostatic steroids
Ceruloplasmin	Retinoids
Fibrin	Vitamin A
Human angiogenic factor	Others
ELR C-X-C chemokines	Nitric oxide
Plasminogen activator	Vitreous fluids
Polyamines	Prostaglandin synthase inhibitor
Substance P	
Urokinase	
Others	
Adenosine	
Angiotropin	
Copper	
Heparin	
Nicotinamide	
Endothelial-cell stimulating angiogenesis factor (ESAF)	

reported for other cell and tissue extracts,[44,46–48] and for a variety of natural and artificial agents, including inhibitors of basement membrane biosynthesis,[16–18,49] placental ribonuclease inhibitor,[50] lymphotoxin,[51] interferons,[52] prostaglandin synthase inhibitors,[53] heparin-binding fragments of fibronectin,[54] protamine,[55] angiostatic steroids,[55a] several antineoplastic and anti-inflammatory agents,[56,56a] platelet factor 4 (PF4),[57] interferon-γ-inducible protein 10 (IP-10),[34,58–60] monokine induced by interferon-γ (MIG),[60] thrombospondin-1,[61–63] angiostatin,[64] and antagonists to $\alpha_V\beta_3$-integrins.[21,22] Although most inhibitors can act directly on the endothelial cell to block migration and/or mitogenesis *in vitro,* their effects *in vivo* may be considerably more complex, involving additional cells and their products.

An important feature of these opposing yet complementary systems is

[46] R. Langer, H. Conn, J. Vacanti, C. C. Haudenschild, and J. Folkman, *Proc. Natl. Acad. Sci. U.S.A.* **77,** 4331 (1980).

[47] S. Brem, I. Preis, R. Langer, H. Brem, J. Folkman, and A. Patz, *Am. J. Ophthalmol.* **84,** 323 (1977).

[48] G. A. Lutty, D. C. Thompson, J. Y. Gallup, R. J. Mello, and A. Fenselau, *Invest. Ophthalmol. Visual Sci.* **24,** 52 (1983).

[49] M. E. Maragoudakis, M. Sarmonika, and N. Panoutscaopoulou, *J. Pharmacol. Exp. Ther.* **244,** 729 (1988).

[50] R. Shapiro and B. L. Vallee, *Proc. Natl. Acad. Sci. U.S.A.* **84,** 2238 (1987).

[51] N. Sato, K. Fukuda, H. Nariuchi, and N. Sagara, *J. Natl. Cancer Inst.* **79,** 1383 (1987).

[52] Y. A. Sidky and E. C. Borden, *Cancer Res.* **47,** 5155 (1987).

[53] H. I. Peterson, *Anticancer Res.* **6,** 251 (1986).

[54] G. A. Homandberg, J. Kramer-Bjerke, D. Grant, G. Christianson, and R. Eisenstein, *Biochim. Biophys. Acta* **874,** 61 (1986).

[55] S. Taylor and J. Folkman, *Nature (London)* **297,** 307 (1982).

[55a] R. Crum, S. Szabo, and J. Folkman, *Science* **230,** 1375 (1985).

[56] P. J. Polverini and R. F. Novak, *Biochem. Biophys. Res. Commun.* **140,** 901 (1986).

[56a] K. Lee, E. Erturk, R. Mayer, and A. T. K. Cockett, *Cancer Res.* **47,** 5021 (1987).

[57] T. E. Maione, G. S. Gray, J. Petro, A. J. Hunt, A. L. Donner, S. I. Bauer, H. F. Carson, and R. J. Sharpe, *Science* **247,** 77 (1990).

[58] R. M. Strieter, S. L. Kunkel, D. A. Arenberg, M. D. Burdick, and P. J. Polverini, *Biochem. Biophysiol. Res. Commun.* **210,** 51 (1995).

[59] A. L. Angiolillo, C. Sgadari, D. D. Taub, F. Liao, J. M. Farber, S. Maheshwari, H. K. Kleinman, G. H. Reaman, and G. Tosato, *J. Exp. Med.* **182,** 155 (1995).

[60] R. M. Strieter, P. J. Polverini, S. L. Kunkel, D. A. Arenberg, M. D. Burdick, J. Kasper, J. Dzuiba, J. VanDamme, A. Walz, D. Marriott, S. Y. Chan, S. Roczniak, and A. B. Shanafelt, *J. Biol. Chem.* **270,** 273 (1995).

[61] F. Rastinejad, P. J. Polverini, and N. P. Bouck, *Cell (Cambridge, Mass.)* **56,** 345 (1989).

[62] D. J. Good, P. J. Polverini, F. Rastinejad, M. M. LeBeau, R. S. Lemons, W. A. Frazier, and N. P. Bouck, *Proc. Natl. Acad. Sci. U.S.A.* **87,** 6624 (1990).

[63] S. S. Tolsma, O. V. Volpert, D. J. Good, W. A. Frazier, P. J. Polverini, and N. Bouck, *J. Cell Biol.* **122,** 497 (1993).

[64] M. S. O'Reilly, L. Holmgren, Y. Shing, C. Chen, R. A. Rosenthal, M. Moses, W. S. Lane, Y. Cao, E. H. Sage, and J. Folkman, *Cell (Cambridge, Mass.)* **79,** 315 (1994).

that with rare exception none of the angiogenic or angiostatic factors are endothelial cell specific nor unique to the process of regulation of angiogenesis. Most of these factors have a wide range of functions and target cells. This is perhaps one of the most essential features of the angiogenic response, the ability of endothelial cells to respond to a variety of mediators. This supports the notion that angiogenesis has evolved as a highly conserved response with a redundant system of factors to assure fruition of neovascularization. For example, during embryonic development, bFGF[65] has been shown to be a principal mediator of vasculogenesis and angiogenesis. In contrast, during wound repair, bFGF may have a more restricted role, whereas an entirely different complement of angiogenic mediators appear to play a role in mediating angiogenesis.[2,3,5,6,38,39,66,67] Although speculation may exist as to whether angiogenic or angiostatic factors are tissue or process specific, the shear redundancy of positive and negative factors that regulate neovascularization affirms the fundamental nature of this process.

Dysregulation of Angiogenesis Leading to Imbalance in Angiogenic and Angiostatic Factors That Contributes to Pathogenesis of Chronic Diseases and Solid Tumor Growth

Several lines of evidence suggest that an imbalance in the production of promoters and inhibitors of angiogenesis contributes to the pathogenesis of several angiogenesis-dependent disorders. For example, in rheumatoid arthritis the unrestrained proliferation of fibroblasts and capillary blood vessels leads to the formation of prolonged and persistent granulation tissue of the pannus whose degradative enzymes contribute to profound destruction of joint spaces.[68–70] Psoriasis, a common genetic skin disease, is a well-known angiogenesis-dependent disorder that is characterized by marked dermal neovascularization. Keratinocytes isolated from psoriatic plaques demonstrate a greater production of angiogenic activity, as compared to normal keratinocytes. Interestingly, this aberrant phenotype is due, in part, to a combined defect in the overproduction of the angiogenic cytokine interleukin-8 (IL-8), and a deficiency in the production of the

[65] W. Risau, *in* "The Development of the Vascular System" (R. N. Feinberg, G. K. Sherer, and R. Auerbach, eds.), p. 58. Karger, Basel, 1991.

[66] C. Sunderkotter, M. Goebeler, K. Schultze-Osthoff R. Bhardwaj, and C. Sorg, *Pharmacol. Ther.* **51,** 195 (1991).

[67] C. Sunderkotter, K. Steinbrink, M. Goebeler, R. Bhardwaj, and C. Sorg, *J. Leukocyte Biol.* **55,** 410 (1994).

[68] E. D. Harris, Jr., *Arthritis Rheum.* **19,** 68 (1976).

[69] A. E. Koch, P. J. Polverini, and S. J. Leibovich, *Arthritis Rheum.* **29,** 471 (1986).

[70] A. E. Koch, P. J. Polverini, S. L. Kunkel, L. A. Harlow, L. A. DiPietro, V. M. Elner, S. G. Elner, and R. M. Strieter, *Science* **258,** 1798 (1992).

angiogenesis inhibitor, thrombospondin-1, resulting in a proangiogenic environment.[71]

For solid tumor growth to succeed, a complex interplay must occur between transformed neoplastic cells and nontransformed resident and recruited immune and nonimmune cells. Although carcinogenesis or neoplastic transformation is dependent on multiple genetic and epigenetic events,[72] the salient feature of all solid tumor growth is the presence of neovascularization.[1,4,7] It appears that tumors are continually renewing and altering their vascular supply.[4] Interestingly, the normal vascular mass of tissue is approximately 20%, whereas, during tumorigenesis, tumor vascular mass may be >50% of the total tumor.[4] In the absence of local capillary proliferation and delivery of oxygen and nutrients, neoplasms cannot grow beyond the size of 2 mm^3.[1,4,9] In addition, the magnitude of tumor-derived angiogenesis has been directly correlated with metastasis of melanoma, prostate cancer, breast cancer, and non-small cell lung cancer (NSCLC).[4,73–78] This would support the notion that tumor-associated angiogenesis is dysregulated in such a manner that a biological imbalance exists that favors either the overexpression of local angiogenic factors or the suppression of endogenous angiostatic factors.[4,42,73] Thus, specific chronic inflammatory/fibroproliferative disorders and growth of solid tumors are associated with an enhanced angiogenic environment. Although the complement of positive and negative regulators of angiogenesis may vary among different physiological and pathological settings, the recognition of this dual mechanism of control is necessary to gain a more thorough understanding of this complex process and its significance in regulating net angiogenesis.

C-X-C Chemokines

The C-X-C chemokine family includes cytokines that in their monomeric forms are less than 10 kDa and are characteristically basic heparin-binding proteins (Table II). The family displays four highly conserved cysteine

[71] B. J. Nickoloff, R. S. Mitra, J. Varani, V. M. Dixit, and P. J. Polverini, *Am. J. Pathol.* **144,** 820 (1994).

[72] P. G. Shields and C. C. Harris, *in* "Lung Cancer" (J. A. Roth, J. D. Cox, and W. K. Hong, eds.), p. 3. Blackwell, Boston, 1993.

[73] J. Folkman, K. Watson, D. Ingber, and D. Hanahan, *Nature (London)* **339,** 58 (1989).

[74] A. Maiorana and P. M. Gullino, *Cancer Res.* **38,** 4409 (1978).

[75] M. Herlyn, W. H. Clark, U. Rodeck, M. L. Mancianti, J. Jambrosic, and H. Koprowski, *Lab. Invest.* **56,** 461 (1987).

[76] N. Weidner, J. P. Semple, W. R. Welch, and J. Folkman, *N. Engl. J. Med.* **324,** 1 (1991).

[77] N. Weidner, P. R. Carroll, J. Flax, W. Blumenfeld, and J. Folkman, *Am. J. Pathol.* **143,** 401 (1993).

[78] P. Macchiarini, G. Fontanini, M. J. Hardin, F. Squartini, and C. A. Angeletti, *Lancet* **340,** 145 (1992).

TABLE II
C-X-C CHEMOKINES

Interleukin-8 (IL-8)
Epithelial neutrophil activating protein-78 (ENA-78)
Growth-related oncogene α (GRO-α)
Growth-related oncogene β (GRO-β)
Growth-related oncogene γ (GRO-γ)
Granulocyte chemotactic protein-2 (GCP-2)
Platelet basic protein (PBP)
Connective tissue activating protein-III (CTAP-III)
β-Thromboglobulin (β-TG)
Neutrophil activating protein-2 (NAP-2)
Platelet factor 4 (PF4)
Interferon-γ-inducible protein (IP-10)
Monokine induced by interferon-γ (MIG)
Stromal cell-derived factor-1 (SDF-1)

amino acid residues, with the first two cysteines separated by one nonconserved amino acid residue. In general, these cytokines appear to have specific chemotactic activity for neutrophils. Because of their chemotactic properties and the presence of the C-X-C cysteine motif, these cytokines have been designated the C-X-C chemokine family. These chemokines are all clustered on human chromosome 4, and they exhibit between 20 and 50% sequence similarity on the amino acid level.[79–83]

Several human C-X-C chemokines have been identified, including PF4, NH_2-terminal truncated forms of platelet basic protein [PBP; connective tissue activating protein-III (CTAP-III), β-thromboglobulin (β-TG), and neutrophil activating protein-2 (NAP-2)], IL-8, growth-related oncogene-α (GRO-α), GRO-β, GRO-γ, IP-10, MIG, epithelial neutrophil activating protein-78 (ENA-78), granulocyte chemotactic protein-2 (GCP-2), and stromal cell-derived factor-1 (SDF-1).[79–86] The NH_2-terminal truncated

[79] M. Baggiolini, B. Dewald, and A. Walz, *in* "Inflammation: Basic Principles and Clinical Correlates" (J. I. Gallin, I. M. Goldstein, and R. Snyderman, eds.), pp. 247–263. Raven, New York, 1992.

[80] M. Baggiolini, A. Walz, and S. L. Kunkel, *J. Clin. Invest.* **84,** 1045 (1989).

[81] K. Matsushima and J. J. Oppenheim, *Cytokine* **1,** 2 (1989).

[82] J. J. Oppenheim, O. C. Zachariae, N. Mukaida, and K. Matsushima, *Annu. Rev. Immunol.* **9,** 617 (1991).

[83] M. D. Miller and M. S. Krangel, *Crit. Rev. Immunol.* **12,** 17 (1992).

[84] A. Walz, R. Burgener, B. Car, M. Baggiolini, S. L. Kunkel, and R. M. Strieter, *J. Exp. Med.* **174,** 1355 (1991).

[85] K. Tashiro, H. Tada, R. Heilker, M. Shirozu, T. Nakano, and T. Honjo, *Science* **261,** 600 (1993).

[86] C. C. Bleul, R. C. Fuhlbrigge, J. M. Casasnovas, A. Aiuti, and T. A. Springer, *J. Exp. Med.* **184,** 1101 (1996).

forms of platelet basic protein are generated when platelet basic protein is released from platelet α-granules and undergoes proteolytic cleavage by monocyte-derived proteases.[87] PF4, the first member of the C-X-C chemokine family to be described, was originally identified for its ability to bind to heparin, leading to inactivation of the anticoagulation function of heparin.[88] Both IP-10 and MIG are interferon-inducible chemokines.[89,90] Although IP-10 appears to be induced by all three interferons (IFN-α, IFN-β, and IFN-γ), MIG is unique in that it appears to be only expressed in the presence of IFN-γ.[89] Although IFN-γ induces the production of IP-10 and MIG, this cytokine attenuates the expression of IL-8, GRO-α, and ENA-78.[91,92] These findings would suggest that members of the C-X-C chemokine family demonstrate disparate regulation in the presence of interferons.

GRO-α, GRO-β, and GRO-γ are closely related C-X-C chemokines, with GRO-α originally described for its melanoma growth stimulatory activity.[93–95] IL-8, ENA-78, and GCP-2 were all initially identified on the basis of their ability to induce neutrophil activation and chemotaxis.[79–84] SDF-1 has been described for its ability to induce lymphocyte migration and prevent infection of T cells by human immunodeficiency virus type 1 (HIV-1).[86,96,97] The C-X-C chemokines have been found to be produced by an array of cells.[79–83,92,98–113] Although numerous *in vivo* and *in vitro*

[87] A. Walz and M. Baggiolini, *J. Exp. Med.* **171,** 449 (1990).

[88] E. Deutsch and W. Kain, *in* "Blood Platelets" (S. A. Jonson, R. W. Monto, J. W. Rebuck, and R. C. Horn, eds.), p. 337. Little, Brown, Boston, 1961.

[89] J. M. Farber, *Biochem. Biophys. Res. Commun.* **192,** 23 (1993).

[90] G. Kaplan, A. D. Luster, G. Hancock, and Z. Cohn, *J. Exp. Med.* **166,** 1098 (1987).

[91] G. L. Gusella, T. Musso, M. C. Bosco, I. Espinoza-Delgado, K. Matsushima, and L. Varesio, *J. Immunol.* **151,** 2725 (1993).

[92] S. Schnyder-Candrian, R. M. Strieter, S. L. Kunkel, and A. Walz, *J. Leukocyte Biol.* **57,** 929 (1995).

[93] A. Ansiowicz, D. Zajchowski, G. Stenman, and R. Sager, *Proc. Natl. Acad. Sci. U.S.A.* **85,** 9645 (1988).

[94] A. Ansiowicz, L. Bardwell, and R. Sager, *Proc. Natl. Acad. Sci. U.S.A.* **84,** 7188 (1987).

[95] A. Richmond and H. G. Thomas, *J. Cell. Biochem.* **36,** 185 (1988).

[96] C. C. Bleul, M. Farzen, H. Choe, C. Parolin, I. Clark-Lewis, J. Sodroski, and T. A. Springer, *Nature (London)* **382,** 829 (1996).

[97] E. Oberlin, A. Amara, F. Bacherlerie, C. Bessia, J. L. Virelizer, F. Arenzana-Seisdedos, O. Schwartz, J. M. Heard, I. Clark-Lewis, D. F. Legler, M. Loetscher, M. Baggiolini, and B. Moser, *Nature (London)* **382,** 833 (1996).

[98] T. Yoshimura, K. Matsushima, J. J. Oppenheim, and E. J. Leonard, *J. Immunol.* **139,** 788 (1987).

[99] K. Matsushima, K. Morishita, T. Yoshimura, S. Lavu, Y. Obayashi, W. Lew, E. Appella, H. F. Kung, E. J. Leonard, and J. J. Oppenheim, *J. Exp. Med.* **167,** 1883 (1988).

[100] R. M. Strieter, S. L. Kunkel, H. J. Showell, and R. M. Marks, *Biochem. Biophys. Res. Commun.* **156,** 1340 (1988).

[101] R. M. Strieter, S. L. Kunkel, H. Showell, D. G. Remick, S. H. Phan, P. A. Ward, and R. M. Marks, *Science* **243,** 1467 (1989).

investigations have shown the importance of C-X-C chemokines in acute inflammation as chemotactic/activating factors for neutrophils and mononuclear cells, only more recently has it become apparent that these C-X-C chemokines may be important in the regulation of angiogenesis.

Assays to Assess Role of C-X-C Chemokines in Regulation of Angiogenesis

Although the process of neovascularization is complex, strategies have been designed to exploit these events and to assess whether various molecules behave as either direct angiogenic or angiostatic factors using both *in vitro* and *in vivo* assay systems. The use of microvascular endothelial cells in culture provides the opportunity to generate bioassays that can specifically determine whether a molecule can induce endothelial cell chemotaxis, proliferation, or tube formation.[4,114] However, a major pitfall with any angiogenesis assay system *in vivo* is the need to separate two processes: inflammation that precedes and contributes to angiogenesis must be distinguished from pure neovascularization.[4,114]

---

[102] R. M. Strieter, S. H. Phan, H. J. Showell, D. G. Remick, J. P. Lynch, M. Genard, C. Raiford, M. Eskandari, R. M. Marks, and S. L. Kunkel, *J. Biol. Chem.* **264,** 10621 (1989).

[103] A. J. Thornton, R. M. Strieter, I. Lindley, M. Baggiolini, and S. L. Kunkel, *J. Immunol.* **144,** 2609 (1990).

[104] V. M. Elner, R. M. Strieter, S. G. Elner, M. Baggiolini, I. Lindley, and S. L. Kunkel, *Am. J. Pathol.* **136,** 745 (1990).

[105] R. M. Strieter, S. W. Chensue, M. A. Basha, T. J. Standiford, J. P. Lynch, and S. L. Kunkel, *Am. J. Respir. Cell. Mol. Biol.* **2,** 321 (1990).

[106] T. J. Standiford, S. L. Kunkel, M. A. Basha, S. W. Chensue, J. P. Lynch, G. B. Toews, and R. M. Strieter, *J. Clin. Invest.* **86,** 1945 (1990).

[107] R. M. Strieter, K. Kasahara, R. Allen, H. J. Showell, T. J. Standiford, and S. L. Kunkel, *Biochem. Biophys. Res. Commun.* **173,** 725 (1990).

[108] Z. Brown, R. M. Strieter, S. W. Chensue, P. Ceska, I. Lindley, G. H. Nield, S. L. Kunkel, and J. Westwick, *Kidney Int.* **40,** 86 (1991).

[109] M. W. Rolfe, S. L. Kunkel, T. J. Standiford, S. W. Chensue, R. M. Allen, H. L. Evanoff, S. H. Phan, and R. M. Strieter, *Am. J. Respir. Cell. Mol. Biol.* **5,** 493 (1991).

[110] B. J. Nickoloff, G. D. Karabin, J. N. W. N. Barker, C. E. M. Giffiths, V. Sarma, R. S. Mitra, J. T. Elder, S. L. Kunkel, and V. M. Dixit, *Am. J. Pathol.* **138,** 129 (1991).

[111] R. M. Strieter, K. Kasahara, R. M. Allen, T. J. Standiford, M. W. Rolfe, F. S. Becker, S. W. Chensue, and S. L. Kunkel, *Am. J. Pathol.* **141,** 397 (1992).

[112] A. E. Koch, S. L. Kunkel, L. A. Harlow, D. D. Mazarakis, G. K. Haines, M. D. Burdick, R. M. Pope, A. Walz, and R. M. Strieter, *J. Clin. Invest.* **94,** 1012 (1994).

[113] A. E. Koch, S. L. Kunkel, M. R. Shah, S. Hosaka, M. M. Halloran, G. K. Haines, M. D. Burdick, R. M. Pope, and R. M. Strieter, *J. Immunol.* **155,** 3660 (1995).

[114] J. Folkman and H. Brem, *in* "Inflammation: Basic Principles and Clinical Correlates" (J. Gallin, I. M. Goldstein, and R. Snyderman, eds.), 2nd Ed., p. 232. Raven, New York, 1992.

The cornea micropocket (CMP) assay using poly(hydroxyethyl methacrylate) (Hydron; Interferon Sciences, New Brunswick, NJ) in rabbit, rat, or mouse corneas, the subcutaneous Matrigel (Collaborative Biomedical, Bedford, MA) implant assay in mice,[115] and the chick embryo chorioallantoic membrane (CAM) assay are *in vivo* assay systems that allow the observation of the full development of neovascularization under controlled conditions.[4,114] In each of these systems, inflammation can be separated from angiogenesis by kinetic analysis and light microscopy.[4,114,115] Because endothelial cell migration is an early event of neovascularization and the ability to separate inflammation from angiogenesis is necessary to determine whether a molecule is a direct angiogenic factor, we have extensively employed two angiogenesis assay systems to assess the role of C-X-C chemokines in the regulation of angiogenesis.[58,60,70,116–119] These assays represent *in vitro* quantitation of chemotaxis of either bovine adrenal or human dermal microvascular endothelial cells and the *in vivo* CMP assay. The advantages of the CMP as compared to the CAM assay are the following: (1) the CMP assay is performed in a mammalian system that avoids potential problems with species specificity of reagents, (2) the age of the chick embryo significantly influences the magnitude of the angiogenic response in the CAM assay,[114] and (3) quantitation of angiogenesis can be readily achieved with the CMP assay by computerized image analysis of the colloidal carbon perfused vessels in the cornea.[114]

Role of ELR Motif of C-X-C Chemokines in Regulation of Angiogenesis

Our laboratory and others have previously found that IL-8 can induce angiogenic activity, independent from inflammation.[70,119,120] Recombinant IL-8 mediates both *in vitro* endothelial cell chemotactic and proliferative activity. In addition, IL-8 induces the full development of neovascularization, independent of inflammation in the CMP assay.[60,70,119] Interestingly, another member of the C-X-C chemokine family, PF4, has been shown to

[115] A. Passaniti, R. M. Taylor, R. Pili, Y. Guo, P. V. Long, J. A. Haney, R. R. Pauly, D. S. Grant, and G. R. Martin, *J. Lab. Invest.* **67,** 519528 (1992).

[116] D. R. Smith, P. J. Polverini, S. L. Kunkel, M. B. Orringer, R. I. Whyte, M. D. Burdick, C. A. Wilke, and R. M. Strieter, *J. Exp. Med.* **179,** 1409 (1994).

[117] D. A. Arenberg, S. L. Kunkel, P. J. Polverini, M. Glass, M. D. Burdick, and R. M. Strieter, *J. Clin. Invest.* **97,** 2792 (1996).

[118] D. A. Arenberg, S. L. Kunkel, P. J. Polverini, S. B. Morris, M. D. Burdick, M. C. Glass, D. T. Taub, M. D. Iannettoni, R. I. Whyte, and R. M. Strieter, *J. Exp. Med.* **184,** 981 (1996).

[119] R. M. Strieter, S. L. Kunkel, V. M. Elner, C. L. Martonyl, A. E. Koch, P. J. Polverini, and S. G. Elner, *Am. J. Pathol.* **141,** 1279 (1992).

[120] D. E. Hu, Y. Hori, and T. P. D. Fan, *Inflammation* **17,** 135 (1993).

have angiostatic properties,[57] as well as to attenuate angiogenesis during the growth of tumors.[121] These findings suggest that members of the C-X-C chemokine family can function as either angiogenic or angiostatic factors in regulating neovascularization. Although it remains unclear whether the COOH terminus (heparin-binding domain) of these chemokines dictates their biological role in regulating angiogenesis, the differences in the function of C-X-C chemokines can be explained by alternative structural domains.

We speculated that members of the C-X-C chemokine family may exert disparate effects in mediating angiogenesis as a function of the presence or absence of the ELR motif (Glu-Leu-Arg) for primarily four reasons. First, members of the C-X-C chemokine family that display binding and activation of neutrophils share the highly conserved ELR motif that immediately precedes the first cysteine amino acid residue, whereas PF4, IP-10, and MIG lack this motif.[122,123] Second, IL-8 (which contains the ELR motif) mediates both endothelial cell chemotactic and proliferative activity *in vitro* and angiogenic activity *in vivo*.[60,70] In contrast, PF4 (lacking the ELR motif) has been shown to have angiostatic properties,[57] and it attenuates growth of tumors *in vivo*.[121] Third, the interferons (IFN-α, IFN-β, and IFN-γ) are all known inhibitors of wound repair, especially angiogenesis.[4,5,114] These cytokines, however, upregulate IP-10 and MIG from a number of cells, including keratinocytes, fibroblasts, endothelial cells, and mononuclear phagocytes.[83,89,90] Finally, we and others have found that IFN-α, IFN-β, and IFN-γ are potent inhibitors of the production of monocyte-derived IL-8, GRO-α, and ENA-78,[91,92] supporting the notion that IFN-α, IFN-β, and IFN-γ may shift the biological balance of ELR- and non-ELR-containing C-X-C chemokines toward a preponderance of angiostatic (non-ELR) C-X-C chemokines.

To evaluate whether C-X-C chemokines display disparate angiogenic activity, endothelial cell chemotaxis is performed in the presence or absence of IL-8, ENA-78, PF4, and IP-10 at concentrations of 50 p*M* to 50 n*M*. Endothelial cell chemotaxis is performed in 48-well chemotaxis chambers (Nucleopore, Cambridge, MA). Bovine adrenal or human dermal microvascular endothelial cells are serum starved for 24 hr, then resuspended at a concentration of 10^6 cells/ml in Dulbecco's modified Eagle's medium (DME) with 0.1% (w/v) bovine serum albumin (BSA) and placed into each

[121] R. J. Sharpe, H. R. Byers, C. F. Scott, S. I. Bauer, and T. E. Maione, *J. Natl. Cancer Inst.* **82,** 848 (1990).

[122] C. A. Hebert, R. V. Vitangcol, and J. B. Baker, *J. Biol. Chem.* **266,** 18989 (1991).

[123] I. Clark-Lewis, B. Dewald, T. Geiser, B. Moser, and M. Baggiolini, *Proc. Natl. Acad. Sci. U.S.A.* **90,** 3574 (1993).

of the bottom wells (25 ml). Nucleopore chemotaxis membranes (5 μm pore size) are first prepared by soaking in 3% (v/v) acetic acid for 12 hr, followed by coating for 2 hr in gelatin (0.1 mg/ml). The membranes are placed over the wells, and chambers are sealed, inverted, and incubated for 2 hr to allow cells to adhere to the membrane. The chambers are then reinverted, 50 ml of sample (containing medium alone, ELR C-X-C chemokines, or non-ELR C-X-C chemokines) is dispensed into the top wells, and chambers are reincubated for an additional 2 hr. Membranes are then fixed and stained with the Diff-Quick staining kit (American Scientific Products, Edison, NJ), and cells that have migrated through the membrane are counted in 10 high power fields (HPF; 400×). Results are expressed as the number of endothelial cells that migrate per HPF after subtracting the background (unstimulated control) to demonstrate specific migration. Each sample is assessed in triplicate. Experiments are repeated at least three times.

Both IL-8 and ENA-78 demonstrate a dose-dependent increase in endothelial migration that is significantly greater than control at concentrations equal to or above 0.1 n*M* and 1 n*M*, respectively.[60] In contrast, neither PF4 nor IP-10 induces significant endothelial cell chemotaxis.[60] Other C-X-C chemokines have been tested for their ability to induce endothelial cell chemotaxis, including ELR C-X-C chemokines IL-8, ENA-78, GCP-2, GRO-α, GRO-β, GRO-γ, PBP, CTAP-III, and NAP-2, and the non-ELR C-X-C chemokines IP-10, PF4, and MIG. In a similar fashion to IL-8 or ENA-78, all of the ELR C-X-C chemokines we have tested demonstrate significant endothelial cell chemotactic activity over the background control, whereas the endothelial cell chemotactic activity induced by MIG is either similar to background control or to the endothelial cell chemotactic activity seen with either PF4 or IP-10.[60] Checkerboard analysis of endothelial cell chemotaxis determines that endothelial cell migration is due to chemotaxis, not chemokinesis. These findings demonstrate that C-X-C chemokines can be divided into two groups with defined biological activities, one which contains the ELR motif and is chemotactic for endothelial cells and the other which lacks the ELR motif and does not induce endothelial chemotaxis.

ELR and Non-ELR C-X-C Chemokines: Angiogenic and Angiostatic Factors

The above studies demonstrate that PF4, IP-10, and MIG are not significant chemotactic factors for endothelial cells, and they suggest that these C-X-C chemokines may be potent inhibitors of angiogenesis. To test this hypothesis, endothelial cell chemotaxis is performed, as above, in the presence or absence of IL-8 (10 nM), ENA-78 (10 nM), or bFGF (5 nM) with or without combining varying concentrations of PF4, IP-10, or MIG from 0 to 10 nM (Fig. 1). Endothelial cell migration in response to either IL-8, ENA-78, or bFGF is significantly inhibited by PF4, IP-10, or MIG in a dose-dependent manner.[60] PF4 and IP-10 at a concentration of 50 pM inhibit either IL-8- or ENA-78-induced endothelial chemotaxis by 50%, whereas PF4 and IP-10 at a concentration of 1 nM attenuate the response to bFGF by 50%. MIG at concentrations of 1 nM, 5 nM, and 10 nM inhibits the endothelial cell chemotactic response to IL-8, ENA-78, and bFGF, respectively, by 50%. Interestingly, although IP-10 and MIG inhibit IL-8-induced endothelial cell chemotactic activity, neither IP-10 nor MIG are effective in attenuating IL-8-induced neutrophil chemotactic activity.[60]

The rat CMP assay of neovascularization[60] is used to determine whether IP-10 or MIG can inhibit the angiogenic activity of either the ELR-containing C-X-C chemokines, bFGF, or vascular endothelial growth factor (VEGF) *in vivo*. Cytokines (IL-8, ENA-78, GRO-α, GCP-2, IP-10, MIG, bFGF, VEGF at a concentration of 10 nM, or combinations of 10 nM each of IL-8 + IP-10, ENA-78 + IP-10, GRO-α + IP-10, GCP-2 + IP-10, IL-8 + MIG, ENA-78 + MIG, bFGF + IP-10, bFGF + MIG, VEGF + IP-10) are combined with sterile Hydron polymer (Interferon Sciences) casting solution, and 5-ml aliquots are air-dried on the surface of polypropylene tubes. Prior to implantation, pellets are rehydrated with normal saline. Animals are anesthetized with an intraperitoneal (i.p.) injection of ketamine (150 mg/kg) and atropine (250 μg/kg). Rat corneas are anesthetized with 0.5% proparacaine hydrochloride ophthalmic solution followed by implantation of the Hydron pellet into an intracorneal pocket (1 to 2 mm from the limbus) created by a cataract knife. Six days after implantation, animals are pretreated i.p. with 1000 U of heparin (Elkins-Sinn, Cherry Hill, NJ), anesthetized with ketamine (150 mg/kg), and perfused with 10 ml of colloidal carbon via the left ventricle. Corneas are then harvested, flattened, and photographed.

No inflammatory response is observed in any of the corneas treated with the above cytokines. Positive neovascularization responses are recorded only if sustained directional ingrowth of capillary sprouts and hairpin loops toward the implant are observed. Negative responses are recorded when either no growth is observed or when only an occasional sprout or hairpin loop displaying no evidence of sustained growth is detected. The ELR C-X-C chemokines (IL-8, ENA-78, GRO-α, and GCP-2), bFGF, and VEGF induce positive corneal angiogenic responses without evidence of significant leukocyte infiltration (Fig. 2). In contrast, Hydron pellets alone, or pellets containing either IP-10 or MIG (10 nM), do not induce a neovascular response in the cornea. When IP-10 is combined with the ELR

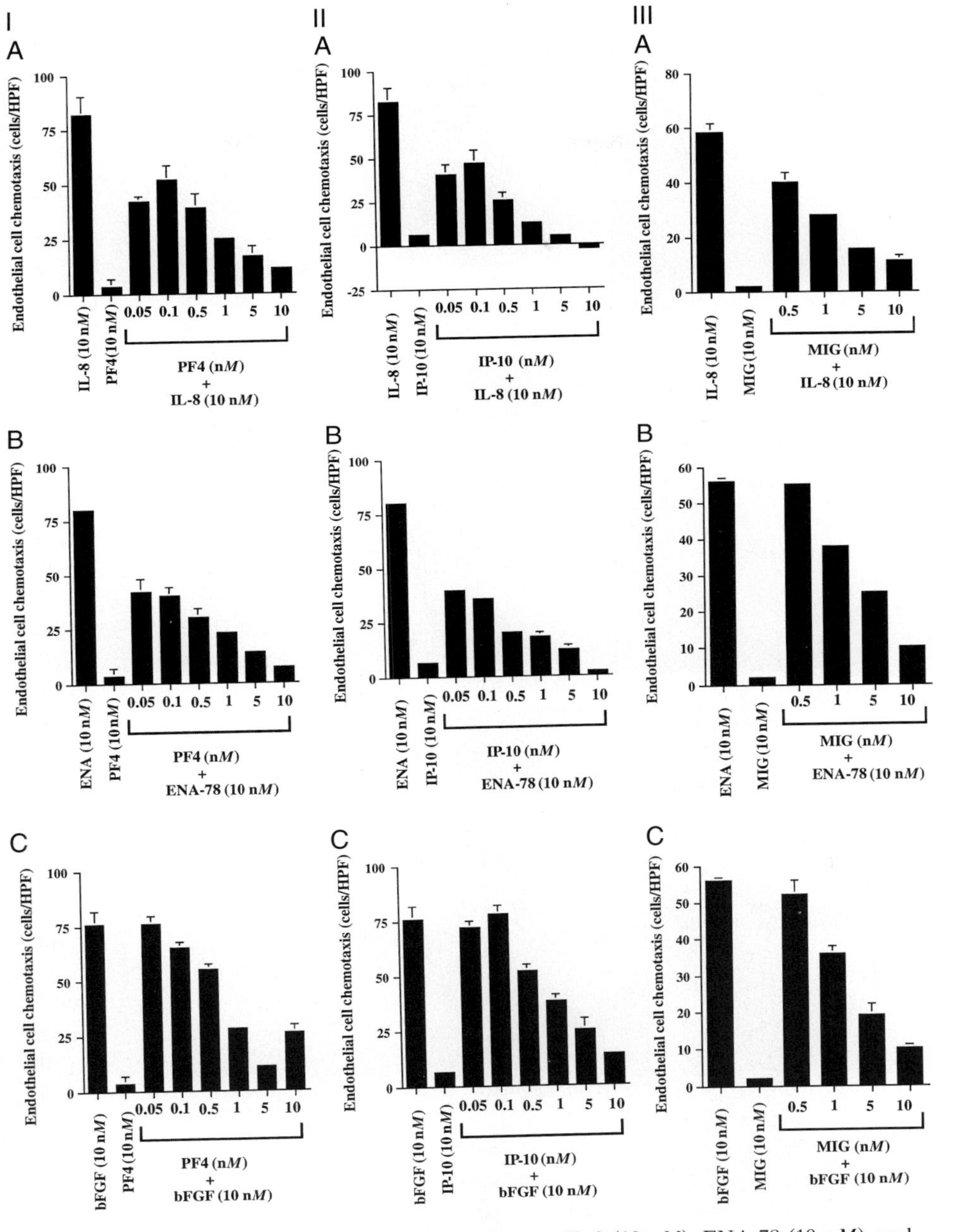

FIG. 1. Endothelial cell chemotaxis in response to IL-8 (10 n*M*), ENA-78 (10 n*M*), and bFGF (5 n*M*) in the presence of varying concentrations PF4 (50 p*M* to 10 n*M*; panel I), IP-10 (50 p*M* to 10 n*M*; panel II), and MIG (500 p*M* to 10 n*M*; panel III). To demonstrate specific migration, background (unstimulated control) migration [cells per high power field (HPF)] was subtracted.

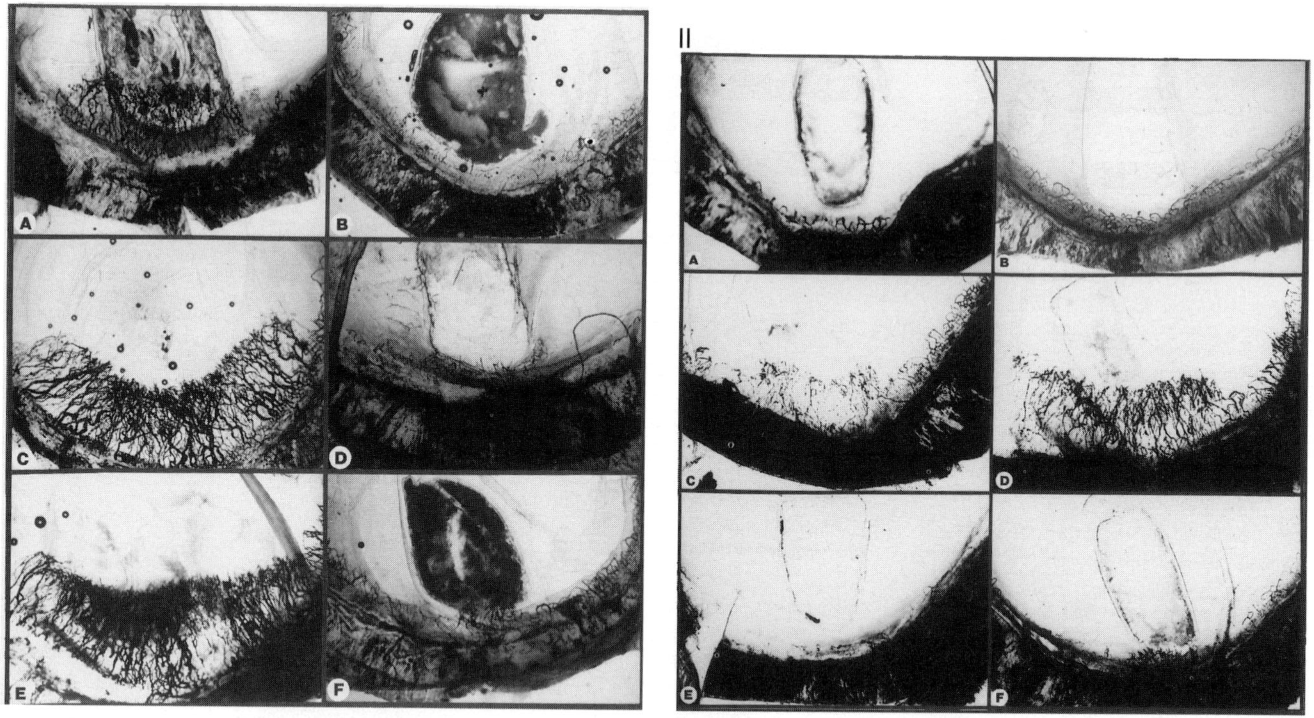

FIG. 2. Rat cornea neovascularization in response to ELR C-X-C chemokines, non-ELR C-X-C chemokines, bFGF, or combinations of compounds. Panel I (A, C, and E) represents the corneal neovascular response to a Hydron pellet containing VEGF (10 n*M*), bFGF (5 n*M*) or ENA-78 (10 n*M*), respectively. Panel I (B, D, and F) represents the corneal neovascular response to a Hydron pellet containing VEGF + IP-10 (10 n*M*), bFGF + IP-10 (10 n*M*), or ENA-78 + IP-10 (10 n*M*), respectively. Panel II (A, B, C, and D) represents the corneal neovascular response to a Hydron pellet containing vehicle control, MIG (10 n*M*), IL-8 (10 n*M*), ENA-78 (10 n*M*), or bFGF (10 n*M*), respectively. Panel II (E and F) represents the corneal neovascular response to the combination of IL-8 with MIG or bFGF with MIG. Magnification: ×25.

C-X-C chemokines (IL-8, ENA-78, GRO-α, and GCP-2), bFGF, or VEGF, IP-10 significantly abrogated the ELR C-X-C chemokine-, bFGF-, or VEGF-induced angiogenic activity. In addition, MIG inhibited IL-8-, ENA-78-, and bFGF-induced corneal angiogenic activity in a similar manner as IP-10.

C-X-C Chemokines: Role of ELR Motif in Regulation of Angiogenesis

To establish whether the ELR motif is the critical structural–functional domain that dictates angiogenic activity for members of the C-X-C chemokine family, muteins are constructed by site-directed mutagenesis of IL-8 that contain either TVR (from IP-10) or DLQ (from PF4) amino acid residue substitutions for the ELR motif, and a mutant of MIG is constructed that contains the ELR motif immediately adjacent to the first cysteine amino acid residue of the primary structure of MIG.[60] The *Escherichia coli* K12 strain DH5aF′ (GIBCO/BRL, Gaithersburg, MD) is used as host for the propagation and maintenance of M13 DNA, and for expression of IL-8 and MIG proteins. Strain CJ236 is used to prepare uracil–DNA for use in site-directed mutagenesis.[124] pGEX 4T-1 (Pharmacia, Piscataway, NJ) is used as the expression vector for all MIG cDNAs.[89] pMAL-c2 (New England Biolabs, Beverly, MA) is used as the expression vector for all IL-8 cDNAs.

Site-directed mutagenesis follows the protocol described by Kunkel *et al.*[124] Individual clones are sequenced using the dideoxynucleotide method[125] with modifications described in the Sequenase (United States Biochemical, Cleveland, OH) protocol. M13 (replicative form) DNA[126] containing confirmed MIG mutations is cleaved with *Bam*HI and *Xba*I (New England Biolabs) and subcloned into pGEX 4T-1. A 197-bp *Sac*I (New England Biolabs) fragment from pMAL.hIL-8 (maltose-binding protein-Ile-Glu-Gly-Arg-human IL-8 fusion protein expression vector) containing the coding sequence for the N-terminal 49 amino acids of the 72-amino acid form of human IL-8 sequence is subcloned to pUC118 (ATCC, Rockville, MD) digested with *Sac*I for site-directed mutagenesis. Clones containing confirmed IL-8 mutations are cleaved with *Sac*I and subcloned into pMAL.hIL-8 digested with *Sac*I.

The open reading frame (ORF) of human MIG[89] is amplified from cDNA generated from interferon-γ-stimulated (1000 U/ml for 16 hr) THP-1

[124] T. A. Kunkel, J. D. Roberts, and R. A. Zakour, *Methods Enzymol.* **154,** 367 (1987).
[125] F. Sanger, S. Nicklen, and A. R. Coulson, *Proc. Natl. Acad. Sci. U.S.A.* **74,** 5463 (1977).
[126] J. Messing, *Methods Enzymol.* **101,** 20 (1983).

cells (ATCC) by PCR (polymerase chain reaction). The 5′ primer used, 5′-CAAGGTGGATCCATGAAGAAAAGTGGTGTTC-3′, encodes a *Bam*HI restriction site immediately upstream of the ATG start site. The 3′ primer, 5′-GCAAGCTCTAGATTATGTAGTCTTCTTTTGACGA-GAACG-3′, encodes a *Xba*I restruction site immediately downstream of the TAA stop codon. The 402-bp fragment is subcloned to M13 mp19 and is confirmed as the human MIG ORF by sequencing. Thr-23 of the ORF sequence is the predicted N-terminal amino acid of the mature, secreted MIG protein,[89] and is hereafter referred to as amino acid position 1. Amino acids Lys-6 and Gly-7 are modified to Glu and Leu, respectively, by site-directed mutagenesis, generating the MIG mutein ELR-MIG. A *Bam*HI restriction site is introduced overlapping Gly-(−1) and Thr-1 by site-directed mutagenesis,[124] resulting in mutein MIG or ELR–MIG cDNAs encoding a Thr-1 to Ser substitution. Then 324-bp fragments obtained from correct M13 RF clones digested with *Bam*HI/*Xba*I are subcloned to pGEX 4T-1 to generate glutathione *S*-transferase–MIG fusion DNAs (GST–MIG or GST–ELR–MIG). The sequence encoded by these DNAs contain the thrombin recognition sequence LVPRGS between the GST and MIG sequences. Digestion of GST–MIG fusion protein with thrombin is predicted to release MIG protein having an N-terminal sequence Gly-Ser-Pro, versus the predicted nonmodified N-terminal sequence Thr-Pro.

Cultures of *E. coli* strain DH5aF′ harboring GST–MIG or GST–ELR–MIG plasmid are grown in 1 liter of LB medium containing 50 μg/ml ampicillin to an OD_{600} of approximately 0.5 at 22° with aeration, and protein expression is induced by the addition of 0.1 m*M* final isopropyl-β-D-thiogalactoside (IPTG) and continued incubation at 22° for 5–6 hr. After induction, the cells are harvested by centrifuging at 6000 *g* for 10 min, and the pellet is washed once in ice-cold phosphate-buffered saline (PBS) and resuspended in 10 ml ice-cold 10 m*M* HEPES, 30 m*M* NaCl, 10 m*M* EDTA, 10 m*M* EGTA, 0.25% (v/v) Tween 20, 1 m*M* phenylmethylsulfonyl fluoride (PMSF; added fresh), pH 7.5 (lysis buffer). The resulting suspension is quick-frozen in liquid nitrogen. After thawing, PMSF is again added to yield a final concentration of 2 m*M*. The suspension is sonciated using a Branson Sonifier 250 equipped with a microtip for 2 min at output setting 5 with a 40% duty cycle. Triton X-100 is added to a final concentration of 1%, and the lysate is nutated for 30 min at room temperature to aid in the solubilization of the fusion protein. The lysate is then centrifuged at 34,500 *g* at 4° for 10 min and the supernatant transferred to a fresh tube.

The GST–MIG protein is purified using the Pharmacia GST purification module essentially as described in the manufacturer's protocol. GST–fusion protein sonicate is passed over a 2-ml glutathione-Sepharose 4B column equilibrated in PBS. After washing with PBS, the GST fusion protein is eluted with 3 column volumes of 10 m*M* reduced glutathione, 50 m*M* Tris-HCl, pH 8.0. Ten units of thrombin per OD_{280} unit of fusion protein is added to the eluted GST–MIG or GST–ELR–MIG fusion protein and incubated at room temperature with occasional gentle mixing for 2–3 hr. MIG or ELR–MIG protein is ≥95% cleaved from the GST protein under these conditions as monitored by sodium dodecyl sulfate–polyacrylamide gel electrophoresis (SDS–PAGE).[127] The MIG-containing solution is adjusted to pH 4.0 using 0.5 *M* sodium acetate, pH 4.0, filtered through a cellulose acetate 0.45-μm filter (Costar, Cambridge, MA), and passed over a Mono S column (Pharmacia) equilibrated with 20 m*M* sodium acetate, pH 4.0. MIG protein is eluted as a single peak using a 0–2 *M* NaCl gradient, and dialyzed against 0.5 m*M* sodium phosphate, 20 m*M* NaCl, pH 7.0. Purified MIG and ELR–MIG are obtained endotoxin-free (<1.0 EU/ml; QCL-1000 test, BioWhittaker, Walkersville, MD), and yields range from 100 to 200 μg/liter [quantitated by amino acid analysis (AAA)] with a purity of >95% (determined by SDS–PAGE, with apparent molecular weight 16,000; AAA accuracy >90%). Mass spectrometry of the purified MIG and ELR–MIG proteins confirms their predicted mass.

The 72-amino acid mature form of IL-8 is amplified using PCR from an IL-8 cDNA in pET3a (kindly provided by I. U. Schraufstatter, Scripps Clinic, La Jolla, CA). The 5′ primer we have used, 5′-AGTGCTAAA-GAACTTAGATG-3′, encodes the beginning reading frame of IL-8, and the 3′ primer, 5′-GGGATCCTCATGAATTCTC-3′, contains a *Bam*HI restriction site immediately after the stop codon. The 220-bp PCR product is purified by gel electrophoresis, digested with *Bam*HI (New England Biolabs), and subcloned into pMal-c2 previously digested with *Xmn*I and *Bam*HI (New England Biolabs) to generate pMal.hIL-8. Clones containing inserts are confirmed by sequencing. Site-directed mutagenesis is used to modify amino acids Glu-4, Leu-5, and Arg-6 to Thr-Val-Arg or Asp-Leu-Gln, generating TVR–IL-8 or DLQ–IL-8, respectively. Correct clones are identified by sequencing, then subcloned as *Sac*I fragments from pUC118 into pMal.hIL-8 digested with *Sac*I.

Cultures of *E. coli* strain DH5aF′ harboring pMal.hIL-8, pMal.TVR–IL-8, or pMal.DLQ-IL-8 are grown in 1 liter of LB medium containing 50 μg/ml ampicillin to an OD_{600} of approximately 0.5 at 37° with aeration, and protein expression is induced by the addition of 0.3 m*M* final IPTG and continued incubation at 37° for 2 hr. Cells are harvested by centrifuging at 5800 *g* for 10 min, and the pellet is washed once in ice-cold PBS and resuspended in 10 ml ice-cold lysis buffer. The resulting suspension is quick-frozen in liquid nitrogen.

[127] U. K. Laemmli, *Nature* (*London*) **227,** 680 (1970).

After thawing, the suspension is sonicated using a Branson Sonifier 250 equipped with a microtip for 2 min at output setting 5 with a 40% duty cycle. The suspension is clarified by centrifugation at 9000 *g* for 10 min at 20°, the supernatant is diluted five-fold in 10 m*M* sodium phosphate, 500 m*M* NaCl, 1 m*M* EGTA, 0.25% Tween 20, pH 7.0 (column buffer), and loaded onto a 10-ml amylose resin (New England Biolabs) affinity column. After extensive washing with column buffer, the MBP fusion protein is eluted with column buffer containing 10 m*M* maltose. Mutein or wild-type IL-8 proteins are released by incubation with 1 μg Factor Xa (New England Biolabs) per OD_{280} unit of MBP fusion protein at room temperature overnight, and they are then passed over a Mono S column (Pharmacia) equilibrated in 10 m*M* sodium phosphate, pH 6.2, and eluted in a 0–1 *M* NaCl gradient. One milliliter of amylose resin is added to fractions containing mutant or wild-type IL-8 protein to remove residual free MBP by incubation for 30 min at room temperature with gentle shaking. The resin is removed by centrifugation, and the supernatant is dialyzed against 0.5 m*M* sodium phosphate, 20 m*M* NaCl, pH 7.5. Yields range from 0.2 to 3.5 mg for wild-type or mutant IL-8 proteins, and samples are $\geq$95% pure as assessed by SDS–PAGE and endotoxin-free ($<$1.0 EU/ml). Proteins are quantitated by amino acid analysis and routinely have accuracies between 88 and 93%.

To determine whether the muteins of IL-8 and MIG behave in a disparate manner in regulating angiogenesis, as compared to wild-type IL-8 and MIG, respectively, endothelial cell chemotaxis and CMP assays are performed. The TVR–IL-8 or DLQ–IL-8 muteins alone fail to induce endothelial cell chemotactic activity, yet these muteins inhibit the maximal endothelial chemotactic activity of wild-type IL-8 by 83 and 88%, respectively.[60] Neither TVR–IL-8 nor DLQ–IL-8 induces neutrophil chemotaxis, nor are they effective in attenuating neutrophil chemotaxis in response to IL-8. Using the *in vivo* rat CMP assay of neovascularization, DLQ–IL-8 (10 n*M*) mutant alone does not induce a positive neovascular response (Fig. 3). However, DLQ–IL-8 (10 n*M*) in combination with either IL-8 (10 n*M*) or ENA-78 (10 n*M*) results in a significant reduction in the ability of either IL-8 or ENA-78 to induce cornea neovascularization.[60] Moreover, the angiostatic activity of the IL-8 muteins is not only unique to inhibition of ELR C-X-C chemokine-induced angiogenic activity, as TVR–IL-8 (10 n*M*) inhibits both bFGF-induced (10 n*M*) maximal endothelial cell chemotaxis and corneal neovascularization. In addition, ELR–MIG (10 n*M*) induces a significant angiogenic response as compared to wild-type MIG.[60] Interestingly, MIG (10 n*M*) inhibits the angiogenic response of ELR–MIG in both endothelial migration and CMP neovascularization assays. These data further support the importance of the ELR motif as a structural domain for angiogenic activity (Table III).

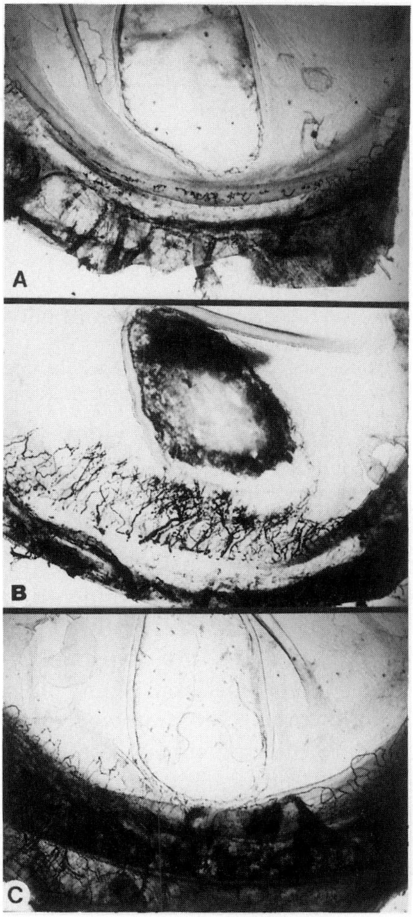

FIG. 3. Rat cornea neovascularization in response to (A) the IL-8 mutant (DLQ–IL-8), (B) wild-type IL-8, or (C) a combination of DLQ–IL-8 + IL-8. Hydron pellets used contained 10 nM DLQ–IL-8, 10 nM IL-8, or DLQ-IL-8 + IL-8. Magnification: ×25.

Role of C-X-C Chemokines in Regulation of Angiogenesis in Non-Small Cell Lung Cancer

To extend these studies to an *in vivo* model system of human tumorigenesis and determine whether C-X-C chemokines regulate tumor-derived angiogenic activity, we employ a human NSCLC/SCID mouse chimera by injecting the human NSCLC cell lines A549 (adenocarcinoma) or Calu 1 (squamous cell carcinoma) into the flanks of SCID mice.[117,118] The A549

TABLE III
C-X-C CHEMOKINES THAT DISPLAY DISPARATE ANGIOGENIC ACTIVITY

Angiogenic C-X-C chemokines containing ELR motif
Interleukin-8 (IL-8)
Epithelial neutrophil activating protein-78 (ENA-78)
Growth-related oncogene α (GRO-α)
Growth-related oncogene β (GRO-β)
Growth-related oncogene γ (GRO-γ)
Granulocyte chemotactic protein-2 (GCP-2)
Platelet basic protein (PBP)
Connective tissue activating protein-III (CTAP-III)
β-Thromboglobulin (β-TG)
Neutrophil activating protein-2 (NAP-2)
Angiostatic C-X-C chemokines that lack ELR motif
Platelet factor 4 (PF4)
Interferon-γ-inducible protein (IP-10)
Monokine induced by interferon-γ (MIG)

and Calu 1 cell lines (ATCC) are maintained in sterile 150-cm^2 tissue culture flasks. Cells are cultured and passaged at 37° in room air/5% CO_2. For inoculation into mice, the cells are trypsinized, harvested, washed, and resuspended in serum-free medium. Four- to six-week-old female CB17-SCID mice (Taconic Farms, Germantown, NY) with serum immunoglobulin (Ig) <1 μg/ml are injected subcutaneously with human NSCLC cells (1×10^6 cells in 100 μl) into each flank. The animals are maintained under sterile conditions in laminar flow rooms and sacrificed in groups of six.

At time of sacrifice, anticoagulated (50 units heparin/500 μl blood) ocular venous plexus blood is collected and centrifuged. The plasma is stored at $-70°$ for later analysis. The lungs are inflated with 4% paraformaldehyde and prepared for histopathological analysis. Tumors are dissected from the mice and measured with a Thorpe caliper (Biomedical Research Instruments, Rockville, MD). A portion of the tumor is fixed in 4% paraformaldehyde for histological analysis and immunohistochemistry. Hematoxylin and eosin (H&E) stained sections are examined under $\times$400 magnification to quantify infiltrating leukocytes. Ten fields are examined in each of the tumor sections from treatment groups. The other portion of the tumor is snap-frozen for subsequent homogenization and sonication in antiprotease buffer (1$\times$ PBS with 2 mM phenylmethylsulfonyl fluoride and 1 μg/ml each of antipan, aprotinin, leupeptin, and pepstatin A), followed by filtration through 0.45-μm filters (Acrodiscs, Gelman, Ann Arbor, MI).

The filtrate is stored at $-70°$ for later analysis of total protein and C-X-C chemokines by specific enzyme-linked immunosorbent assay (ELISA).

For ELISA determination, antigenic IL-8 or IP-10 is quantitated using a modification of a double-ligand method. Flat-bottomed 96-well microtiter plates (Nunc Immuno-Plate I 96-F, Naperville, IL) are coated with 50 μl/well of the polyclonal anti-IL-8 or anti-IP-10 antibody (1 ng/μl in 0.6 *M* NaCl, 0.26 *M* H_3BO_4, and 0.08 *N* NaOH, pH 9.6) for 24 hr at 4° and then washed with PBS, pH 7.5, 0.05% Tween 20 (wash buffer). Microtiter plate nonspecific binding sits are blocked with 2% BSA in PBS, and plates are incubated for 60 min at 37°. Plates are rinsed three times. A 50-μl sample (neat, and 1:10) is added, followed by incubation for 1 hr at 37°. Plates are washed three times, 50 μl/well of biotinylated polyclonal rabbit anti-IL-8 or anti-IP-10 antibody [3.5 ng/μl in PBS, pH 7.5, 0.05% Tween 20, and 2% fetal calf serum (FCS)] is added, and plates are incubated for 45 min at 37°. Plates are washed three times, streptavidin–peroxidase conjugate (Bio-Rad, Richmond, CA) is added, and the plates are incubated for 30 min at 37°. Plates are washed again and chromogen substrate (Bio-Rad) added. The plates are incubated at room temperature to the desired extinction, and the reaction is terminated with 50 μl/well of 3 *M* H_2SO_4 solution. Plates are read at 490 nm in an automated microplate reader (Bio-Tek Instruments, Winooski, VT). Standards are dilutions of recombinant IL-8 or IP-10 from 100 ng/ml to 1 pg/ml (50 μl/well). This method consistently detects IL-8 or IP-10 concentrations greater than 50 pg/ml in a linear fashion. Tumor samples are run in parallel for total protein (TP) content (Pierce, Rockford, IL), and results are expressed as ng of IL-8 or IP-10 per mg total protein. All tumor homogenates are normalized to total protein prior to lyophilization (SpeedVac, Savant, Farmingdale, NY) and used in the CMP assay to assess angiogenic activity.

In the IL-8 or IP-10 depletion studies, SCID mice receive intraperitoneal injections of 500 μl of either neutralizing rabbit anti-human IL-8, neutralizing rabbit anti-human IP-10, control (preimmune) serum, or no treatment, every 48 hr for 6 to 10 weeks, starting at the time of cell inoculation. Polyclonal rabbit anti-human IL-8 or anti-human IP-10 sera is produced by immunization of rabbits with IL-8 or IP-10 (Peprotech, Rocky Hill, NJ) in multiple intradermal sites with Freund's complete adjuvant. The specificity of the IL-8 or IP-10 antiserum is confirmed by ELISA and Western blot analysis, and the reagent is not cross-reactive with a panel of 12 other recombinant human cytokines or the murine chemokines KC and MIP-2. The antiserum at a dilution of 1:1000 inhibits 30 ng of recombinant CXC chemokine in the CMP assay. In the IP-10 treatment studies, mice receive intratumor injections of either human recombinant IP-10 (1 μg in 20 μl of sterile saline every other day) or an equimolar concentration of an irrelevant protein (human serum albumin, HSA) beginning at the time of tumor inoculation. Tumor specimens from these mice are processed as described above.

Interleukin-8 as Endogenous Angiogenic Factor That Promotes Neovascularization during Tumor Growth of Non-Small Cell Lung Cancer

Although previous studies have demonstrated that IL-8 is significantly elevated in human NSCLC,[116] we postulate that IL-8 may be acting *in vivo* to support tumorigenesis by promoting angiogenesis.[117] To test this hypothesis, SCID mice are inoculated with either A549 or Calu 1 cells (10^6 cells/flank) and sacrificed weekly beginning with the second and third week of tumor growth, respectively. The experiments using A549 cells are terminated at 8 weeks due to morbidity noted in the animals secondary to tumor burden. There is a progressive increase in tumor size in A549-bearing animals beginning at week 2 through week 8. In contrast, animals bearing Calu 1 tumors demonstrate little growth until week 8. The production of IL-8, as measured by ELISA, from A549 tumors increases in direct correlation with tumor size ($r^2 = 0.90$), and levels are maximal at 6 to 8 weeks of growth. In contrast, the production of IL-8 by Calu 1 tumors is delayed, yet correlated with tumor size ($r^2 = 0.76$) ($n = 12$ tumors). The size of the A549 and Calu 1 tumors is significantly correlated with their mass ($r^2 = 0.91$). The A549 (adenocarcinoma) tumors produce markedly greater levels of IL-8 (1.2 ± 0.2 versus about 0 ng/mg total protein) and are 50-fold larger in size (279 ± 26 versus 5.6 ± 1.4 mm^2) than Calu 1 (squamous cell carcinoma) tumors by 8 weeks. Plasma levels of IL-8 from both A549 and Calu 1 tumor-bearing animals parallel the production of IL-8 from the primary tumors and are maximal in A549 tumor-bearing animals (1.14 ± 0.21 ng/ml) at 7 weeks. The H&E staining of A549 tumors demonstrates a paucity of infiltrating neutrophils. These results demonstrate that the magnitude of IL-8 production directly correlates with tumor growth and not with infiltrating of leukocytes.

To delineate the role of IL-8 during tumorigenesis of A549 cells in SCID mice, animals are subjected to a strategy of IL-8 depletion. SCID mice injected with A549 cells are treated at the time of inoculation, and every 48 hr for a period of 6 weeks, in one of the following ways: passive immunization with neutralizing IL-8 antibodies, immunization with control antibodies, or untreated. The A549 tumor-bearing animals treated with neutralizing antibodies to IL-8 demonstrate a >40% reduction (101 ± 12.8 mm^2) in tumor growth at 6 weeks, as compared to animals bearing A549 tumors that are either untreated (193 ± 20.7 mm^2) or treated with control antibodies (173 ± 23.7 mm^2, $n = 12$ tumors in each group; $p < 0.01$ for anti-IL-8-treated versus control antibody or untreated tumors). The degree of neutrophil infiltration in the tumors is small, with no significant difference noted between anti-IL-8-treated mice (3.2 ± 0.4 neutrophils per HPF) and control antibody-treated mice (4.1 ± 0.5 neutrophils/HPF; $p = 0.14$). There is no significant difference in tumor size between the control antibody-treated and untreated groups.

To take advantage of the fact that A549 cells express human VLA-2,[128] and to assess spontaneous metastases to the lungs of tumor-bearing mice, fluorescence activated cell sorting (FACS) analysis of human VLA-2 is performed on single-cell suspensions of mouse lungs. At the time of sacrifice, lungs from human A549 tumor-bearing animals are perfused with normal saline and dissected free of the thoracic cavity. The right lung is minced and then incubated for 1 hr in digestion medium (RPMI with 0.02% collagenase type IV and 0.1 mg of bovine pancreas grade II DNase I). Cells are further separated by repeatedly aspirating the cell suspension through a 20-ml syringe. Cells are then pelleted at 600 *g* for 10 min at 4°, resuspended in sterile water for 30 sec to lyse remaining red blood cells (RBCs), washed in 1× PBS, and resuspended in complete medium with 5% FCS. Cells are counted, transferred at a concentration of 5×10^6 cells/ml to fluorescent antibody buffer [1% FA buffer (Difco, Detroit, MI), 1% FCS, and 0.1% azide], and maintained at 4° for the remainder of the staining procedure. One hundred microliters of cells are labeled with fluorescein isothiocyanate (FITC)-conjugated rat anti-human CD49b (1 μg, Pharmingen, San Diego, CA). This antibody recognizes the human a_2 portion of the β_1-integrin, VLA-2. FITC-conjugated rat IgG is used as a control antibody. Unbound antibody is washed with FA buffer, and the cell suspension is analyzed with FACS (Becton Dickinson, Lincoln Park, NJ). The data are expressed as the percentage of cells staining positively with anti-human CD49b. We have demonstrated a trend toward a reduction in the number of metastatic cells in animals treated with neutralizing antibodies to IL-8.

To further determine the mechanism of growth inhibition, we directly evaluate angiogenic activity from A549 tumors of animals that have been treated *in vivo* with either control or neutralizing anti-IL-8 antibodies for 6 weeks. Tumor homogenates are normalized to total protein, incorporated into Hydron pellets, and embedded into the rat corena. Five of the six A549 tumor samples from control antibody-treated animals induce positive corneal angiogenic responses. In contrast, four of six A549 tumor samples from anti-IL-8-treated animals induce no corneal neovascular response,

[128] S. A. Mette, J. Pilewski, C. A. Buck, and S. M. Albelda, *Am. J. Respir. Cell. Mol. Biol.* **8,** 562 (1993).

with the remaining two inducing only weak angiogenic activity. Importantly, there is no infiltration of the corneal tissue by inflammatory cells in any of the test samples, suggesting that the angiogenic responses are mediated entirely by factors present in tumor tissue, rather than by products of infiltrating inflammatory cells.

To further confirm that decreased angiogenic activity correlates with a reduction in tumor vascularity, vessel density is quantified from A549 tumors of SCID mice treated with either control or neutralizing IL-8 antibodies. Tumor tissue sections are dewaxed with xylene and rehydrated through graded concentrations of ethanol. Slides are blocked with normal rabbit serum (BioGenex, San Ramon, CA) and overlaid with 1:500 dilution of either control (goat) or goat anti-Factor VIII-related antigen antibodies. Slides are then rinsed and overlaid with secondary biotinylated rabbit anti-goat IgG (1:35) and incubated for 60 min. After washing twice with Tris-buffered saline, slides are overlaid with a 1:35 dilution of alkaline phosphatase conjugated to streptavidin (BioGenex), and incubated for 60 min. Fast Red (BioGenex) reagent is used for chromogenic localization of Factor VIII-related antigen. After optimal color development, sections are immersed in sterile water, counterstained with Mayer's hematoxylin, and coverslipped using an aqueous mounting solution. A549 tumor specimens from anti-IL-8- and control antibody-treated SCID mice are examined in a blinded fashion for the presence of Factor VIII-related antigen immunolocalization. Sections are first scanned at low magnification (×40) to identify vascular "hot spots." Areas of greatest vessel density are then examined under high magnification (×400) and counted. A distinct area of positive staining for Factor VIII-related antigen is counted as a single vessel. Results are expressed as the mean number of vessels ± SEM per high power field (HPF; ×400). A total of 30 HPFs are examined and counted from three tumors of each of the treatment groups. Tumor vessel density in animals passively immunized with neutralizing IL-8 antibodies is significantly lower than in tumors of control antibody-treated animals (4.5 ± 0.4 versus 10.7 ± 0.4 vessels per HPF, $p < 0.0001$). These studies demonstrate that a primary angiogenic signal for A549 tumor neovascularization *in vivo* is directly mediated by tumor-associated IL-8.

Interferon-γ-Inducible Protein as Endogenous Angiostatic Factor That Inhibits Neovascularization during Tumor Growth of Non-Small Cell Lung Cancer

To determine whether IP-10 protein is present in human NSCLC, freshly isolated specimens of bronchogenic tumors are assessed by specific IP-10

ELISA.[118] The levels of IP-10 from tumor specimens are significantly higher than in normal lung tissue (1.11 ± 0.39 versus 0.23 ± 0.07 ng/mg total proteins, $p < 0.05$). To ascertain whether the presence of IP-10 protein varies by histologic cell type, results are further subdivided by cell type (SCCA versus adenocarcinoma). The increase in IP-10 from NSCLC tissue is entirely attributable to the higher levels of IP-10 present in squamous cell carcinoma (SCCA) as compared to adenocarcinoma (2.25 ± 0.83 versus 0.19 ± 0.06 ng/mg total protein, $p < 0.05$).

Although these experiments demonstrate that IP-10 protein is significantly elevated in specimens of freshly isolated SCCA, we postulate that IP-10 may be acting *in vivo* to regulate tumor-derived angiogenesis.[118] To test this hypothesis, we preincubate specimens of human SCCA normalized to total protein in the presence of either control or neutralizing antibodies to IP-10 and assess their angiogenic activity using either *in vitro* endothelial cell chemotaxis or CMP assay. SCCA samples preincubated with neutralizing IP-10 antibodies, as compared to control antibodies, demonstrate a significant increase in their endothelial cell chemotactic activity (85.0 ± 8.0 versus 48.0 ± 3.0 cells/HPF, $p < 0.05$). These findings are further confirmed using the rat CMP assay, as SCCA specimens preincubated in the presence of neutralizing antibodies to IP-10, as compared to control antibodies, demonstrate an augmented neovascular response in the cornea.

The above findings suggest that IP-10 represents an important endogenous angiostatic factor in freshly isolated NSCLC (SCCA) of the lung. However, to determine if this angiostatic activity is physiologically relevant during the course of *in vivo* tumor growth, a human NSCLC/SCID mouse model of human NSCLC tumorigenesis is employed. SCID mice are inoculated with either A549 (adenocarcinoma) of Calu 1 (SCCA) cells in a similar manner as the experiments outlined above for assessing IL-8. The production of IP-10 from A549 and Calu 1 tumors is inversely correlated with tumor growth ($r = -0.648$ and -0.688 for A549 and Calu 1 tumors, respectively, $p < 0.05$). In addition, IP-10 levels are significantly higher in the Calu 1 (SCCA) tumors compared to A549 tumors. Plasma IP-10 levels from tumor-bearing SCID mice parallel the findings from the primary tumors. Furthermore, the appearance of spontaneous lung metastases in SCID mice bearing A549 tumors occur after IP-10 levels from either the primary tumor or plasma reaches a nadir. To determine whether IP-10 *in vitro* is an autocrine growth factor for these cell lines, A549 and Calu 1 cells are cultured in the presence or absence of recombinant IP-10 for 24 and 48 hr. The presence of exogenous IP-10 does not alter proliferation, as compared to appropriate controls ($p > 0.2$). These findings suggest that IP-10 functions as neither an autocrine growth factor nor an inhibitor of cellular proliferation of human NSCLC cell lines.

Because IP-10 is found to be a potent endogenous angiostatic molecule in SCCA, the reduced expression of IP-10 in A549 (adenocarcinoma) tumors, as compared to Calu 1 (SCCA) tumors, may contribute to their more aggressive behavior. We hypothesize that restoration of tumor-associated IP-10 in A549 tumors can lead to inhibition of tumorigenesis via an IP-10-dependent decrease in tumor-associated angiogenic activity and neovascularization. SCID mice bearing A549 tumors are injected (intratumor) with either recombinant human IP-10 (1 in 20 μl of normal saline) or 20 μl of an equimolar concentration of an irrelevant human protein, HSA, every 48 hr for a period of 8 weeks beginning at the time of tumor cell inoculation. The intratumor administration of IP-10 results in a 40 and 42% reduction in tumor size and mass, respectively, as compared to tumors treated with intratumor HSA (0.79 $\pm$ 0.14 versus 1.37 $\pm$ 0.23 g and 136 $\pm$ 17 versus 217 $\pm$ 27 mm^2 at 8 weeks, respectively, $p \leq 0.03$). To exclude the possibility that IP-10 inhibits tumor growth by recruiting tumoricidal leukocytes, quantitation of tumor-infiltrating leukocytes is performed by light microscopy and morphometric analysis. The A549 tumors from SCID mice treated for 8 weeks with IP-10, as compared to HSA, reveal no evidence for alterations in intratumor leukocyte populations.

We have sought to determine whether IP-10 treatment of the primary tumor also reduces spontaneous lung metastases.[118] Six H&E-stained lung sections from each lung of SCID mice treated with either intratumor IP-10 or HSA for 8 weeks (n = 6 animals per group) are examined under low magnification (×40) for evidence of spontaneous metastases and counted. In addition, using an Olympus BH-2 microscope coupled to a Sony 3CCD camera and a Macintosh IIfx computer, the total area of metastatic tumor burden per lung section is quantitated using NIH Image 1.55 software. Data are expressed as either the number of metastases per lung section or the area of metastatic tumor per section (square pixels at ×40 magnification). The number of metastases is significantly reduced in mice treated with IP-10 as compared with HSA (3.5 $\pm$ 0.6 versus 8.7 $\pm$ 0.9 metastases per lung section, respectively, $p < 0.001$). In addition, the size (area) of the lung metastases per section is also dramatically reduced in the IP-10- as compared to the HSA-treated mice (37 $\pm$ 13 $\times$ 10^3 versus 142 $\pm$ 29 $\times$ 10^3 square pixels, respectively, $p < 0.01$). To further demonstrate the importance of endogenous IP-10 in the regulation of human NSCLC (SCCA) tumor growth, we passively immunize SCID mice bearing Calu 1 tumors with either neutralizing rabbit anti-human IP-10 or control antibodies for 10 weeks. Calu 1 tumors from animals that are passively immunized with neutralizing antibodies to IP-10 for 10 weeks demonstrate a 1.8- to 2.9-fold increase in tumor size, as compared to tumors from animals that have received control antibodies. However, there is no evidence of lung metastases in either group.

To further determine the mechanism of growth inhibition by intratumor administration of IP-10, we directly evaluate angiogenic activity from A549 tumors of animals that have been treated *in vivo* with either IP-10 or HSA for 8 weeks. Tumor homogenates are normalized to total protein, incorporated into Hydron pellets, and embedded into the rat cornea. Nine of 12 A549 tumor samples from IP-10-treated tumors induce no significant corneal neovascular response, with the remaining 3 inducing only weak angiogenic activity. In contrast, 11 of 12 A549 tumor samples from HSA-treated tumors induce positive corneal angiogenic responses. Importantly, there is no infiltration of the corneal tissue by inflammatory cells in any of the test samples, suggesting that the angiogenic response is mediated entirely by factors present in tumor tissue, rather than by products of infiltrating inflammatory cells.

To further confirm that the decreased angiogenic activity correlates with a reduction in tumor vasculature, vessel density by FACS analysis of Factor VIII-related antigen-expressing endothelial cells from the primary tumors is quantified from A549 tumors of SCID mice treated with either intratumor IP-10 or HSA. A portion of tumor is minced into <1 mm^3 sections and incubated for 1 hr in protease digestion medium (Dispase, Collaborative Biomedical Products, Two Oak Park, MA). Cells are then pelleted at 600 *g* for 10 min, red blood cells are lysed, and the sample is washed in PBS and resuspended in complete medium with 5% FCS. Cells are counted, and transferred to fluorescent antibody buffer (1% FA buffer Difco, 1% FCS, and 0.1% (w/v) azide, 5×10^6 cells/ml) and maintained at 4° for the remainder of the staining procedure. One hundred microliters of tumor cells are labeled with rabbit anti-Factor VIII-related antigen antibodies to recognize tumor-associated endothelial cells. Preimmune rabbit IgG is used as a control. FITC-conjugated goat anti-rabbit IgG is used as a secondary antibody. FACS analysis is then employed to detect Factor VIII-related antigen-expressing cells. Tumor-derived endothelial cells are expressed as the percentage of cells from the tumor that are positive for Factor VIII-related antigen. Tumor-derived Factor VIII-related antigen-expressing endothelial cells are markedly reduced in primary tumors treated with IP-10, as compared to HSA (7.8 ± 1.3 versus 32.4 ± 8.7%, $p < 0.05$). These studies demonstrate that IP-10 behaves as a potent angiostatic factor for the attenuation of tumor-derived neovascularization leading to reduced tumorigenicity and spontaneous metastases.

Conclusion

Angiogenesis is regulated by a dual, yet opposing, balance of angiogenic and angiostatic factors. For example, the magnitude of the expression of angiogenic and angiostatic factors in a primary tumor correlates with both tumor growth and potential of spontaneous metastases. The above studies using both *in vitro* and *in vivo* systems have demonstrated that, as a family, the C-X-C chemokines behave as either angiogenic or angiostatic factors, depending on the presence of the ELR motif that immediately precedes the first cysteine amino acid residue of the primary structure of these cytokines. Moreover, in the context of an *in vivo* angiogenesis-dependent model system, C-X-C chemokines are important endogenous factors that regulate tumor growth, tumor-derived angiogenic activity and neovascularization, and the potential for spontaneous metastases. These findings support the notion that both *in vitro* or *in vivo* assay systems are useful to assess the behavior of angiogenic and angiostatic C-X-C chemokines to delineate their role in the regulation of angiogenesis.

Acknowledgments

This work was supported, in part, by National Institutes of Health Grants CA72543 (D. A. A.); CA66180, P50 HL56402, and P50 CA69568 (R. M. S.); HL39926 (P. J. P.); and HL31693 and HL35276 (S. L. K.).

[15] Role of Chemokines in Antibacterial Host Defense

By Theodore J. Standiford, Steven L. Kunkel, and Robert M. Strieter

Introduction

Overview of Bacterial Pneumonia

Bacterial pneumonia is the second most common cause of hospital-acquired infection, and it is the leading cause of death among all nosocomial infections. Despite the development of new broad spectrum antibiotics, nosocomial and even community-acquired bacterial pneumonia continues to be a major cause of morbidity and mortality in the United States. The emergence of multidrug-resistant microbes in the immunocompromised host has made the treatment of these infections increasingly difficult, underscoring the importance of immune host defense in determining the eventual outcome of severe bacterial infection. Effective pulmonary host defense against bacterial invasion is primarily dependent on the rapid clearance of the etiologic agent from the respiratory tract. Innate, or natural immunity, is the principal pathway for effective elimination of bacterial organisms

0076-6879/97 $25.00

from the lung. The three phagocytic cells that constitute innate immunity in the lung include resident alveolar macrophages, recruited neutrophils [polymorphonuclear leukocytes (PMN)], and recruited mononuclear phagocytes. Although effective control of bacterial organisms can occur independently of acquired immune responses, bacterial phagocytosis and killing are greatly enhanced by the presence of specific opsonizing antibody secreted by sensitized B lymphocytes.

Cytokine Mediators of Lung Antibacterial Host Defense

The recruitment and/or activation of leukocytes in the setting of bacterial challenge is a complex and dynamic process that is dependent on the coordinated expression of pro- and anti-inflammatory cytokines. Several cytokine mediators, including tumor necrosis factor-α (TNF-α) and interferon-γ (IFN-γ), represent critical components of both acquired and innate immune responses directed against a variety of infectious agents. Tumor necrosis factor has been shown to enhance natural immunity by directly augmenting the ability of PMN and macrophages to phagocytose and kill gram-positive and gram-negative bacterial organisms. In addition, TNF appears to mediate the influx of PMN to the lung in mice after intrapulmonary challenge with either *Staphylococcus aureus* or *Pseudomonas aeruginosa.* Likewise, IFN-γ can stimulate PMN and macrophage microbicidal activity *in vitro.* Moreover, the intratracheal (i.t.) administration of recombinant IFN-γ can augment PMN influx and bacterial clearance in rats challenged with *P. aeruginosa* intratracheally, which appears to be mediated by endogenously produced TNF.

In addition to TNF and IFN-γ, several lines of evidence suggest that chemokines, particularly C-X-C chemokines, are integral components of antibacterial host defense. First, C-X-C chemokines exert potent PMN stimulatory and chemotactic activities both *in vitro* and *in vivo.* Specific activating effects include enhancement of PMN phagocytosis of immunoglobulin A (IgA)-coated microspheres or IgG-coated erythrocytes, as well as enhanced killing of bacterial, mycobacterial, and fungal organisms by PMN. Furthermore, chemokines, in particular interleukin-8 (IL-8), have been detected in increased amounts within the lungs of patients with pulmonary and pleural bacterial infections. Finally, we have identified the compartmentalized expression of murine macrophage inflammatory protein-2 (MIP-2), the functional murine homolog of IL-8, within the lungs of mice with experimental *Klebsiella* pneumonia, and the inhibition of MIP-2 bioactivity *in vivo* can result in substantial attenuation of bacterial clearance in animals challenged with *Klebsiella.*

Assessment of Effect of Chemokines on Polymorphonuclear Leukocyte Phagocytosis in Vitro

To begin to assess the role of chemokines in antibacterial host defense, *in vitro* systems are used to establish direct effects of chemokines on specific leukocyte populations. The PMN is instrumental in both the phagocytosis and killing of bacterial organisms. The efficacy of microbial phagocytosis by PMN is greatly enhanced by opsonization of bacteria with specific (immunoglobulin) and nonspecific (C3b) opsonins, as well as the expression of Fc and complement receptors on the surface of PMN. In addition, β_2-integrins, particularly CD11b, have been shown to mediate attachment of microbes or foreign substances to PMN, facilitating effective phagocytosis. Once internalized, the killing of bacteria occurs by oxidative and nonoxidative pathways.

To study the effect of C-X-C chemokines on PMN phagocytic and bactericidal activity, human PMN are isolated from blood by Ficoll–Hypaque density centrifugation, sedimentation in 5% dextran/0.9% saline, and separated from erythrocytes by hypotonic lysis. Human PMN (10^5 cells) are incubated with 5% fetal calf serum (FCS) (as a source of opsonin) for 5 min at 37° in sterile polypropylene tubes. The bacterium used in phagocytic and microbicidal studies is *Escherichia coli*, rough strain NCTC 86 (ATCC 4157, Rockville, MD). This strain is ideal for phagocytic assays because it lacks a capsule and is therefore readily ingested. Other organisms, including gram-positive and gram-negative bacterial species, can also be used in phagocytic assays. However, the ingestion of heavily encapsulated organisms, such as *Klebsiella pneumoniae* and *P. aeruginosa*, is poor in the absence of specific opsonizing antibody. *Escherichia coli* (10^6 bacteria) are added to PMN and incubated for 30 min at 37° on a rocker plate, then centrifuged at 300*g* for 10 min at 4°. The supernatants are removed and extracellular bacteria removed by washing the cells three times with cold Hanks' balanced-salt solution (HBSS). Cells are resuspended in 250 μl of HBSS plus 5% FCS (v/v), cytospins performed, and slides allowed to air dry. Diff-Quick staining is performed, and 200 cells per slide are counted to determine the percentage of PMN containing bacteria and the number of intracellular *E. coli* per PMN.

Because it is sometimes difficult to determine if an organism has truly been internalized, alternative approaches have been described.[1,2] One such approach is to label bacteria with fluorescein isothiocyanate (FITC) by incubating organisms with 0.1 mg/ml FITC isomer (Sigma, St. Louis, MO)

[1] D. A. Drevets and P. A. Campbell, *J. Immunol. Methods* **42,** 31 (1991).

[2] S. O'Neill, E. Lesperance, and D. J. Klass, *Am. Rev. Respir. Dis.* **130,** 225 (1984).

in 0.1 *M* $NaHCO_3$ at 25° for 60 min. At the conclusion of the assay, cells are mixed with ethidium bromide (50 mg/ml final concentration) and visualized with a fluorescence microscope. In the presence of ethidium bromide, internalized bacteria will remain bright green in color, whereas extracellular bacteria will turn from green to orange. A second approach to distinguish intracellular from extracellular organisms can be used when studying the phagocytosis of gram-positive organisms, in particular *aureus.* In these studies, lysostaphin (20 μg/ml) can be added for 15 min at the end of phagocytosis to lyse all extracellular but not intracellular organisms.

The C-X-C chemokines IL-8 and murine MIP-2 have been used as stimulants in PMN phagocytic assays. Interleukin-8 represents a likely participant in antibacterial host defense in humans, as this chemokine is a potent PMN activating and chemotactic factor, and IL-8 is expressed within the lungs of patients with lung bacterial infections. Murine MIP-2 represents the functional murine homolog of IL-8 and has previously been shown to be a potent inducer of chemotaxis and degranulation of human PMN. Treatment of human PMN with either human recombinant IL-8 (hrIL-8) or murine recombinant MIP-2 (mrMIP-2) results in a dose-dependent increase in PMN phagocytosis, with a maximal 1.7- and 2.1-fold increase in phagocytic index over that observed in unstimulated PMN.

Assessment of Effect of Chemokines on Polymorphonuclear Leukocyte Complement, Immunoglobulin, and β_2-Integrin in Vitro

Having demonstrated that hrIL-8 and mrMIP-2 enhance PMN phagocytic activity, flow cytometry techniques are employed to determine if these chemokines augment phagocytic activity by regulating the cell surface expression of C3b (CR1), Fc, or CD11b receptors on human PMN. The following monoclonal antibodies (Pharmingen, San Diego, CA) are used as directly FITC-labeled conjugates: anti-FcγRII (CD32), mouse IgG_2; anti-FcγRIII (CD16), mouse IgG_1; anti-CR1 (CD35), mouse IgG_1; anti-Mac-1 (CD11b), mouse IgG_1; and control mouse IgG_1 and IgG_2. Aliquots of purified PMN (5×10^5 cells in 0.1 ml) are stained with 1 μg of monoclonal antibody (MAb) for 30 min on ice in the dark, washed, fixed in 2% paraformaldehyde (w/v) in phosphate-buffered saline (PBS), and stored at 4° in the dark until analyzed. Samples are analysed on an EPICS C flow cytometer with accompanying software (Coulter, EPICS Division, Hialeah, FL), examining at least 20,000 events per sample. Treatment of PMN with hrIL-8 or mrMIP-2 results in a significant increase in the cell surface expression of CD11b. In addition, IL-8, but not mrMIP-2, induces a modest increase in PMN CR1 expression. However, neither hrIL-8 nor mrMIP-2 induces appreciable changes in either FcγII or FcγIII receptor expression.

Assessment of Effect of Chemokines on Polymorphonuclear Leukocytes Microbicidal Activity in Vitro

To determine if chemokines enhance the ability of PMN to kill ingested microbes, bactericidal assays are performed. For these studies, human PMN (10^5 cells) are incubated with 5% FCS for 5 min at 37° in sterile polypropylene tubes. *Escherichia coli* (10^6 bacteria) and chemokines or equal volume of HBSS are added and incubated for an additional 1 hr at 37° on a rocker plate. The supernatants are removed and the cells washed three times with HBSS. The cells are then lysed by adding 1 ml ice-cold sterile water and incubating on ice for 10 min. One milliliter of 2× HBSS is added per sample, then undiluted and serial dilutions are plated to 1×10^{-3} on blood agar plates. Plates are incubated for 18 hr at 37°, and colony counts performed (cfu, colony-forming units). Percent survival of intracellular bacteria is calculated by the following formula:

$$\% \text{ Survival} = \frac{\text{number } E.\ coli \text{ cfu/ml PMN lysate} \times 100}{\text{total number intracellular } E.\ coli}$$

Total intracellular *E. coli* is the product of total number of PMN times the mean number of intracellular *E. coli* per PMN. As shown in Fig. 1, incubation of human PMN with mrMIP-2 or hrIL-8 results in a substantial increase in killing of intracellular *E. coli*, with augmented killing observed at mrMIP-2 or hrIL-8 concentrations of 10 ng/ml and above.

Murine Model of *Klebsiella* Pneumonia

The use of a murine model to assess the role of chemokines in gram-negative bacterial pneumonia has a number of attractive features. Specifically, the advantages of this model are severalfold: (1) the model has been shown to reproduce many of the histological and immunologic features as those found in human and other animal models of gram-negative bacterial pneumonia; (2) manipulation of the *K. pneumoniae* inoculum results in significant alterations in the magnitude of cellular recruitment, lung injury, and lethality; and (3) reagents are available to characterize the biological effects of specific chemokines *in vivo*, either through specific cytokine "depletion" experiments or by transient overexpression of cytokines in a compartmentalized fashion. The use of *K. pneumoniae* in *in vivo* studies is particularly attractive because (1) this organism is a common cause of both nosocomial and community-acquired pneumonia; (2) *K. pneumoniae* is one of several gram-negative bacteria that can be difficult to treat clinically due to novel mechanisms of antibiotic resistance (i.e., extended spectrum β-lactamases); and (3) the mice chosen for our studies readily develop a

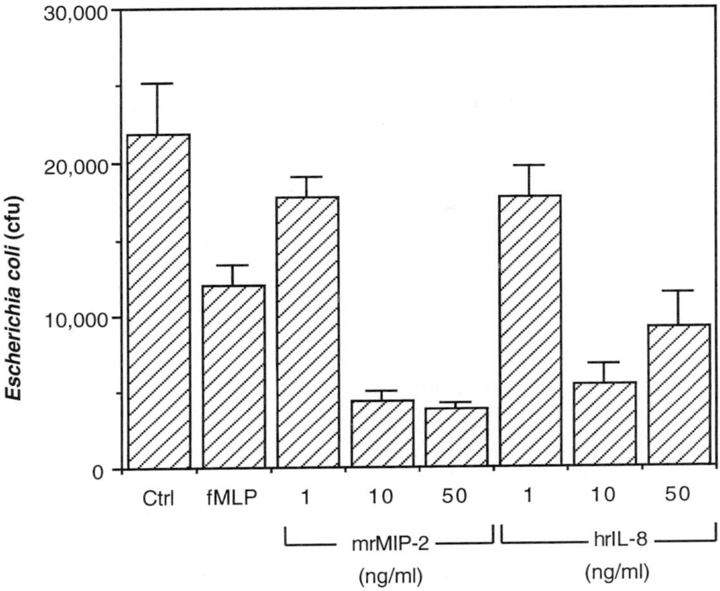

FIG. 1. Effect of mrMIP-2 and hrIL-8 on killing of intracellular *E. coli* by resting and chemokine-treated human PMN. Resting or cytokine-stimulated PMN (10^5) were incubated with 5% FCS and *E. coli* (10^6) in HBSS for 1 hr and then the number of viable bacteria in cell lysates determined. Decreases in *E. coli* colony-forming units reflect increased PMN microbicidal activity.

lethal pneumonia in response to intratracheal inoculation with this organism.

It is important to realize that considerable genetic heterogeneity in susceptibility to *K. pneumoniae* exists. For example, the inbred strains CBA/J and C3H, as well as the outbred strain CD-1, readily develop pneumonia in response to intratracheally administered *K. pneumoniae*, whereas B57B/6 are more resistant to *K. pneumoniae* when administered intratracheally. Furthermore, the virulence of *K. pneumoniae* is strain-specific, and is determined by the expression of a number of virulence factors, including pili and capsules.

Klebsiella pneumoniae Inoculation

We chose to use *K. pneumoniae* strain 43816, serotype 2 (ATCC) in our studies, as this strain has been shown to induce an impressive inflammatory response in mice.[3] *Klebsiella pneumoniae* is grown in tryptic soy broth (Difco, Detroit, MI) for 18 hr at 37°. The concentration of bacteria in broth is determined by measuring the amount of absorbance at 600 nm. A standard of absorbancies based on known colony-forming units is used to calculate the inoculum concentration. Growth for this period results in a concentration of approximately 10^9 cfu/ml medium. Bacteria are then pelleted by centrifugation at 10,000 rpm for 30 min, washed two times in saline, and resuspended at the desired concentration. Animals are anesthetized with approximately 1.8–2 mg pentobarbital per animal intraperitoneally. A midline incision is made through the skin overlying the trachea. The trachea is then exposed by blunt dissection, and 30 μl of bacterial inoculum or saline is administered via a sterile 26-gauge needle. The skin incision is closed with a single surgical staple.

The pathophysiological event that occur in response to intratracheal challenge with *K. pneumoniae* are clearly dependent on the dose of bacteria administered. Administration of 10^2 cfu in naive CBA/J or CD-1 mice results in an approximately 20–50% mortality, with the majority of animals clearing the infection without long-term sequela. Using a dose of 10^3 cfu, survival is observed in less than 50% of animals, whereas inoculum doses of 10^4 cfu or greater are universally fatal within 4 days of intratracheal challenge. All animals inoculated with *K. pneumoniae* at doses of 10^3 cfu or greater develop signs of systemic toxicity, including lethargy, decreased food intake, and ruffled fur by 24 hr. Gross examination of the lung 48 hr after inoculation reveals dense consolidation of the lung, which occurs in a lobar distribution, similar to that observed in humans with *Klebsiella* pneumonia. Histological examination of the lungs at that time reveals the presence of substantial quantities of intraalveolar PMN (Fig. 2B,C), as well as a moderate influx of airspace macrophages. In addition, *K. pneumoniae* organisms are present in alveolar spaces and intracellularly within alveolar macrophages and PMN (Fig. 2D).

Assessment of Lung Leukocyte Influx in Klebsiella Pneumonia

As discussed above, lung inflammation after the intratracheal administration of *K. pneumoniae* can be assessed grossly by histological examination. To more precisely quantitate numbes of leukocytes in the lung alveolar spaces, total cell counts from bronchoalveolar lavage (BAL) fluid are determined. At various time points, animals are sacrificed and the right ventricle perfused with 1 ml (PBS) containing 5 mm EDTA. The trachea is then exposed and cannulated with a 1.7 mm o.d. polyethylene catheter. Bron-

[3] I. A. J. M. Bakker-Woudenberg, A. F. Lokerse, M. T. Ten Kate, J. W. Mouton, M. C. Woodle, and G. Storm, *J. Infect. Dis.* **168,** 164 (1993).

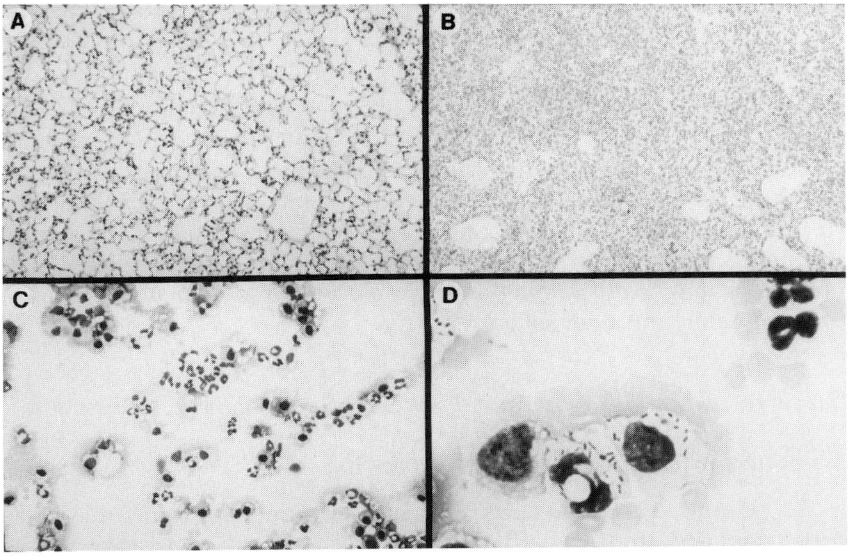

FIG. 2. Composite histological sections and bronchoalveolar lavage (BAL) differentials from saline- or *K. pneumoniae*-challenged mice. (A and B) Histological sections of lung 48 hr after the intratracheal administration of saline or *K. pneumoniae* (10^3 cfu), respectively. (C and D) BAL differentials from *K. pneumoniae*-challenged animals at 48 hr at magnification ×40 and ×100, respectively.

cholalveolar lavage is performed by instilling PBS containing 5 m*M* EDTA in 1-ml aliquots. Approximately 5 ml of lavage fluid is retrieved per mouse, total cells are counted after hypotonic lysis, and cytospins prepared for determination of BAL cell differential counts. Bronchoalveolar lavage cell differentials in mice 48 hr after inoculation with *K. pneumoniae* 10^3 cfu reveal 35.4 ± 5.1 macrophages, 63.6 ± 4.9% PMN, and 1.0 ± 0.7 lymphocytes, whereas BAL cell differentials in saline-challenged control animals consist of 97.9 ± 0.4% macrophages, 0.9 ± 0.3% PMN, and 1.3 ± 0.4% lymphocytes.

Numbers of PMN in the lung can be indirectly quantitated by assessing whole lung myeloperoxidase (MPO) activity.[4] Myeloperoxidase is an enzyme that is present in large quantities within PMN but not other leukocyte or stromal cell populations. At various times, animals are sacrificed and the right ventricle perfused with 1 ml PBS to remove blood from the pulmonary vasculature. The heart and lungs are removed enbloc, then the heart, vascular structures, and major airways are dissected away. Whole

[4] S. E. Goldblum, K. M. Wu, and M. Jay, *J. Appl. Physiol.* **59,** 1978 (1985).

lungs are homogenized in 2 ml of 50 mM potassium phosphate, pH 6.0, hexadecyltrimethylammonium bromide (5 g/liter) and 5 mM EDTA. The resultant homogenate is sonicated and centrifuged at 12,000g for 15 min at 4°. The supernatant is mixed 1 : 15 with assay buffer containing 100 mM potassium phosphate, pH 6.0, 0.083 ml H_2O_2 (30% stock solution diluted 1 : 100, v/v), and 0.834 ml *o*-dianisidine hydrochloride (10 mg/ml) and read at 490 nm in an enzyme-linked immunosorbent assay (ELISA) reader. MPO units are calculated as the change in absorbance over time. Inoculation of animals with 10^3 cfu *K. pneumoniae* results in a greater than 3-fold increase in lung MPO activity at 48 hr, as compared to matched animals receiving saline intratracheally.

Characterization of Chemokine Expression in *Klebsiella* Pneumonia

Assessment of Chemokine mRNA Expression

To determine the relevance of specific cytokines to events that occur in response to intratracheal bacterial challenge, it is imperative to first characterize the time course and the cellular sources of cytokine expression during the evolution of *Klebsiella* pneumonia. Measurement of both cytokine mRNA and protein production is assessed in multiple organs to determine if the expression of cytokines is compartmentalized. Cytokine mRNA in organ homogenates is determined by either Northern blot analysis or reverse transcription–polymerase chain reaction (RT–PCR). In our experience, Northern blot analysis is often not sufficiently sensitive to detect specific cytokine message in organ homogenate, unless mRNA is concentrated by oligo(dT) purification. RT–PCR represents a much more sensitive method of detecting cytokine mRNA.

Lungs are harvested at specific times postinoculation with *K. pneumoniae*, immediately snap-frozen in liquid nitrogen, and stored at −70°. Total cellular RNA from the lungs is isolated by homogenizing the lungs in a solution containing 25 mM Tris, pH 8.0, 4.2 M guanidine isothiocyanate, 0.5% Sarkosyl, and 0.1 M 2-mercaptoethanol.[5] After homogenization, the suspension is added to a solution containing an equal volume of 100 mM Tris, pH 8.0, 10 mM EDTA, and 1.0% sodium dodecyl sulfate (SDS). The mixture is extracted twice each with phenol–chloroform and chloroform–isoamyl alcohol. The RNA is isopropyl alcohol precipitated and the pellet dissolved in diethyl pyrocarbonate (DEPC) water. Total RNA is determined by spectrometric analysis at 260 nm wavelength.

[5] J. M. Chirgwin, A. E. Przybyca, R. J. MacDonald, and W. J. Rutter, *Biochemistry* **18,** 5294 (1979).

Five micrograms of total RNA is reverse-transcribed into cDNA utilizing a reverse transcription kit (BRL, Gaithersburg, MD) and oligo(dT)12–18 primers. The cDNA is then amplified using the specific primers for murine MIP-2, with β-actin primers serving as a control. The MIP-2 sense and antisense primers used have the sequences 5′-GCTGGCCACCAACCACCAGG-3′ and 5′-AGCGAGGCACATCAGGTACG-3′, respectively, giving an amplified product of 350 base pairs. The β-actin sense and antisense primers used have the sequences 5′-ATGGATGACGATATCGCTC-3′ and 5′-GATTCCATACCCAGGAAGG-3′, respectively, giving an amplified product of 812 base pairs. The amplification buffer contains 50 m*M* KCl, 10 m*M* Tris-HCl (pH 8.3), and 2 m*M* $MgCl_2$. Specific oligonucleotide primers are added (200 ng/sample) to the buffer, along with 5μl of the reverse-transcribed cDNA samples. The mixture is incubated for 5 min at 94°, then cycled 35 times at 93° for 45 sec, 52° for 45 sec, and elongated at 72° for 80 sec.

After amplification, the sample (8 μl) is separated on a 2% agarose gel containing 0.3 mg/ml (0.003%) of ethidium bromide and bands visualized and photographed using UV transillumination. The photographic negative is then processed using video densitometry and an image analysis program (Scion, Frederick, MD). The specificity of the MIP-2 PCR product is confirmed using Southern blot analysis, transfer to nylon membranes, and hybridization with a digoxigenin-labeled internal murine MIP-2 oligonucleotide probe. As shown in Fig. 3, the intratracheal administration of *K. pneumoniae* results in the induction of MIP-2 mRNA in lung homogenates by 24 hr postinoculation, with maximal expression by 48 hr after bacterial challenge. MIP-2 mRNA was not detected in the lungs of untreated animals, with minimal expression noted in the lungs of animals 48 hr after the intratracheal administration of saline.

Quantitation of Chemokine Protein Production in Klebsiella Pneumonia

Whereas Northern blot analysis or RT–PCR allows for semiquantitative assessment of cytokine mRNA expression, quantitation of chemokine levels requires measurement of cytokines using specific and sensitive ELISA. At designated time points, whole lungs are harvested and homogenized in 2 ml of lysis buffer containing 0.5% Triton X-100, 150 m*M* NaCl, 15 m*M* Tris, 1 m*M* $CaCl_2$, and 1 m*M* $MgCl_2$ (pH 7.40). Homogenates are incubated on ice for 10 min, then centrifuged at 2500 rpm for 10 min at 4°. Supernatants are collected, passed through a 0.45-μm filter (Gelman, Ann Arbor, MI), then stored at −20° for assessment of cytokine levels.

Murine MIP-2 (or other cytokines) can be quantitated using a modification of a double-ligand method. Briefly, flat-bottomed 96-well microtiter

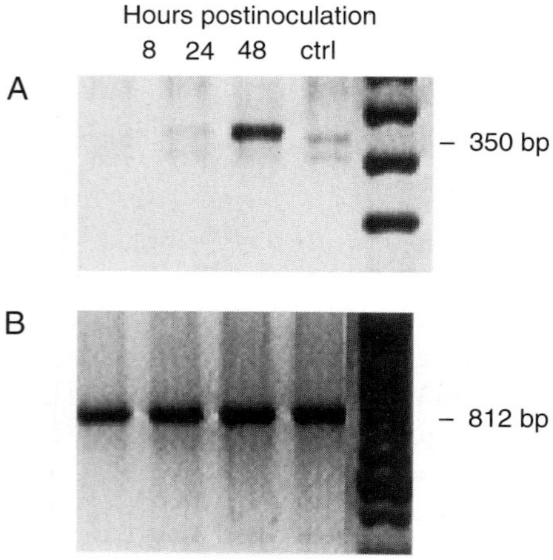

FIG. 3. Time-dependent production of MIP-2 mRNA in lung homogenates after inoculation with *K. pneumoniae* (10^3 cfu). (A) PCR product of MIP-2 (35 cycles), (B) PCR product of β-actin (35 cycles). Molecular weight markers are found to the right. The lungs from three animals were combined at each specific time point assayed.

plates (Nunc Immuno-Plate I 96-F, Naperville, IL) are coated with 50 μl/well of rabbit anti-MIP-2 antibody (1 μg/ml in 0.6 *M* NaCl, 0.26 *M* H_3BO_4, and 0.08 *N* NaOH, pH 9.6) for 16 hr at 4° and then washed with PBS, pH 7.5, 0.05% Tween 20 (wash buffer). Microtiter plate nonspecific binding sites are blocked with 2% bovine serum albumin (BSA) in PBS and incubated for 90 min at 37°. Plates are rinsed four times with wash buffer and diluted (neat and 1:10). Cell-free supernatants (50 μl) in duplicate are added, followed by incubation for 1 hr at 37°. Plates are washed four times, followed by the addition of 50 μl/well biotinylated rabbit anti-MIP-2 antibodies [3.5 μg/ml in PBS, pH 7.5, 0.05% Tween 20, and 2% FCS (v/v)], and plates incubated for 30 min at 37°. Plates are washed four times, streptavidin–peroxidase conjugate (Bio-Rad, Richmond, CA) added, and the plates incubated for 30 min at 37°. Plates are washed again four times and chromogen substrate (Bio-Rad) added. The plates are incubated at room temperature to the desired extinction, and the reaction is terminated with 50 μl/well of 3 *M* H_2SO_4 solution. Plates are read at 490 nm in an ELISA reader. Standards are 1/2 log dilutions of recombinant murine MIP-2 from 1 pg/ml to 100 ng/ml.

This ELISA method consistently detects murine MIP-2 concentrations above 25 pg/ml. The ELISA does not cross-react with TNF, IL-2, IL-4, IL-6, or IFN-γ. In addition, the ELISA does not cross-react with other members of the murine chemokine family, including murine JE/MCP-1,

MIP-1α, RANTES, KC, growth related oncogene α (GROα), and epithelial neurophil activating protein-78 (ENA-78). Measurement of chemokines in *Klebsiella* pneumonia reveals an increase in MIP-2 protein in lung homogenates by 24 hr postinoculation, with maximal levels at 48 hr and a return to baseline levels by 6 days after *Klebsiella* challenge (Fig. 4). The induction of MIP-2 appears to be compartmentalized to the lung, as no increases in MIP-2 levels are noted in plasma or liver homogenates. Using a similar sandwich ELISA technique, we have also noted increases in the C-C chemokine MIP-1α in the lungs after administration of *K. pneumoniae*, with the time course of expression similar to that observed with MIP-2. No significant increases in the expression of the chemokines GROα, MIG (monokine induced by IFN-γ), or RANTES was noted. Tumor necrosis factor and IL-12 are also expressed during the evolution of pneumonia. However, levels of these cytokines remain elevated to 6 days postinoculation with *K. pneumoniae*.

Immunolocalization of Chemokines in Pneumonia

To identify cellular sources or reservoirs of cytokines in the lung in *Klebsiella* pneumonia, immunohistochemistry is performed using a purified rabbit anti-murine antibody. Lungs are inflated with 1 ml of 4% paraformaldehyde in PBS to improve resolution of anatomic relationships, then excised *en bloc.* Paraffin-embedded tissue is processed for immunohistochemical localization of cytokines using a peroxidase staining technique (Vectastain

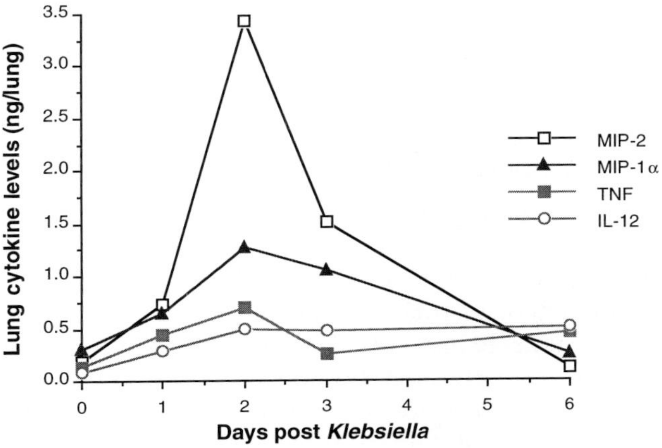

FIG. 4. Time-dependent production of cytokines in lung homogenates from CBA/J mice after the intratracheal administration of saline or *K. pneumoniae* (10^2 cfu).

Elite ABC kit, Vector, Burlingame, CA). In addition, immunolocalization of cytokines can be performed on BAL cytospin preparations. For paraffin-embedded tissue, tissue sections are dewaxed with xylene, and rehydrated through graded concentrations of ethanol (100 through 50% ethanol). BAL cytospins are incubated with 1:1 methanol/3% H_2O_2 (v/v) for 15 min at room temperature prior to staining. Tissue or BAL cytospin nonspecific binding sites are blocked using blocking solution (Vector). Tissue sections or cytospins are washed with PBS and incubated with a 1:1000 to 1:5000 (v/v) dilution of purified rabbit anticytokine antibodies (1 mg/ml) or equivalent dilutions of purified rabbit IgG for 24 hr at 4°. The slides are washed, then incubated for 30 min with anti-rabbit biotinylated antibodies (Vector) at room temperature. Next, the sections are washed twice in Tris-buffered saline and incubated with Vectastain ABC reagent (Vector). True blue chromogenic substrate (BioGenex, San Ramon, CA) is used for localization of specific cytokines. After optimal color development (usually between 5 and 10 min), tissue sections are rinsed in sterile water, counterstained with Fast Red (KPL Laboratories, Gaithersburg, MD), and coverslipped using Permount mounting solution.

Immunohistochemical staining of lung tissue reveals the presence of cell-associated MIP-2 within alveolar macrophages by 24 hr postadministration of *K. pneumoniae.* By 48 hr, alveolar macrophages continue to express MIP-2. However, staining of MIP-2 is also noted within PMN at this time. These studies suggest that although the alveolar macrophage appears to initiate and maintain the influx of PMN in pneumonia, recruited PMN can further amplify the inflammatory cascade via the elaboration of chemotactic and activating cytokines. A similar staining pattern has been noted for TNF, with early expression of TNF by alveolar macrophages followed by immunolocalization of TNF to PMN, corresponding to a bimodel pattern of TNF protein production within the lung after *Klebsiella* challenge.

Assessment of Chemokine Function in *Klebsiella* Pneumonia

In the previous series of experiments, we have characterized the inflammatory response that occurs in response to intratracheal administration of *K. pneumoniae*, as well as identified specific cytokines that are produced during the evolution of *Klebsiella* pneumonia. To establish causal relationships between cytokines and specific biological effects, *in vivo* cytokine depletion studies are performed. End points that can be examined in this murine model include lung inflammatory cell influx, cytokine expression, lung bacterial clearance, and survival. The performance of chemokine neutralization studies requires the administration of specific neutralizing antibodies or receptor antagonists. We have successfully generated potent rabbit anti-murine neutralizing antibodies for a variety of chemokines, including MIP-2. Neutralization of MIP-2 is confirmed *in vitro* by the inhibition of human PMN chemotaxis to murine recombinant MIP-2. *In vivo*, MIP-2 production in lung in response to *Klebsiella* challenge intratracheally can be inhibited by the intraperitoneal administration of rabbit anti-murine MIP-2 serum or purified antibodies.

To generate these antibodies, specific pathogen-free rabbits are immunized intradermally and intraperitoneally with 20 μg of cytokine in Freund's complete adjuvant (CFA). A total volume of 1.5 ml is given intradermally (less than 100 μl/injection site) and 0.5 ml is given intraperitoneally. The animals are boosted every 2 weeks two times with cytokine and Freund's incomplete adjuvant, starting 2 weeks after the primary immunization. Two weeks after the third injection, the rabbits are bled and the sera accessed for antibody titers. The animals are boosted every 2–3 months or as necessary to keep the titers at or greater than 10^6. To purify polyclonal antibodies, immune serum is passed through a sterile protein A column, and immunoglobulin is eluted. The anti-MIP-2 antibodies produced have a half-life of 48 hr *in vivo*, which requires the repeated administration of purified antibody every 48 hr for studies lasting longer than that time. The advantage of purified polyclonal antibodies rather than whole serum is the potential avoidance of nonspecific effects of serum components other than IgG. However, we have not noted significant toxicity due to administration of immune serum, even when given for a 2-week period.

Assessment of Role of Chemokines as Mediators of Inflammation in Klebsiella Pneumonia

To assess the role of MIP-2 as a mediator of PMN influx in bacterial pneumonia, mice are passively immunized with either rabbit anti-murine MIP-2 serum (0.5 ml) or purified polyclonal rabbit anti-MIP-2 antibody (5 mg/animal) intraperitoneally, followed 2 hr later by the intratracheal administration of *K. pneumoniae* (10^3 cfu). At various time points, lungs are harvested and assessed for total lung MPO activity. In addition, BAL cell counts and differentials are performed to quantitate numbers of airspace inflammatory cells. Because lung MPO activity is greatest at 48 hr postchallenge, this time point is examined. As shown in Fig. 5, the intratracheal inoculation with *K. pneumoniae* in animals receiving preimmune serum results in a nearly three-fold increase in lung MPO activity at 48 hr, as compared to animals administered saline intratracheally. Importantly, lung MPO activity in anti-MIP-2-treated animals inoculated with *Klebsiella* at 48 hr was decreased by 50% as compared to animals receiving preimmune serum, whereas anti-MIP-2 antibodies did not alter lung MPO activity in saline-challenged animals. Furthermore, passive immunization with anti-MIP-2 serum resulted in an approximately 45% reduction in percent BAL

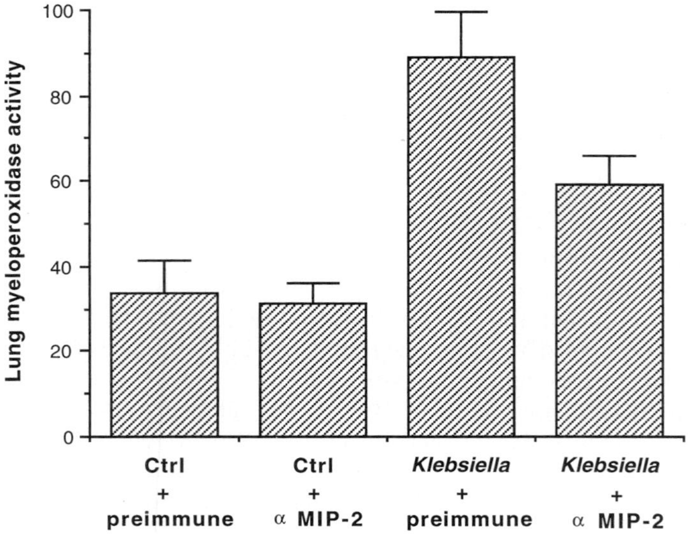

FIG. 5. Lung MPO activity 48 hr after intratracheal inoculation with either saline or *K. pneumoniae* (10^3 cfu) in animals pretreated intraperitoneally with either anti-MIP-2 antibody or preimmune serum. Error bars represent standard error of the mean (SEM).

PMN and total PMN without influencing numbers of BAL macrophage or lymphocyte populations (data not shown). These observations indicate that MIP-2 is a relevant mediator of PMN influx in gram-negative bacterial pneumonia.

Assessment of Role of Chemokines in Lung Bacterial Clearance

To determine if MIP-2 represents a critical signal in effective lung bacterial clearance *in vivo*, the effect of MIP-2 neutralization on lung *K. pneumoniae* colony-forming units is determined. Mice are administrated either preimmune serum or anti-MIP-2 serum 2 hr prior to *K. pneumoniae* inoculation, and lungs are harvested at 24 and 48 hr postinoculation. Lungs are placed in 1 ml of sterile saline, then homogenized with a tissue homogenizer under a vented hood. After the lung homogenates are placed on ice, serial 1 : 10 dilutions are made in sterile 96-well plates (Costar, Cambridge, MA). Ten microliters of each dilution is plated on soy base blood agar plates, plates are incubated for 18 hr at 37°, and then colonies are counted. As shown in Table I, animals receiving anti-MIP-2 serum have a 6- and 4.4-fold greater number of *K. pneumoniae* colony-forming units isolated from lung homogenates at 24 and 48 hr postinoculation, respectively, as

TABLE I
EFFECT OF ANTI-MIP-2 SERUM ON *Klebsiella pneumoniae* COLONY-FORMING UNITS IN LUNGS[a]

Time postinoculation (hr)	Treatment	*K. pneumoniae* colony-forming unit
24	Preimmune	$1.74 \pm 0.10 \times 10^3$
	Anti-MIP-2	$10.1 \pm 0.04 \times 10^3$
48	Preimmune	$8.32 \pm 0.27 \times 10^6$
	Anti-MIP-2	$36.3 \pm 0.12 \times 10^6$

[a] Animals were pretreated with either 0.5 ml rabbit anti-MIP-2 antibody or rabbit preimmune serum intraperitoneally 2 hr prior to intratracheal inoculation with 10^3 cfu *K. pneumoniae.*

compared to control animals. To assess for early dissemination of *Klebsiella*, *K. pneumoniae* colony-forming units are determined in both plasma and liver homogenates at 24 hr postinoculation. As shown in Table II, animals receiving anti-MIP-2 antibodies have a 3.3- and 2.4-fold increase in *K. pneumoniae* colony-forming units in liver and plasma, as compared to animals receiving preimmune serum.

Assessment of Role of Chemokines in Regulating Proinflammatory Cytokine Expression in Pneumonia

One potential mechanism whereby chemokines may modulate the host response to bacterial pathogens is by directly regulating the expression of

TABLE II
EFFECT OF ANTI-MIP-2 SERUM ON *Klebsiella pneumoniae* COLONY-FORMING UNITS IN PLASMA AND LUNG[a]

Site	Treatment	*K. pneumoniae* CFU
Plasma	Preimmune	$1.17 \pm 0.21 \times 10^1$
	Anti-MIP-2	$2.82 \pm 0.45 \times 10^1$
Liver	Preimmune	$3.47 \pm 0.35 \times 10^1$
	Anti-MIP-2	$1.12 \pm 0.18 \times 10^2$[b]

[a] At 24 hr postinoculation. Animals were pretreated with either 0.5 ml rabbit anti-MIP-2 antibody or rabbit preimmune serum intraperitoneally 2 hr prior to intratracheal inoculation.
[b] $p < 0.05$ as compared to animals receiving preimmune serum.

other relevant inflammatory cytokines. To address this issue, animals are passively immunized with rabbit anti-murine MIP-2 antiserum or control serum 2 hr prior to inoculation with *K. pneumoniae.* Plasma and lung are then harvested at the time of maximal cytokine expression (~48 hr postinoculation). As compared to animals receiving control serum, treatment with anti-MIP-2 serum does not alter the levels of TNF, IFN-γ, or IL-12 in plasma or lung (data not shown). Moreover, levels of the C-C chemokine MIP-1α are not different in anti-MIP-2-treated animals as compared to controls. These studies suggest that the biological effects of MIP-2 are direct and are not dependent on MIP-2-mediated regulation of other activating and/or chemotactic cytokines.

Tumor necrosis factor has been shown to be expressed early in several animal models of endotoxemia, and *in vivo* neutralization studies indicate that TNF serves as a proximal inducer of more distal proinflammatory cytokines, including members of the C-C chemokine family. To determine if the expression of MIP-2 is dependent on the endogenous production of TNF in pneumonia, TNF neutralization studies are performed. To inhibit TNF *in vivo*, we have employed a soluble human TNF receptor–immunoglobulin fusion protein (sTNFR:Fc, Immunex, Seattle, WA). For these studies, 100 μg of sTNFR:Fc is given intraperitoneally 2 hr prior to *K. pneumoniae* administration. The sTNFR:Fc is composed of soluble dimeric human p80 TNF receptor linked to the Fc region of human IgG_1.[6] The sTNFR:Fc has been used because this construct results in 50–1000 times greater efficacy in neutralizing TNF bioactivity as compared to monomeric soluble TNF receptor or anti-TNF antibody. Like observations made in endotoxin-challenged mice, treatment with sTNFR:Fc does not alter MIP-2 levels in plasma or lungs after the intratracheal administration of *K. pneumoniae*, indicating that the production of MIP-2 in *Klebsiella* pneumonia is not dependent on a TNF-driven cytokine cascade.

Assessment of Contribution of Chemokines to Survival in Klebsiella Pneumonia

The use of animal models of human disease allow for a number of clinical end points to be assessed, with the most important of these outcomes being survival. The necessity of specific cytokines for effective bacterial clearance resulting in host survival can again be determined with *in vivo* cytokine depletion studies. To determine if cytokines are necessary components of antibacterial host defense, we have utilized sublethal doses of *K. pneumoniae* ranging from LD_{20} to LD_{50}. Conversely, if detrimental effects

[6] K. D. Peppel, M. Crawford, and B. Beutler, *J. Exp. Med.* **174,** 1483 (1991).

of specific cytokines are anticipated, LD_{50}–LD_{100} doses of *Klebsiella* are used. To assess the contribution of MIP-2 to ultimate survival in *Klebisella* pneumonia, mice are administered either 0.5 ml rabbit preimmune control serum or rabbit anti-murine MIP-2 serum intraperitoneally 2 hr prior to *K. pneumoniae* inoculation (10^2 or 10^3 organisms). For mortality studies, 0.25 ml of preimmune or anti-MIP-2 serum is administered at 48-hr intervals following the initial administration. As shown in Fig. 6A, no mortality is observed in animals passively immunized with control serum until 48 hr postinoculation with 10^3 cfu, at which time mortality increases substantially, with 100% lethality noted by 5 days postinoculation. A significant increase

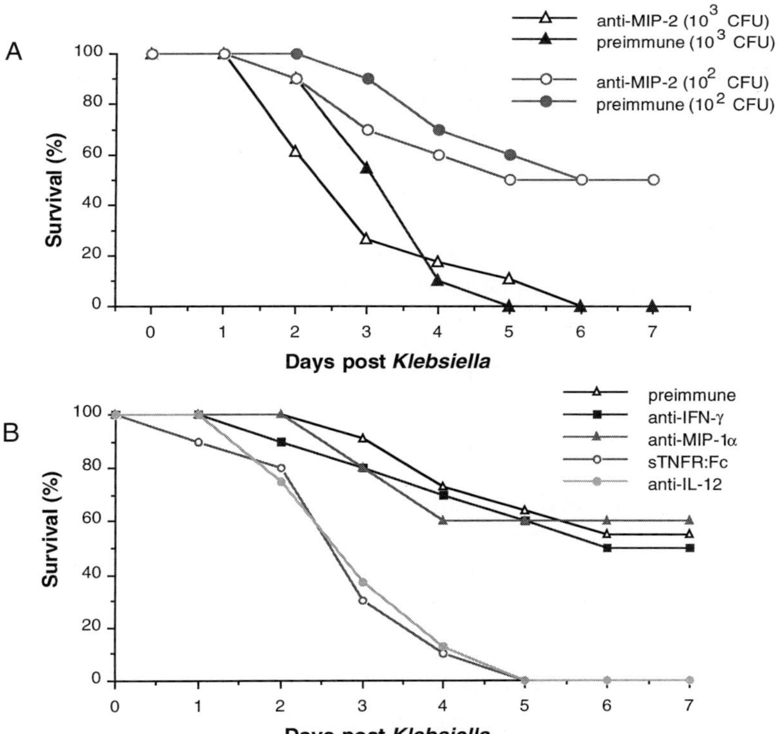

FIG. 6. Effect of cytokine depletion on survival in *Klebsiella* pneumonia. (A) CBA/J mice were passively immunized with rabbit preimmune or rabbit anti-murine MIP-2 serum, then inoculated with either 10^2 or 10^3 cfu *K. pneumoniae.* (B) CBA/J mice were passively immunized with either rabbit preimmune or rabbit anti-murine MIP-1α serum, anti-IFN-γ serum, anti-IL-12 serum, or sTNFR:Fc, followed by the intratracheal administration of 3×10^2 cfu *K. pneumoniae.*

in early mortality is noted in the anti-MIP-2-treated group at 48 and 72 hr as compared to controls, whereas no differences in mortality are noted at time points after 72 hr. Using an inoculum of 10^2 cfu *K. pneumoniae*, a similar trend toward increased early mortality is observed in anti-MIP-2 treated-animals. However, long-term survival in animals treated with anti-MIP-2 serum is not different than animals receiving control serum.

To understand the relative importance of other cytokines in effective lung antibacterial host defense, cytokine neutralization studies are performed using a *K. pneumoniae* inoculum of 3×10^2 cfu. As compared to CBA/J mice receiving control serum intraperitoneally, treatment with either sTNFR:Fc (100 μg i.p., then 50 μg i.p. every 48 hr) or rabbit anti-murine IL-12 serum (0.5 ml i.p., followed by 0.25 ml i.p. every 48 hr) results in substantial decreases in both early and long-term survival (Fig. 6B). In contrast, no increased lethality is observed in animals treated with either rabbit antimurine IFN-γ serum or rabbit antimurine MIP-1α serum. These studies indicate that the expression of TNF, MIP-2, and IL-12 is required for early and/or late survival in murine *Klebsiella* pneumonia, whereas depletion of IFN-γ or MIP-1α is not by itself sufficient to alter outcome in animals infected with *K. pneumoniae.* It is possible that chemokines such as MIP-1α may participate in events other than bacterial clearance, including lung remodeling and repair. The role of MIP-1α and other C-C chemokines in *Klebsiella* pneumonia is the focus of ongoing studies.

New Approaches to Assessing Role of Chemokines in Antibacterial Host Defense

Intratracheal Cytokine Gene Therapy

Advances in our understanding of cytokine biology have made possible the implementation of novel strategies targeted at augmentation or neutralization of specific cytokines and/or cytokine receptors. However, immunotherapy is often complicated by significant toxicity, especially when cytokines or cytokine antagonists are given systemically. Technological breakthroughs in gene therapy using adenoviral, retroviral, and liposomal vectors have provided powerful tools to study the biological effects of specific cytokine mediators, as well as develop novel and clinically applicable therapies in certain disease states.

We have employed an intratracheal gene therapy approach using recombinant human type 5 adenoviral vectors in the treatment of murine *Klebsiella* pneumonia. The use of intratracheal adenoviral gene therapy to over-express specific cytokines within the lung has a multitude of desirable features. The adenoviral vector utilized results in very efficient infectivity of the airway epithelium and subsequent genetic transfer of the viral genome. This results in high levels of cytokine expression in lung, with little increase in plasma cytokine levels posttransfection. Furthermore, because the E1 region of the viral genome has been deleted, the recombinant adenoviruses used are replication-deficient, resulting in only transient expression of cytokines within the lung. Although relatively short-lived transgene expression represents a major drawback of adenoviral gene therapy in chronic inheritable diseases such as cystic fibrosis, the transient nature of cytokine expression in bacterial pneumonia may result in early beneficial effects (i.e., rapid bacterial clearance) without the detrimental effects of prolonged cytokine overexpression. Hence, the transient and compartmentalized expression of cytokines using intratracheally administered recombinant adenoviral vectors appears not only to represent an ideal approach to investigating the effect of specific cytokines, but also to provide a clinically applicable therapeutic intervention.

Our initial studies have focused on transient lung IL-12 transgene expression using a recombinant human type 5 adenoviral vector containing the murine IL-12 p35 and IL-12 p40 cDNAs.[7,8] This Ad5-based recombinant contains an expression cassette for the p35 subunit cDNA of murine IL-12 inserted in E1 in place of nucleotides 342–3523, and an expression cassette for the p40 subunit inserted in E3 in place of nucleotides 28,133–30,818. Expression of each IL-12 subunit cDNA is driven by the human cytomegalovirus (CMV) immediate early promoter and terminated by the polyadenylation signal of SV40. Transcription of both cDNAs is in the same direction as the E1 and E3 transcription units that they replace. Administration of this recombinant adenovirus results in biologically active IL-12 expression both *in vivo* and *in vitro.*

In *in vivo* studies, animals are administered 5×10^8 plaque-forming units (PFUs) of either the IL-12 vector (Ad5mIL-12) or a control virus (Ad5LacZ) intratracheally. The administration of Ad5mIL-12 results in the expression of both the p35 and p40 PCR products by 24 hr, with continued expression to 7 days postadministration. Furthermore, administration of Ad5IL-12 results in an approximate nine-, three-, and six-fold increase in IL-12 levels at days 1, 2, and 7, respectively, as compared to animals receiving Ad5LacZ. Immunohistochemical studies indicate that the expression of cell-associated IL-12 is localized predominantly to airway

[7] A. J. Bett, W. Haddara, L. Prevec, and F. L. Graham, *Proc. Natl. Acad. Sci. U.S.A.* **91,** 8802 (1994).

[8] J. M. Bramson, M. Hitt, W. S. Gallichan, K. L. Rosenthal, J. Gauldie, and F. L. Graham, *Hum. Gene Ther.* **7,** 333 (1996).

epithelial cells posttransfection. However, alveolar macrophages and alveolar epithelial cells also express IL-12 after Ad5IL-12 administration.

To determine the effect of transient lung IL-12 transgene expression on survival in *Klebsiella* pneumonia, animals are administered *K. pneumoniae* and adenovirus concomitantly (Fig. 7). In animals that received *K. pneumoniae* (10^3 cfu) alone without adenoviral administration, mortality is noted by 48 hr after bacterial administration, with only 10% of animals surviving to 5 days and no animals surviving long term. Similarly, animals that received *K. pneumoniae* immediately followed by Ad5LacZ (5×10^8 pfu) intratracheally experience significant mortality by 48 hr, with no animals surviving to 5 days. In contrast, treatment of *K. pneumoniae*-infected animals with Ad5IL-12 (5×10^8 pfu) results in a significant shift in the survival curve to the right, with approximately 45% of animals clearing the infection and surviving long term.

Similar gene therapy approaches are being used to assess the effect of chemokine overexpression on outcome in murine bacterial pneumonia. A human Ad5-based recombinant containing an expression cassette for the rat MIP-2 cDNA inserted in the E1 region of the viral genome (Ad5rMIP-2) has been developed. The intratracheal administration of this recombinant adenoviral vector at a concentration of 4×10^8 pfu results in the expression of bioactive rMIP-2 for 10–14 days, which is temporally associated with an impressive influx of PMN into the airspace of both rats and mice. In preliminary studies using a *K. pneumoniae* inoculum of 3×10^2 cfu, treatment with Ad5rMIP-2 paradoxically decreases survival in *Klebsiella* pneu-

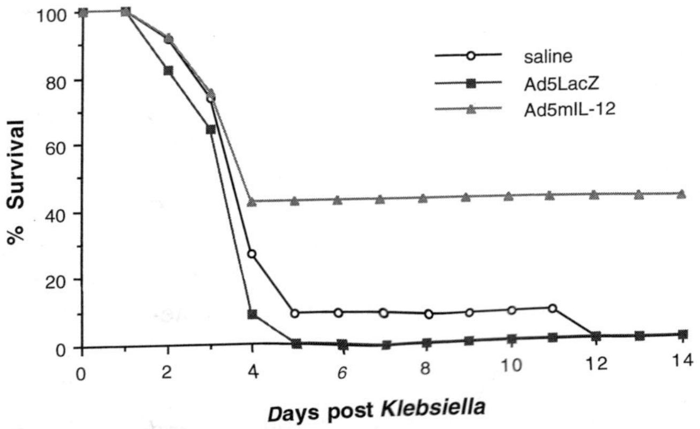

FIG. 7. Effect of transient IL-12 transgene expression on survival in *Klebsiella* pneumonia. CBA/J mice were inoculated with 10^3 cfu *K. pneumoniae*, immediately followed by administration of 5×10^8 PFU Ad5LacZ or Ad5mIL-12.

monia from 50% in animals receiving control vector to no survival in animals receiving Ad5rMIP-2. Mechanisms by which transient overexpression of rMIP-2 results in increased rather than decreased lethality in pneumonia have yet to be defined. However, a likely explanation is that MIP-2 overexpression results in exuberant PMN influx and/or activation culminating in excessive lung injury. Studies are ongoing to identify mechanisms involved.

Conclusion

The difficulty in treating lung infections caused by gram-negative bacteria underscores the need to explore novel approaches to therapy in the treatment of patients with this disease. By using a murine model of *Klebsiella* pneumonia, we can determine effects and consequences of specific immune interventions as may occur in human gram-negative pneumonia. Furthermore, by understanding the roles of specific cytokines in the setting of bacterial pneumonia, we can then develop novel therapeutic strategies to inhibit or augment these cytokines, which may translate into improved outcomes in patients with this devastating disease.

Acknowledgments

This research was supported in part by National Institutes of Health Grants 1P50HL46487, HL50057, HL31693, HL35276, HL58200, HL57243, CA66180, and AA10571.

[16] Animal Models of Asthma: Role of Chemokines

By David A. Griffiths-Johnson, Paul D. Collins, Peter J. Jose, and Timothy J. Williams

Introduction

Chemokines are potent chemoattractants for leukocytes, and this has led to speculation that chemokines might be responsible for the influx of inflammatory cells into the lung seen in asthma.[1,2] In this chapter, we discuss the pathogenesis of the disease, focusing on the inflammatory component, how it has been modeled in different animal species, and how this has

[1] K. B. Bacon and T. J. Schall, *Int. Arch. Allergy Immunol.* **109,** 97 (1996).
[2] R. Alam, J. York, M. Boyars, S. Stafford, J. A. Grant, J. Lee, P. Forsythe, T. Sim, and N. Ida, *Am. J. Respir. Crit. Care Med.* **153,** 1398 (1996).

0076-6879/97 $25.00

led to the discovery of specific chemokines. We review the evidence that chemokines are involved in cell recruitment *in vivo* in animal models.

Pathogenesis of Asthma

Symptoms and Prevalence

Asthma is characterized by airway hyperreactivity (AHR) manifest by episodic symptoms of acute bronchoconstriction with shortness of breath. Underlying this AHR is airway inflammation, the modulation of which is a key target in asthma therapy. The incidence of asthma is widely believed to be rising, and epidemiological studies support this for childhood asthma in Europe, Australasia, the United States, and in urban communities in Africa.

Experimental Asthma

To understand the mechanisms of the disease, an asthmatic episode may be induced experimentally by instillation of antigen directly into individual lobes of the lung of an asthmatic. This results in an immediate-onset airway response (IAR) characterized by bronchoconstriction that occurs within minutes and lasts for up to 1–2 hr. This is followed by a late-onset airway response (LAR) in 30–50% of cases associated with the development of airway inflammation and AHR. LAR typically has an onset of 5–6 hr and lasts 12–24 hr. There is frequently a transient increase in the number of neutrophils during IAR, but if the response progresses to an LAR there is an influx of eosinophils, monocytes–macrophages, and lymphocytes in parallel with the development of AHR.[3,4]

Histology

The pathological changes in chronic asthma have been assessed by taking biopsies of airway tissues and by performing bronchial washes.[5] Biopsies frequently show a damaged epithelium in the large and medium-size airways, with thickening of the basement membrane and edema due to microvascular leakage. Increased mucus secretion occurs, and mucus can mix with leaked plasma proteins to form mucus plugs in the airways.

[3] J. B. Sedgwick, W. J. Calhoun, G. J. Gleich, H. Kita, J. S. Abrams, L. B. Schwartz, B. Volovitz, M. Ben-Yaakov, and W. W. Busse, *Am. Rev. Respir. Dis.* **144,** 1274 (1991).

[4] W. J. Metzger, D. Zavala, H. B. Richerson, P. Moseley, P. Iwamota, M. Monick, K. Sjoerdsma, and G. W. Hunninghake, *Am. Rev. Respir. Dis.* **135,** 433 (1987).

[5] P. K. Jeffery, *Br. Med. Bull.* **48,** 23 (1992).

One of the most consistent features, even in mild asthmatics, is the presence of inflammatory cells in the tissues and in the bronchial washings. These include activated eosinophils, lymphocytes, and immature macrophages.[6,7] As discussed below, there is now evidence that chemokines may be involved not only in accumulation but also in activation of leukocytes.[8]

The severity of clinical symptoms correlates with several different measures of the inflammatory response. For example, both the number of eosinophils and the levels of major basic protein and eosinophil cationic protein, both released by activated eosinophils, are increased in the bronchoalveolar lavage (BAL) fluid of asthmatics.[9,10] There is *in vitro* evidence that these proteins, which are highly cationic, produce changes to the integrity of the respiratory epithelium and that this may be involved in the generation and maintenance of AHR.[11]

Animal Models of Asthma

A number of different species have been used to model acute asthma; however, chronic asthma is less well modeled. In sensitized guinea pigs, there is a rapid and intense contraction of the bronchial smooth muscle after antigen challenge both *in vitro* and *in vivo*. Cellular infiltration after antigen challenge was demonstrated by Dunn *et al.*[12] Margination of eosinophils and neutrophils was observed in the peribronchial vasculature only 8 min after antigen challenge. Six hours after antigen challenge, eosinophils were the predominant leukocyte to have migrated into the smooth muscle layer around the large airways. Between 6 and 24 hr eosinophils migrated to the epithelium where they persisted for up to 7 days. There was a parallel increase in eosinophils in BAL fluid. Bronchoconstriction occurred during

[6] J. Bousquet, P. Chanez, A. M. Campbell, A. M. Vignola, and P. Godard, *Clin. Exp. Allergy* **25**(2), 39 (1995).

[7] J. B. Sedgwick, W. J. Calhoun, R. F. Vrtis, M. E. Bates, P. K. McAllister, and W. W. Busse, *J. Immunol.* **149,** 3710 (1992).

[8] N. W. Lukacs, R. M. Strieter, S. W. Chensue, and S. L. Kunkel, *J. Leukocyte Biol.* **59,** 13 (1996).

[9] P. Diaz, M. C. Gonzalez, F. R. Galleguillos, P. Ancic, O. Cromwell, D. Shepherd, S. R. Durham, G. J. Gleich, and A. B. Kay, *Am. Rev. Respir. Dis.* **139,** 1383 (1989).

[10] J. Bousquet, P. Chanez, J. Y. Lacoste, G. Barneon, M. N. Ghavanian, I. Enander, P. Venge, S. Ahlstedt, J. Simony-Lafontaine, P. Godard, and P.-B. Michel, *N. Engl. J. Med.* **323,** 1033 (1990).

[11] G. J. Gleich, N. A. Flavahan, T. Fujisawa, and P. M. Vanhoutte, *J. Allergy Clin. Immunol.* **81,** 776 (1988).

[12] C. J. Dunn, G. A. Elliott, J. A. Oostveen, and I. M. Richards, *Am. Rev. Respir. Dis.* **137,** 541 (1988).

both IAR and LAR, and this was associated with AHR[13] that was 2 to 3-fold above baseline. This is a much smaller degree of AHR than that seen in humans (approximately 10-fold), and this is often cited as a criticism of using the guinea pig to model human asthma. A second difference is the preferential production of immunoglobulin G_1 (IgG_1) in the guinea pig rather than IgE, which is the cell-fixing homocytotropic isotype seen in most species including humans. Nevertheless, the guinea pig has proved an invaluable model in studying the acute allergic reaction.

Rats[14] and mice[15] are increasingly being used to model asthma as the technology to measure lung function in very small animals has been developed. An advantage of both rats and mice is the large range of cytokines, antibodies, and more recently chemokines available for mechanistic studies.

Rabbits are sometimes used to model asthma, but effective immunization requires injection of antigen 24 hr after birth,[16] which can serve as a drawback for many laboratories. Alternative sensitization regimes have been described; however, these are frequently quite protracted.[17] A further drawback to using rabbits is that, in addition to eosinophil recruitment, antigen induces a strong neutrophil influx that is not seen in humans.

Both sheep and dogs are used to a limited extent as some are naturally sensitive to *Ascaris suum.*[18,19] Basenji greyhounds have a spontaneous and hereditary AHR although the mechanism underlying this is unknown.

Allergic airways responses in nonhuman primates have been studied since before the mid-1970s. A large number are naturally sensitive to *Ascaris suum,* including cynomolgus monkeys, which have an approximate 30-fold leftward shift in the bronchoconstriction dose–response curve to methacholine challenge. This is similar to that seen in humans.[20] No other species used to model human asthma shows such levels of AHR.[21] Monkeys have

[13] E. Boichot, V. Lagente, C. Carre, P. Waltmann, J. M. Mencia-Huerta, and P. Braquet, *Clin. Exp. Allergy* **21,** 67 (1991).

[14] W. Elwood, J. O. Lotvall, P. J. Barnes, and K. F. Chung, *J. Allergy Clin. Immunol.* **88,** 951 (1991).

[15] P. S. Foster, S. P. Hogan, A. J. Ramsay, K. I. Matthaei, and I. G. Young, *J. Exp. Med.* **183,** 195 (1996).

[16] W. R. Marsh, C. G. Irvin, K. R. Murphy, B. L. Behrens, and G. L. Larsen, *Am. Rev. Respir. Dis.* **131,** 875 (1985).

[17] W. J. Metzger, K. Sjoerdsma, L. Brown, T. Coyle, C. Page, and C. Touvay, *in* "Ginkgolides—Chemistry, Biology, Pharmacology and Clinical Perspectives" (P. Braquet, ed.), p. 313. J. R. Prous Science, 1988.

[18] W. M. Abraham, *Am. Rev. Respir. Dis.* **135,** S49 (1987).

[19] K. F. Chung, A. B. Becker, S. C. Lazarus, O. L. Frick, J. A. Nadel, and W. M. Gold, *Appl. Physiol.* **58**(4), 1347 (1985).

[20] R. H. Gundel, C. D. Wegner, and L. G. Letts, *Am. Rev. Respir. Dis.* **146,** 369 (1992) (abstract).

[21] M. Pretolani and B. B. Vargaftig, *Biochem. Pharmacol.* **45,** 791 (1993).

been used to assess the role of proinflammatory mediators such as platelet activating factor (PAF), leukotriene B_4 (LTB_4), and interleukin-5 (IL-5). Pretreatment with a single dose of an anti-IL-5 antibody has been shown to suppress antigen-induced eosinophilia for up to 3 months.[22]

Allergic Mechanisms

Current theories of allergic asthma suggest that it results from the body mounting an inappropriate immunologic response to foreign moieties (antigens) encountered usually by inhalation. During the induction phase the antigen is processed by antigen-presenting cells, including macrophages and dendritic cells.[23] These cells migrate to the regional lymph nodes where the antigen is re-presented at the surface of the cell in conjunction with major histocompatibility complex (MHC) class II molecules. Interaction with naive T cells ($CD45RA^+$) leads to development of memory ($CD45RO^+$) T-cell clones that circulate and act as a surveillance system for antigen. These cells have a long life span and recirculate between lymphoid and mucosal tissue. Further antigen stimulation of the $CD45RO^+$ T-cell clones leads to rapid clonal expansion, release, and homing of the cells to target organs. *In vitro* studies on highly purified T cells have demonstrated that a number of chemokines including macrophage inflammatory protein-1α (MIP-1α), MIP-1β, RANTES, and monocyte chemotactic protein-1 (MCP-1) stimulate T-lymphocyte proliferation, endogenous production of IL-2, and increased expression of its cell surface receptor CD25 indicating cell activation.[24]

Antigen-stimulated T cells produce a number of cytokines. Interleukin-4 causes isotype switching of B lymphocytes to produce antigen-specific IgE. Part of this switch mechanism may be due to the action of chemokines. When B cells were stimulated with either RANTES or MIP-1α in combination with IL-4 and MHC class II antigen complex, there was increased production of antigen-specific IgE and IgG_4.[25]

Antigen-specific IgE circulates and binds to mast cells, monocytes–macrophages, basophils, and eosinophils through FcεRI (the high affinity IgE receptor) or FcεRII. Subsequent exposure to antigen results in cross-linking of IgE molecules and triggering of cells, especially mast cells, to

[22] P. J. Mauser, A. Pitman, X. Fernandez, S. K. Foran, G. K. Adams, W. Kreutner, R. W. Egan, and R. W. Chapman, *Am. Rev. Respir. Dis.* **152,** 467 (1995).

[23] A. Bellini, E. Vittori, M. Marini, V. Ackerman, and S. Mattoli, *Chest* **103,** 997 (1993).

[24] D. D. Taub, S. M. Turcovski-Corrales, M. L. Key, D. L. Longo, and W. J. Murphy, *J. Immunol.* **156,** 2095 (1996).

[25] H. Kimata, A. Yoshida, C. Ishioka, M. Fujimoto, I. Lindley, and K. Furusho, *J. Exp. Med.* **183,** 2397 (1996).

release preformed mediators (e.g., histamine), which rapidly cause bronchoconstriction, vasodilation, and edema. Cross-linking also stimulates the synthesis and release of LTB_4, LTC_4, LTD_4, prostaglandin D_2 (PGD_2), and PAF,[26] all of which are proinflammatory and/or bronchoconstrictory. In addition, there is evidence for the production of the C-C chemokine MARC by murine mast cells *in vitro*.[27] The human mast cell leukemic cell line HMC-1 is a source of multiple chemokines including MCP-1, MIP-1α, MIP-1β, RANTES, and IL-8.[28]

Other proinflammatory cytokines whose synthesis and release are triggered by antigen include IL-1α, IL-1β, IL-3, IL-4, IL-5, granulocyte–macrophage colony-stimulating factor (GM-CSF), and tumor necrosis factor-α (TNF-α). IL-3, IL-5, and GM-CSF increase differentiation of eosinophil precursors, release eosinophils from the bone marrow, and prolong their survival in tissues.[29,30] Some of these cytokines have been shown to stimulate the release of chemokines. In a murine model, IL-4-transfected tumor cells were transplanted into the skin of syngeneic mice. Eighteen hours later there was a remarkable eosinophilia at these skin sites accompanied by an increase in eotaxin mRNA.[31]

In 1986, Mosmann *et al.*[32] hypothesized the existence of two subpopulations of murine $CD4^+$ lymphocytes, the Th_1 and Th_2 subsets, based on their profile of cytokine release. Th_1 cells produce IL-2 and interferon-γ (IFN-γ), whereas Th_2 cells produce IL-4 and IL-5. Both Th_1 and Th_2 cells produce IL-3 and GM-CSF. In humans, separation of T cells into distinct populations is more difficult, but there is evidence for this.[33] In a segmental antigen challenge study Robinson *et al.*[34] demonstrated an increase in $CD4^+$ cells with mRNA of a Th_2 cytokine profile in the BAL fluid. This has been supported by further studies measuring mRNA in biopsy tissues[35] and

[26] S. T. Holgate, *Clin. Exp. Allergy* **21,** 11 (1991).

[27] P. A. Kulmburg, N. E. Huber, B. J. Scheer, M. Wrann, and T. Baumruker, *J. Exp. Med.* **176,** 1773 (1992).

[28] R. S. Selvan, J. H. Butterfield and M. S. Krangel, *J. Biol. Chem.* **269,** 13893 (1994).

[29] Y. Yamaguchi, Y. Hayashi, Y. Sugama, Y. Miura, T. Kasahara, S. Kitamura, M. Torisu, S. Mita, A. Tominaga, K. Takatsu, and T. Suda, *J. Exp. Med.* **167,** 1737 (1988).

[30] P. F. Weller, *Clin. Immunol. Immunopathol.* **62,** S55 (1992).

[31] M. E. Rothenberg, A. D. Luster, and P. Leder, *Proc. Natl. Acad. Sci. U.S.A.* **92,** 8960 (1995).

[32] T. R. Mosmann, H. Cherwinski, M. W. Bond, M. A. Giedlin, and R. L. Coffman, *J. Immunol.* **136,** 2348 (1986).

[33] S. Romagnani, *Immunol. Today* **13,** 379 (1992).

[34] D. Robinson, Q. Hamid, A. M. Bentley, S. Ying, A. B. Kay, and S. R. Durham, *J. Allergy Clin. Immunol.* **92,** 313 (1993).

[35] V. Ackerman, M. Marini, E. Vittori, A. Bellini, G. Vassali, and S. Mattoli, *Chest* **105,** 687 (1994).

cytokine levels in the BAL fluid of allergic subjects.[36,37] There is evidence[38] that the Th_2 cytokines IL-4, IL-10, and IL-13 all inhibit cytokine-induced (TNF-α and IFN-γ) RANTES release from human airway smooth muscle cells. It is therefore likely that the different pattern of cytokines produced in conditions with a preponderance of either Th_1 or Th_2 lymphocytes will affect the local chemokine profile *in vivo*.

Purification of Chemokines from Inflammatory Lavage Fluids

Our aims have been to develop animal models in which we use *in vivo* bioassays to identify leukocyte chemoattractants present in inflammatory lavage fluids. Here we concentrate on our experience with a guinea pig model of the eosinophilic response to inhaled antigen, where we discovered a novel C-C chemokine, eotaxin,[39,40] and earlier work where we identified the neutrophil chemoattractants in a rabbit model of zymosan-induced peritonitis as IL-8, melanoma growth stimulating activity (MGSA), and the complement activation product C5a.[41–43] The amount and nature of the sample will govern chromatography conditions, which need to be adjusted for each new type of sample, so the emphasis here is placed on the reasons for the particular steps used. Examples of successful strategies and further details are given in Fig. 1.

Lavage fluids obtained from the airways or peritoneum generally contain high protein concentrations and, even after centrifugation and filtration, can quickly block or damage high-performance liquid chromatography (HPLC) columns. Thus, the first step should be a cleanup, to permit subsequent HPLC, which retains and concentrates all the bioactivity

[36] G. Krishnaswamy, M. C. Liu, M.-C. Su, M. Kumai, H.-Q. Xiao, D. G. Marsh, and S.-K. Huang, *Am. J. Respir. Cell Mol. Biol.* **9,** 279 (1993).

[37] S. R. Durham, S. Ying, V. A. Varney, M. R. Jacobson, R. M. Sudderick, I. S. Mackay, A. B. Kay, and Q. A. Hamid, *J. Immunol.* **148,** 2390 (1992).

[38] J. Matthias, S. J. Hirst, P. J. Jose, A. Robichaud, N. Berkman, C. Witt, C. H. C. Twort, P. J. Barnes, and K. F. Chung, *J. Immunol.* **158,** 1841 (1997).

[39] D. A. Griffiths-Johnson, P. D. Collins, A. G. Rossi, P. J. Jose, and T. J. Williams, *Biochem. Biophys. Res. Commun.* **197,** 1167 (1993).

[40] P. J. Jose, D. A. Griffiths-Johnson, P. D. Collins, D. T. Walsh, R. Moqbel, N. F. Totty, O. Truong, J. J. Hsuan, and T. J. Williams, *J. Exp. Med.* **179,** 881 (1994).

[41] P. D. Collins, P. J. Jose, and T. J. Williams, *J. Immunol.* **146,** 677 (1991).

[42] B. C. Beaubien, P. D. Collins, P. J. Jose, N. F. Totty, M. D. Waterfield, J. Hsuan, and T. J. Williams, *Biochem. J.* **271,** 797 (1990).

[43] P. J. Jose, P. D. Collins, J. A. Perkins, B. C. Beaubien, N. F. Totty, M. D. Waterfield, J. Hsuan, and T. J. Williams, *Biochem. J.* **278,** 493 (1991).

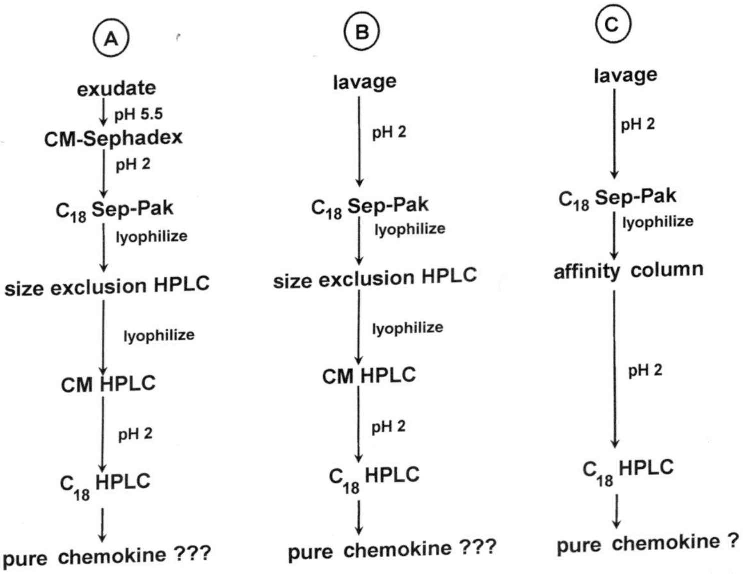

FIG. 1. Flow diagrams for purifying chemokines from exudate and lavage fluids. (A) When the samples are very proteinaceous (e.g., zymosan-induced peritoneal exudates), carboxymethyl(CM)-Sephadex C-25 soft gel chromatography is recommended as the first step because Sep-Pak cartridges (Waters/Millipore, Marlborough, MA) quickly become overloaded or blocked. (B) BAL fluids are less proteinaceous, and C_{18} reversed-phase Sep-Pak cartridges can be used as the first step. (C) When affinity columns are available, the number of chromatography steps can be reduced, and the chemokine is more likely to be pure after C_{18} reversed-phase HPLC. The details below supplement those in the text.

Exudate is adjusted to pH 5.5 with 0.15 *M* HCl, centrifuged, and applied to CM-Sephadex C-25 equilibrated in 0.15 *M* NaCl/10 m*M* sodium phosphate, pH 5.5. Either columns or a batch procedure may be used. After washing the gel in equilibration buffer, elution of bound proteins is achieved by increasing the salt concentration and/or the pH. The eluate is then adjusted to pH 2 and applied to C_{18} Sep-Pak cartridges as described in the legend to Fig. 2. The dried sample is suitable for any form of chromatography.

Size-exclusion columns are equilibrated in 0.08% (v/v) trifluoroacetic acid (TFA) (pH 2.0), and the sample is applied in a small volume (approximately 1% column volume) of the same solvent. For bioassay aliquots of the collected fractions are mixed with carrier protein [bovine serum albumin (BSA), <0.1 ng endotoxin/mg] and lyophilized. The remainder of the bioactive fractions are lyophilized without carrier protein for the next step in purification.

Cation-exchange HPLC can often be performed at pH values higher than that used for the soft gel cleanup step because of the lower ionic strength of the start buffer (e.g., 10 m*M* buffer without added salt). Although we have used ammonium acetate gradients (because this salt has the advantage of being somewhat volatile during lyophilization), we now prefer the use of sodium chloride gradients (0–1.0 *M*), which gives better resolution. Bioactive fractions from ion exchange or samples eluted from affinity chromatography can be subjected directly to reversed-phase HPLC after adjusting the pH to 2.0 with 20% TFA.

Reversed-phase HPLC is performed on wide bore (300 Å) fully end-capped C_{18} columns (usually 4 × 250 mm) using gradients of acetonitrile in 0.08% TFA. Sometimes we use a second reversed-phase step on a narrow bore C_{18} column (2 × 150 mm) with a shallower gradient and smaller fraction size to enhance purification.

while removing large quantities of unwanted material. Because chemokines are generally small cationic proteins present in iow concentration, cation exchange is a good first step cleanup. We have used carboxymethyl Sephadex C-25 because much of the ionic charge is accessible only to proteins of less than about 30 kDa. This means that a small amount of gel will bind virtually all the chemokines while excluding the majority of the larger cationic proteins as well as the anionic proteins such as the major contaminant, albumin. To avoid the need for dialysis of large lavage volumes, we choose the highest pH (e.g., pH 5.5) at which all the activity will bind at the ionic strength of the lavage medium (usually 0.15 *M* NaCl, sometimes with 10 m*M* EDTA). This strategy usually discards >90% of the unwanted proteins while retaining virtually all the bioactivity that, after washing the gel, is then eluted by increasing the ionic strength and/or the pH.

An alternative first step cleanup for BAL fluid is the use of C_{18} reversed-phase Sep-Pak cartridges prior to HPLC. Although reversed-phase has a lower protein capacity than ion-exchange chromatography, chemokines, perhaps because they are smaller, are concentrated in preference to the unwanted proteins so that relatively large volumes of fluid may be processed (Fig. 2). Proteins bind to the reversed-phase cartridges whereas salts pass through. Thus, elution with a volatile reagent followed by evaporation under vacuum in a centrifuge gives a sample that is amenable to any subsequent form of chromatography.

After the initial cleanup has been completed, standard HPLC methods for proteins (size exclusion, ion exchange, reversed phase) are suitable (Fig. 1). Size exclusion requires a small sample volume, whereas ion-exchange and reversed-phase chromatographies are amenable to larger loading volumes. Reversed-phase HPLC has the further advantage of accepting, and indeed desalting, samples in high salt buffers (e.g., from ion exchange or hydrophobic interaction). Chemokines obtained from natural sources may separate into more than one peak on reversed-phase HPLC, perhaps as a result of differential glycosylation.[40] Bioactive fractions are assessed for purity by sodium dodecyl sulfate–polyacrylamide gel electrophoresis (SDS–PAGE) with silver staining, mass spectrometry, and Edman degradation. The latter two methods are also a considerable aid to identification.

If the bioactivity in the lavage or exudate has been identified and sufficient antibodies are available, affinity columns of solid-phase antibody are particularly useful after an initial cleanup. Once the chemokine is eluted from the antibody column, reversed-phase HPLC serves to desalt, concentrate, and often to achieve complete purification.

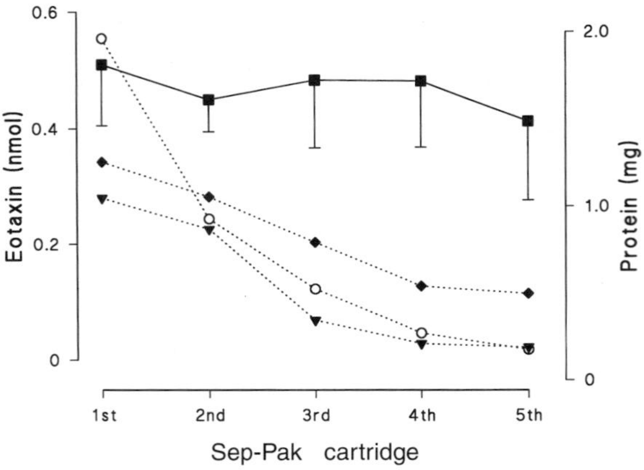

FIG. 2. Preferential concentration of chemokines, relative to total protein, on C_{18} reversed-phase Sep-Pak cartridges. BAL fluid (270 ml) obtained from 10 guinea pigs, challenged to induce lung eosinophilia, was centrifuged to remove cells and debris, adjusted to pH 2 with 20% TFA, and centrifuged again. Five Sep-Pak cartridges (1.6 ml C_{18} Environmental cartridges obtained from Waters/Millipore, Marlborough, MA) joined in series were wetted with a minimum of 5 column volumes (40 ml) of acetonitrile and washed with 5 column volumes of 0.1% TFA. After sample application and washing with a further 5 column volumes of 0.1% TFA, the cartridges were separated and eluted individually with 3 column volumes (4.8 ml) of acetonitrile/0.1% TFA. The eluates were dried under vacuum using a SpeedVac concentrator (Savant, Farmingdale, NY). The results are from an analysis of 2% of the samples eluted from each Sep-Pak (the remaining 98% being used for an affinity chromatography purification step) in three separate preparations (3 × 10 guinea pigs). Protein levels (solid line) were similar in each of the cartridges, indicating that the cartridges were saturated with the majority of the proteins in the BAL fluid. In contrast, most of the eotaxin (dotted lines) was retained by the early Sep-Pak cartridges, suggesting preferential concentration of this small protein. In practice, material eluted from the first three cartridges is used for affinity purification.

Role of Chemokines in Selective Cell Recruitment

Interleukin-8

Interleukin-8, a member of the C-X-C chemokine family, has been purified from the supernatant of stimulated human monocytes.[44–46] It has been shown to be a potent activator of, and chemoattractant for, neutrophils

[44] T. Yoshimura, K. Matsushima, S. Tanaka, E. A. Robinson, E. Appella, J. J. Oppenheim, and E. J. Leonard, *Proc. Natl. Acad. Sci. U.S.A.* **84,** 9233 (1987).

[45] J.-M. Schröder, U. Mrowietz, E. Morita, and E. Christophers, *J. Immunol.* **139,** 3474 (1987).

[46] A. Walz, P. Peveri, H. Aschauer, and M. Baggiolini, *Biochem. Biophys. Res. Commun.* **149,** 755 (1987).

in vitro[47] and *in vivo* in rabbit skin.[48,49] In primates, intravenous infusion of IL-8 causes an increase in blood neutrophils and their margination in the lung, liver, and spleen.[50] Antigen instillation into the trachea of ragweed-sensitized dogs leads to the production of IL-8 and an increase in the number of neutrophils in BAL fluid.[51] In humans, increased levels of IL-8 have been reported in the BAL fluid of asthmatics 4 hr after segmental antigen challenge.[52]

There is evidence that IL-8 also acts on cell types other than neutrophils. Intraperitoneal or aerosol administration of human IL-8 into guinea pigs causes a local influx of lymphocytes by 4 hr and eosinophil accumulation by 24 hr. Surprisingly, IL-8 fails to cause neutrophil recruitment in these experiments.[53] Intradermal injection of IL-8 into guinea pigs causes a dose-dependent eosinophil accumulation, although IL-8 is a weak agonist when compared with a number of C-C chemokines.[54] Sehmi *et al.*[47] have demonstrated that human peripheral venous eosinophils purified from individuals with a blood eosinophilia, but not from those of less eosinophilic controls, migrate to IL-8 in *in vitro* assays of chemotaxis. This suggests that an *in vivo* priming mechanism might exist in people with a blood eosinophilia. To test this, eosinophils are purified from placental cord blood and cultured with IL-3 and IL-5. These cells have a similar IL-8 responsiveness to those purified from the blood of eosinophilic subjects. Incubation of eosinophils from normal donors with IL-5 also enhances the response to IL-8. Thus, although IL-8 may normally not be a potent eosinophil chemoattractant, it is far more effective if cells have been previously primed by prior exposure to IL-5. As we know that eosinophils are frequently in an activated state in asthma and therefore primed, IL-8 may have a role in inflammation in asthma.

[47] R. Sehmi, O. Cromwell, A. J. Wardlaw, R. Moqbel, and A. B. Kay, *Clin. Exp. Allergy* **23,** 1027 (1993).

[48] M. Rampart, J. Van Damme, L. Zonnekyn, and A. G. Herman, *Am. J. Pathol.* **135,** 1 (1989).

[49] S. Nourshargh, J. A. Perkins, H. J. Showell, K. Matsushima, T. J. Williams, and P. D. Collins, *J. Immunol.* **148,** 106 (1992).

[50] K. J. Van Zee, E. Fischer, A. S. Hawes, C. A. Hebert, T. G. Terrell, J. B. Baker, S. F. Lowry, and L. L. Moldawer, *J. Immunol.* **148,** 1746 (1992).

[51] T. Kaneko, P. R. Massion, M. Hara, and J. A. Nadel, *Am. J. Respir. Crit. Care Med.* **153,** 136 (1996).

[52] L. M. Teran, M. Carroll, A. J. Frew, S. Montefort, L. C. K. Lau, D. E. Davies, I. Lindley, P. H. Howarth, M. K. Church, and S. T. Holgate, *Int. Arch. Allergy Immunol.* **107,** 374 (1995).

[53] D. Smith, L. Burrows, and J. Westwick, "Chemotactic Cytokines." Plenum, New York, 1991.

[54] P. D. Collins, V. B. Weg, L. H. Faccioli, M. L. Watson, R. Moqbel, and T. J. Williams, *Immunology* **79,** 312 (1993).

C-C Chemokines

Shortly after the discovery of IL-8 two groups simultaneously purified what at the time was believed to be a selective monocyte chemoattractant.[55,56] This activity is now known as MCP-1 and was the first human member of the C-C family of chemokines to be described. The cDNA of IL-8 and MCP-1 have been used to screen genetic libraries derived from human and animal tissues. This procedure has identified the genetic sequence of related but previously unidentified chemokines.

The C-C chemokines are potent *in vitro* chemoattractants for eosinophils, lymphocytes, monocytes, macrophages, and to a lesser extent for basophils, typically over a concentration range of 0.1–100 n*M*. Given the evidence for the involvement of eosinophils, lymphocytes, basophils, and mast cells in asthma, the C-C chemokines are suspected of playing an important role *in vivo*. There is currently significant interest in the pharmaceutical industry in developing chemokine antagonists for the treatment of asthma.

RANTES. The C-C chemokine RANTES (regulated on activation in normal T cells expressed and secreted) was first described by Schall *et al.*,[57] who demonstrated that it is produced by T lymphocytes and causes the selective migration of blood monocytes and $CD4^+$, $CD45RO^+$ memory lymphocytes *in vitro*.[58] As discussed earlier, memory lymphocytes are critical to the development of sensitization and thus to the development of the asthmatic response.

In 1992, Kameyoshi *et al.* showed that a potent eosinophil chemoattractant was released from human platelets when stimulated with thrombin. The activity was purified, sequenced, and found to be RANTES.[59] This was a particularly important discovery because all the other known human C-C chemokines at the time, namely, MCP-1, MIP-1α, and MIP-1β, had only been reported to act on monocytes and lymphocytes. Subsequent *in vitro* and *in vivo* studies have supported the ability of RANTES to act as a chemoattractant for eosinophils.[60–62] RANTES also up-regulates the

[55] K. Matsushima, C. G. Larsen, G. C. Dubois, and J. J. Oppenheim, *J. Exp. Med.* **169,** 1485 (1989).

[56] T. Yoshimura, E. A. Robinson, S. Tanaka, E. Appella, J.-I. Kuratsu, and E. J. Leonard, *J. Exp. Med.* **169,** 1449 (1989).

[57] T. J. Schall, J. Jongstra, B. J. Dyer, J. Jorgensen, C. Clayberger, M. M. Davis, and A. M. Krensky, *J. Immunol.* **141,** 1018 (1988).

[58] T. J. Schall, K. Bacon, K. I. Toy, and D. V. Goeddel, *Nature (London)* **347,** 669 (1990).

[59] Y. Kameyoshi, A. Dorschner, A. I. Mallet, E. Christophers, and J.-M. Schröder, *J. Exp. Med.* **176,** 587 (1992).

[60] A. Rot, M. Krieger, T. Brunner, S. C. Bischoff, T. J. Schall, and C. A. Dahinden, *J. Exp. Med.* **176,** 1489 (1992).

expression of EG2 on eosinophils, an indicator of cellular activation.[63] Eosinophils express constitutive RANTES mRNA and thus are not only activated by RANTES but also have the potential to produce it.[64]

RANTES is also known to act on basophils and mast cells. It is a potent chemoattractant for these cells and causes the release of mediators, although the latter response is weaker than that produced by MCP-1 or MCP-3.[61]

RANTES is therefore not only produced by a key cell in an allergy, the T lymphocyte, but it also is chemotactic for memory T lymphocytes, eosinophils, mast cells, and basophils and is strongly implicated in asthma. Increased levels of RANTES have been reported in the BAL fluid of asthmatics after segmental antigen challenge[65,66] and in the airways of mild asthmatics.[67] There is now good evidence that, like many chemokines, RANTES is produced by numerous cell types. It is particularly interesting that bronchial epithelial cells produce RANTES as they are in an ideal location to establish a chemoattractant gradient in the airways where eosinophilia is frequently reported.[68]

As discussed earlier, large amounts of RANTES are released from platelets when stimulated with thrombin.[59] Two animal models may support the role of the platelet-derived RANTES in inflammation and AHR. In the first, a guinea pig model of antigen- and PAF-induced lung eosinophilia, platelet depletion decreases the eosinophil accumulation by more than 50%.[69] In the second, platelet depletion in a rabbit model has a similar inhibitory effect and also blocks LAR bronchoconstriction and AHR.[70]

---

[61] C. A. Dahinden, T. Geiser, T. Brunner, V. Von Tscharner, D. Caput, P. Ferrara, A. Minty, and M. Baggiolini, *J. Exp. Med.* **179,** 751 (1994).

[62] R. Alam, S. Stafford, P. Forsythe, R. Harrison, D. Faubion, M. A. Lett-Brown, and J. A. Grant, *J. Immunol.* **150,** 3442 (1993).

[63] T. Kakazu, J. Chihara, A. Saito, I. Higashimoto, T. Yamamoto, D. Kurachi, and S. Nakajima, *Int. Arch. Allergy Immunol.* **108**(Suppl. 1), 43 (1995).

[64] S. Ying, Q. Meng, L. Taborda-Barata, C. J. Corrigan, J. Barkans, B. Assoufi, R. Moqbel, S. R. Durham, and A. B. Kay, *Eur. J. Immunol.* **26,** 70 (1996).

[65] S. Sur, H. Kita, G. J. Gleich, T. C. Chenier, and L. W. Hunt, *J. Allergy Clin. Immunol.* **97,** 1272 (1996).

[66] L. M. Teran, N. Noso, M. Carroll, D. E. Davies, S. Holgate, and J.-M. Schroder, *J. Immunol.* **157,** 1806 (1997).

[67] N. Berkman, V. L. Krishnan, T. Gilbey, R. Newton, B. O'Conner, P. J. Barnes, and K. Fan Chung, *Am. J. Respir. Crit. Care Med.* **154,** 1804 (1997).

[68] J. H. Wang, J. L. Devalia, C. Xia, R. J. Sapsford, and R. J. Davies, *Am. J. Respir. Cell Mol. Biol.* **14,** 27 (1996).

[69] A. Lellouch-Tubiana, J. Lefort, M.-T. Simon, A. Pfister,and B. B. Vargaftig, *Am. Rev. Respir. Dis.* **137,** 948 (1988).

[70] A. J. Coyle, C. P. Page, L. Atkinson, R. Flanagan, and W. J. Metzger, *Am. Rev. Respir. Dis.* **142,** 587 (1990).

RANTES protein has not been measured in these experiments, however, and therefore its role in the contribution of platelets is only speculative. What is clearly needed in these models is to pretreat animals with a neutralizing anti-RANTES antibody.

The effects of RANTES on eosinophilia appear to be species dependent. In both dogs and monkeys intradermal injection of human RANTES causes a local eosinophilia, although both studies also have reported a marked monocyte–macrophage recruitment.[71,72] Intradermal injection of human RANTES in guinea pigs fails to recruit ^{111}In-labeled eosinophils *in vivo,*[40] and guinea pig RANTES activates macrophages but not eosinophils *in vitro.*[73] In a rat model in which RANTES is overexpressed in bronchial epithelial cells, there is a pronounced increase in mononuclear cells primarily of monocytes and not of eosinophils.[74]

Macrophage Inflammatory Protein-1α and -1β. MIP-1α and MIP-1β are closely related chemokines that have been shown to cause the chemotaxis of lymphocytes and, at higher concentrations of MIP-1α, eosinophils. They also activate basophils and mast cells. Alam *et al.*[75] have demonstrated the release of MIP-1α from the basophils of 14 out of 20 asthmatics compared with much lower levels produced by only 2 out of 10 controls.

MIP-1α is more potent than MIP-1β and acts on all phenotypes of lymphocytes in a concentration-dependent manner. At low concentrations it is chemotactic for B cells and CD8$^+$ T cells. At higher concentrations it has preferential chemotactic activity on CD4$^+$ cells. However, it does not discriminate between different phenotypes of CD4$^+$ cells as has been shown for RANTES and MIP-1β, both of which have chemotactic activity only on CD45RO$^+$ T cells.[76,77]

[71] R. Meurer, G. van Riper, W. Feeney, P. Cunningham, D. Hora, M. S. Springer, D. E. MacIntyre, and H. Rosen, *J. Exp. Med.* **178,** 1913 (1993).

[72] P. D. Ponath, S. Qin, D. J. Ringler, I. Clark-Lewis, J. Wang, N. Kassam, H. Smith, X. Shi, J.-A. Gonzalo, W. Newman, J.-C. Gutierrez-Ramos, and C. R. Mackay, *J. Clin. Invest.* **97,** 604 (1996).

[73] E. M. Campbell, M. L. Watson, A.E.I. Proudfoot, T. N. C. Wells, T. Yoshimura, and J. Westwick, *Br. J. Pharmacol.* **119,** 50P (1997).

[74] T. A. Braciak, K. Bacon, Z. Xing, D. J. Torry, F. L. Graham, T. J. Schall, C. D. Richards, K. Croitoru, and J. Gauldie, *J. Immunol.* **157,** 5076 (1996).

[75] R. Alam, P. A. Forsythe, S. Stafford, M. A. Lett-Brown, and J. A. Grant, *J. Exp. Med.* **176,** 781 (1992).

[76] T. J. Schall, K. Bacon, R. D. R. Camp, C. Herbert, and D. V. Goeddel, *J. Exp. Med.* **177,** 1821 (1993).

[77] D. D. Taub, K. Conlon, A. R. Lloyd, J. J. Oppenheim, and D. J. Kelvin, *Science* **260,** 355 (1996).

In one study, Lukacs *et al.*,[78] have sensitized mice with *Schistosoma mansoni* egg antigen and later challenged them by intratracheal administration of the antigen. This leads to an inflammatory influx primarily of neutrophils and eosinophils, which is maximal by 8 hr postchallenge. This is accompanied by an increased expression of MIP-1α mRNA extracted from the cell pellet of the lavage fluid. Immunohistochemical staining shows MIP-1α staining on epithelial cells, alveolar macrophages, and mononuclear cells. Pretreatment with a neutralizing anti-MIP-1α antibody causes a significant reduction in lung eosinophilia but not neutrophilia. This at first appears surprising given the relatively low efficacy of MIP-1α on human eosinophils. However, murine MIP-1α is far more potent on mouse eosinophils than human MIP-1α is on human eosinophils.[78] In another murine model of antigen-induced pulmonary eosinophilia,[79] the increased MIP-1α message has been shown not to be associated with the pulmonary eosinophilia. In conclusion, there is good evidence for a role of MIP-1α in allergic responses in the mouse, but this is not necessarily directly linked with the eosinophilia, suggesting that other chemokines may be important in this component.

Monocyte Chemotactic Proteins. MCP-1 was first described in 1989[55,80,81] and shown to be a monocyte chemoattractant effective at subnanomolar concentrations, with no effect on neutrophils.[82] It is now known that MCP-1 is also a chemoattractant for $CD4^+$ (particularly the $CD45RO^+$ subset) and $CD8^+$ T cells, but not for B cells.[83] Although MCP-1 is not a good chemoattractant for basophils, it is a potent stimulator of histamine and LTC_4 release.[84–86] It has little activity on eosinophils.

[78] N. W. Lukacs, R. M. Strieter, C. L. Shaklee, S. W. Chensue, and S. L. Kunkel, *Eur. J. Immunol.* **25,** 245 (1995).

[79] J. A. MacLean, R. Ownbey, and A. D. Luster, *J. Exp. Med.* **184,** 1461 (1996).

[80] T. Yoshimura, N. Yuhki, S. K. Moore, E. Appella, M. I. Lerman, and E. J. Leonard, *FEBS Lett.* **244,** 487 (1989).

[81] T. Yoshimura, E. A. Robinson, S. Tanaka, E. Appella, and E. J. Leonard, *J. Immunol.* **142,** 1956 (1989).

[82] B. J. Rollins, A. Walz, and M. Baggiolini, *Blood* **78,** 1112 (1991).

[83] M. W. Carr, S. J. Roth, E. Luther, S. S. Rose, and T. A. Springer, *Immunology* **91,** 3652 (1994).

[84] S. C. Bischoff, M. Krieger, T. Brunner, and C. A. Dahinden, *J. Exp. Med.* **175,** 1271 (1992).

[85] R. Alam, M. A. Lett-Brown, P. A. Forsythe, D. J. Anderson-Walters, C. Kenamore, C. Kormos, and J. A. Grant, *J. Clin. Invest.* **89,** 723 (1992).

[86] P. Kuna, S. R. Reddigari, D. Rucinski, J. J. Oppenheim, and A. P. Kaplan, *J. Exp. Med.* **175,** 489 (1992).

Given that there is a clear increase of macrophages, particularly during the early phase of an asthmatic episode, Sousa *et al.*[87] have investigated whether there is an altered expression of MCP-1 in asthmatic airways. Using anti-MCP-1 antibodies they have demonstrated an increase in MCP-1 in bronchial biopsy tissues from asthmatics. Staining is most intense in the epithelium and subepithelium. Furthermore, increased MCP-1 has been detected in the lavage fluid[2] and sputum[88] of asthmatics.

In a rat model of immune complex-mediated acute inflammatory lung injury, Brieland *et al.*[89] have shown a time-dependent release of MCP-1 into the lavage fluid. Alveolar macrophages purified from these animals have a markedly increased expression of MCP-1 mRNA, and cell culture supernatants of these cells contain MCP-1. This may represent a further cell source and establish a positive feedback loop whereby MCP-1 release from alveolar macrophages leads to the infiltration of further monocytes–macrophages into the lung. Another source of MCP-1 may be the eosinophil. When a human leukemic eosinophil cell line is stimulated with the proinflammatory cytokine TNF-α, there is an increased production of MCP-1 that could be inhibited by pretreatment with glucocorticoids.[90]

To date, three further members of the human MCP family have been described based on sequence similarities, namely, MCP-2, MCP-3,[91] and MCP-4.[92] A mouse MCP-5 has also been described.[93] The human MCPs are good chemoattractants of monocytes and lymphocytes but not neutrophils.[91,94] When injected intradermally into the rabbit, MCP-1, -2, and -3 cause a selective monocyte recruitment.[91]

When the MCPs are tested in a Boyden chamber assay it is shown that MCP-3 is a good eosinophil chemoattractant. MCP-2 is slightly less potent,

[87] A. R. Sousa, S. J. Lane, J. A. Nakhosteen, T. Yoshimura, T. H. Lee, and R. N. Poston, *Am. J. Respir. Cell Mol. Biol.* **10,** 142 (1994).

[88] K. Kurashima, N. Mukaida, M. Fujimura, J.-M. Schröder, T. Mutsuda, and K. Matsushima, *J. Leukocyte Biol.* **56,** 313 (1996).

[89] J. K. Brieland, M. L. Jones, S. J. Clarke, J. B. Baker, J. S. Warren, and J. C. Fantone, *Am. J. Respir. Cell Mol. Biol.* **7,** 134 (1992).

[90] L. A. Goldstein, R. M. Strieter, H. L. Evanoff, S. L. Kunkel, and N. W. Lukacs, *Mediators of Inflammation* **5,** 218 (1996).

[91] J. Van Damme, P. Proost, J.-P. Lenaerts, and G. Opdenakker, *J. Exp. Med.* **176,** 59 (1992).

[92] M. Uguccioni, P. Loetscher, U. Forssmann, B. Dewald, H. Li, S. H. Lima, Y. Li, B. Kreider, G. Garotta, M. Thelen, and M. Baggiolini, *J. Exp. Med.* **183,** 2379 (1996).

[93] G.-Q. Jia, J.-A. Gonzalo, C. Lloyd, L. Kremer, L. Lu, C. Martinez-A., B. K. Wershil, and J.-C. Gutierrez-Ramos, *J. Exp. Med.* **184,** 1939 (1997).

[94] M. Uguccioni, M. D'Apuzzo, M. Loetscher, B. Dewald, and M. Baggiolini, *Eur. J. Immunol.* **25,** 64 (1995).

and MCP-1 has no activity[95] although deletion of the amino terminal residue from MCP-1 converts it to a potent eosinophil chemoattractant. MCP-4 is equipotent with eotaxin (see below) as a chemoattractant for eosinophils.[92]

Finally, MCP-1 and MCP-3 are potent activators of basophils and stimulate release of their granule contents. For activation, MCP-3 = RANTES $\gg$ MCP-1, whereas MCP-1 and MCP-3 are more potent at causing mediator release (MCP-1 = MCP-3 $\gg$ RANTES > MIP-1α).[61,84,86] MCP-4 has also been shown to be a potent activator of basophil histamine release.[96]

Eotaxin. In 1992, we embarked on an investigation of endogenous eosinophil chemoattractants generated in an allergic airways model *in vivo.*[39,40] We used a pharmacological protocol whereby guinea pigs are sensitized to ovalbumin over a 3-week period and then challenged by ovalbumin aerosol under the cover of an antihistamine to prevent fatal anaphylaxis. At different times after antigen challenge animals are sacrificed and the airways lavaged. Using an *in vivo* bioassay system measuring the accumulation of ^{111}In-labeled eosinophils in guinea pig skin,[97] eosinophil chemoattractant activity of the lavage fluid is assessed (Fig. 3a).

The eosinophil chemoattractant activity peaks at 3–6 hr after the allergen challenge, returns to baseline by 24 hr (Fig. 3b), and is purified by three sequential HPLC steps: cation exchange, sizing, and reversed-phase (see the section Purification of Chemokines from Inflammatory Lavage Fluids). At each step, fractions are bioassayed and the active fractions pooled for further purification. Three fractions from the final purification step are active. These three fractions have the same amino-terminal sequence but small differences in molecular size. Finally, 70 out of 73 amino acids have been positively identified, and we suggest the molecular size differences reflect differential O-glycosylation (Fig. 4). The novel protein that we identified is a C-C chemokine, and, given its eosinophil selectivity, we called it eotaxin. *In vivo,* eotaxin induces the selective accumulation of ^{111}In-labeled eosinophils when injected in the skin at only 1–2 pmol per site. *In vitro,* eotaxin increases the intracellular concentration of calcium in both guinea pig and human eosinophils at 2–4 n*M*, and it induces the aggregation of guinea pig eosinophils at 10–40 n*M*.[39,40] Comparison of its sequence with other C-C chemokines indicates closest homology with the human MCPs, namely, MCP-2 (54%), MCP-1 (53%), and MCP-3 (51%),

[95] N. Noso, P. Proost, J. Van Damme, and J.-M. Schröder, *Biochem. Biophys. Res. Commun.* **200,** 1470 (1994).

[96] C. Stellato, P. Collins, H. Li, J. White, P. D. Ponath, W. Newman, D. Soler, G. LaRosa, L. M. Schwiebert, C. Bickel, M. Liu, B. Bochner, T. J. Williams, and R. Schleimer, *J. Clin. Invest.* **99,** 926 (1997).

[97] L. H. Faccioli, S. Nourshargh, R. Moqbel, F. M. Williams, R. Sehmi, A. B. Kay, and T. J. Williams, *Immunology* **73,** 222 (1991).

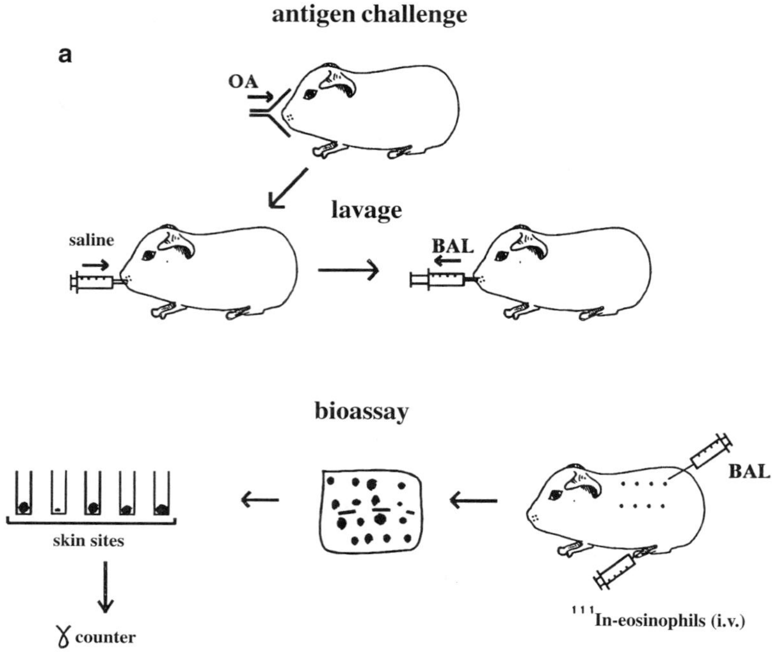

FIG. 3. (a) Male Dunkin Hartley guinea pigs were sensitized with intraperitoneal ovalbumin (OA, 1 mg/kg) on day 1 followed by exposure to aerosolized antigen (2%, w/v, OA, 5 min) on day 8 and challenge 7–10 days later (1%, w/v, OA, 20 min) under antihistamine cover to protect against fatal anaphylaxis (pyrilamine 10 mg/kg, intraperitoneally, 30 min before challenge). At different times after antigen challenge BAL was performed with 4 ml of saline. Samples were centrifuged and the supernatant stored frozen prior to bioassay. BAL samples were assayed by intradermal injection into guinea pigs previously given intravenous injections of 5×10^6 ^{111}In-labeled eosinophils. After 4 hr, assay animals were sacrificed and the skin sites punched out for gamma counting.

and lower homology with MIP-1α (31%) and RANTES (26%).The homology with guinea pig MCP-1[98] is only 44%.

Using PCR (polymerase chain reaction) with primers based on the protein sequence of guinea pig eotaxin to obtain probes, guinea pig lung cDNA libraries have been screened and the full-length eotaxin cDNA cloned.[99,100] This cDNA encodes a 96-amino acid protein with a 23-amino

[98] T. Yoshimura, *J. Immunol.* **150,** 5025 (1993).

[99] P. J. Jose, I. M. Adcock, D. A. Griffiths-Johnson, N. Berkman, T. N. C. Wells, T. J. Williams, and C. A. Power, *Biochem. Biophys. Res. Commun.* **205,** 788 (1994).

[100] M. E. Rothenberg, A. D. Luster, C. M. Lilly, J. M. Drazen, and P. Leder, *J. Exp. Med.* **181,** 1211 (1995).

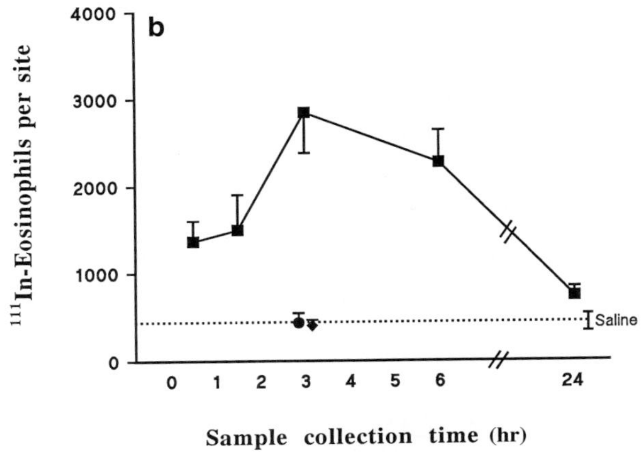

FIG. 3. (*continued*). (b) Time course of generation of eosinophil chemoattractant activity in sensitized animals after antigen challenge (■, n = 4–10). Significant activity was detected in samples taken at 0.5, 1.5, 3, and 6 hr after antigen challenge. No significant activity was detected in 24-hr samples nor at 3 hr after sham (saline) challenge of sensitized animals (●, n = 5) or antigen challenge of sham sensitized animals (◆, n = 5). Data are presented as means ± SEM. The dotted line shows the ^{111}In-labeled eosinophil content of saline-injected skin sites. (Figure 3b is reproduced from *The Journal of Experimental Medicine,* 1994, **179,** 881–887, by copyright permission of The Rockefeller University Press.)

acid leader sequence. In sensitized guinea pigs there is a high constitutive expression of eotaxin mRNA in the lung, which increases to a peak by 3 hr and returns to baseline by 6 hr. In naive guinea pigs, constitutive eotaxin message is highest in the lung. It is detected at lower levels in the intestines, stomach, heart, spleen, liver, thymus, and kidneys but not in the skin, brain, or bone marrow.[100]

Rothenburg *et al.*[31] have used the cDNA encoding guinea pig eotaxin to screen a mouse genomic library. They identified a murine eotaxin gene and prepared recombinant murine eotaxin. Murine eotaxin cDNA has 78% homology with guinea pig eotaxin. Murine eotaxin is extremely potent as a chemoattractant for eosinophils *in vitro* and has no effect on macrophages or neutrophils. Eotaxin mRNA is constitutively expressed in mucosal tissues of mice including the skin, lung, and intestines, although levels are greatest in the thymus and skin. Gonzalo *et al.*[101] have used a murine model of asthma to demonstrate that the pronounced lung eosinophilia peaking at

[101] J.-A. Gonzalo, G.-Q. Jia, V. Aguirre, D. Friend, A. J. Coyle, N. A. Jenkins, G.-S. Lin, H. Katz, A. Lichtman, N. Copeland, M. Kopf, and J.-C. Gutierrez-Ramos, *Immunity* **4,** 1 (1996).

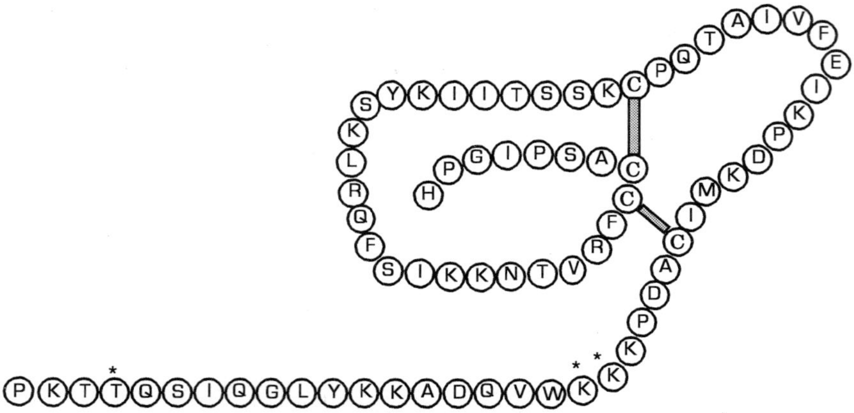

FIG. 4. Amino acid sequence of guinea pig eotaxin. The mature secreted molecule has 73 residues of which 70 were positively identified by protein sequencing. The three remaining residues (indicated with asterisks) were identified using the cloned full-length eotaxin cDNA (derived by screening a guinea pig lung library).

3 hr after antigen challenge is concomitant with an increase in murine eotaxin expression in lung tissue.

The role of T lymphocytes and eotaxin has been investigated in a murine model of antigen-induced pulmonary eosinophilia.[79] The authors demonstrated a longer time course of eotaxin expression than in the guinea pig. Eotaxin mRNA is increased at 3 hr, reaches maximal levels by 6 hr, and is still elevated above constitutive expression at 24 hr. There is an increase in MIP-1α mRNA at 3 and 6 hr but no change in RANTES message. To investigate the role of T lymphocytes, animals are pretreated (twice) with an anti-CD3 monoclonal antibody before the antigen challenge. This reduces the $CD4^+$ and $CD8^+$ cell population by 95% (as assessed by flow cytometry) and blocks both the antigen-induced eosinophilia and the increase in eotaxin message. In contrast, T-cell depletion prolongs MIP-1α expression up to 24 hr and induces RANTES message production. This effect is also seen in sham sensitized animals, suggesting that the induction of MIP-1α and RANTES mRNA is not associated with pulmonary eosinophilia. These results suggest that eotaxin, and not MIP-1α or RANTES, is the key regulator of eosinophilia in this model.

We have showed that guinea pig eotaxin is highly potent as a stimulator of human eosinophils, implying the existence of a human eotaxin homolog and an eotaxin receptor on human eosinophils.[40] Subsequently, in 1996, two papers were published demonstrating a human equivalent to guinea pig eotaxin. Ponath *et al.*[72] showed that human eotaxin has 61.8% homology (at the amino acid level) with guinea pig eotaxin and 63.2% homology with murine eotaxin. Again eotaxin is potent and selective for eosinophils *in vitro*. When injected into the skin of rhesus monkeys it produces a significant eosinophilia at 10 pmol per injection site. Infiltrating cells counted by histology are >90% eosinophils. In another study, Garcia-Zepeda *et al.*[102] also cloned human eotaxin describing an identical sequence and similar *in vitro* activity. They also demonstrated that both endothelial, epithelial, and peripheral blood eosinophils can be stimulated to produce eotaxin.

The discovery of eotaxin in animal models of allergy and now in humans is significant in that a selective eosinophil chemoattractant has been identified in a number of different species. Furthermore, an eotaxin receptor has been identified on eosinophils and designated CCR3.[103,104] This is the only known receptor to which eotaxin has been shown to bind. On CCR3-transfected cells eotaxin has a high binding affinity, namely, a K_D of approximately 0.1 nM. RANTES and MCP-3 also bind to CCR3 (K_D of 2.7 and 3.1 nM, respectively), but MIP-1α does not. Receptor expression of CCR3 on eosinophils is high when compared with, for example, CCR1 and CCR2 on monocytes and T lymphocytes.[103,104] We have taken advantage of a species difference between the eotaxin receptor in guinea pig and humans to show that eosinophil responses can be blocked by agents that bind to but do not activate the eotaxin receptor.[105] Blocking anti-eotaxin antibody pretreatment significantly reduces antigen-induced lung eosinophilia in mice.[106]

Interleukin-5 and Eotaxin

Following the description of IL-5 as an eosinophilopoietic factor, numerous studies have demonstrated the central role of this cytokine in the control of eosinophil accumulation associated with allergic inflammation.[107] Several *in vivo* studies have used neutralizing anti-IL-5 antibodies to suppress

[102] E. A. Garcia-Zepeda, M. E. Rothenberg, R. T. Ownbey, J. Celestin, P. Leder, and A. D. Luster, *Nature Med.* **2,** 449 (1996).

[103] P. D. Ponath, S. Qin, T. W. Post, J. Wang, L. Wu, N. P. Gerard, W. Newman, C. Gerard, and C. R. Mackay, *J. Exp. Med.* **183,** 2437 (1996).

[104] B. L. Daugherty, S. J. Siciliano, J. DeMartino, L. Malkowitz, A. Sirontino, and M. S. Springer, *J. Exp. Med.* **183,** 2349 (1996).

[105] S. Marleau, D. A. Griffiths-Johnson, P. D. Collins, Y. S. Bakhle, T. J. Williams, and P. J. Jose, *J. Immunol.* **157,** 4141 (1996).

[106] J. A. Gonzalo, C. M. Lloyd, L. Kremer, E. Finger, C. Martinez-A., M. H. Siegelman, M. I. Cybulsky, and J.-C. Gutierrez-Ramos, *J. Clin. Invest.* **98,** 2332 (1996).

[107] C. J. Sanderson, *Blood* **79,** 3101 (1992).

antigen or parasite-induced eosinophil infiltration.[22,108,109] Other investigations have detected IL-5 protein in BAL fluid and plasma, as well as increased IL-5 mRNA expression in lung tissue, from asthmatic individuals.[110] IL-5-deficient mice fail to produce a lung eosinophilia following antigen or parasite challenge.[15]

These observations, coupled with *in vitro* data showing eosinophil chemotaxis and/or chemokinesis, have implied that IL-5 might have an important role as an eosinophil chemoattractant when generated locally at sites of allergic inflammation such as the asthmatic lung. In contrast, we and others have been unable to demonstrate substantial eosinophil chemotactic or chemoattractant activity associated with IL-5. We find that the intravenous administration of low amounts of IL-5 (18 pmol/kg) to guinea pigs results in an acute (30 min to 1 hr) and profound (10- to 20-fold increase) blood eosinophilia that is sustained for at least 6 hr. In the presence of this blood eosinophilia the local infiltration of eosinophils in skin in response to intradermally injected eotaxin is considerably enhanced, and the number of cells accumulating correlates with the circulating eosinophil levels.[111] We have identified the bone marrow as the source of this rapidly mobilizable pool of eosinophils.

On the basis of these observations we have proposed the following mechanism to explain the essential role of IL-5 in the regulation of eosinophil accumulation (Fig. 5). Exposure to an appropriate stimulus, for example, antigen challenge of a sensitized individual, is followed by the local generation of IL-5 and eotaxin (i). Eotaxin acts close to its site of production in the lung, where it induces the migration of eosinophils from the microvasculature into the tissue (ii). In contrast, IL-5 passes into the circulation, acting remotely in the bone marrow where it releases a rapidly mobilizable pool of eosinophils into the circulation (iii). These cells are then available to enhance the local accumulation of cells in response to eotaxin (iv). This novel action of IL-5 is in addition to its effects on eosinophilopoiesis, which over a longer time would be expected to replenish the bone marrow reserve (v). It is also possible that the eosinophils released by IL-5 form a distinct circulating pool of primed cells (vi) that respond preferentially to chemoattractants produced locally in tissues. Finally, the presence of IL-5 in the lung may additionally prolong the survival of eosinophils in the tissues (vii).

[108] N. Chand, J. E. Harrison, S. Rooney, J. Pillar, R. Jakubicki, K. Nolan, W. Diamantis, and R. D. Sofia, *Eur. J. Pharmacol.* **211,** 121 (1992).
[109] E. Coeffier, D. Joseph, and B. B. Vargaftig, *Br. J. Pharmacol.* **113,** 749 (1994).
[110] D. H. Broide, M. Lotz, and A. J. Cuomo, *J. Allergy Clin. Immunol.* **89,** 958 (1992).
[111] P. D. Collins, S. Marleau, D. A. Griffiths-Johnson, P. J. Jose, and T. J. Williams, *J. Exp. Med.* **182,** 1169 (1995).

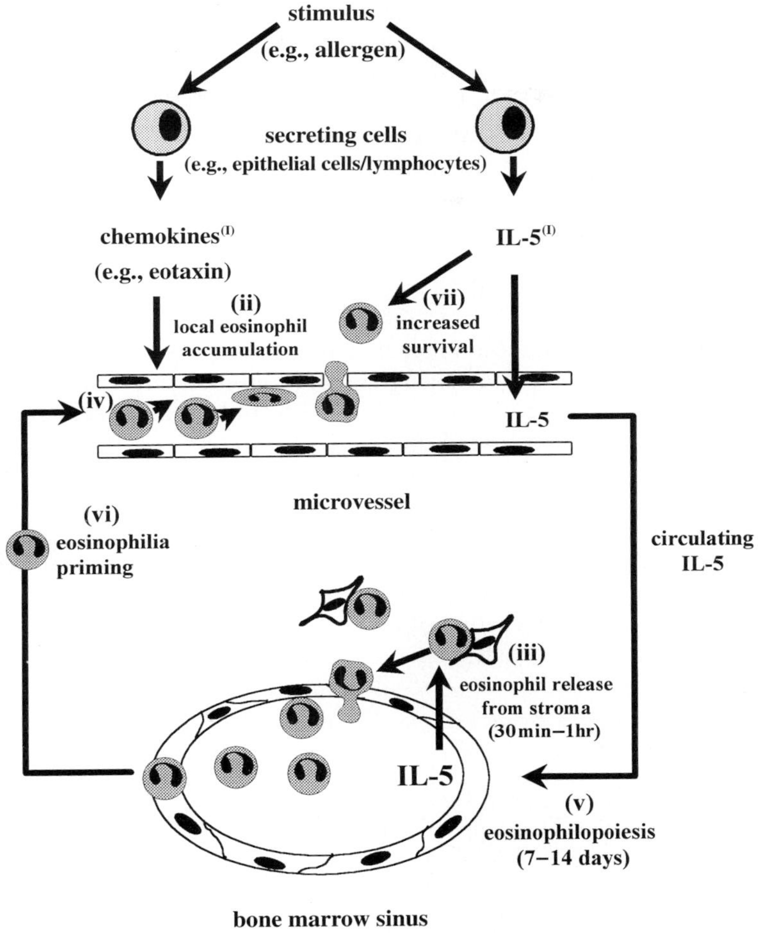

FIG. 5. Proposed mechanism for cooperation between chemokines and IL-5 in the local recruitment of eosinophils in allergic inflammation (numbers in parentheses relates to the text in the section Interleukin-5 and Eotaxin).

Leukocyte Trafficking

Adhesion Molecules

The accumulation of leukocytes in tissues is believed to take place in a sequence of different phases, that is, rolling along the microvessel wall, tethering, firm adhesion and flattening, and emigration through endothelial cell junctions.[112] This process can be visualized using, for example, intravital microscopy of the exposed mesentery of rats.[113] Rolling involves a light tethering of the cell to the vascular endothelium through binding of surface adhesion molecules on the leukocyte to their counterreceptors on the vascular endothelium (e.g., L-selectin on eosinophils binds to CD34 on endothelial cells). This rolling may be transient or, given appropriate stimuli (discussed below), may resolve into a firmer adhesion through a different set of molecules (e.g., the $\alpha_4\beta_1$-integrin VLA4 expressed on eosinophils binds to its counterreceptor VCAM-1 on the endothelium). The adhesion molecules and counterreceptors used by the inflammatory cells in asthma have been reviewed in detail.[114]

Selective cell accumulation may be partly explained through the differential expression of adhesion molecules on certain cells. For example, VLA4 is expressed on eosinophils but not on neutrophils. However, additional mechanisms are involved. IL-4 and IL-13 levels increase in asthma, and both have been shown to increase expression of vascular cell adhesion molecule-1 (VCAM-1) on endothelial cells. Anti-VCAM-1 antibodies decrease antigen-induced eosinophil infiltration in both the guinea pig[115,116] and sheep.[117] Similarly, IL-1 and TNF-α increase the expression of ICAM-1 (intercellular adhesion molecule-1, the counterligand for β_2-integrins) and the adhesion and migration of leukocytes including eosinophils, which is blocked by anti-ICAM antibodies.

Involvement of Chemokines in Interactions between Leukocytes and Endothelial Cells

It is now believed that a key mechanism for stimulating firm adhesion of the rolling cell is through its encountering an area with a high local concentration of a particular chemokine. This signal is believed to be maintained on the luminal surface for sampling by the rolling leukocytes, a theory first proposed for the neutrophil chemoattractant C5a.[118] This may be achieved by binding of the chemokines to surface glycoproteins to prevent rapid removal in the bloodstream. In 1993, Rot demonstrated *in*

[112] T. A. Springer, *Cell (Cambridge, Mass.)* **76,** 301 (1994).

[113] S. Nourshargh, S. Larkin, A. Das, and T. J. Williams, *Blood* **85,** 2553 (1995).

[114] S. Nourshargh, *in* "New Drugs for Asthma" (P. J. Barnes, ed.), p. 220. IBC Technical Services, London, 1992.

[115] V. B. Weg, T. J. Williams, R. R. Lobb, and S. Nourshargh, *J. Exp. Med.* **177,** 561 (1993).

[116] M. Pretolani, C. Ruffie, J. R. Lapa-e-Silva, D. Joseph, R. R. Lobb, and B. B. Vargaftig, *J. Exp. Med.* **180,** 795 (1994).

[117] W. M. Abraham, M. W. Sielczak, A. Ahmed, A. Cortes, I. T. Lauredo, J. Kim, B. Pepinsky, C. D. Benjamin, D. R. Leone, R. R. Lobb, and P. F. Weller, *J. Clin. Invest.* **93,** 776 (1994).

[118] M. G. Tonneson, D. C. Anderson, T. A. Springer, A. Knelder, N. Avdi, and P. M. Henson, *Fed. Proc.* **45,** 379 (1986).

vitro that IL-8 will bind to glycosaminoglycan proteins.[119] Both MIP-1α and RANTES have been shown to bind to surface proteins.[120] This interaction must be dynamic with shedding or internalization of the chemokine to remove the chemoattractant when appropriate. Shed chemokines may bind to passing red blood cells that express clearance receptors for C-C and C-X-C chemokines.[121] Once the rolling leukocyte recognizes the bound chemokine there is an increased affinity and/or increased expression of surface integrins leading to firmer adhesion. The cell then crawls along the surface of the endothelium and undergoes diapedesis and emigration via interendothelial cell junctions.

A number of control mechanisms are involved in chemokine-mediated leukocyte accumulation. The production of chemokines is regulated by transcription factors sometimes activated by other cytokines or chemokines.[122] Different cytokines/chemokines may be generated together or in sequence, and they may act synergistically to produce leukocyte recruitment (e.g., with different mediators activating the leukocyte and endothelium). In addition, mechanisms exist to act as negative regulators (e.g., regulating secretion action or metabolism of chemokines). Thus, a complex regulatory system is involved in the control of leukocyte trafficking *in vivo,* which is only superficially modeled by *in vitro* systems such as the Boyden chamber.

Summary and Conclusion

In studies of disease processes, increasing knowledge leads to an increased awareness of the complexity of the underlying mechanisms. The intense research activity in the chemokine field has made this acutely manifest. Numerous chemokines have been discovered through the use of (1) bioassay of *in vitro* cell culture supernatants and *in vivo* exudates from animal models of inflammation and (2) molecular biology techniques. Any one chemokine can often be produced by a number of different cell types and exert its effects on different target cells. This has been interpreted by some as implying a high degree of redundancy. Although this is understandable, in disease processes parallel and sequential mechanisms are possible, and potentially important therapeutic targets have emerged.

There is compelling evidence from animal and clinical studies that eosinophils are important effector cells in asthma, but this relationship is as yet

[119] A. Rot, *Eur. J. Immunol.* **23,** 303 (1993).

[120] Y. Tanaka, D. H. Adams, S. Hubscher, H. Hirano, U. Siebenlist, and S. Shaw, *Nature (London)* **361,** 79 (1993).

[121] W. C. Darbonne, G. C. Rice, M. A. Mohler, T. Apple, C. A. Hebert, A. J. Valente, and J. B. Baker, *J. Clin. Invest.* **88,** 1362 (1991).

[122] A. Manni, J. Kleimberg, V. Ackerman, A. Bellini, F. Patalano, and S. Mattoli, *Biochem. Biophys. Res. Commun.* **220,** 120 (1996).

unproven in the human disease. Two possible targets to prevent eosinophil recruitment to the lung are IL-5 and its receptor, which are important in several aspects of eosinophil biology, and eotaxin and its receptor, CCR3. The eotaxin receptor is particularly attractive as a target as it is expressed in high numbers on eosinophils, but not other leukocytes, and appears to be the major detector of the eosinophil for eotaxin and other chemokines such as MCP-4. Eotaxin and CCR3 knockout mice are being developed, and animal models will continue to be invaluable when antagonists are available. In the shape of receptor antagonists, the chemokine field may yet provide the final proof of concept for the long-established eosinophil theory of asthma in humans.

[17] Identification and Structural Characterization of Chemokines in Lesional Skin Material of Patients with Inflammatory Skin Disease

By JENS-MICHAEL SCHRÖDER

Introduction

A number of cutaneous inflammatory disorders are histologically characterized by a disease- and activity-dependent presence of different leukocyte types and subpopulations. The appearance of leukocytes in inflamed skin led to the working hypothesis that apart from panleukotactic factors, which attract all different leukocyte forms in a similar manner, cell-specific attractants are generated having more cell-selective chemoattractive properties. Chemokines represent a family of rather cell-selective or specific leukocyte chemoattractants, which consist of mostly basic and heparin-binding proteins.

In contrast to well-known leukoattractants such as the plasma-derived complement split product C5a, the majority of chemokines seem to be generated under inflammatory conditions mainly in affected tissues and to a lesser extent in the blood. Therefore, the highest concentrations of chemokines [except those released by activated platelets, which are a rich source of the chemokines β-thromboglobulin, platelet factor 4 (PF4), and RANTES][1] can be expected and observed in tissue rather than in blood plasma.

[1] Y. Kameyoshi, A. Dörschner, A. I. Mallet, E. Christophers, and J.-M. Schröder, *J. Exp. Med.* **176,** 587 (1992).

METHODS IN ENZYMOLOGY, VOL. 288

0076-6879/97 $25.00

For proof of the biological relevance of each chemokine in diseases, it is essential to get as much evidence as possible that this working hypothesis is true. To gain insight into mechanisms that are responsible for tissue leukocyte immigration by a particular leukocyte form or subset, biological material should be analyzed for the presence or absence of leukocyte chemotactic factors showing biochemical properties known for chemokines (or other groups of leukoattractants). Because a number of inflammatory skin diseases have a characteristic pattern and localization of different leukocytes in skin, we were interested to know which factors are of particular importance for leukocyte infiltration; therefore, we have optimized biochemical methods to identify the factors responsible for leukocyte tissue infiltration.

This chapter concerns the methodology for isolating and characterizing chemokines obtained from lesional scales of patients with inflammatory skin diseases with a major focus on psoriasis. It includes as main sections bioassays for the detection of chemoattractants, strategies of chemokine purification by modern high-performance liquid chromatography (HPLC) techniques, results in different materials and discussion of the advantages and limitations, some pitfalls, as well as potential directions for future research in this area.

Methods of Chemokine Extraction from Biological Material

To successfully characterize biologically active chemokines from biological material, one must optimize its extraction. For lesional skin material the following procedure is useful. Lesional scale material (50 g) is suspended in 350 ml of 0.1 *M* citric acid. After adding 350 ml ethanol (96%), the suspension is homogenized using a homogenizer at 2000 rpm for 60 min and chilling in an ice–water mixture. Homogenized scales are centrifuged at 1000 *g* for 30 min at 4°, and the opalescent supernatant is then concentrated using ultrafilters (Amicon, Danvers, MA, YM2 filters, cutoff of 2 kDa), followed by diafiltration against the buffer required for further experiments. After diafiltration samples should be centrifuged after one freezing and thawing cycle. Otherwise problems can occur when HPLC is performed as the next step.

The addition of ethanol is essential for optimal extraction of lesional scale material. Omission of this organic solvent leads to strongly increased amounts of fine material in the supernatants, which cannot be filtered and thus impair subsequent HPLC separations. Alternatively, scale lipids can first be extracted using ethyl acetate followed by treatment of the remaining scales with ethanolic citric acid and subsequent extraction of chemokines and other polypeptides.

Extraction of skin material with acidic ethanolic buffers allows direct application of samples to HPLC columns. High molecular mass proteins, which are present in huge amounts when extraction is performed at neutral pH in the absence of ethanol, are present only in low concentrations. In addition, the use of acidic buffers leads to a better extraction of extracellular matrix-bound chemokines, whereas the stickiness of the very hydrophobic chemokines will be reduced by adding organic solvents such as ethanol.

Assays for Detection of Chemokines

For purification and biochemical characterization of chemokines present in biological skin material, it is essential to have assays at hand that allow its detection to follow the chemokine during purification. For detection of chemokines and followup studies during purification by HPLC we have used several different assay systems with different sensitivities. The procedures for each assay system are described in the following sections.

Measurement of Polymorphonuclear Leukocyte Chemotaxis

For detection of chemotactic activity in HPLC fractions we use single-blind well Boyden chambers. Multiwell chambers cannot be used with the method described below. Boyden chambers (Costar, Bodenheim, Germany) consist of two parts. The lower compartment, usually consisting of a volume of approximately 100 μl, is filled with an appropriate stimulus dissolved in chemotaxis medium. This lower part is covered by a filter, the upper part is screwed on the lower part, and finally the cell suspension is added to the upper chamber. For assaying polymorphonuclear leukocyte (PMN, neutrophil), chemotactic activity freshly purified (human) neutrophils need to be used.

Isolation of Human Neutrophils. Freshly taken venous blood (100 ml) is immediately mixed with 10 ml of an acidic dextran containing anticoagulant solution [65 m*M* citric acid, 85 m*M* sodium citrate, and 20 g/liter dextran T-70 (Sigma, Munich, Germany]. The blood is centrifuged for 20 min at 500 *g* at room temperature; the plasma as well as the white buffy coat are pipetted away. The remaining blood cell sediment is mixed with the same volume of a gelatin solution [2.5% (w/v) in 0.9% aqueous saline (0.9% NaCl)], previously prewarmed to 37°, and stored at 37° for 30 min. The yellowish supernatant containing a few red blood cells is collected and centrifuged (400 *g*) at room temperature for 10 min. Now the supernatant, except for 0.5 ml, is removed by pipetting, and the sediment is mixed carefully with 0.15 *M* ammonium chloride in water, pH 7, and stored for 7 min at room temperature followed by centrifugation at 4° for 10 min at

150 *g*. The cell sediment is washed twice with phosphate-buffered saline (PBS), pH 7.2, containing 128 m*M* NaCl, 8.1 m*M* Na_2HPO_4, 1.5 m*M* KH_2PO_4, 2.7 m*M* KCl, and 0.1% (w/v) bovine serum albumin (BSA) and stored at 4° until the experiments are started. For chemotaxis experiments, cells should be stored no longer than 4 hr after isolation.

Using this technique final cell preparations contain more than 90% PMN (>85% neutrophils), with a viability greater than 97% as assessed by trypan blue exclusion. The main contaminants of these preparations are eosinophils (2–7%), T lymphocytes (3–5%), and monocytes (2–4%).

Measurement of Chemotaxis. For chemotaxis experiments PMN are suspended in cold (4°) PBS containing $CaCl_2$ (0.6 m*M*) and $MgCl_2$ (0.8 m*M*) [complete PBS, (c)PBS]. The lower parts of Boyden chambers are filled with fractions to be tested. Because HPLC solvents are usually not compatible with the bioassay, several special procedures need to be used for testing biological activity. In the case that acidic solvents are used for HPLC, the acids should be volatile as well as present at low concentrations. Usually 5–10 μl of each fraction (~0.1% of the whole content of each fraction) is sufficient for testing. In most cases fractions can be used directly by diluting them in 300 μl (c)PBS. In the case that the pH of the final solution is below pH 6.9 or the content of organic compounds is too high, the HPLC fractions (10–40 μl) should be frozen and subsequently lyophilized after adding 10 μl (c)PBS to the aliquots of each fraction in a microtiter plate for 30 min. Subsequently 300 μl of (c)PBS is added to each well of the microtiter plate containing the lyophilizate and immediately transferred to the lower part of two Boyden chambers.

Thereafter the lower part is covered with self-punched (7-mm punch) polyvinylpyrrolidone-containing polycarbonate filters (pore size 3 μm, Costar; note that these are not identical with chemotaxis filters distributed by the manufacturer), which previously were washed with 1 *M* NaOH in 50% (v/v) aqueous ethanol for 7 min followed by three washes in water. The washing procedure is essential to get sufficient numbers of cells migrating through the filter (possibly by enhancing anionic sites on the filter, which enhance binding of cationic chemokines).

The cover is then screwed into the chamber, and in each Boyden chamber the fluid present in the upper part is carefully withdrawn. In the case that air bubbles are present in the lower part, the chamber should be opened and refilled. Thereafter, 100 μl of the prewarmed (37°) cell suspension [2×10^6 cells/ml (c)PBS] is added to the upper part of prewarmed (37°) Boyden chambers, and each chamber is covered with thin slides (to avoid loss of water) and then incubated for 2 hr at 37°. Incubation in moist incubators containing carbon dioxide, which is very common for chemotaxis experiments, is not necessary when this procedure is used.

After incubation remaining cells and fluids in the upper part of Boyden chambers are drawn away followed by the opening of each chamber. Filters are carefully removed from the lower chamber, and to each chamber 10 μl of 0.1% (v/v) Triton X-100 (Sigma) in water is added and incubated for 5 min. Then the complete contents of each chamber are transferred to a microtiter plate and to each well 100 μl of the β-glucuronidase (EC 3.2.1.31) substrate *p*-nitrophenyl-β-D-glucuronide (Sigma; 10 m*M* in 0.1 *M* aqueous sodium acetate, pH 4.0) is added followed by overnight (18 hr) incubation at 37°. The enzymatic reaction is terminated by adding 100 μl of 0.1 *M* glycine buffer, pH 10.0, and *p*-nitrophenolate is measured using a microtiter plate photometer at 405 nm.

For calibration of chemotactic migration defined numbers of cells are lysed in 110 μl of Triton X-100 (Sigma) in (c)PBS, and lysates are measured for β-glucuronidase content. A calibration curve, as shown in Fig. 1, can be established.

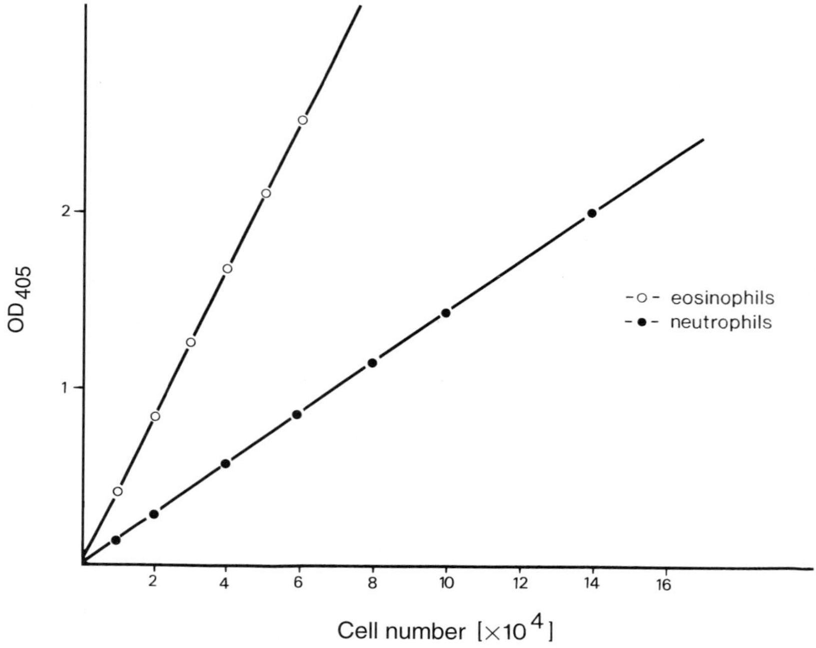

FIG. 1. Eosinophils contain more β-glucuronidase than neutrophils. Defined numbers of neutrophils or eosinophils were lysed with Triton X-100 and investigated for β-glucuronidase activity (formation of *p*-nitrophenolate, measured as OD_{405}).

In a few cases where crude tissue extracts are assayed for the presence of attractants false-positive results can be obtained due to free β-glucuronidase activity present in the samples. This can be easily identified by incubating the HPLC fractions with β-glucuronidase substrate in the absence of cells. We avoid this problem by treating each biological sample prior to testing at pH 2 for 10 min and subsequently restoring both pH and osmolarity. β-Glucuronidase is acid-unstable, and thus the presence of endogenous enzyme in crude extracts presents no problems when reversed-phase HPLC is used for separation of proteins. In our experience none of the known proteinaceous attractants are affected under these conditions.[2]

In each test series positive as well as negative controls need to be included at appropriate concentrations (i.e., chemotaxins that are expected in the HPLC fractions). In the case of chemokine analyses, interleukin-8 (IL-8, Peprotech, London, U.K., 10^{-9} *M*) as well as *N*-formyl-Met-Leu-Phe (fMLP, Sigma, 10^{-8} *M*) serve as positive controls for chemotactic responsiveness of the PMN preparation, whereas (c)PBS is used as a negative control.

Measurement of Eosinophil Chemotaxis

When eosinophil (Eo) chemotactic factors have to be purified from biological material, purified Eo preparations are necessary to determine Eo chemoattractants with a similar chemotaxis method.

Isolation of Human Eosinophils. For Eo purification we use the following procedure. The PMN-rich supernatants obtained after gelatin sedimentation (see above) are washed twice with ice-cold PBS (pH 7.4), and eosinophils are separated by centrifugation in a gradient of isotonic Percoll (Pharmacia, Freiburg, Germany), similar to the method described by Gärtner.[3] A discontinuous gradient is prepared using 1.5 ml of Percoll (Pharmacia, Uppsala, Sweden) solutions with the densities 1.075, 1.077, 1.0785, 1.080, 1.0825, and 1.085 g/ml in 15-ml polystyrene tubes.

The crude PMN preparation is layered on the top of the gradient and centrifuged at 900 *g* for 10 min at room temperature. Cells are collected from each interface, and aliquots are stained with eosin to identify the purity of each eosinophil preparation. Fractions containing more than 90% eosinophils are collected. Contaminating erythrocytes are lysed with NH_4Cl in water [0.84% (w/v), pH 7.0]. The recovery of human eosinophils is rather low. From healthy donors, usually $1–5 \times 10^6$ pure eosinophils are obtained from 100 ml of blood.

[2] J.-M. Schröder, U. Mrowietz, E. Morita, and E. Christophers, *J. Immunol.* **139,** 3474 (1987).
[3] I. Gärtner, *Immunology* **40,** 133 (1980).

Measurement of Chemotaxis. Chemotaxis experiments are performed as described for neutrophils. Because of the higher β-glucuronidase content of human eosinophils (see Fig. 1), 10^5 cells per chamber or less is sufficient for determination of chemotactic migration. Incubation times of the cells in the Boyden chamber need to be elongated to 1.5 hr instead of 1 hr in the case of neutrophils. When very low numbers of eosinophils are available (this is often the case) the number per chamber can be decreased down to 5×10^4 cells. As shown in Fig. 1, 2×10^3 cells can be easily assayed. Contaminating neutrophils (<30%) usually are not problematic because the low numbers migrating under experimental conditions contribute to <10% of the extinction measured. For control of maximum migration by contaminating neutrophils IL-8 (10^{-9} *M*) can be included, because IL-8 does not stimulate eosinophil chemotaxis.[2,4]

Polymorphonuclear Leukocyte Enzyme Release as Bioassay for Chemokine Identification

Chemotactic cytokines are capable of inducing extracellular degranulation of neutrophils when pretreated with the fungal metabolite cytochalasin B. Thus this property can be used for the identification of chemokines in HPLC fractions.[2,5] Elastase, β-glucuronidase, and myeloperoxidase are most commonly used for these purposes.[2,5,6]

Measurement of Enzyme Release. For determination of enzyme releasing activity in HPLC fractions, PMN [10^7 cells/ml (c)PBS] are preincubated with cytochalasin B [Sigma, 5 μg/ml from a stock (1 mg/ml) in dimethyl sulfoxide (DMSO)] for 5 min at 37°. One hundred microliters of sample in (c)PBS (HPLC fractions are treated as described for chemotaxis) is added to the wells of a microtiter plate and prewarmed to 37° (5 min). Thereafter to each well 100 μl of the cell suspension is added and incubated for 30 min at 37°. After centrifugation 100 μl of supernatants are carefully transferred to another microtiter plate and incubated with 100 μl of substrate solution in appropriate buffers. For β-glucuronidase determination 10 m*M* *p*-nitrophenyl-β-D-glucuronide in 0.1 *M* sodium acetate, pH 4.0, serves as substrate. In this case the enzymatic reaction is terminated after overnight (18 hr) incubation by adding 200 μl of 0.4 *M* glycine buffer, pH 10, and *p*-nitrophenolate is determined using a microtiter plate photometer.

[4] T. Yoshimura, K. Matsushima, S. Tanaka, E. A. Robinson, E. Appella, J. J. Oppenheim, and E. J. Leonard, *Proc. Natl. Acad. Sci. U.S.A.* **84,** 9233 (1987).

[5] J.-M. Schröder, N. Persoon, and E. Christophers, *J. Exp. Med.* **171,** 1091 (1990).

[6] A. Walz, P. Peveri, H. Aschauer, and M. Baggiolini, *Biochem. Biophys. Res. Commun.* **149,** 755 (1987).

As a total control one well contains instead of the stimulus 100 μl of 0.2% (v/v) Triton X-100 for lysis of the cells.

In cases where fast detection of PMN enzyme-releasing activity is necessary, myeloperoxidase (MPO) (EC 1.11.1.7) can be determined. One hundred microliters of supernatant is incubated with 100 μl of 10 m*M* *o*-phenylenediamine dihydrochloride (Sigma) in 0.1 *M* citrate/phosphate buffer, pH 5, containing 0.001% (v/v) hydrogen peroxide for 10 min at room temperature in the dark. The enzymatic reaction is stopped by adding 100 μl of 2 *M* H_2SO_4, and the brownish color is examined in a microtiter plate at 485 nm using a multichannel photometer (SLT 210, Kontron, Salzburg, Austria). As controls we use fMLP (4×10^{-7} *M*) and 0.1% (w/v) aqueous hexadecyltrimethylammonium bromide (Sigma, used instead of Triton X-100, which destroys MPO activity) as the 100% control. Enzyme releasing activity can be expressed as either OD_{405} units (β-glucuronidase), OD_{486} units (MPO), or percentages of the total enzyme content (total control).

Desensitization of Chemotaxis and Degranulation as Method for Detection of Chemokines

The following assay system is a modification of the chemotaxis methods and enzyme-release methods described above. The lower compartment in a Boyden chamber is filled with HPLC fractions previously lyophilized and dissolved in (c)PBS or appropriately diluted in (c)PBS and is covered with a polycarbonate filter (pore size 3 μm, Costar) as described for chemotaxis experiments. Unlike the chemotaxis experiments 50 μl of a 3×10^{-9} *M* IL-8 solution in (c)PBS [alternatively, 10^{-8} *M* *N*-formylnorleucylleucylphenylalanine (fNLP, Sigma) or 10^{-9} *M* rC5a (Sigma)] is added to the upper chamber followed by adding 50 μl of a PMN suspension [4×10^6 cells/ml (c)PBS] and covering the chambers with glass slides. Thereafter, the chambers are incubated for 2 hr at 37° and treated as described for neutrophil chemotaxis.

In a different approach aliquots of HPLC fractions are either lyophilized and dissolved in (c)PBS or appropriately diluted in (c)PBS. Then 50 μl is added to the upper chamber, which previously was filled with either 2×10^{-9} *M* IL-8, 10^{-8} *M* fNLP, or 10^{-9} *M* rC5a into the prewarmed (37°) lower chamber. Thereafter 50 μl of the PMN suspension (4×10^6 cells/ml) is added to the upper chamber, and the experiment is performed as described.

In another approach to achieve desensitization of enzyme release, cells are preincubated with either 10^{-8} *M* IL-8, 10^{-8} *M* fNLP, 5×10^{-9} *M* rC5a, or (c)PBS (for control) at a density of 10^7 cells/ml PBS without Ca^{2+} and

Mg^{2+} for 15 min at room temperature. Thereafter, a stock solution of cytochalasin B in DMSO is added (final concentration 5 μg/ml) to the cell suspension and incubated for 5 min at 37°. Stock solutions of $CaCl_2$ and $MgCl_2$ are then added to give the required final concentration of Ca^{2+} and Mg^{2+} in (c)PBS. One hundred microliters of PMN suspension (10^7 cells/ml) is added to 100 μl of properly treated HPLC fractions in a microtiter plate and incubated for additional 30 min. Detection of enzymatic activity is performed as described for degranulation.

Strategies of Chemokine Characterization and Purification

It is essential to use acidic buffers containing organic solvents for extraction of chemokines from skin samples. Extraction with physiological buffers such as saline or PBS usually was found to be insufficient. Furthermore, drastic losses of activity, possibly by binding of the highly cationic and hydrophobic chemokines to tissue glycosaminoglycans and/or hydrophobic structures, were seen. After extraction all samples should first be clarified by centrifugation (1000 *g* or more) followed by filtration through 0.1-μm cellulose nitrate filters.

The combination of chemotaxis experiments and desensitization experiments after HPLC separation of biological samples (1) allows for an early decision of how many peptidelike leukocyte attractants are present and (2) permits a partial evaluation of the receptor type, which is involved in eliciting locomotion by attractants present in biological samples. Because of the high sensitivity of the locomotion assays (the detection limit is usually near 20–100 pg chemokine per chamber), these methods allow the partial biological characterization of chemotactic mediators present in biological samples obtained from diseased skin.

Instead of chemotaxis desensitization, alternatively the desensitization of degranulation can be used for detection of chemokines. In our hands this assay is only useful when sufficient biological material is available, because PMN degranulation induced by chemokines requires 10- to 30-fold higher amounts of chemokines than are necessary for chemotaxis.

For purification of chemokines from lesional scales of diseased skin, we use two strategies to obtain pure chemokines for detailed biological and biochemical characterization.

Purification of Chemokines: Strategy I

Acidic and ethanolic extracts obtained from 5–50 g of lesional scale material are concentrated to a volume of less than 5 ml and applied to a Superdex column (2.6 × 79 cm, Pharmacia) previously equilibrated with

0.1 *M* ammonium formate, pH 5.0, at 8°. Proteins are eluted with a flow of 10 ml/hr and separated as 5-ml fractions. Figure 2 shows a characteristic chromatogram. Aliquots (10 μl) of each fraction are tested in either the chemotaxis assay system, the degranulation assay system, or a degranulation

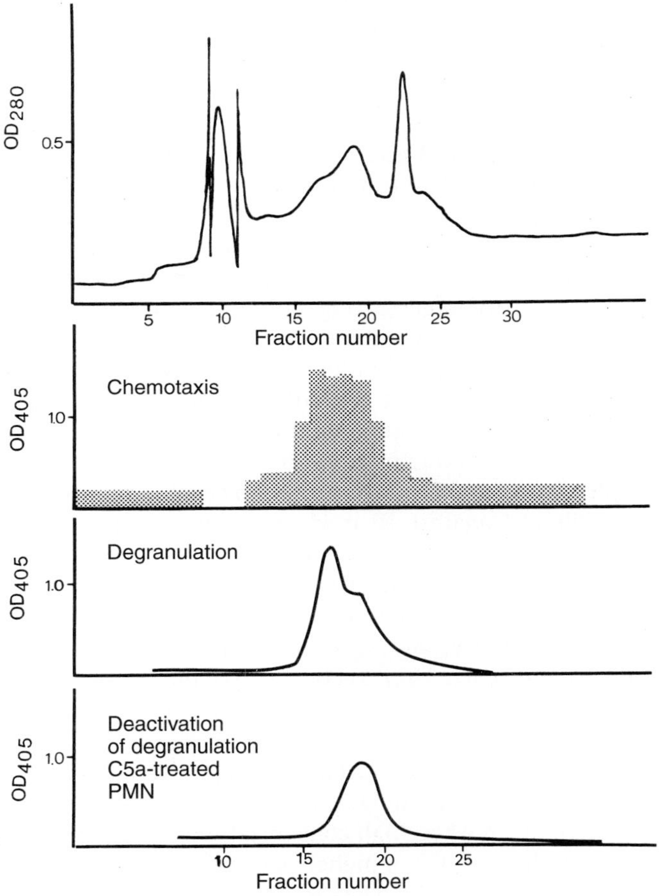

FIG. 2. Size-exclusion chromatography of a crude psoriatic scale extract. A psoriatic scale extract was separated on a Superdex column, and fractions were assayed for neutrophil chemotaxis (shaded area). Degranulation of PMN (marker enzyme myeloperoxidase) is shown. Note the presence of a peak and a shoulder. Degranulation experiment performed with neutrophils, which were deactivated with C5a, is shown as well. Note the absence of the earlier eluting peak, which corresponds to $C5a_{des\text{-}Arg}$ present in psoriatic scales.

desensitization assay system[7] (Fig. 2). These bioassays provide information about the content of chemotactically active peptides in the samples as well as the presence of chemotaxis-inhibiting fractions. Using this method as the first step, these inhibitors are separated from the biological activity due to an earlier elution time (Fig. 2). Thus, further characterization of biological activity is not affected by the presence of chemotaxis inhibitors,[8] which can easily impair the detection of chemotaxins that coelute in other separation systems.

The use of ammonium formate as elution buffer also allows lyophilization of biologically active fractions prior to further purification by HPLC techniques, owing to volatility of this salt. We prefer to concentrate samples using Amicon filters (YM2, cutoff 2 kDa) under acidic conditions to prevent losses of bioactive material. When only low amounts of proteins are present we use acidic buffers for diafiltration (whenever possible), which contain 20% (v/v) acetonitrile.

The use of a desensitization assay provides early information about the presence or absence of $C5a_{des\text{-}Arg}$ in the samples (Fig. 2). Thus, in principle, for purification of chemokines acting on neutrophils fractions that elute later than $C5a_{des\text{-}arg}$ can be combined (Fig. 2). For further analyses we combine all fractions containing biological activity in the 6- to 20-kDa area, because it is possible to separate the different chemotaxins during further purification by either ion-exchange or reversed-phase HPLC.

For separation of chemotactic peptides a cation-exchange HPLC column is used. Samples from Superdex size-exclusion chromatography are concentrated and diafiltered against 10 m*M* ammonium formate, pH 5.0, and applied to a TSK-CM-3SW HPLC column (LKB, Bromma, Sweden, 125 × 12 mm). Proteins are eluted with a gradient of increasing concentrations of ammonium formate (final concentration 0.5 *M*, flow rate 1 ml/min) and finally with 0.5 *M* formate, acidified with formic acid to pH 3.0. A portion (10–30 μl) of each 1-ml fraction is lyophilized in the presence of 10 μl (c)PBS and tested for PMN stimulating activity (chemotaxis, MPO release).

Figure 3 shows a representative HPLC run. It should be remarked that the slope of maximum degranulation-inducing activity and chemotactic activity are not identical when high concentrations of the chemoattractants are present, as can happen when higher percentages of the whole fractions are used for testing (Fig. 3). When less material is used for testing a different pattern of chemotactic activity can be seen (Fig. 3). Reasons for this appar-

[7] J.-M. Schröder and E. Christophers, *J. Invest. Dermatol.* **87,** 53 (1986).

[8] J.-M. Schröder, *in* "Molecular Basis of Inflammation" (J. Navarro, ed.), p. 239. Ares Serono Symposia Publications, Rome, 1994.

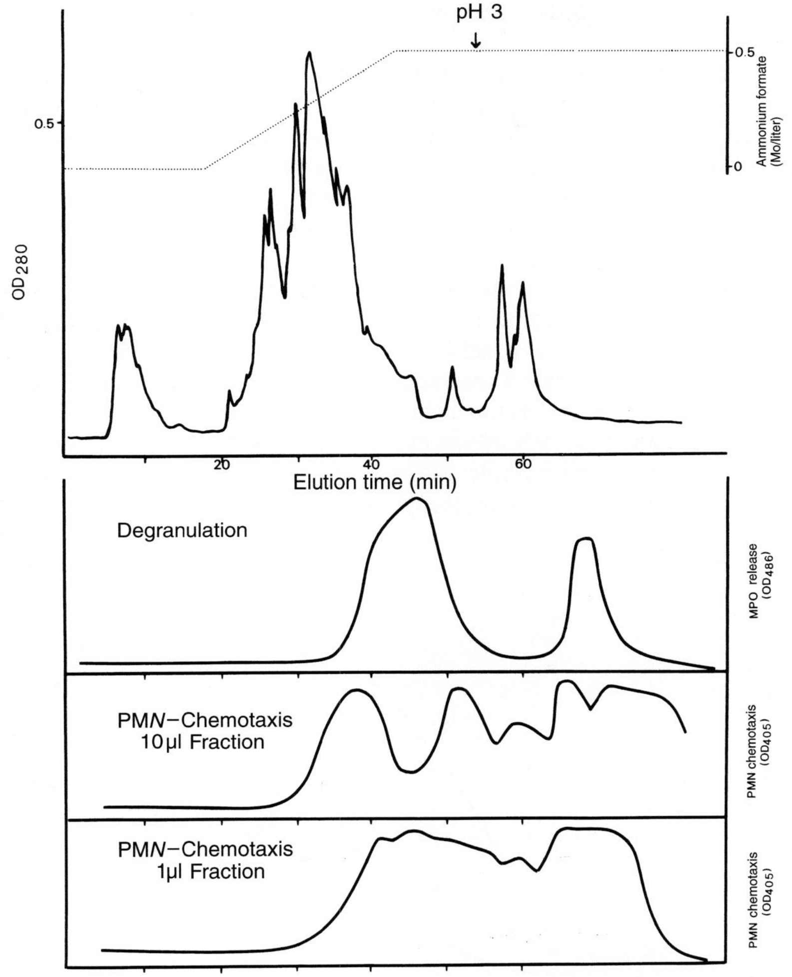

FIG. 3. Cation-exchange HPLC of size-exclusion chromatography-purified psoriatic scale-derived neutrophil attractants. Pooled neutrophil attractant-containing size-exclusion HPLC fractions were separated on a TSK-CM-3SW cation-exchange HPLC column, and fractions were assayed for myeloperoxidase-releasing activity (*upper*) and neutrophil chemotactic activity using 10-μl fractions (*middle*) and 1-μl fractions (*lower*). Note the changes of activity in the chemotaxis assay system when different amounts of the fractions were tested.

ent discrepancy are differences in dose–response curves of chemotactic activity and degranulating activity. Whereas half-maximum chemotactic activity occurs at lower doses than that of degranulation, chemotaxis dose–response curves are bell-shaped (Fig. 4), in contrast to those of degranulation, which are sigmoid. Thus it is important to investigate fractions at different dilutions to obtain information about the fraction that contains the highest number of units of PMN chemotactic activity (1 unit is defined as the amount of the chemotaxin, per milliliter, which elicits half-maximum responses).

As the next purification step we use a preparative wide pore reversed-phase (RP-8) HPLC column (7 μm particle size, C_8 Nucleosil with end capping, 250 × 12.6 mm, Macherey and Nagel, Düren, Germany), previously equilibrated with water/acetonitrile/trifluoroacetic acid (TFA), 90 : 10 : 0.1 (v/v/v). Fractions from cation-exchange HPLC are acidified with TFA to pH 2.4 and then, whenever possible, the whole volume of combined fractions (using a 5-ml sample loop) is directly applied to the column. Alternatively, samples can be concentrated using ultrafilters (Amicon YM2,

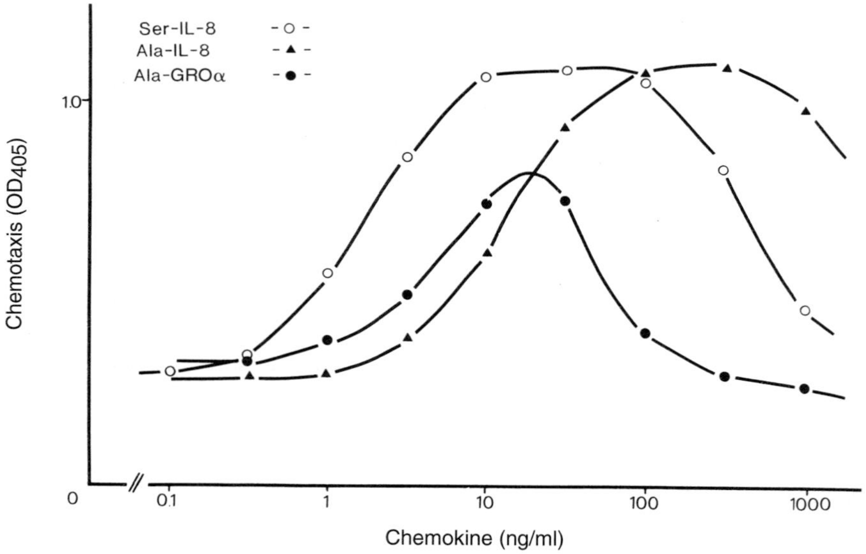

FIG. 4. Neutrophil chemotactic activity of two different IL-8 forms and GRO-α. Purified [Ser-IL-8]^{72}, [Ala-IL-8]^{77}, and [Ala-Gro-α]^{73} were tested in the Boyden chamber system for neutrophil chemotactic activity. Note the different ED_{50} values for [Ser^{72}]- [Ala^{77}]IL-8 and the higher efficacy (percentage of input migrating cells) of the IL-8 variants compared to GRO-α. GRO-α shows a narrow, bell-shaped dose–response curve.

cutoff 2 kDa) after acidification of the samples and the addition of 10% acetonitrile (v/v) to the final volume.

Whereas on cation-exchange HPLC the protein content can be recorded by measuring absorbance at 280 nm (215 nm should not be used owing to the strong absorption of ammonium formate buffer), proteins eluting on preparative RP-8 HPLC can be detected by absorbance at 215 nm, which is much more sensitive than that at 280 nm. Proteins are then eluted with an increasing gradient of acetonitrile similar to that shown in Fig. 9 at a flow rate of 3 ml/min. We also have tried *n*-propanol instead of acetonitrile as eluent. *n*-Propanol gives less resolution of peaks and thus was not used further for RP-8 HPLC.

For testing of bioactivity aliquots of fractions (10–40 μl) are mixed with 10 μl of (c)PBS and then lyophilized in a microtiter plate. Because of the use of volatile solvents, testing of an increasing volume of each fraction is possible without any interference with the assays.

Gel Electrophoresis of Chemokines. Fractions showing biological activity are investigated using sodium dodecyl sulfate–polyacrylamide gel electrophoresis (SDS–PAGE). The method of Schägger and von Jagow[9] with Tricine instead of glycine is used (in the presence of 8 *M* urea).

Fractions to be tested (10 μl) are mixed with 10 μl of sample buffer [50 m*M* Tris, 4% (w/v) SDS, 12% (w/v) glycerol, pH 6.8] and boiled for 10 min. Samples are then loaded on the stacking gel and separated electrophoretically (gel dimension 130 × 100 × 1 mm) for 18 hr at 10 mA current, 30 V power (power limit 10 W) at room temperature. Then the gel is fixed for 30 min with aqueous 2-propanol [30% (v/v)] containing 10% (v/v) acetic acid and 0.3% (v/v) glutaraldehyde, followed by washing three times with deionized water. Proteins are stained by a 30-min incubation with 0.03% (w/v) aqueous silver nitrate solution followed by developing with a solution of 10% saturated aqueous Na_2CO_3 solution containing 0.1% (v/v) of saturated aqueous formaldehyde [40% (w/v)]. Development is terminated by incubation with acetic acid [3% (w/v) in water].

Alternatively, commercially available "high density gels" can be used with the Phast electrophoresis system (Pharmacia). This permits faster detection of proteins, but resolution is not as high. Figure 5 shows an example of the analysis of purified IL-8 preparations obtained from psoriatic scales and different N-terminal variants of IL-8 by this technique.

Further Purification of Chemokines. When activity-containing fractions from preparative RP-8 HPLC show bands representing relative molecular masses exceeding 30 kDa, size-exclusion HPLC is useful as the next purification step. Samples are lyophilized, and the residue is dissolved in 50–200

[9] H. Schägger and G. V. von Jagow, *Anal. Biochem.* **166,** 368 (1987).

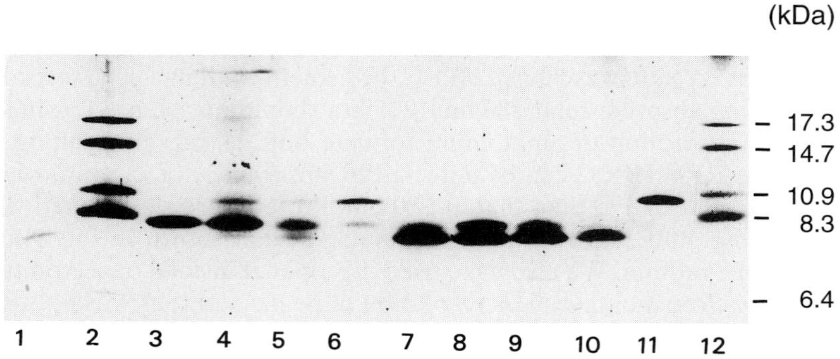

FIG. 5. Tricine–SDS gel electrophoresis of different IL-8 forms and psoriasis-derived IL-8. Samples were analyzed using the Tricine–SDS gel electrophoresis system in the presence of 8 *M* urea. Whereas lanes 2 and 12 contained molecular weight markers, in lane 3 [Ser^{72}]IL-8 was applied. Lane 7 contained [Glu^{69}]IL-8, and in lane 11 [Ala^{77}]IL-8 was applied. Lanes 4–6 contained different psoriasis-derived IL-8 preparations, which mainly contained [Ser^{72}]IL-8 and to a lesser extent [Ala^{77}]IL-8 (lanes 4 and 6). In lanes 8–10 the major psoriasis-derived IL-8 form [Glu^{69}]IL-8 was applied. Lane 1 contained small amounts of psoriasis-derived GRO-α1.

μl of 0.1% aqueous TFA and applied to a size-exclusion HPLC column (TSK-2000-3SW HPLC column, 0.75 × 60 cm, containing a 0.75 × 12 cm precolumn), which previously has been equilibrated with 0.1% TFA in water (pH ~2). Proteins are eluted at a flow rate of 1 ml/min with 0.1% TFA in water, and proteins are detected in the effluent by recording the absorbance at 215 nm. Peak biological activity of chemokines [IL-8, growth-related oncogene-α (GRO-α), RANTES] appears in fractions corresponding to a relative molecular mass of approximately 10 kDa (Fig. 6).

The acidic conditions used (aqueous TFA) allow optimal separation of low molecular mass proteins (1–20 kDa) from high molecular mass proteins.[2,10] It should be noted that by the use of rather neutral and salt-containing solutions the resolution in the area 5–20 kDa is far less than that seen under the conditions given. Furthermore, it is our experience that the recovery of chemokines is better when acidic conditions are used for elution. As a positive side effect we noted a drastic increase of the half-life of the column when acidic conditions are used for elution. Other size-exclusion columns we tested, even those based on zirconium oxide as matrix, did not show the high resolution of the TSK-2000 column. The use of 0.1% TFA as eluent in this HPLC system allows the direct application of whole fractions containing biologically active material on a reversed-phase HPLC column for the subsequent separation procedure.

[10] G. B. Irvine and C. Shaw, *Anal. Biochem.* **155,** 141 (1986).

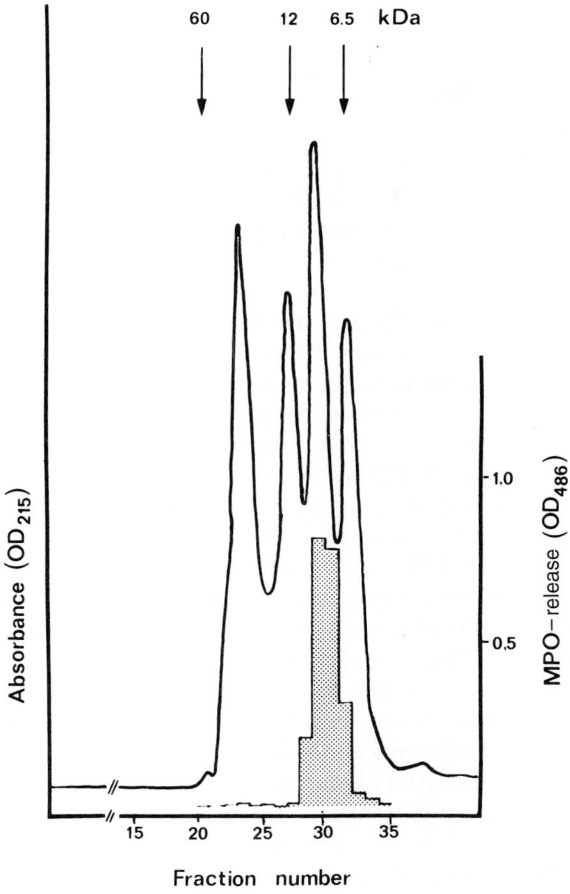

FIG. 6. TSK-2000 size-exclusion HPLC of psoriasis-derived IL-8 preparations. Partially purified psoriatic scale-derived IL-8 was separated on a TSK-2000 HPLC column using 0.1% aqueous trifluoroacetic acid as solvent. Fractions were assayed for neutrophil enzyme-releasing activity with myeloperoxidase as marker enzyme.

As the next step we originally used an analytical RP-18 HPLC column for final purification and found biological activity eluting in a single, sharp, symmetrical peak. Amino-terminal sequencing of the protein peak containing biological activity did not reveal a chemokine. Instead we often obtained the sequence of ubiquitin, which is not an attractant for neutrophils and thus indicated that a chemotaxin is present as a contaminant (<3% of the whole amounts).

To separate the attractant from ubiquitin we tried several HPLC systems. Although cation exchange is useful for separation, it can be used only as an initial step. Unfortunately, when it is used early in purification, resolution is not high. When this HPLC step was used again later, we noted a nearly complete loss of biological activity, possibly by its binding to the matrix. An HPLC column that allows separation of ubiquitin from chemotactic factors is the Poly F (fluorocarbon) HPLC column (Dupont de Nemours, Bad Nauheim, Germany). We used this column for separation of ubiquitin, after the preparative RP-8 HPLC run, with a gradient of acetonitrile as eluent.

Later it became clear that a cyano(CN)propyl HPLC column, used in the reversed-phase mode, is very useful for separating ubiquitin and other molecules we often misinterpreted as neutrophil attractants. This column has been shown to be useful for purification of different chemokines from different cellular sources.[11–13] Therefore, psoriatic scale-derived neutrophil attractant preparations, which often reveal by SDS–PAGE only low contamination with high molecular mass components, are directly applied (when fractions from size-exclusion HPLC are used) to an end-capped wide-pore CN-propyl HPLC column (5 μm, 250 $\times$ 4.6 mm; J. T. Baker, Gross Gerau, Germany), previously equilibrated with 0.1% TFA (v/v) in water. Proteins are eluted with a gradient of *n*-propanol containing 0.1% TFA. It is important that samples applied to this column do not contain organic solvents. We noted that in the presence of organic solvents (from previous reversed-phase HPLC steps) biological activity often eluted in the void volume, without any separation. Again, biological activity can be tested after complete lyophilization of aliquots for biological activity as described earlier. A typical chromatogram is shown in Fig. 7 for the major neutrophil chemotactic protein present in lesional psoriatic scales. Often we found single peaks representing nearly pure chemokines.

Biologically active fractions are finally purified on a narrow pore (125 nm) reversed-phase (RP-18) HPLC column (Nucleosil, 5 μm octadecyl silica, Bischoff, Leonberg, Germany), previously equilibrated with 0.1% (v/v) TFA in water. Proteins are eluted using a gradient of increasing concentration of acetonitrile, and aliquots of fractions are tested for biological activity as described. The earlier eluting component (peak I) usually shows from a major component a band corresponding to 8.0 kDa together with a minor band at 8.2 kDa (Fig. 5). On the other hand the later eluting

[11] A. Walz and M. Baggiolini, *Biochem. Biophys. Res. Commun.* **159,** 969 (1989).

[12] Y. Kameyoshi, A. Dörschner, A. I. Mallet, E. Christophers, and J.-M. Schröder, *J. Exp. Med.* **176,** 587 (1992).

[13] J.-M. Schröder, H. Gregory, J. Young, and E. Christophers, *J. Invest. Dermatol.* **98,** 241 (1992).

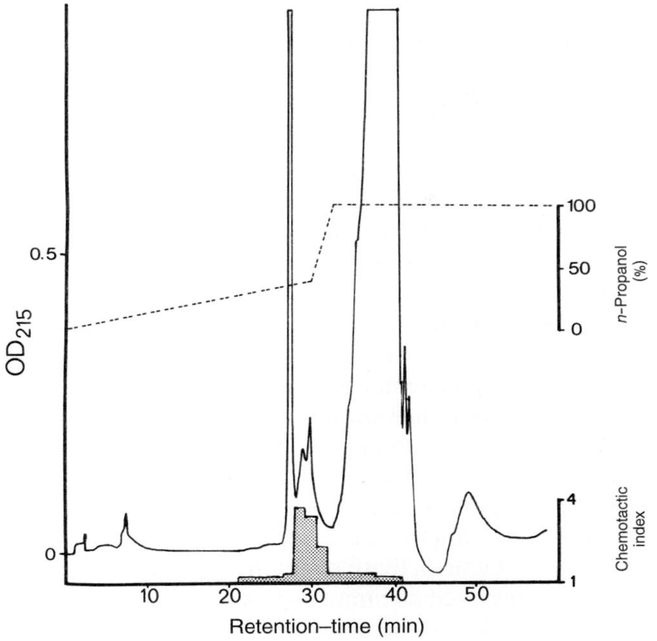

FIG. 7. Cyanopropyl HPLC of partially purified psoriasis-derived IL-8. Prepurified psoriatic scale-derived IL-8 preparations were separated on a cyanopropyl HPLC column using *n*-propanol as eluent, and fractions were tested for myeloperoxidase-releasing activity.

and major peak reveals a major band at 8.2 kDa with an additional band at 8.0 kDa. Because both peaks are not completely separated, compounds of both peaks can be detected on SDS–PAGE. Sometimes an additional line corresponding to an M_r near 9000 can be seen in peak I. Amino-terminal sequencing of both peaks consistently reveals the sequences shown in Table I for peak I and peak II, which are identical with those known for [Ser-IL-8][72], [Ala-IL-8][77], and [Glu-IL-8][69].

TABLE I
AMINO-TERMINAL AMINO ACID SEQUENCE ANALYSES OF PSORIASIS-DERIVED NEUTROPHIL ATTRACTANTS

Sample	N-terminal sequence
Peak I	90% SAKELRXQXIKTYXKP
	10% AVLPRSAKEL
Peak II	ELRXQXIKTYSKP
[Ala-IL-8][77]	AVLPRSAKELRCQCIKTYSKP . . .

The recovery of the IL-8 species from lesional psoriatic scales is low despite optimizing each step. Usually 2–10 μg of pure IL-8 cytokine variants can be obtained from 20–50 g of scales.

When the CN-propyl HPLC column is tested with several chemokines [IL-8, GRO-α, neutrophil-activating peptide-2 (NAP-2), RANTES, and macrophage chemotactic protein-1 (MCP-1)], all chemokines elute with very low amounts of propanol,[11–13] and thus many contaminating proteins can be separated using this HPLC system. These properties usually allow a high recovery at this step. In the majority of our experiments we saw in both CN-propyl and RP-18 HPLC two peaks close together, which both showed high chemotactic activity.

Table II indicates the losses during purification of IL-8 from lesional scale extracts. When PMN chemotaxis is used for estimation, in each purification step the number of remaining units of chemotactic activity, defined as the number of units (half-maximum chemotactic response eliciting doses), should be determined. We used this technique in the initial phase, when an enzyme-linked immunosorbent assay (ELISA) for IL-8 was unavailable. With this bioassay it is impossible to determine losses during the initial purification of chemotactic activity owing to the presence of a number of other, structurally unrelated chemotaxins, namely, GRO-α and $C5a_{des\text{-}Arg}$, which are both strong PMN chemotaxins.

Starting with the preparative RP-8 HPLC step the losses of IL-8 can be calculated because both $C5a_{des\text{-}Arg}$ as well as GRO-α elute earlier from the RP-8 HPLC column and therefore do not interfere with the estimation of IL-8. The total amount of free IL-8 present in lesional scales is underestimated because of the losses after the cation-exchange HPLC. By the use

TABLE II
PURIFICATION OF INTERLEUKIN-8 FROM PSORIATIC SCALES

Separation procedure	Protein[a] (μg)	IL-8[b] (μg)	Yield (%)	Purity (%)
Crude extract	17400	123.9	100	0.7
CM-TSK HPLC	2567	86.7	80	3.4
Preparative RP-8 HPLC	512	21.9	17.7	4.3
TSK-2000 HPLC	138	14.7	11.9	10.7
CN-propyl HPLC	14.3	9.4	8.6	65.7
RP-18 HPLC	2.9	2.8	2.3	96.5

[a] Protein content was measured using either aliquots or the whole samples, which were in part diafiltered, after RP-18 HPLC and integration of peaks absorbing at 215 nm using ubiquitin for calibration.

[b] IL-8 was determined in aliquots of the samples at appropriate dilution using a specific sandwich ELISA.

TABLE III
BIOCHEMICAL AND BIOLOGICAL PROPERTIES OF DIFFERENT PSORIASIS SCALE-DERIVED INTERLEUKIN-8 SPECIES

IL-8 variant[a]	Molecular mass[b] (kDa)	Retention time (min) RP HPLC[c]	Retention time (min) CM-TSK HPLC[d]	ED_{50} (chemotaxis)[e] (ng/ml)	N-terminal sequence[f]	Relative amount[g] (%)
1	8.3	33.3	21.4	4.4	n.d.[h]	2
2	7.9	34.1	23.2	2.9	n.d.	3
3	8.7	33.5	25.7	13.6	AVLPRSAK	13
4	8.3	33.4	31.3	2.8	SAKELR	34
5	7.9	34.2	29.6	2.1	ELRXQX	100
6	4.6	34.1	n.d.	>300	SAKELR	21
7	4.4	33.1	n.d.	>300	ELRXQX	44

[a] IL-8 variants are purified according to strategy I and strategy II.
[b] Relative molecular masses were determined using a Tricine-containing gel in the presence of 8 *M* urea.[9]
[c] RP-18 HPLC was performed with a narrow pore HPLC column using identical conditions.
[d] Retention time was obtained from TSK-CM HPLC reinvestigations.
[e] Concentration to elicit half-maximum neutrophil chemotaxis was determined using the method described in the text.
[f] Single-letter code for amino acids is used.
[g] Amounts are relative to the major component (=100).
[h] n.d., Not determined.

of an IL-8 ELISA we obtained evidence that approximately 30–40% of the bulk of immunoreactive IL-8 (which by Western blotting represents bands near 8 kDa) elutes at lower ionic strength from the cation-exchange column than the major peak, indicating the presence of more anionic IL-8 species in lesional scales. The existence of less cationic chemokines in lesional scale material is not unexpected. When we analyzed IL-8 preparations obtained from culture supernatants of stimulated monocytes, apart from the cationic IL-8 forms, several species with isoelectric points less than pH 7.5 are visible.[14] These less cationic species may indicate a posttranslational modification of IL-8, most likely hydrolysis of glutamine and asparagine residues to the corresponding acids, which add additional negative charges, as has been postulated for IL-1.[15]

Principally, it is possible to characterize the other IL-8 related cytokines by HPLC. In Table III biochemical and biological properties of different IL-8 forms obtained from lesional psoriatic scales are summarized. Purifica-

[14] J.-M. Schröder, U. Mrowietz, and E. Christophers, *Biochem. Biophys. Res. Commun.* **152,** 277 (1988).
[15] T. P. Hopp, S. K. Dower, and C. J. March, *Immunol. Res.* **5,** 271 (1986).

tion of the less abundant IL-8 forms present in psoriatic scales require at least 50–100 g of lesional scale material to obtain pure products. The final yield of these minor IL-8 forms is low. Often we lost the whole activity during purification. To optimize the yield of different psoriasis-derived IL-8 forms and to characterize other neutrophil chemotactic chemokines present in lesional psoriatic skin we developed a completely different strategy for isolation of chemokines from biological samples.

Purification of Interleukin-8 and Growth Related Oncogene-α: Strategy II

Lesional psoriatic scale material (1–50 g) is extracted with ethanolic citrate buffer, as already described in detail, diafiltrated against 10 m*M* Tris/10 m*M* citrate buffer, pH 8.0, using an Amicon YM2 filter (cutoff 2 kDa), and clarified after freezing/thawing by centrifugation. This extract is passed through an anti-IL-8 affinity column, which is prepared as follows. Anti-IL-8 monoclonal antibodies are coupled to a small preactivated Sepharose column. A commercially available cartridge filled with *N*-hydroxysuccinimide (NHS)-coupled Sepharose (10 × 5 mm, Pharmacia) is activated according to the instructions of the manufacturer and then incubated with 1 ml of a monoclonal anti-IL-8 antibody solution (1 mg/ml PBS, pH 7.4, without BSA) for 30 min. The column is washed with PBS and then treated with 0.1 *M* ethanolamine, pH 9, to block remaining NHS residues. After equilibrating the column with 10 m*M* Tris/citrate buffer, pH 8, the scale extract (10 ml) is passed two times through the column with a flow rate of 1 ml/min. Thereafter the column is washed with 2 ml Tris/citrate buffer, pH 8. Low affinity bound material is washed from the column by the use of 3 ml PBS, pH 7.4, containing 2 *M* NaCl. This material is collected for further chromatography as described later. The high affinity bound material is stripped from the column by the use of 3 ml of glycine buffer, pH 2.0.

Both low and high affinity bound materials are separately concentrated and first separated by preparative RP-8 HPLC as described. Biologically active fractions are diafiltered against equilibration buffer [50 m*M* ammonium formate, pH 4.0, containing 20% (v/v) acetonitrile] and then separated by cation-exchange HPLC. Samples (100–400 μl) are applied to a microbore Mono S PC 1.65 HPLC column (Pharmacia, Freiburg, Germany), attached to a microbore Smart HPLC apparatus (Pharmacia, Freiburg, Germany), previously equilibrated with equilibration buffer. Proteins are eluted with a gradient of increasing concentrations of NaCl (maximum, 1 *M*) in equilibration buffer using a flow rate of 100 μl/min. During the HPLC runs, absorbance at 215 and 280 nm as well as conductivity are monitored. Figure 8 shows a typical HPLC run with investigation of PMN chemotactic activity (Fig. 8A) and IL-8 immunoreactivity in HPLC fractions (Fig. 8B).

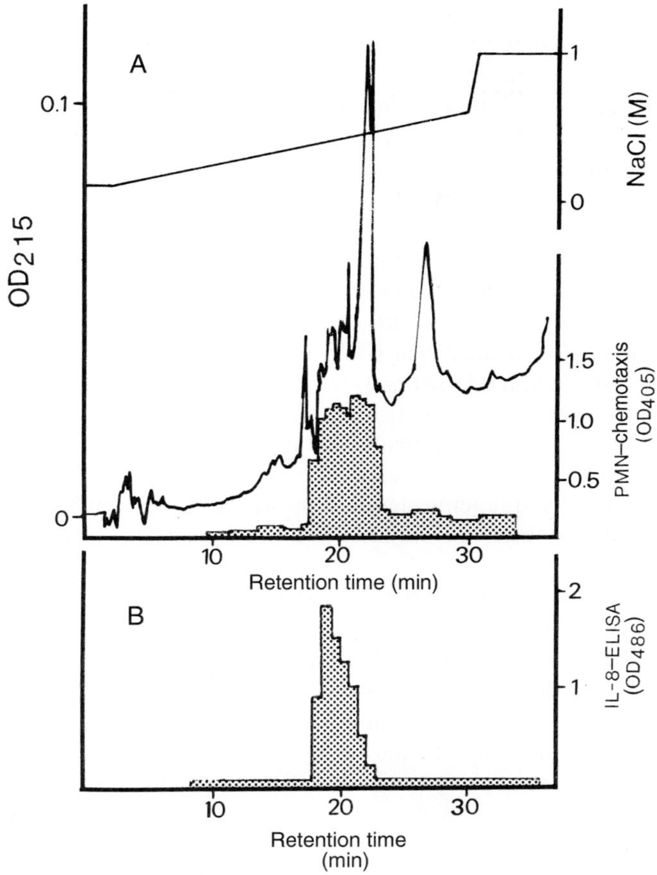

FIG. 8. Micro-cation-exchange HPLC of affinity-purified psoriasis-derived IL-8. Anti-IL-8 affinity column-bound psoriasis-derived proteins were separated on a microbore Mono S HPLC column and eluted with increasing concentrations of NaCl. Fractions containing neutrophil chemotactic activity are shaded in (A). (B) Fractions were also tested for IL-8 immunoreactivity with a solid-phase ELISA.

Interleukin-8 Solid-Phase Enzyme-Linked Immunosorbent Assay. For detection of IL-8-like immunoreactivity a 96-well ELISA microtiter plate (Nunc, Roskilde, Denmark) is filled with HPLC fractions (1–30 μl), which are lyophilized when acetonitrile is present. Subsequently 100 μl of coating buffer (0.1 *M* Na_2CO_3, pH 10.5) is added to each well, and the plate is incubated overnight at 4°. The plate is washed and then incubated with monoclonal antibodies against IL-8 (or other chemokines, when necessary) at an appropriate concentration (usually ~10 μg/ml) followed by addition of a peroxidase-labeled anti-mouse immunoglobulin (IgG) antibody (Dianova, Hamburg, Germany) at the appropriate dilution and *o*-phenylenediamine as substrate for detection of peroxidase activity. Because this procedure is only used to detect peaks that show the highest immunoreactivity, a complete calibration curve for the estimation of the amount of IL-8 is not necessary.

The use of a solid-phase ELISA is limited to low protein amounts in each well. When the sample contains more protein than can be bound to the ELISA plate surface (usually when the extinction at 215 nm is higher than 0.3–0.5), false-negative results can occur. In such a case, either samples should be diluted or a sandwich ELISA should be used.

Further Purification of Psoriasis-Derived Interleukin-8. The presence of many protein peaks, even when high affinity IL-8 affinity column-bound material is chromatographed, means that it is necessary to further chromatograph the IL-8-like material. Final purification by RP-18 HPLC and subsequent N-terminal sequencing again support the initial finding that [Glu-IL-8]^{69} and its less cationic variants are the major IL-8 form present in psoriatic skin.

The HPLC run of material bound with low affinity to the affinity column consistently reveals the presence of both PMN chemotactic as well as immunoreactive material. A detailed analysis of the proteins is lacking so far. Some of the proteins, which by SDS–PAGE (Tricine methods) showed bands near 4 kDa (Fig. 5), revealed N-terminal sequences corresponding to that of both major IL-8 forms in psoriatic skin [Glu-IL-8]^{69} and [Ser-IL-8].^{72} Because the mobility on Tricine SDS–PAGE is faster than that of authentic 69- and 72-residue forms, it is likely that C-terminal truncation products are present.

Careful SDS–PAGE analyses of IL-8 preparations obtained from lesional scales as well as different cell cultures show similar results (Fig. 5). Because in Western blot experiments using different anti IL-8 monoclonal antibodies no bands are visible (data not shown), it is likely that apart from C-terminal truncation cysteine bridges might be disrupted in these IL-8 variants. This hypothesis is supported by data obtained from electrospray ionization–mass spectrometry (ESI–MS) that are consistent with the existence of N-acetylated, C-terminally truncated forms of IL-8 with opened disulfide bridges.[16]

Thus the compounds with low affinity for the anti-IL-8 column, obtained from lesional psoriatic scale material, represent a tool for investigating the catabolism of chemokines *in vivo.* The use of anti-IL-8 affinity chromatogra-

[16] J.-M. Schröder and A. I. Mallet, unpublished observation (1996).

phy prior to a detailed chromatographic analysis of neutrophil chemotactic chemokines in lesional psoriatic scales for depletion of IL-8 also allows the analysis of neutrophil attractants different from IL-8, which otherwise are difficult to detect because of the similar biological and biochemical properties.

Purification of Growth-Related Oncogene-α

The depletion of psoriatic scale extracts from IL-8 allows the isolation of natural GRO-α, which represents the second major neutrophil attractant in psoriatic scales.[13] The following procedure is used to purify GRO forms present in lesional scales. IL-8-depleted scale extract (5–10 ml) (the effluent of the anti-IL-8 affinity column chromatography step) is applied to a heparin-Sepharose cartridge (HiTrap, 10 × 5 mm, Pharmacia, Freiburg, Germany) and passed through the column with a flow rate of 0.5 ml/min. This procedure is repeated twice with the effluent and the same column. Thereafter, the column is washed with 2 ml of a 0.1 *M* Tris/citrate buffer, pH 8. Proteins bound to the column are then eluted using a gradient of increasing concentrations of NaCl (0–2 *M* within 20 min) in 0.1 *M* Tris/citrate buffer, pH 8.0, monitoring the absorbance at 280 nm. Fractions (1 ml) are collected and tested for PMN chemotactic activity.

Heparin-bound proteins usually elute in three major and several minor peaks. Chemotactic activity for neutrophils can be found in nearly all fractions containing protein. Attempts to purify PMN attractants from each peak revealed a multiplicity of biological activities, which finally give amounts too small for further analyses. Therefore, we always pool the heparin-binding proteins and use them in the next step for preparative RP-8 HPLC.

Proteins stripped from the heparin column are acidified to pH 2–3 with formic acid, concentrated and diafiltered against 0.1% TFA using Amicon YM2 filters (cutoff 2 kDa), and then applied (1–5 ml) to a preparative RP-8 HPLC column (C_8 Nucleosil with end capping, 250 × 12.6 mm, 7 μm particle size, Macherey and Nagel) previously equilibrated with water/acetonitrile/TFA (90:10:0.01, v/v/v). UV absorbance is monitored at 215 nm, and fractions are taken manually according to peaks or shoulders of UV absorbance at a flow rate of 3 ml/min. Although the automatic collection of fractions is more convenient, we often had to rechromatograph fractions because of unwanted shoulders or peaks within the biologically active fractions.

Figure 9 shows a representative chromatogram of an IL-8-depleted psoriatic scale extract. The major peak of PMN chemotactic activity originates from GRO-α (Fig. 9A). Solid-phase ELISAs for other chemokines

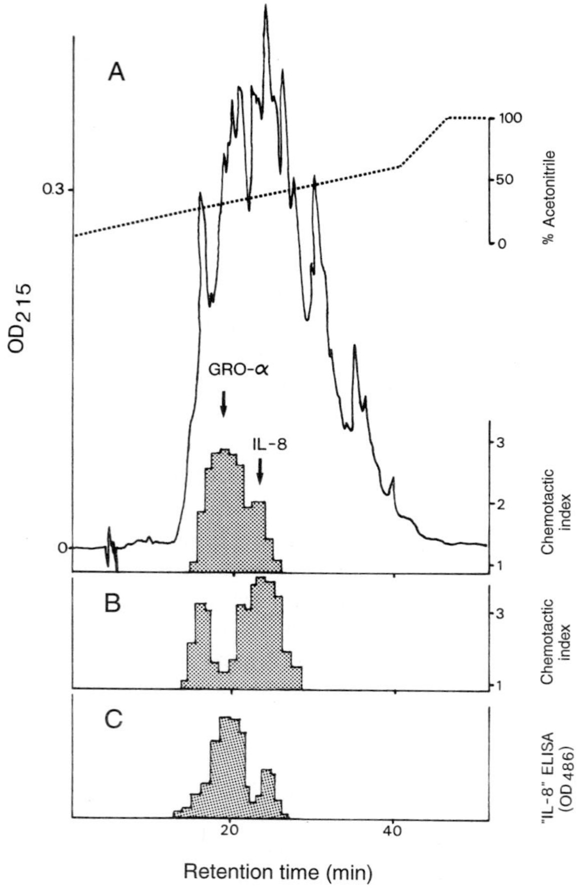

FIG. 9. Preparative reversed-phase HPLC of IL-8-depleted psoriatic scale extracts. Psoriatic scale extracts, depleted from IL-8, were chromatographed on a heparin-Sepharose column, and bound proteins were separated by preparative reversed-phase (RP-8) HPLC. Whereas (A) shows PMN chemotactic activity when 3-μl aliquots of HPLC fractions were used for testing, (B) shows neutrophil chemotactic activity when 30-μl aliquots were used. (C) Presence of "IL-8 immunoreactivity" using the cross-reacting monoclonal antibody 52E8 (this antibody cross-reacts with GRO-α).

such as epithelial neutrophil activating protein-78 (ENA-78), NAP-2, and related compounds, as well as RANTES, in our hands do not show significant peaks in HPLC fractions, whereas a monoclonal antibody that recognizes both IL-8 and GRO-α reveals a major peak and a minor one (Fig. 9C). The major peak consistently contains the highest chemotactic activity, which can be detected even at 1:10,000 dilution of the fraction. For detection of GRO-α by the chemotaxis assay it is important to use appropriately diluted fractions. When high amounts of GRO-α are present in the samples and 30-μl aliquots of the fractions are used, two peaks of biological activity are present (Fig. 9B), which correspond to a single peak of activity when 3-μl aliquots of HPLC fractions are used for testing instead (Fig. 9A).

Micro-Cation-Exchange High-Performance Liquid Chromatography of Growth-Related Oncogene-α. To purify GRO-α from a lesional psoriatic scale extract, fractions from RP-8 HPLC eluting early (see Fig. 9) are further separated by microbore Mono S HPLC. To bioactive fractions from RP-8 HPLC, aqueous ammonia is added until the pH is near 4.0. Then the whole sample is applied to the column, which previously was equilibrated with 50 m*M* ammonium formate, pH 4.0, containing 20% (v/v) acetonitrile. Proteins are eluted with a gradient of increasing concentrations of NaCl (maximum 1 *M*) in equilibration buffer using a flow rate of 100 μl/min. UV absorbance is monitored as described. Manually taken fractions are tested at optimal dilution for PMN chemotactic activity.

Figure 10 shows a representative run. Usually two to three peaks of PMN chemotactic activity are detectable. Because one peak will not be recognized by PMN desensitized with recombinant human C5a, it seems to represent $C5a_{des\text{-}Arg}$. Purification of this material is possible, but this is not within the scope of this article. Both of the other peaks are purified separately to apparent homogeneity by microbore RP-18 HPLC as demon-

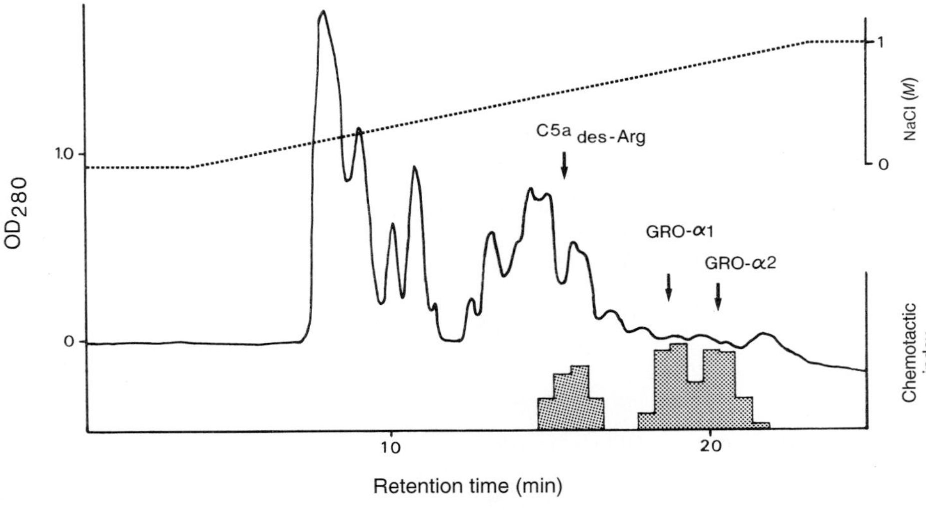

FIG. 10. Micro-cation-exchange HPLC of psoriasis-derived GRO-α. Partially reversed-phase HPLC purified psoriasis-derived GRO-α was further purified by microbore Mon S HPLC. Chemotactic activity in fractions is shown in the shaded area.

strated before for purification of IL-8. Representative runs are shown in Fig. 11. To fractions containing biological activity from microbore Mono S HPLC, TFA is added to give a final concentration of 1%. This sample is applied to a microbore RP-18 HPLC column (Sephasil C_{18}, 2.1 × 100 mm, 5 μm particle size, Pharmacia, Freiburg, Germany), previously equilibrated with 10% (v/v) acetonitrile in 0.1% (v/v) aqueous TFA. Proteins are eluted with a gradient of increasing concentrations of aceonitrile.

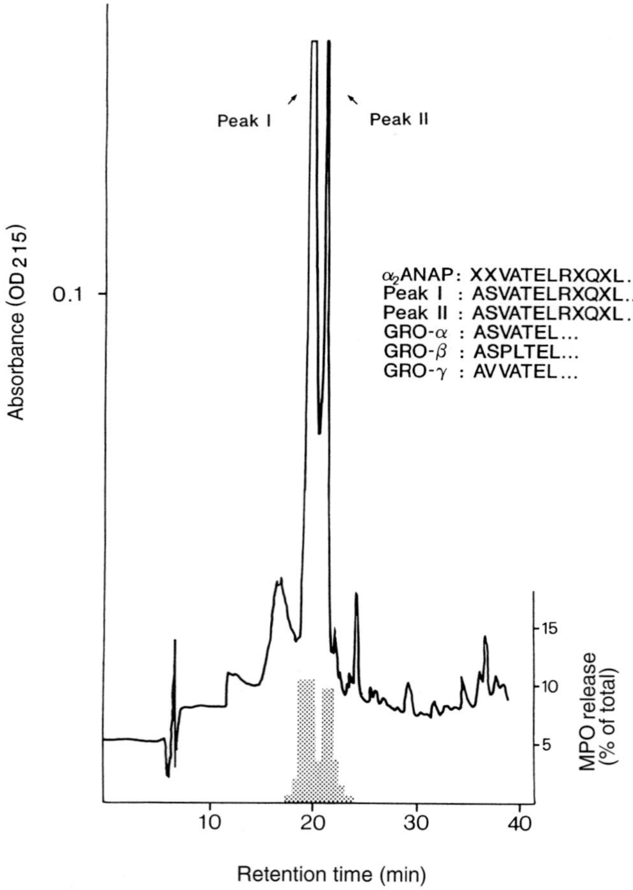

FIG. 11. Reversed-phase HPLC of psoriasis-derived GRO-α. Fractions from Mono S HPLC containing GRO-α2 were purified by the use of an analytical RP-18 HPLC column with acetonitrile as eluent. Fractions were tested for myeloperoxidase-releasing activity (shaded area). Inset shows N-terminal sequence data of both peaks as well as GRO-α1 (termed α_2ANAP) and the deduced sequences of GRO-α, GRO-β, and GRO-γ.

Three biochemically distinct forms of GRO-α can be isolated. The first one, termed GRO-α1 shows an N-terminal sequence that does not allow identification of the first two amino acids, whereas other residues are identical with those deduced from the GRO-α cDNA sequence (Fig. 11, inset). The second and the third GRO-α species elute together from the cation-exchange column but can be separated from one another by the use of RP-18 HPLC (Fig. 11). Again, both show an identical N-terminal amino acid sequence. Electrospray ionization–mass spectrometry analyses of psoriasis-derived GRO-α1, -α2, and -α3 unexpectedly show for one GRO-α species exactly the mass deduced from the sequence together with higher masses, which may come from O-glycosylated forms.[16] Two other GRO-α species show masses that differ from theoretical values for C-terminal truncation products with or without glycosylation. The recovery of purified GRO-α species usually is found to be in the range of 0.5–7 μg per 50 g of pooled lesional scales for each species.

Purification of additional and minor neutrophil chemotactic cytokines from psoriatic scales revealed no other chemokines, including GRO-β, GRO-γ, ENA-78, or GCP-2. Thus, it can be concluded that lesional psoriatic scales contain only different posttranslationally modified variants of IL-8 as well as GRO-α as neutrophil chemotactic cytokines. Initial attempts to further analyze different GRO-α forms by peptide mapping revealed differences in fingerprinting. Thus it is possible that different GRO-α forms exist having slightly modified amino acid sequences, which could be the result of a GRO-α gene polymorphism. Because we always used pooled scales from up to 30 different patients, this would explain the unexpected heterogeneity of GRO-α, which is clearly not the result of posttranslational modifications or the presence of GRO-β or GRO-γ.

Purification of C-C Chemokines from Lesional Scale Material

A number of C-C chemokines are attractive candidates to explain the disease-specific histological pattern of leukocyte tissue infiltration. Whereas C-X-C chemokines are mainly neutrophil attractants and do not show any eosinophil chemotactic activity, C-C chemokines express activity in monocytes, lymphocytes, and eosinophils (Eo's).

A number of skin diseases including parasitic infections have as a characteristic the presence of Eo's but not neutrophils. Thus it is interesting to speculate whether Eo-chemotactic C-C chemokines are responsible for this infiltration. To investigate whether C-C chemokines are released as bioactive cytokines we have analyzed lesional scale material obtained from patients with drug reactions and atopic dermatitis. In both dermatoses a dermal but not epidermal infiltrate by eosinophils (proven by the presence of immunoreactive major basic protein as a marker protein of eosinophils rather than by direct staining for intact eosinophils) is a characteristic feature.

The major problem is the recruitment of sufficient biological material for investigation. Another disadvantage is the localization of Eo's within the dermis and not in the epidermis. Scales originate from epidermis, and thus only cytokines produced within the epidermis or spilled over from the dermis will principally appear in scales. Because it is likely that the spillover effect is limited, the absence of any activity in extracts obtained from scales of such dermatoses does not mean that the cytokine to be investigated is not produced within the dermis.

Despite these limitations we are able to obtain in some materials sufficient evidence for the production of bioactive C-C chemokines, whereas attempts at discovery and purification of MCP-1 and RANTES from psoriatic scales completely failed. This negative finding may reflect the absence of infiltrating monocytes and eosinophils into the active psoriasis lesion rather than methodical problems of its identification. In scale extracts obtained from patients with drug reactions we were successful in identifying bioactive RANTES.[17]

Purification of RANTES from Scale Extracts

Pooled lesional scales (0.5–3 g) obtained from patients with drug reactions or atopic dermatitis are suspended in 40 ml of 0.1 *M* citric acid containing 50% (v/v) ethanol. After homogenization the material is treated as described (see methods for chemokine extraction). Clarified protein extracts are further separated according to strategy II (see above). Extracts previously diafiltered against 10 m*M* Tris/10 m*M* citrate buffer, pH 8.0, using an Amicon YM2 filter (cutoff 2 kDa) are concentrated to 5 ml and then passed three times at a flow rate of 1 ml/min through a heparin-Sepharose cartridge (10 × 5 mm, Pharmacia, Freiburg, Germany), previously equilibrated with the diafiltration buffer. Thereafter the column is washed with 2 ml of 0.1 *M* Tris/citrate buffer, pH 8.0. Proteins bound to the column are eluted with 2 ml of 2 *M* NaCl in 0.1 *M* Tris/citrate buffer. Initially we used a gradient of NaCl to strip proteins from the column. The low amounts of material eluting from the column led to the experience that bulk elution appears to be more efficient, giving a lower volume for further separation.

The entire amount of heparin-bound material is then diafiltered against 0.1% (v/v) TFA and applied to a preparative RP-8 HPLC column (C_8

[17] J.-M. Schröder, N. Noso, M. Sticherling, and E. Christophers, *J. Leukocyte Biol.* **59,** 1 (1996).

Nucleosil with end capping, 250 × 12.6 mm, particle size 7 μm, Macherey and Nagel) previously equilibrated with water/acetonitrile/TFA (90 : 10 : 0.01, v/v/v). UV absorbance is monitored at 215 nm, and fractions are taken manually according to peaks or shoulders of UV absorbance. Ten microliters of each fraction is pipetted into microtiter plates and mixed with 10 μl of (c)PBS. After freezing at −70° samples are lyophilized, and lyophilizates present in each well are dissolved in 300 μl (c)PBS. Boyden chambers (in duplicate) are filled with samples and tested for Eo chemotactic activity using the system described under Measurement of Eosinophil Chemotaxis.

Figure 12 shows a typical example of drug reaction-derived scales. A single peak of Eo chemotactic activity can be detected. When 10 μl aliquots of the fraction are tested for RANTES immunoreactivity by the use of a solid-phase ELISA (for detection we use a mixture of two in-house monoclonal antibodies at a final concentration of 10 μg/ml and 100 μl per well; for details, see the section on IL-8 solid-phase ELISA), immunoreactive RANTES appears exactly in the same fractions containing Eo chemotactic activity (Fig. 12).

For further purification and biochemical characterization it is essential to use microbore HPLC methods. All of our attempts to further purify this activity by ion-exchange HPLC or size-exclusion HPLC using normal analytical HPLC columns led to a complete loss of activity. Therefore, as the next step for purification micro-cation-exchange HPLC was seen to be optimal. For isolation of the Eo attractant, biologically active fractions from RP-8 HPLC are directly applied to a Smart Mono S HPLC column, previously equilibrated with 50 m*M* ammonium formate, pH 4.0, containing 20% (v/v) acetonitrile, and proteins are eluted with a gradient of increasing concentrations of NaCl (maximum concentration 1 *M*) in equilibration buffer using a flow rate of 100 μl/min. For good recovery it is essential to add 20% acetonitrile to the eluent. When proteins are eluted with buffers in the absence of acetonitrile losses, possibly due to nonspecific absorption, are extremely high, even when the microbore HPLC column is used.

In our hands RANTES never could be obtained on ion-exchange HPLC as a single peak absorbing at 280 nm. When the peak containing Eo-chemotactic activity from Mono S-HPLC is purified to apparent homogeneity by microbore RP-18 HPLC (see strategy II) sometimes (but rarely) biological activity elutes in a single peak absorbing at 215 nm. With a solid-phase RANTES ELISA, immunoreactive RANTES again appears in fractions containing Eo-chemotactic activity. Moreover, a single band near 8 kDa can be detected on Western blot analysis using anti-RANTES antibody. We did not find any fractions that contained immunoreactive RANTES but no Eo-chemotactic activity or vice versa. The methods we

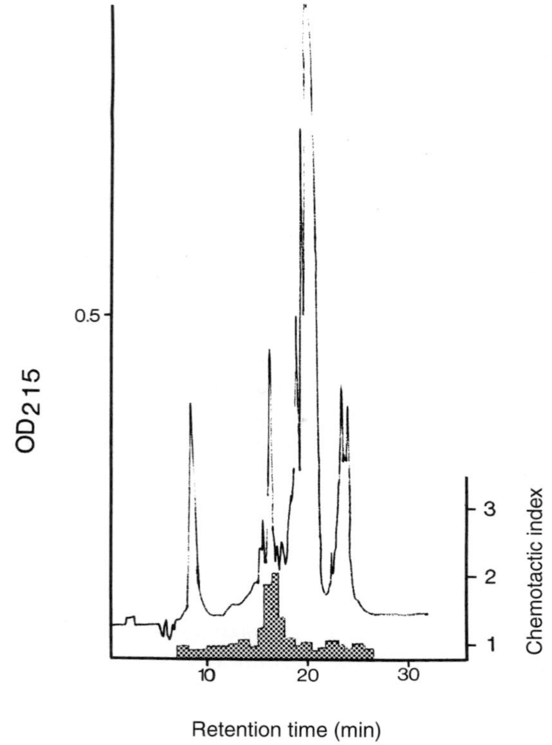

FIG. 12. Reversed-phase HPLC analysis of lesional scales for eosinophil chemotactic chemokines. Scale extracts obtained from patients with drug reactions were separated by heparin-Sepharose chromatography, and heparin-bound proteins were chromatographed on a preparative RP-8 HPLC column. Eosinophil chemotactic activity in HPLC fractions is shown in the shaded area.

use for purification of Eo-chemotactic cytokines allow a clear discrimination between known Eo-chemotactic chemokines. When applied as a mixture together to the microbore RP-18 column these chemokines elute in the following order: natural eotaxin < glycosylated natural MCP-3 < natural RANTES < recombinant MCP-3. Thus, the presence of a single peak of Eo-chemotactic activity in scales points toward the absence of Eo-chemotactic chemokines other than RANTES.

Outlook

The intent of this article is to show ways of detection, purification, and further biochemical characterization of chemokines obtained directly from diseased skin. We have optimized the methods of chemokine purification by the use of modern microbore HPLC techniques together with the elimination of procedures that usually led to dramatic losses and thus affect the recovery of these cytokines. The methods described should be easily adaptable to biological samples obtained from sources other than skin. We have also extracted chemokines from nasal polyps as well as bronchoalveolar lavage fluid.

Miniaturization of the techniques by the use of microbore HPLC technology together with highly sensitive analytical techniques [silver staining of proteins on SDS–PAGE, enhanced chemiluminescence Western blot, and matrix-associated laser desorption ionization–mass spectrometry (MALDI)] should provide information about biological and biochemical properties of chemokines present in minute amounts of biological samples (i.e., biopsy material) obtained from patients. It can be expected that in the near future these techniques will supplement the information so far obtained only from immunohistochemical investigations as well as *in situ* hybridization studies.

Acknowledgments

This work was supported by the Deutsche Forschungsgemeinschaft. I thank Dr. Michael Sticherling for providing several chemokine monoclonal antibodies, Christine Gerbrecht-Gliessmann, Jutta Quitzau, Marlies Brandt, and Birgit Radtke for technical assistance, and Gabriele Tams for editorial help.

Section III

Signal Transduction of Chemokines

[18] Calcium Mobilization Assays

By SHAUN R. MCCOLL and PAUL H. NACCACHE

Introduction

Activation of leukocytes by chemotactic factors can modulate the intracellular levels of several ions, including sodium, potassium, hydrogen, and calcium. The latter, in particular, is used as an index of leukocyte activation in response to stimulation by chemotactic factors. In general, activation of leukocytes by chemotactic factors will lead to a transient increase in the intracellular level of free calcium that is dependent to varying degrees on both an influx of calcium from outside the cell and release of calcium from intracellular stores. As discussed later in this chapter, the intracellular release and influx may be linked.

General Model of Calcium Mobilization

The fundamental feature of calcium as a signaling element is its maintenance in an extreme state of disequilibrium. Indeed, mammalian cells, whose total concentration of calcium is about 1–2 m*M*, maintain their intracellular concentration of free cytoplasmic calcium in the neighborhood of 0.1 to 0.2 μM in the face of a very large unfavorable electrochemical gradient. This allows for the required large (in relative terms), rapid, and transient changes in the concentration of free cytoplasmic calcium that characterize the activation of most cells. Intracellular calcium is distributed among several pools that include free cytoplasmic calcium, calcium bound to various lipids or proteins, and calcium that is accumulated in various storage compartments. For the purposes of this discussion, one can assume that the free cytoplasmic calcium is regulated by exchanges with the intracellular storage pools and with the extracellular milieu, which typically contains about 1–2 m*M* of calcium.

The mechanism responsible for the release of calcium from intracellular storage pools in nonmuscle cells has been linked to the phospholipase C-mediated generation of inositol trisphosphate [Ins(1,4,5)P_3]. Cell stimulation is also accompanied by increases in the plasma membrane permeability to calcium and the diffusion into the cytoplasm (and into the calcium storage organelles) of extracellularly derived calcium through calcium channels that differ from other (voltage–dependent) calcium channels. These specialized channels are known as calcium release-activated calcium channels or

0076-6879/97 $25.00

CRAC.[1] The nature of the linkage between the emptying of intracellular stores and the activation of CRAC is still conjectural. Cells sense that the calcium storage pools have been depleted and respond by allowing calcium to flow in from the external milieu (for a detailed review of the current state of knowledge concerning this phenomenon, termed capacitative calcium influx, see the comprehensive review by Berridge).[2] Thus, there are at least three parameters of calcium mobilization that can be monitored as indices of cell activation by chemokines (as well as by other agonists): (1) changes in the concentration of free cytoplasmic calcium, (2) discharge of calcium from the intracellular storage pools, and (3) influx of calcium.

Methods for Measuring Intracellular Calcium

Fluorescent Dyes

Various methods for monitoring the levels of intracellular calcium have been developed. Some are based on the distribution of ^{45}Ca, others on the use of calcium electrodes, and still others on the use of aequorin or aequorin-based indicators. Although each of these methods is valid, they all tend to be either more difficult to interpret (radioactivity measurements), more technically demanding (calcium electrodes), or more limited in their applicability (difficulties in incorporating aequorin into cells) than the method based on the use of fluorescent indicators. These assays employ the fluorescent calcium-binding dyes originally developed by R. Y. Tsien.[3,4] The second-, and third-, generation dyes Fura-2, Indo-1, and Fluo-3, have circumvented to a great extent the problems with Quin-2 (a first-generation indicator), which acted as a buffer of intracellular calcium.[3,5,6] Other calcium indicators are also available (Molecular Probes, Eugene, OR). Our laboratories have considerable experience with each of the first three indicators, but we generally find Fura-2 to be satisfactory for most uses. The choice of dye is based on an empirical determination of which works best in a particular cellular system, and although we found that Fura-2 is satisfactory for most uses, some cells dictate the use of other dyes.

The basic principle of the use of the dyes is the chemical attachment of an acetoxymethyl ester group that renders them hydrophobic. This en-

[1] R. Penner, C. Fasolato, and M. Hoth, *Curr. Opin. Neurobiol.* **3,** 368 (1993).
[2] M. J. Berridge, *Biochem. J.* **312,** 1 (1995).
[3] G. Grynkievicz, M. Poenie, and R. Y. Tsien, *J. Biol. Chem.* **260,** 3440 (1985).
[4] R. Y. Tsien, *Nature* (*London*) **290,** 527 (1981).
[5] R. Y. Tsien, T. J. Rink, and M. Poenie, *Cell Calcium* **6,** 145 (1985).
[6] R. Y. Tsien, *Methods Cell Biol.* **30,** 127 (1989).

ables the modified dyes to cross the plasma membrane into the cytoplasm where nonspecific cellular esterases cleave the acetoxymethyl ester moiety from the molecule, trapping the free acid form in the cytoplasm. Only the free acid form and not the esterified counterparts act as calcium indicators. This principle enables significant accumulation of the indicators inside cells. The amount that accumulates depends on several factors and varies from cell type to cell type. Such factors include the concentration of esterified dye, the length of the incubation period as well as the level of intracellular esterases associated with a particular cell type, and the relative permeability of the plasma membrane to the free acid form, as this will affect retention of the free acid.[6] If necessary, the solubility and therefore the uptake of the dyes into cells can be increased using dispersing agents such as Pluronic F-127, serum, or bovine serum albumin. Additionally, retention of dye in cells can be enhanced using the anion-exchange inhibitor probenicid.[7] These procedures are rarely needed with hematopoietic cells.

The chemical structures of Flura-2, Indo-1,[3] and Fluo-3[8] have been adequately reported previously. Their most important feature is that their spectral characteristics (fluorescence intensity or excitation or emission spectra) change on binding calcium. Each of the three probes has different spectral characteristics that may confer specific advantages under certain experimental conditions or with various types of recording apparatus.

Loading

Our major experience with these dyes has been with the use of Fura-2 to assay intracellular calcium mobilization in human neutrophils, although we have also used both Fura-2 and Indo-1 in monocytes and the human embryonic kidney 293 cell line (HEK293). Some of the details that follow may differ from cell type to cell type.

In general, loading of the indicator into neutrophils is accomplished by incubating a cell suspension of $1–2 \times 10^7$ cells/ml in Hanks' balanced salt solution (HBSS) containing 1.6 μM calcium at 37° for 30 min to 1 hr with 1–2 μM Fura-2. As far as neutrophils are concerned, it is advantageous to have the cells resuspended in magnesium-free buffers and swirling gently during this loading process. We have found that this reduces the tendency for neutrophils to aggregate and form clumps. To achieve this, we use a gyratory water bath. The cell suspension is then washed to remove extracellular indicator. Two washes are usually sufficient to achieve this. The cells should then be allowed to equilibrate at 37° for 5–10 min to stabilize the

[7] J. E. Merritt, S. A. McCathy, M. P. A. Davies, and K. E. Moores, *Biochem. J.* **269,** 513 (1990).
[8] A. Minta, J. Kao, and R. Y. Tsien, *J. Biol. Chem.* **264,** 8171 (1989).

baseline fluorescence. The loading as well as the preincubation can be carried out at room temperature. This results, at least in the case of neutrophils, in a cell population that is more stable and whose ability to respond to various agonists persists longer in time than that of cells maintained at 37°. However, we have also observed changes in the behavior of the cells under these conditions, in particular with respect to their ability to be primed by various growth factors such as granulocyte–macrophage colony-stimulating factor (GM-CSF) and tumor necrosis factor (TNF) (S. R. McColl and P. H. Naccache, unpublished observations). In our laboratories, we assay calcium mobilization using SLM or Aminco-Bowman luminescence spectrophotometers (Spectronic Instruments Inc., Rochester, NY). These are highly sensitive fluorescence spectrophotometers that possess excellent analytical software. Other fluorescence spectrophotometers have also been used.

Monitoring

Once the indicator has been loaded, the cells are divided into aliquots in cuvettes and placed into heated cuvette compartments in the spectrofluorimeters. Although quartz cuvettes will provide a sensitivity advantage (especially at the near-UV excitation wavelength of Fura-2), disposable plastic cuvettes (Sarstedt, Newton, NC, acrylic cuvettes) are adequate for most uses owing to the high fluorescence quantum yield of the calcium indicators. The cuvettes also contain small magnetic fleas that serve to keep the cells in suspension. Fluorescence of Fura-2 is monitored at 340 and 510 nm, excitation and emission, respectively. Fluorescence of Indo-1 and Fluo-3 is monitored at 350 and 488 nm excitation and 405 and 520 nm emission, respectively. The cells may then be stimulated by the addition of a chemotactic factor, which is added directly to the cuvette via an injection port. On stimulation of G-protein-coupled receptors, an increase in fluorescence should be observed almost immediately. It should reach a maximum level within 5–15 sec of injection and then gradually (within the next 1–5 min) return to baseline levels.

Quantitation

Calibration of the fluorescence can be achieved most simply by measuring the fluorescence of the cell suspensions following lysis of the cells with 0.1% Triton X-100 (v/v) in the presence of saturating (1 m*M*) and extremely low (10 n*M*) concentrations of calcium. The latter is achieved by adding excess amounts of calcium chelators (EGTA) and by increasing the pH of the medium to greater than pH 8 by the addition of NaOH. This allows one to increase the affinity of EGTA for calcium and to decrease the free calcium in the medium to the nanomolar range. The free calcium is then calculated according to the following formula:

$$\text{Calcium} = K_d[(F - F_{min})/(F_{max} - F)]$$

Alternatively, F_{max} and F_{min} can also be determined without lysis of the cells. Probe saturation (F_{max}) is achieved following the addition of a calcium ionophore such as ionomycin (A23187 cannot be used as its fluorescence profile interferes with the assay). F_{min} can then be monitored by quenching the intracellular probe by the addition of Mn^{2+} to the ionophore-treated cells. This procedure must be carried out immediately after the measurement of changes in intracellular calcium levels in order to minimize the amount of leakage from the cells.

Calibration of the fluorescence can also be accomplished by comparing the fluorescence of the samples measured as the ratio of 340/380 or 400/490 nm in the case of Fura-2 and Indo-1, respectively, to a standard curve obtained in the presence of known amounts of free calcium.[3] This procedure has the advantage of making the determination of free calcium independent of the levels of dye loading and/or leaking.

Assessing Calcium Influx

Monitoring the fluorescence levels of a cell loaded with a calcium indicator will give an index of the changes in the concentration of free cytoplasmic calcium regardless of the source of calcium. Several approaches exist that allow one to estimate whether some or all of the additional calcium is derived from extracellular sources. These can be divided into two experimental groups: inhibition of calcium influx and direct monitoring of calcium influx.

Calcium can be prevented from entering into cells either by removing it from the extracellular milieu or by blocking the calcium entry mechanisms (calcium channels). Removal of extracellular calcium can be accomplished using chelating agents such as EGTA or by resuspending the cells in "calcium-free" medium. There are two specific precautions to keep in mind during these manipulations. The first is that the intracellular free calcium concentration is only on the order of 0.1 μM and that contaminating calcium in nominally calcium-free buffers can be as high as 1–5 μM. Second, this approach assumes that the only effect of calcium chelation (or removal) is extracellular. This is only approximately correct for a short period as, given sufficient time, the extracellular chelating agents will eventually act as a sink for intracellular calcium. Experimentally, this requires that chelating agents only be added a few seconds before stimulation of the cells. In addition, chelating agents are likely to also affect plasma membrane-bound calcium and in so doing change in some uncontrolled manner its fundamental biophysical and biochemical properties.

Calcium influx may also be abrogated by inhibiting the calcium channels through which it diffuses into the cells (CRAC). Although the nature of these channels in nonexcitable cells (also termed receptor-operated channels or ROC) remains to be adequately characterized, it has become clear that they are distinct from voltage-gated channels. As mentioned above, the elucidation of the linkage between the emptying of the intracellular calcium stores and the opening of the calcium channels is also an area of intense investigation and debate.[9,10] CRAC are inhibited by trivalent and divalent ions with a characteristic potency order: $La^{3+} > Zn^{2+} > Cd^{2+} > Be^{2+} = Co^{2+} = Mn^{2+} > Ni^{2+} > Sr^{2+} > Ba^{2+}$. CRAC are also inhibited by a series of organic compounds of which the most commonly used are SK&F96365 (about 10–20 *M*; see however Leung *et al.*[11] for a word of caution about high concentrations of SK&F96365) and econazole (10–20 *M*). The effects of the multivalent cations or the CRAC inhibitors on the increase in cytoplasmic free calcium induced by a particular agonist (e.g., a chemokine) can therefore be used to estimate the proportion of the total increase that derives from the extracellular medium.

The final and probably most direct means of monitoring the effects of a chemokine on the permeability of the plasma membrane to calcium using fluorescent dyes depends on the ability of Mn^{2+} to quench their fluorescence.[12] The permeability to Mn^{2+} under resting conditions is quite low. On the other hand, it is generally accepted that Mn^{2+} ions can diffuse through CRAC, once these are open, and then bind to the fluorescent calcium dyes with high affinity. In contrast to the effect of calcium, the binding of Mn^{2+} leads to a quenching of the fluorescence of the dyes, which can be monitored most conveniently at the (calcium-independent) isobestic point of the particular dye being used. Fura-2 has been most extensively used for this purpose at 360 and 510 nm (excitation and emission wavelengths, respectively), Indo-1 can also be used in a similar fashion (350 and 455 nm, excitation and emission wavelengths, respectively).

Modulation of Intracellular Calcium Levels

The ability to monitor and modulate the filling state of the intracellular calcium stores is limited to some extent by the partial knowledge of the

[9] M. J. Berridge, *Biochem. J.* **314,** 1055 (1996).

[10] M. B. Hallet, E. J. Pettit, and E. V. Davies, *Biochem. J.* **314,** 1054 (1995).

[11] Y. M. Leung, C. Y. Kwan, and T. T. Loh, *Biochem. Pharmacol.* **51,** 605 (1996).

[12] S. Sozzani, M. Molino, M. Locati, W. Luini, C. Cerletti, A. Vecchi, and A. Mantovani, *J. Immunol.* **150,** 1544 (1993).

nature, distribution, and function of the stores. However, the presence of calcium storage pools, the filling of which is controlled by ATP-dependent calcium pumps, is well established.[2] Inhibitors of the calcium pump that is responsible for the accumulation of calcium into these stores have been described. Their addition to cells induces the leakage of calcium from the storage pools into the cytoplasm. The best characterized of the calcium pump inhibitors is thapsigargin; others include cyclopiazonic acid and 2,5-di(tert-butyl)-1,4-benzohydroquinone. Their addition to cells induces the rapid emptying of calcium stores. However, it should be kept in mind that, as discussed above, the emptying of the calcium stores leads to the influx of calcium through the CRAC. An alternative method to modulate the responses of the calcium stores in permeabilized cell preparations involves the use of antagonists of the Ins(1,4,5)P_3 receptors (e.g., heparin) or of blocking antibodies directed against the receptor.

The filling state of intracellular calcium stores can also be monitored directly using fluorescent calcium indicators that have the propensity to accumulate in these organelles (e.g., Mag-Fura-2).[13]

The above methods to modulate the state of filling of the calcium stores are of most use for the determination of their functional role in responses to chemokines, as opposed to their role as indicators of responsiveness to chemokines. However, one may envisage examining the calcium responses of cells to a particular chemokine following the depletion of the calcium storage pools in order to characterize the cellular mechanisms involved in this response.

Calcium Mobilization as Index of Monitoring Chemokine Functions

At first glance, calcium mobilization studies would appear to be a good index for monitoring chemokine function. The calcium assay is rapid, reproducible, and robust, and as long as the specialized equipment is available it can be used in a routine manner. The increase in the level of intracellular calcium is causally linked to other neutrophil functions including lipid mediator synthesis, degranulation, and the respiratory burst.[14–16] On the other hand, it is true to say that if we consider the major function of chemokines to be as chemotactic factors, then the dissociation of cell movement (chemotaxis, cytoskeletal rearrangement) from alterations in the level

[13] A. M. Hofer and T. E. Machen, *Proc. Natl. Acad. Sci. U.S.A.* **90,** 2598 (1993).

[14] C. A. Rouzer and B. Samuelsson, *Proc. Natl. Acad. Sci. U.S.A.* **84,** 7393 (1987).

[15] C. A. Rouzer and S. Kargman, *J. Biol. Chem.* **263,** 10980 (1988).

[16] P. H. Naccache, S. Therrien, A. C. Caon, N. Liao, C. Gilbert, and S. R. McColl, *J. Immunol.* **142,** 2438 (1989).

of intracellular calcium[16,17] lends weight to a line of thinking that assessment of intracellular calcium, at least in isolation, may not be an ideal measure of leukocyte activation by chemokines.

One also needs to consider the question of potency of action. In this context, *in vitro* chemotaxis assays clearly hold the upper hand. With several chemokines, chemotactic effects are observed with as little as 10^{-12}–10^{-11} M, and maximal effects are often observed at 10^{-10}–10^{-9} M. This contrasts with assessment of calcium mobilization in which detectable increases in the level of intracellular free calcium are usually observed at about 10^{-10} M and are maximal at between 10^{-7} and 10^{-6} M. These observations imply that the measurement of the induction of cell movement by chemokines is significantly more sensitive than is that of calcium mobilization.

A further question that arises is whether calcium mobilization assays offer a useful means of assessing the different types of chemokine receptors present on the surface of leukocytes, in terms of both the number of different types to which a particular ligand will bind as well as which ligands may share the same receptor. This form of assessment takes advantage of the transient desensitization of chemotactic factor receptors.[18,19] A browse through the literature suggests the existence of at least two basic forms of desensitization: homologous and heterologous. The former occurs when a cell is stimulated by a ligand that renders the cell unresponsive to further exposure to the same ligand. On the other hand, heterologous desensitization occurs when exposure to one ligand renders the cell unresponsive to exposure to other ligands that apparently bind to different receptors. A well-documented example is that which occurs when neutrophils are stimulated with fMet-Leu-Phe. This renders the cells unresponsive to addition of fMet-Leu-Phe (homologous desensitization) and a variety of other chemotactic factors including interleukin-8 (IL-8), leukotriene B_4 (LTB_4), and platelet-activating factor (PAF) (heterologous desensitization).

As far as studies of desensitization go, the calcium mobilization assay is at least as effective as, and probably more useful than, any other assay. Certainly, the manipulations required to observe desensitization are more easily conducted using the calcium mobilization assay than any other assay of leukocyte activation. However, interpretation of the results of these experiments can be a major stumbling blocking. Put simply, how does one distinguish between heterologous and homologous desensitization when using this phenomenon to identify (a) the number of receptors for a specific

[17] B. Hendey and F. R. Maxfield, *Blood Cells* **19,** 143 (1993).

[18] J. D. Winkler, H. M. Sarau, J. J. Foley, S. Mong, and S. T. Crooke, *J. Pharmacol. Exp. Ther.* **246,** 204 (1988).

[19] P. H. Naccache, S. R. McColl, A. C. Caon, and P. Borgeat, *Br. J. Pharmacol.* **97,** 461 (1989).

ligand that are present on a particular leukocyte or (b) the number of ligands that interact with an individual receptor?

In summary, although calcium mobilization is a very useful measure of leukocyte activation by chemokines, and may provide information regarding the relative distribution of various chemokine-binding molecules on the leukocyte surface, it is likely to be most useful when assessed in the context of other functional responses.

Acknowledgment

This study was supported by a Project Grant from the NH&MRC (Australia) and by grants from the MRC and the Arthritis Society (Canada).

[19] G-Protein Activation by Chemokines

By SUZANNE K. BECKNER

Introduction

Heterotrimeric (α, β, γ) G proteins serve as molecular switches to transduce information from occupied (activated) receptors to appropriate intracellular effectors.[1] Transmembrane signaling mechanisms regulated by heterotrimeric G proteins include adenylyl cyclase (EC 4.6.1.1, adenylate cyclase), phosphatidylinositol-specific phospholipases A and C, guanylate cyclase, cGMP phosphodiesterase, and ion channels, specifically those that regulate Ca^{2+} and K^{+}. Other processes shown to be regulated by heterotrimeric G proteins include the mitogen-activated protein (MAP) kinase cascade activated by growth factors[2] and vesicular trafficking.[3] Both of these pathways, as well as others, are also regulated by single monomeric G proteins that exhibit GTP binding and hydrolysis activities and are also considered part of the G-protein superfamily.[4] With the exception of Rho, these monomeric G proteins are not considered further. This chapter focuses on the receptor-coupled heterotrimeric G proteins.

The heterotrimeric G proteins consist of three distinct subunits: α (39–46 kDa), β (37 kDa), and γ (8 kDa). The α subunit has a single, high affinity binding site for the guanine nucleotides GTP and GDP and possesses

[1] A. G. Gilman, *Annu. Rev. Biochem.* **56,** 615 (1987).
[2] M. H. Cobb and E. J. Goldsmith, *J. Biol. Chem.* **270,** 14843 (1995).
[3] C. Nuoffer and W. E. Balch, *Annu. Rev. Biochem.* **63,** 949 (1994).
[4] A. Hall, *Annu. Rev. Cell Biol.* **10,** 31 (1994).

0076-6879/97 $25.00

TABLE I
FAMILIES OF MAMMALIAN G PROTEINS

Family	Toxin	Receptors[a]	Effector[b]
G_s	Cholera	Glucagon, β-adrenergic, TSH, LH, VIP	Adenylyl cyclase (↑), Ca^{2+} channels (↑), Na^+ channels (↓)
G_i/G_o	Pertussis	α-Adrenergic, muscarinic, opiate, IGF-II	Adenylyl cyclase (↓), phospholipase C (↑), Ca^{2+} channels (↓)
		Rhodopsin	cGMP phosphodiesterase (↑)
G_q	None	α-Adrenergic, muscarinic	Phospholipase (C-β_{1-4} (↑)
α_{15}		B cells, myeloid cells	Phospholipase C-β_{1-4} (↑)
α_{16}		T cells, myeloid cells	Phospholipase C-β_{1-4} (↑)
G_{12}	None	??	??

[a] TSH, Thyroid-stimulating hormone; LH, luteinizing hormone; VIP, vasoactive intestinal peptide; IGF-II, insulin-like growth factor type II.
[b] ↑, Stimulation; ↓, inhibition.

intrinsic GTPase activity. The β and γ subunits exist as a tightly associated complex that functions as a unit. There is increasing evidence that the $\beta\gamma$ complex and the free α subunit as well as the heterotrimeric $\alpha\beta\gamma$ complex can all interact with and regulate effector proteins and may, in fact, act synergistically.

The initial identification of G proteins was defined by the function of the α subunit. The use of sensitive molecular techniques has resulted in the discovery of additional G-protein family members based on sequence homology. Newly identified G-protein α subunits are designated by numbers. However, there are currently four major subfamilies of α subunits classified on the basis of sequence relationships represented by $G\alpha_s$, $G\alpha_i$, $G\alpha_q$, and $G\alpha_{12}$ (Table I).

The G_s family was the first to be identified because of its ability to stimulate adenylyl cyclase, which generates cAMP from ATP. More recently, $G\alpha_s$ has been shown to stimulate dihydropyridine-sensitive voltage-gated Ca^{2+} channels in skeletal muscle and inhibit cardiac Na^+ channels.[5] Similarly, the G_i family was initially classified by its ability to inhibit adenylyl cyclase (which may be a direct effect of $\beta\gamma$ on adenylyl cyclase[6]) and the rhodopsin-sensitive cGMP phosphodiesterase, but it has since been implicated in the stimulation of phospholipase C (PLC) (initially designated G_o but α subunits have since been shown to be members of the G_i family). The hydrolysis of phosphatidylinositol 4,5-bisphosphate (PIP_2) by PLC generates two intracellular messengers: inositol 1,4,5-trisphosphate (IP_3),

[5] B. Schubert, A. M. J. VanDongen, G. E. Kirsch, and A. M. Brown, *Science* **245,** 516 (1989).
[6] W. J. Tang and A. G. Gilman, *Science* **254,** 1500 (1991).

which increases intracellular Ca^{2+}, and diacylglycerol (DAG), which stimulates protein kinase C (PKC).

Members of the G_s and G_i families are nearly ubiquitous in all cell types. Less is known about G_q and G_{12}. Most members of these families are also ubiquitous, with the exceptions of α_{15} and α_{16} (both members of the $G\alpha_q$ family), which seem restricted to B and T lymphocytes, respectively, as well as certain myeloid cells.[7] The effector roles of G_q and G_{12} are still undetermined, but both may stimulate phospholipase C-β_1, -β_2, -β_3, and -β_4.

At least 21 distinct G-protein α subunits, 5 β subunits, and 9 γ subunits have been identified.[8] Most of the well-characterized G-protein subunits have been cloned, and specific antibodies have been developed that recognize these specific subunits.

Methods to study receptor regulation of G-protein activity are highly variable and are dependent on the ligand, the specific G protein, and the cell type. The most extensively studied receptor-coupled G-protein systems are the hormone [glucagon, β-adrenergic, thyroid-stimulating hormone (TSH), follicle-stimulating hormone (FSH), vasoactive intestinal peptide (VIP), etc.]-sensitive (stimulatory) adenylyl cyclase systems and the neurotransmitter (α-adrenergic, muscarinic, cholinergic, opiate, etc.)-responsive (inhibitory) adenylyl cyclase and phospholipase A and C effectors. Although some lymphokine and cytokine G-protein-coupled systems have been studied,[9] little is known about chemokine receptor regulation of G proteins. The methodology described in this chapter summarizes studies with the well-characterized receptor-coupled G-protein systems.

G-Protein Regulation of Transmembrane Signaling

In the inactive state, the G protein exists as a heterotrimeric complex with GDP bound to the α subunit. On ligand binding to the specific cell surface receptor, the complex is activated by GDP exchange for GTP on the α subunit, causing a conformational change that dissociates the complex from the receptor. The α subunit possesses intrinsic GTPase activity that converts it back to the inactive GDP-bound form.

A wide variety of cell surface receptors have been shown to exert their intracellular effects via activation of G proteins, including small peptide and large glycoprotein hormones, neurotransmitters, lymphokines, and more recently chemokines. The activation of G proteins is initiated by binding of the ligand to a specific cell surface receptor, which results in a conforma-

[7] M. I. Simon, M. P. Strathmann, and N. Gautan, *Science* **252,** 802 (1991).

[8] B. R. Conklin and H. R. Bourne, *Cell* (*Cambridge, Mass.*) **73,** 631 (1993).

[9] W. L. Farrar, J. L. Cleveland, S. K. Beckner, E. Bonvini, and S. W. Evans, *Immunol. Rev.* **92,** 49 (1986).

tional change of the receptor and allows the high affinity interaction of the occupied receptor with the membrane-bound G-protein complex. It is not clear if the Gα subunits or heterotrimeric G-protein complexes are associated with effector systems or only activate effectors after ligand binding and activation. Ligand–receptor interaction catalyzes guanine nucleotide exchange on the α subunit and effector activation. The GTP-bound form of the α subunit dissociates from the receptor and activates effector proteins that modulate the activity of specific intracellular proteins. The dissociation of the G protein from the receptor also reduces the affinity of the receptor for its ligand.

The structure and function of well-characterized G-protein-coupled receptors have been reviewed.[10] Despite the wide variety of agonists that activate a wide variety of cellular functions, these receptors exhibit remarkable homology. Cloning and sequencing studies have shown that these G-protein-coupled receptors are characterized by seven hydrophobic stretches of 20–25 amino acids predicted to form transmembrane α helices connected by alternating extracellular and intracellular loops.[10] It is clear that a single receptor can activate multiple pathways within a cell, although the predominant pathway may vary by cell type.

Chemokine Receptors Coupled to G Proteins

The receptors for chemokine and chemoattractant receptors belong to the serpentine superfamily of G-protein-coupled receptors, yet these receptors mediate unique and specific intracellular effects compared to the better characterized peptide, hormone, and neurotransmitter receptors. Chemokine receptors that have been cloned and shown to be homologous to G-protein-coupled receptors include interleukin-8 (IL-8), RANTES, macrophage inflammatory protein-1α (MIP-1α), monocyte chemoattractant protein-1 (MCP-1), growth-related oncogene (GRO), *N*-formylmethiony leucyl phenylalanine (fMLP), C5a and platelet-activating factor (PAF).[11] Most of these receptors activate PLC and regulate IP_3 and intracellular Ca^{2+} mobilization[12,13] and in some cases (IL-8) activate the MAP kinase cascade.[14] These findings suggest that chemokine receptors interact primarily with the G_i family of G proteins. In fact, studies with transfected

[10] C. D. Strader, T. M. Fong, M. R. Tota, and D. Underwood, *Annu. Rev. Biochem.* **63,** 101 (1994).

[11] P. M. Murphy, *Annu. Rev. Immunol.* **12,** 593 (1994).

[12] A. Ben-Baruch, D. F. Michiel, and J. J. Oppenheim, *J. Biol. Chem.* **270,** 11703 (1995).

[13] K. B. Bacon, B. A. Premack, P. Gardner, and T. J. Schall, *Science* **269,** 1727 (1995).

[14] S. A. Jones, B. Moser, and M. Thelen, *FEBS Lett.* **364,** 211 (1995).

cells confirmed that IL-8, fMLP, and C5a interact with members of both the G_i and G_q families of G proteins.[15–18]

The complexities of chemokine regulation of G proteins is evidenced by studies of the MCP-1 receptor. Dubois *et al.* demonstrated that MCP-1 mediated intracellular Ca^{2+} mobilization without detectable increases in inositol phosphates and also regulated tyrosine phosphorylation of specific 42- and 44-kDa substrates in a T-cell hybrid.[19] Myers *et al.*[20] demonstrated that MCP-1 mobilizes intracellular Ca^{2+} and also inhibits adenylyl cyclase in embryonic kidney cells transfected with MCP-1 receptors. It remains to be determined whether the effects on Ca^{2+} mobilization are mediated by adenylyl cyclase and which G proteins interact with one another in this system. However, MIP-1α increased cAMP levels in MO7e cells, a human factor-dependent line,[21] implicating interaction with the G_s family of G proteins.

The involvement of the G_i family of G proteins in chemokine function is supported by the findings that a number of chemokine responses are *Bordetella pertussis* sensitive (see below). The signal transduction by the MCP-1 receptor both in a T-cell hybrid[19] and in kidney cells transfected with the receptor[20] was blocked by treatment with pertussis toxin. Similarly, pertussis toxin abrogated IL-8-mediated intracellular Ca^{2+} mobilization and chemotaxis in Jurkat cells transfected with IL-8R1 and IL-8R2 receptors,[14] fMLP-induced actin polymerization and cytosolic Ca^{2+} elevation in polymorphonuclear leukocytes,[22] and IL-2-activated natural killer (NK) chemotaxis induced by RANTES and MCP-1.[23]

In addition to regulation of the heterotrimeric G proteins, fMLP and IL-8 receptors have also been shown to couple to the small GTP binding protein RhoA, which mediates rapid activation of leukocyte integrin adhesion.[24] The Rho family of G proteins belongs to a group of cytosolic regulatory proteins whose activity, like the heterotrimeric G proteins, is regulated

[15] D. Wu, G. J. LaRosa, and M. I. Simon, *Science* **261,** 101 (1993).

[16] P. Gierschik, D. Sidiropoulos, and K. H. Jacobs, *J. Biol. Chem.* **264,** 21470 (1989).

[17] T. T. Amatruda, N. P. Gerard, C. Gerard, and M. I. Simon, *J. Biol. Chem.* **268,** 10139 (1993).

[18] M. A. Buhl, B. J. Eisfelder, G. H. Worthen, G. L. Johnson, and M. Russell, *FEBS Lett.* **323,** 132 (1993).

[19] P. A. Dubois, D. Palmer, M. L. Webb, J. A. Ledbetter, and R. A. Shapiro, *J. Immunol.* **156,** 1356 (1996).

[20] S. J. Myers, L. M. Wong, and I. F. Charo, *J. Biol. Chem.* **270,** 5786 (1995).

[21] C. Mantel, S. Aronica, Z. Luo, M. S. Marshall, Y. J. Kim, S. Cooper, N. Hague, and H. E. Broxmeyer, *J. Immunol.* **154,** 2342 (1995).

[22] G. M. Omann, J. M. Harter, N. Hassan, P. J. Mansfield, S. J. Suchard, and R. R. Neubig, *J. Immunol.* **149,** 2172 (1992).

[23] A. A. Maghazachi, A. al-Aoukaty, and T. J. Schall, *J. Immunol.* **153,** 4969 (1994).

[24] C. Laudana, J. J. Campbell, and E. C. Butcher, *Science* **271,** 981 (1996).

by GTP binding and hydrolysis.[4] This family of small G proteins also includes Arf, Ras, Sarl, Rab, and Rac.

Methods to Study G Proteins

The initial discovery of a role for G proteins in receptor transmembrane signaling was by Rodbell and colleagues,[25] who were studying the glucagon-sensitive adenylyl cyclase system in liver membranes. Like other polypeptide hormones and neurotransmitters, this complex consists of a receptor, G proteins, and the effector adenylyl cyclase, which converts ATP to cAMP. This second messenger then activates a specific protein kinase (PKA) that initiates a cascade of protein phosphorylation that regulates the activity of a number of key regulatory enzymes.

The involvement of G proteins in receptor-mediated signal transduction is traditionally determined by analysis of guanine nucleotide effects on ligand binding and regulation of second messenger systems, regulation by bacterial toxin-catalyzed ADP-ribosylation, direct measurement of ligand activation of membrane GTP binding and hydrolysis activity, and immunohistochemistry with specific G-protein subunit antibodies. More sophisticated techniques include direct receptor/G-protein reconstitution or transfection studies with purified or cloned proteins.

Identification of Receptors as G Protein Coupled

The initial studies of Rodbell[25] demonstrated that guanine nucleotides are absolutely required for glucagon activation of the adenylyl cyclase system. A multitude of studies in highly diverse ligand-linked adenylyl cyclase systems has demonstrated that basal as well as ligand stimulation and inhibition of adenylyl cyclase is enhanced/abrogated by GTP and more effectively by the nonhydrolyzable analog Gpp(NH)p.

The binding of ligands that are G protein coupled should also be guanine nucleotide sensitive in that receptor affinity is decreased by guanine nucleotides. It has long been known that certain families of G proteins, specifically the G_s and G_i families, are ADP-ribosylated by bacterial toxins, namely, cholera and *B. pertussis* toxins, respectively. Such ADP-ribosylation by these toxins affects the activities of the respective G protein.

Modification of Receptor Affinity by G Proteins. Following the observations that adenylyl cyclase activation was stimulated by GTP,[25] studies by Lin *et al.*[26] demonstrated that the incubation of rat liver plasma membranes

[25] M. Rodbell, *Nature* (*London*) **284,** 17 (1980).

[26] M. C. Lin, S. Nicosia, P. M. Lad, and M. Rodbell, *J. Biol. Chem.* **252,** 2790 (1977).

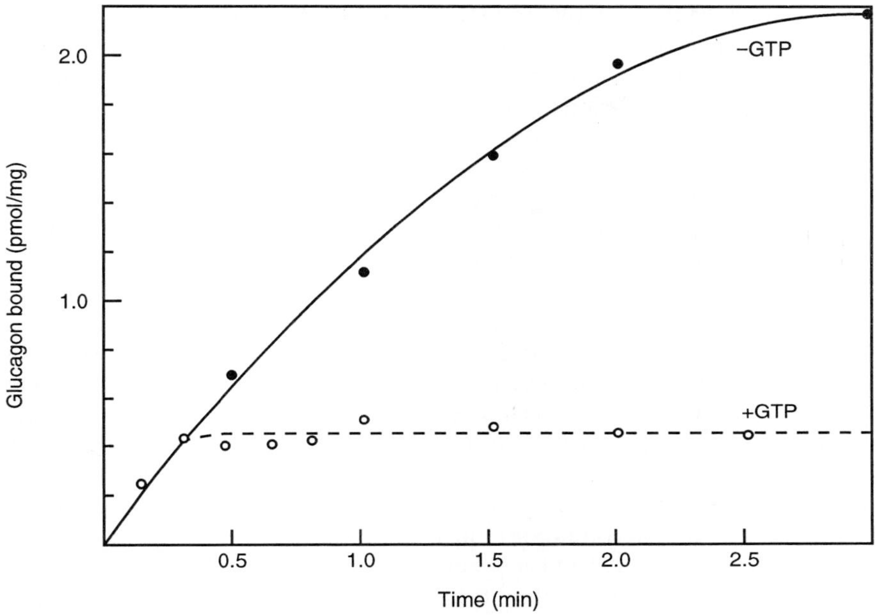

FIG. 1. Time course of glucagon binding in the presence and absence of GTP. Rat liver plasma membranes were incubated with 3 n*M* ^{3}H-labeled glucagon for the indicated times in the presence or absence of 1 μM GTP. Nonspecific binding was determined by the addition of 1 μM native glucagon as described. (Reproduced with permission from Lin *et al.*[26])

with GTP significantly decreased the binding of radiolabeled glucagon (Fig. 1). Additional studies demonstrated that micromolar concentrations of guanine nucleotides increased the rate of ^{125}I-labeled glucagon binding but decreased binding affinity without altering the number of receptor sites.[27]

To determine an effect of guanine nucleotides on chemokine binding, incubate membranes [GTP and Gpp(NH)p cannot penetrate intact cells] of responsive cells with chemokine in the presence and absence of 1 μM GTP and assess the picomoles bound per milligram protein at different time points (0–10 min) as in Fig. 1. Binding conditions should be optimized for the chemokine being studied. More rigorous determination of changes in receptor affinity can be accomplished by Scatchard analysis of binding in the presence and absence of GTP.

Effect of Bacterial Toxins. During the study of G proteins in the mid-1980s, it was observed that certain bacterial toxins catalyze the ADP-

[27] M. Rodbell, H. M. J. Krans, S. L. Pohl, and L. Birnbaumer, *J. Biol. Chem.* **246,** 1872 (1971).

ribosylation of the α subunits of G proteins. Cholera toxin catalyzes the transfer of the ADP-ribose moiety of NAD^+ to a specific arginine residue of certain G proteins, primarily those of the G_s family (45,000 Da). Similarly, pertussis toxin ADP-ribosylates those α subunits that contain a specific cysteine residue near the carboxyl terminus, primarily those of the G_i family (40,000 Da). It is therefore possible to determine the presence of these α subunits in cell membranes by incubating membranes in the presence and absence of these toxins with NAD^+ and monitoring the radioactivity incorporated into the appropriate proteins.

Add 100–200 μg membrane protein to the reaction mix to achieve a final concentration of 50 m*M* Tris-HCl (pH 7.5), 100 μM GTP, 2 m*M* dithiothreitol (DTT), 10 μM ^{3}H-, ^{14}C-, or ^{32}P-labeled NAD^+ (3500 cpm/pmol), and 1 μg/ml cholera or pertussis toxin. Incubate 10 min at 37° and stop the reaction with ice-cold Tris-HCl (pH 7.5) and immediate centrifugation of the membrane pellet in a microcentrifuge. Wash the pellet with Tris buffer, solubilize in polyacrylamide gel buffer containing 1% (v/v) sodium dodecyl sulfate (SDS), and separate proteins electrophoretically.[28] Radioactivity incorporated into the 45-kDa $G\alpha_s$ (cholera toxin) or 40-kDa $G\alpha_i$ (pertussis toxin) subunit can then be quantitated by slicing and counting gels or by Western blot with anti-$G\alpha$ antibodies. Alternatively, solubilize membranes in immunoprecipitation buffer containing 1% (v/v) Triton, immunoprecipitate with appropriate $G\alpha$ antibodies, separate proteins electrophoretically, and quantitate by autoradiography.[29]

Modification of $G\alpha$ by cholera toxin constitutively activates these proteins by inhibiting GTPase, whereas modification by pertussis toxin prevents receptor-mediated activation of G proteins. Therefore, pretreatment of cells or plasma membrane preparations with cholera toxin results in enhanced activation of second messenger systems by G_s-coupled ligands, and pretreatment with pertussis toxin abrogates transmembrane signaling by G_i-coupled ligands.

To measure the effect of toxins on G-protein-mediated transmembrane signaling, measure the desired second messenger in cell membranes or intact cells, in the presence or absence of cholera or pertussis toxin. It is not recommended to measure ultimate physiological responses in membranes or cells after toxin pretreatment, given the complexity of transmembrane signaling systems and the interactions between various G-protein systems.[30]

[28] S. K. Beckner and M. Blecher, *Biochim. Biophys. Acta* **673,** 477 (1981).
[29] S. K. Beckner, S. Hattori, and T. Y. Shih, *Nature* (*London*) **317,** 71 (1985).
[30] D. J. Roof, M. L. Applebury, and P. C. Sternweis, *J. Biol. Chem.* **260,** 16242 (1985).

Both cholera and pertussis toxins are composed of β (binding) subunits and the active ADP-ribosylating α subunits(s). The cholera toxin A-1 and A-2 α subunits are connected by a disulfide bond, so it is necessary to include DTT in the reaction mix to dissociate the active enzyme in membrane preparations. In intact cells, binding of the β subunits to the cell membrane will release the active α subunits. The activity of both toxins also requires NAD^+ as a substrate for ADP-ribosylation of the Gα subunits. It is therefore necessary to include 1 m*M* NAD^+ in membrane experiments. In intact cells, endogenous NAD^+ is sufficient.

To measure the effect of cholera or pertussis toxin on second messenger systems in membranes, incubate the membranes with and without 10 μg/ml toxin, 1 m*M* NAD^+, 2 m*M* DTT, and 10–100 μM GTP with the appropriate reaction mix for the desired second messenger system. Compare the effect of toxin treatment to nontreated control reactions to determine the extent of toxin stimulation or inhibition of second messenger generation.

To measure the effect of toxins on second messenger systems in intact cells, pretreat cells with 5 μg/ml of the appropriate toxin in phosphate-buffered saline (PBS, pH 7.4) for 2–3 hr to allow the toxin to dissociate and ADP-ribosylate Gα. Immediately monitor intracellular messenger levels. Figure 2 demonstrates the effect on intracellular cAMP levels in response to epinephrine following cholera toxin pretreatment.[31] Cholera toxin pretreatment significantly increased the rate and magnitude of cAMP production in hepatocytes.

Measurement of G-Protein Activity

Given that G proteins possess intrinsic guanine nucleotide binding and hydrolysis activities, it is possible to directly measure these enzymatic activities and their regulation by specific ligands in membrane preparations and reconstituted systems. With the availability of specific anti-G-protein antibodies, it is also possible to quantitate different G-protein levels in cells, membranes, and tissues by routine immunohistochemistry techniques.

GTP Binding. The measurement of ligand-stimulated GTP binding has been fully reviewed in a previous volume of *Methods in Enzymology.*[32] Reaction conditions are highly dependent on the cell type and must be optimized for incubation temperature, membrane concentration, NaCl, Mg^{2+}, and guanine nucleotide concentrations as reviewed.[32]

A general method is described here that was used to measure the interleukin-2 (IL-2) responsive system in lymphocyte membranes. Il-2 in-

[31] S. K. Beckner and M. Blecher, *Biochim. Biophys. Acta* **673,** 467 (1981).

[32] T. Wieland and K. H. Jacobs, *Methods Enzymol.* **237,** 3 (1994).

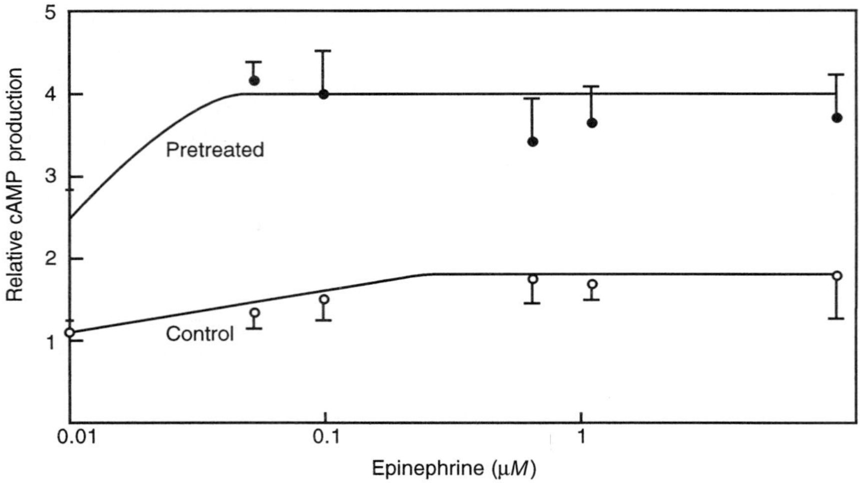

FIG. 2. Epinephrine responsiveness following cholera toxin pretreatment. Confluent monolayers of RL-PR-C hepatocytes were incubated with (●) or without (○) 5 μg/ml cholera toxin in phosphate-buffered saline, pH 7.4, for 3 hr at 37°. Monolayers were washed well at 22° to remove excess cholera toxin, exposed to the indicated concentration of epinephrine in phosphate-buffered saline (pH 7.4), 1.5% bovine serum albumin for 10 min at 37°, and cAMP determined. Data are expressed as picomoles cAMP produced relative to control, in the absence of epinephrine, ±SD of triplicate determinations from two separate experiments. Control values were 2.58 and 29.7 pmol cAMP/60-mm petri dish for basal and cholera toxin, respectively. [Reprinted from *Biochim. Biophys. Acta* **673,** S. K. Beckner and M. Blecher, Regulatory states of adenylate cyclase in RL-PR-C cloned rat hepatocytes, pp. 467–476. Copyright 1991 with kind permission of Elsevier Science–NL, Sara Burgerhartstraat 25, 1055 KV Amsterdam, The Netherlands.]

hibits adenylyl cyclase,[33] activates protein kinase C,[34] and stimulates IL-2-dependent GTP binding and hydrolysis.[35] Both the biochemical and cellular effects are pertussis toxin sensitive,[35] suggesting that IL-2 is coupled to a member of the G_i/G_o family of G proteins.

To prepare membranes, homogenize cells (4°) in 25 m*M* Tris-HCl (pH 7.6), 5 m*M* $MgCl_2$, 1 m*M* phenylmethylsulfonyl fluoride (PMSF, a protease inhibitor). Centrifuge at 150*g* for 5 min to remove debris and collect membranes by centrifugation of the supernatant at 8000*g* for 10 min (4°). The final reaction buffer consists of 10 m*M* Tris-HCl (pH 7.8), 10 m*M* $MgCl_2$, 1 m*M* EDTA, 0.2% (v/v) bovine serum albumin (BSA), 0.5 m*M* ascorbic

[33] S. K. Beckner and W. L. Farrar, *J. Immunol.* **140,** 208 (1988).
[34] S. K. Beckner and W. L. Farrar, *J. Biol. Chem.* **261,** 3043 (1986).
[35] S. W. Evans, S. K. Beckner, and W. L. Farrar, *Nature* (*London*) **325,** 166 (1987).

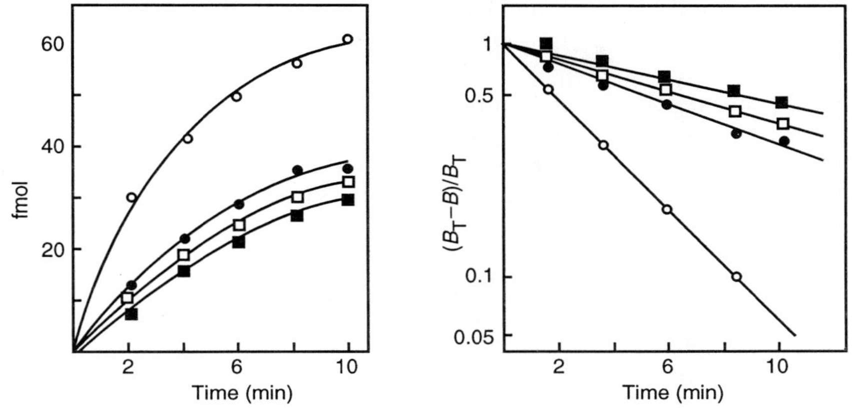

FIG. 3. Stimulation of [^{35}S]GTPγS binding to lymphocyte membranes by IL-2. (*Left*) Time course of binding; (*right*) binding data transformed to compare binding rates. The assay mixture included either membrane and IL-2 (○), membrane, IL-2, and GDP (●), membrane alone (□), or membrane and GDP (■). B_T, Maximum binding; B, observed binding; t, time (min); k rate constant; and $(B_T - B)/B_T = e^{-kt}$. [Reprinted with permission from *Nature,* Evans *et al.*, **325,** 166 (1987). Copyright 1987 Macmillan Magazines Limited.]

acid, 2 m*M* adenylyl imidodiphosphate, 100 n*M* [^{35}S]GTPγS with and without 1 μM GDP, and appropriate concentrations of ligand or diluent in 50 μl. Initiate the reaction with 50 μg of membrane protein (1 mg/ml) and incubate for the optimal time at the optimal concentration. Stop the reaction with 450 μl of cold buffer containing 25 μM unlabeled GTPγS, 100 m*M*, NaCl, 10 m*M* Tris-HCl (pH 7.4), and 0.1% Lubrol. Separate free nucleotide from bound by filtration through a 0.45-μm nitrocellulose membrane and washing with 20 ml of stopping buffer minus GTPγS.

Typical GTP binding kinetics can be seen in Fig. 3. IL-2 stimulated GTP binding over a 10-min period compared to membrane controls and controls with excess GDP to inhibit GTP binding.[35] Such binding data can be transformed to obtain rate constants (Fig. 3).

Kinetic analysis of GTP binding to purified $G\alpha_s$[36] and $G\alpha_i$[37] proteins has been demonstrated. A photoreactive GTP analog, azidoanilido-GTP, has been used to compare GTP binding activities of different purified G-protein α subunits.[38] Okamoto *et al.*[39] have developed a novel method to quantitate GTP binding to specific G proteins in membrane preparations

[36] J. K. Northrup, M. D. Smigel, and A. G. Gilman, *J. Biol. Chem.* **257,** 11416 (1982).

[37] T. Okamoto and I. Nishimoto, *Proc. Natl. Acad. Sci. U.S.A.* **88,** 8020 (1991).

[38] T. A. Fields, M. E. Linder, and P. J. Casey, *Biochemistry* **33,** 6877 (1994).

[39] T. Okamoto, T. Ikezu, Y. Murayama, E. Ogata, and I. Nishimoto, *FEBS Lett.* **305,** 125 (1992).

by loading membranes with GTPγS and allowing them to bind to specific G-protein antibodies on an enzyme-linked immunosorbent assay (ELISA) plate.

It is also possible to determine the presence of GTP binding proteins by measuring GTP binding to proteins separated via SDS–polyacrylamide gel electrophoresis (SDS–PAGE) by electroblotting and direct binding with [α-^{32}P]GTP. Following electrophoretic separation, transfer proteins to nitrocellulose sheets. Incubate nitrocellulose blots for 30 min at room temperature in 0.3% (w/v) BSA, 0.25% (w/v) Tween 20, 2 m*M* DTT, 50 μM $MgCl_2$, 20 m*M* Tris-HCl (pH 7.7), and 5 μCi of 1 n*M* [α-^{32}P]GTP. Wash twice with buffer without BSA and [α-^{32}P]GTP, air-dry, and detect ^{32}P by autoradiography.[40]

GTPase Activity. A general method for the measurement of receptor-stimulated hydrolysis of GTP by membrane preparations was reviewed in a previous volume of *Methods in Enzymology.*[41] As with GTP binding, conditions must be optimized for each specific cell type. It should be noted that this assay measures the release of P_i as an indirect measurement of GTPase activity. The G-protein α subunits hydrolyze GTP and GDP/P_i remain associated with the α subunit. The affinity of the α subunit for P_i is low, and it is rapidly released.

The IL-2-sensitive system in lymphocytes is again used as an example of measuring GTPase activity.[35] The reaction mix is identical to that used for measuring GTP binding except that 100 n*M* [γ-^{32}P]GTP replaces GTPγS and an ATP regenerating system is added (0.1 m*M* ATP, 3 m*M* creatine phosphate, and 75 units/ml creatine phosphokinase). Stimulation of GTP hydrolysis by IL-2 is measured by the release of $^{32}PO_4$ from [^{32}P]GTP. Controls include membranes alone and reactions with saturating amounts of GDP. Initiate the reaction by adding 100 μg of membrane protein and stop by adding 500 μl of 5% (w/v) Norit A in potassium phosphate buffer (pH 7.0). After centrifugation in a microcentrifuge, count an aliquot of the supernatant for $^{32}PO_4$. As seen in Fig. 4, the release of P_i was linear over 15 min, suggesting that GTP was hydrolyzed at a rate of 44 pmol/mg protein/min, similar to the rate of fMLP stimulation of hydrolysis in neutrophil membranes.[42]

The GTPase activity of purified G proteins has been characterized and the active sites required for this activity identified by mutational[42–46] and

[40] J. P. Doucet, S. Fournier, M. Parulekar, and J. M. Trifaro, *FEBS Lett.* **247,** 127 (1989).

[41] P. Gierschik, T. Bouillon, and K. H. Jacobs, *Methods Enzymol.* **237,** 13 (1994).

[42] F. Okajima, T. Katada, and M. Ui, *J. Biol. Chem.* **260,** 6761 (1985).

[43] D. R. Brandt and E. M. Ross, *J. Biol. Chem.* **261,** 1656 (1986).

[44] D. R. Brandt and E. M. Ross, *J. Biol. Chem.* **260,** 266 (1985).

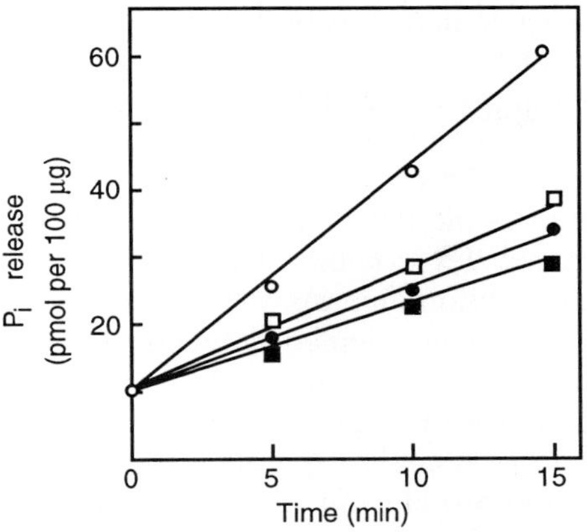

FIG. 4. Stimulation of [γ-^{32}P]GTP hydrolysis by IL-2. GTP hydrolysis was determined by measuring the release of [^{32}P]P_i from [γ-^{32}P]GTP. The reaction mixture included IL-2 (○), IL-2 and GDP (●), membrane alone (□), or membrane and GDP (■). [Reprinted with permission from *Nature*, Evans *et al.*, **325,** 166 (1987). Copyright 1987 Macmillan Magazines Limited.]

structural analysis.[47] Methods to study GTP hydrolysis and nucleotide exchange rates for the Rho family of small GTP binding proteins have been reviewed in a previous volume of *Methods in Enzymology.*[48]

Immunochemistry Techniques. The availability of specific G-protein antibodies allows immunochemical detection and quantitation of specific G proteins. Specific methodology depends on the cell type, antibody affinity, and protein concentration. Immunoblotting and detection kits are available from several manufacturers, including Promega (Madison, WI), Pierce (Rockford, IL), Amersham (Arlington Heights, IL), Life Technologies (Gaithersburg, MD), and others. A general method used for the detection of G_q is described below.[49]

To determine the presence of specific G proteins, separate membrane proteins electrophoretically. Transfer to nitrocellulose membranes electrophoretically as described by the manufacturer. Block immunoblots for 2 hr

[45] C. Kleuss, A. S. Raw, E. Lee, S. R. Sprang, and A. G. Gilman, *Proc. Natl. Acad. Sci. U.S.A.* **91,** 9828 (1994).

[46] M. Freissmuth and A. G. Gilman, *J. Biol. Chem.* **264,** 21907 (1989).

[47] M. B. Mixon, E. Lee, D. E. Coleman, A. M. Berghuis, A. G. Gilman, and S. R. Sprang, *Science* **270,** 954 (1995).

[48] A. J. Self and A. Hall, *Methods Enzymol.* **256,** 67 (1995).

[49] G. Milligan, *J. Neurochem.* **61,** 845 (1993).

at 37° with 5% (w/v) gelatin in Tris-buffered saline (TBS; 20 m*M* Tris-HCl, pH 7.5, 500 m*M* NaCl). Add primary G-protein antibody appropriately diluted in TBS containing 1% gelatin, incubate overnight, and wash extensively with distilled water and finally with TBS containing 0.1% (v/v) Tween 20 and then again with TBS.

There are numerous methods available to identify reactive antibodies, including radioactive ^{125}I[50,51] or biotinylated (Life Technologies) second antibody detected by autoradiography and horseradish peroxidase,[49] respectively, or other enzyme-coupled detection systems amplified as described by the manufacturer and measured colorimetrically. It is possible to then quantitate the levels of G proteins by immunoblotting varying amounts of membrane lysates in parallel with defined amounts of the appropriate purified G protein and measuring band density by scanning.[49] Another technique is to use specific G-protein antibodies to immunoprecipitate G proteins[29] and quantitate levels by dot-blot.[52]

Using immunochemistry techniques, the levels of G_s, G_i/G_o, and G_q have been measured in guinea pig uterine membranes at different stages of pregnancy.[50] Levels of G_q and G_{11} have been measured in rat central nervous system tissues[49] and G_s, G_i, and G_q measured in myocardial membranes.[51] Monoclonal antibodies to specific G proteins are available from a variety of commercial sources. Alternatively, anti-G-protein peptide antisera can be developed. The specificity and functional application of such antisera for the detection of G proteins have been reviewed in a previous volume of this series.[53] The application of polyclonal G-protein antisera to determine levels of G proteins in hematopoietic cells has been described.[54]

Protein/Membrane Reconstitution Studies

Definitive evidence for the involvement of a particular G protein in receptor function is ultimately demonstrated by reconstitution of purified, cloned, or transfected receptors and G proteins in a defined system. Although complicated, such studies provide a clear indication of the direct interaction of receptors and G proteins with a specific signaling pathway. However, it must be considered that most reconstitution systems use receptors, G proteins, or effector systems at much higher levels than exist physio-

[50] H. Warsop, A. Khouja, D. P. Wichelhaus, and C. T. Jones, *J. Dev. Physiol.* **19,** 91 (1993).
[51] M. C. Michel, O. E. Brodde, and P. A. Insel, *J. Hypertens.* **11,** 355 (1993).
[52] H. Qadota, C. P. Python, S. B. Inoue, M. Arisawa, Y. Anraku, T. Watanabe, D. E. Levin, and Y. Ohya, *Science* **272,** 279 (1996).
[53] G. Milligan, *Methods Enzymol.* **237,** 268 (1994).
[54] T. M. Wilkie, P. A. Scherle, M. P. Strathmann, V. Z. Slepak, and M. I. Simon, *Proc. Nat. Acad. Sci. U.S.A.* **88,** 10049 (1991).

logically, and the *in vivo* interactions are much more complex. However, direct interaction of various components can be demonstrated. The specific methodologies for G protein and receptor purification, cloning, and reconstitution vary greatly depending on the systems and are referenced below. Traditionally, bovine or rabbit brain and liver have served as the richest sources of purified G proteins.

Gilman and co-workers have purified, cloned, and reconstituted numerous G proteins. $G\alpha_s$ is the most well characterized and has been purified, cloned,[55] and reconstituted[56] to demonstrate ligand regulation, GTP binding,[57] and hydrolysis.[58] Other G proteins are less well characterized. Members of the $G\alpha_i$ family have been purified, characterized,[59–61] cloned,[62] and reconstituted into phospholipid vesicles.[63] Slepak *et al.*[64] have cloned $G\alpha_o$ and used mutational analysis to study GTP binding and association of α with $\beta\gamma$ subunits. $G\alpha_q$ has been purified from turkey erythrocytes[65] and reconstituted with purified PLC-β_1.[66] Purified bovine cerebral $G\alpha_q$ has been reconstituted with recombinant muscarinic receptors and agonist-stimulated GTP binding measured.[67] A method for separating $G\alpha_i$ and $G\alpha_o$ subunits by affinity chromatography with $\beta\gamma$ subunits has also been described.[68]

Many different methods have been used to reconstitute receptors with G-protein subunits (see above), and the most recent methodology is described.[67] Coreconstitute purified or recombinant receptors and G proteins by gel filtration of the protein–lipid mixture on a Ultrogel AcA-34 column, which allows the resolution of reconstituted vesicles free from proteins, detergent, and other small molecules.[67] Add receptor (0.3 to 1.2 μM) in 500 mM phosphate buffer (pH 7.0), 0.1% digitonin to $G\alpha$ protein in 20 mM

[55] M. P. Grazino, M. Freissmuth, and A. G. Gilman, *J. Biol. Chem.* **264,** 409 (1989).
[56] D. R. Brandt, T. Asano, S. E. Pedersen, and E. M. Ross, *Biochemistry* **22,** 4357 (1983).
[57] T. Asano, S. E. Pedersen, C. W. Scott, and E. M. Ross, *Biochemistry* **23,** 5460 (1984).
[58] T. Asano and E. M. Ross, *Biochemistry* **23,** 5467 (1984).
[59] G. M. Bokoch, T. Katada, J. K. Northup, M. Ui, and A. G. Gilman, *J. Biol. Chem.* **259,** 3560 (1984).
[60] T. Katada, G. M. Bokoch, J. K. Northup, M. Ui, and A. G. Gilman, *J. Biol. Chem.* **259,** 3568 (1984).
[61] P. C. Sternweis and J. D. Robishaw, *J. Biol. Chem.* **259,** 13806 (1984).
[62] R. Graff, R. Mattera, J. Condina, M. K. Estes, and L. Birnbaumer, *J. Biol. Chem.* **267,** 24307.
[63] P. C. Sternweis, *J. Biol. Chem.* **261,** 631 (1986).
[64] V. Z. Slepak, T. M. Wilkie, and M. I. Simon, *J. Biol. Chem.* **268,** 1414 (1993).
[65] G. L. Waldo, J. L. Boyer, A. J. Morris, and T. K. Harden, *J. Biol. Chem.* **266,** 14217 (1991).
[66] A. V. Smrcka, J. R. Hepler, K. O. Brown, and P. C. Sternweis, *Science* **251,** 804 (1991).
[67] G. Berstein, J. L. Blank, A. V. Smrcka, T. Higashijima, P. C. Sternweis, J. H. Exton, and E. M. Ross, *J. Biol. Chem.* **267,** 8081 (1992).
[68] I. Pang and P. C. Sternweis, *Proc. Nat. Acad. Sci. U.S.A.* **86,** 7814 (1989).

Na–HEPES (pH 7.2), 3 m*M* EGTA, 1 m*M* DTT, 0.25% sodium cholate, 5 m*M* $MgCl_2$, and 1 μM GDP. Add the lipid mix such that the final concentrations are 0.4 m*M* 1-palmitoyl-2-oleoylphosphatidylethanolamine, 0.18 mg/ml bovine brain phosphatidylserine, and 45 μM cholesteryl hemisuccinate. Add the $\beta\gamma$ subunits in 20 m*M* Tris-HCl (pH 8.0), 1 m*M* EDTA, 0.05% (v/v) Lubrol 12A9. Apply the mixture to the column and run in 20 m*M* Na–HEPES (pH 8.0), 2 m*M* $MgCl_2$, 1 m*M* EDTA, and 100 m*M* NaCl. Collect fractions and monitor for reconstituted vesicles as described.[67]

Okamoto and Nishimoto have developed a technique to reconstitute receptors and G proteins in a soluble system.[37] They analyzed the kinetics of GTP binding in a system that consisted of a peptide fragment of the insulin-like growth factor II/mannose 6-phosphate receptor and purified $G\alpha_i$ and $\beta\gamma$.

Conclusion

Since the initial discovery of the important role of G proteins in transmembrane signaling by Rodbell and colleagues in the 1970s, biochemical, molecular, and immunological techniques have greatly enhanced our understanding of this complicated transduction system. A wide variety of distinct G proteins have been purified and cloned, and specific families of G proteins that couple to well-characterized intracellular messenger systems have been identified.

Although the methods described above are useful to examine the involvement of G proteins in ligand signaling, it is likely that multiple systems regulate the activity of one another. Such redundancy is not unexpected, given the complexities of cellular function. The determination of an involvement of G proteins in chemokine regulation of cell function requires more than studies with bacterial toxins. As noted above, reconstitution of purified components in a defined system with the technology available today, is the best technique to identify which chemokine receptors directly interact with a specific G protein.

The regulation of receptor activation of cellular processes is very complex. Receptors coupled to G proteins stimulate GTP binding to $G\alpha$ subunits, which decreases receptor affinity for ligands. Phosphorylation of G-protein-coupled receptors[69] provides an additional level of regulation. Similarly, G-protein activity is also regulated by covalent modification including phosphorylation[70,71] and lipid modification to regulate cellular local-

[69] R. T. Premont, J. Inglese, and R. J. Lefkowitz, *FASEB J.* **9,** 175 (1995).

[70] M. D. Houslay, *Cellular Signaling* **3,** 1 (1991).

[71] T. L. Z. Jones, *in* "G Proteins" (A. M. Spiegel, T. L. Z. Jones, W. F. Simonds, and L. S. Weinstein, eds.), p. 49. R. G. Landes, Austin, Texas, 1994.

ization and function[72] in addition to toxin-catalyzed ADP-ribosylation. Heterotrimeric G-protein activity is also regulated by accessory proteins, such as AFR, which is a monomeric G protein,[73] and other monomeric G proteins such as Rho.[4]

The availability of improved cloning techniques, recombinant receptors, and G proteins and the development of defined reconstitution systems will now allow more rigorous studies to better understand specific signaling pathways for signal transduction. Rubenstein *et al.*[74] have addressed the biochemical basis of receptor/G-protein selectivity by measuring regulation of different G-protein activities by β-adrenergic receptors reconstituted into phospholipid vesicles. Chidiac and Wells[75] have used reconstitution systems to evaluate the cooperative interactions among varius G proteins. Traynor and Nahorski[76] have adapted a GTP binding assay as a functional measure to determine the efficacy and intrinsic activity of opioid agonists.

Kuang *et al.*[77] have examined the pathways by which the C-C chemokine receptors activate phospholipase C in COS-7 cells cotransfected with different receptors and G proteins. Their findings demonstrated that the CRK-1 receptor can couple to $G\alpha_{14}$ but not $G\alpha_{16}$, suggesting that some of the C-C chemokine receptors, unlike the C-X-C chemokine receptors, discriminate against $G\alpha_{16}$, a hematopoietic-specific $G\alpha$ subunit.

New techniques are being developed to identify and quantitate G proteins and their activities. Chen *et al.*[78] have developed a colorimetric assay for measuring the activation of G_s and G_q by specific receptors that requires only 30,000 cells and has a rapid throughput to allow screening of numerous ligand systems. The assay uses the β-galactosidase (*lacZ*) gene fused to five copies of the cAMP response element (CRE) to detect the activation of CRE binding protein that results from an increase in intracellular cAMP (through melanocortin stimulation of G_α) or calcium (through bombesin stimulation of G_q). Cell lines were stably transfected with the melanocortin or bombesin receptor and transiently transfected with the pCRE/β-Gal plasmids. Cells were then stimulated with hormone and β-galactosidase activity measured. The increase (-fold) in β-galactosidase activity was similar in magnitude to increases in cAMP and adenylyl cylase activity measured by traditional methods.

[72] P. B. Wedegaertner, P. T. Wilson, and H. R. Bourne, *J. Biol. Chem.* **270,** 503 (1995).
[73] J. Moss and M. Vaughan, *J. Biol. Chem.* **270,** 12327 (1995).
[74] R. C. Rubenstein, M. E. Linder, and E. M. Ross, *Biochemistry* **30,** 10769 (1991).
[75] P. Chidiac and J. W. Wells, *Biochemistry* **31,** 10908 (1992).
[76] J. R. Traynor and S. R. Nahorski, *Mol. Pharmacol.* **47,** 848 (1995).
[77] Y. Kuang, Y. Wu, H. Jiang, and D. Wu, *J. Biol. Chem.* **271,** 3975 (1996).
[78] W. Chen, T. S. Shields, P. J. Stork, and R. D. Cone, *Anal. Biochem.* **226,** 349 (1995).

Current studies with the better characterized G-protein-coupled systems will undoubtedly identify still more unique G proteins and second messenger systems. Clearly there is still much work to be done, and, given that chemokine receptors are only relatively recently being identified and characterized, the role of G proteins in chemokine signaling will be more fully characterized in the future.

Acknowledgments

The author is grateful to Dr. Richard Horuk and Dr. Victor Rebois for helpful discussion and critical review of the manuscript.

[20] Adenylate Cyclase Assays to Measure Chemokine Receptor Function

By Richard Horuk

Introduction

At the latest count, eight different chemokine receptors have been identified and isolated by molecular cloning techniques.[1] These receptors belong to a large family of proteins, characterized by a presumed heptahelical structure,[2,3] that couple to and signal via a family of heterotrimeric G proteins composed of α, β, and γ subunits.[4] A variety of effector systems are activated by the G-protein complex; these include the phospholipases (C, A_2, and D), protein kinase C, adenylate cyclase, and phosphatidylinositol 3-kinase.[2,3] Some of these enzymes are discussed in other chapters of this volume, and we limit ourselves here to a discussion of chemokine activation of adenylate cyclase.

Analysis of the activity of adenylate cyclase has been widely used for studying the function of G-protein-linked receptors. Adenylate cyclase activity is regulated in either a positive or negative manner by receptor

[1] R. Horuk, *Trends Pharmacol. Sci.* **15,** 159 (1994).
[2] H. G. Dohlmam. J. Thorner, M. G. Caron, and R. J. Lefkowitz, *Annu. Rev. Biochem.* **60,** 653 (1991).
[3] W. C. Probst, L. A. Snyder, D. I. Schuster, J. Brosius, and S. C. Sealfon, *DNA Cell Biol.* **11,** 1 (1992).
[4] M. Rodbell, *Nature (London)* **284,** 17 (1980).

0076-6879/97 $25.00

agonists, depending on whether the receptor is coupled to stimulatory (G_s) or inhibitory (G_i) G proteins.[4] A number of receptors including the β-adrenergic, glucagon, and vasoactive intestinal peptide (VIP) receptors couple to G_s[2,3] and thus stimulate adenylate cyclase, whereas chemoattractant receptors such as *N*-formylmethionylleucylphenylalanine (fMLP), C5a, and some chemokine receptors [RANTES and macrophage chemotactic protein-1 (MCP-1)] couple to G_i and thus inhibit adenylate cyclase.[2,3,5,6] Interestingly, not all chemokine receptors coupled to G_i mediate an inhibitory adenylate cyclase effect in neutrophils.[7] In addition to their activation by hormones, discussed above, G proteins can also be directly modulated by aluminum fluoride and by ADP-ribosylation of their α subunits by cholera and pertussis toxins.[4] Finally, the catalytic subunit of adenylate cyclase can also be directly stimulated by the diterpene forskolin.[4,8]

Principle

Adenylate cyclase is a membrane-bound enzyme that catalyzes the conversion of ATP to cAMP and pyrophosphate (Fig. 1). The cAMP produced by the cells is a second messenger molecule that acts primarily by activating cAMP-dependent protein kinases before it is itself degraded by phosphodiesterases to AMP (Fig. 1). Thus, as we have seen the activation of adenylate cyclase by receptor–hormone complexes leads to changes in cAMP concentrations, and these changes in the levels of the cyclic nucleotide can be used to measure adenylate cyclase activity. However, it is important to note at this juncture that cAMP levels can also be modulated by receptor-independent mechanisms. For example, the diterpene analog forskolin (Fig. 2) mimics the hormonal activation of adenylate cyclase by binding to the catalytic subunit of the enzyme. This direct activation also results in an increase in cAMP levels. In addition, the methylxanthines caffeine and theophylline (Fig. 2) also increase cAMP levels by inhibiting the enzyme phosphodiesterase (Fig. 1) that degrades cAMP and thus indirectly increases cAMP concentrations.

A variety of methods, mainly radioisotopic, have been used to measure adenylate cyclase activity by measuring changes in the concentration of cAMP. In this chapter, we describe four methods that can be used to measure cyclase activity either in cell membranes and tissue homogenates or in tissue slices and cultured cells. The method that we currently use in

[5] T. T. Armatruda, N. P. Gerard, C. Gerard, and M. I. Simon, *J. Biol. Chem.* **268,** 10139 (1993).
[6] S. J. Myers, L. M. Wong, and I. F. Charo, *J. Biol. Chem.* **270,** 5786 (1995).
[7] A. M. Spiegel, *in* "G Proteins" (A. M. Spiegel, ed.), 1st Ed. R. G. Landes, Austin, Texas, 1994.
[8] K. B. Seamon and J. W. Daly, *Adv. Cycl. Nucl. Prot. Phosphor. Res.* **20,** 1 (1986).

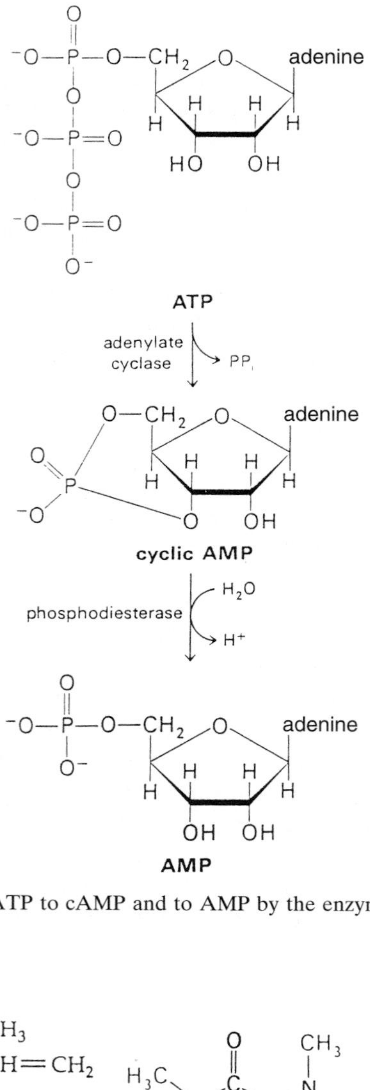

FIG. 1. Conversion of ATP to cAMP and to AMP by the enzymes adenylate cyclase and phosphodiesterase.

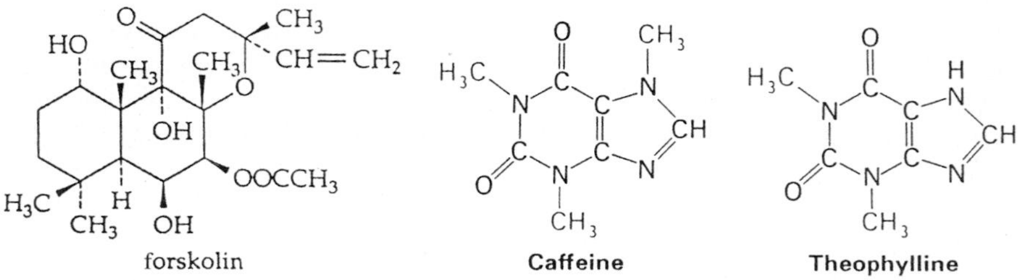

FIG. 2. Structures of the diterpene forskolin, caffeine, and theophylline.

our laboratory depends on the conversion of ^{32}P-labeled ATP to cAMP and is used to measure enzyme activity in cell homogenates or membranes. The second method that we describe can be used to measure cyclase activity in tissue slices and in cultured cells and involves prelabeling intracellular pools with [^{3}H] adenine, which is then converted to [^{3}H]cAMP. Both methods rely on chromatographic techniques to separate the radiolabeled cAMP from the reaction mixture, and a dual-column separation technique developed by Salomon *et al.*[9] that efficiently achieves this separation is described in some detail. The final two methods of determining adenylate cyclase activity are based on the detection of cAMP by radioimmunoassay (RIA) and by enzyme immunoassay (EIA), and brief discussions of some of the methods and procedures currently in use are given.

Adenylate Cyclase Activity: Conversion of [^{32}P] ATP to [^{32}P] cAMP

The principle of the assay depends on the conversion of the substrate, usually [α-^{32}P]ATP, to [^{32}P]cAMP and the subsequent separation of the radioactively labeled cAMP from unconverted substrate and other metabolites by a two-step sequential column chromatography procedure.[9] The first step of the purification involves separation with a Dowex cation-exchange resin, and this is then followed by a second separation step over an aluminum oxide (alumina) column. This procedure offers a number of advantages over existing assays, not the least of which is its increased sensitivity.

The assay is influenced by a variety of considerations. First, it has an absolute requirement for divalent cations (Mg^{2+} or Mn^{2+}), and these need to be present throughout the assay. Second, enzymes such as phosphohydrolases and nucleotidases that degrade ATP and phosphodiesterases that convert cAMP to AMP are present in the membrane preparations and can influence the accurate determination of enzyme activity. To circumvent problems with ATP degradation, an ATP regenerating system comprising creatine kinase and creatine phosphate is routinely added. Problems with degradation of cAMP by phosphodiesterases are overcome by adding phosphodiesterase inhibitors to the assay. A variety of such inhibitors are available, and they include papaverine, IBMX (3-isobutyl-1-methylxanthine), Ro 20-1724 [4-(3-butoxy-4-methoxybenzyl)-2-imidazolidinone], and Rolipram [4-(3-cyclopentyloxy-4-methoxyphenyl)-2-pyrrolidinone]. Any one or more of those reagents can be used to inhibit the activity of the enzymes. Rolipram, in particular, is a fairly effective inhibitor, and we routinely add it to the assay at concentrations of 1 μM. Third, GTP is absolutely required for hormone-induced activation and inhibition of adenylate cyclase and is

[9] Y. Salomon, C. Londos, and M. Rodbell, *Anal. Biochem.* **58,** 541 (1974).

usually present at a final concentration of around 2.5 μM. Alternatively, the nonhydrolyzable GTP analog GTPγS [guanosine 5'-*O*-(3-thiotriphosphate)] can also effectively substitute for GTP in this manner. Finally, the time and temperature of the incubation are important in the assay. We routinely measure cyclase activity at 30° for 10 min and have shown that cAMP formation is fairly linear under these circumstances.

Materials

Where possible we have listed the materials for the assay together with a recommended list of sources. It is hoped that this information will be helpful to investigators trying to set up this assay.

Chemical	Supplier
[α-^{32}P]ATP, 2 mCi/ml	New England Nuclear (Boston, MA)
[^{3}H]cAMP	New England Nuclear
ATP	Sigma (St. Louis, MO)
cAMP	Sigma
Creatine phosphokinase	Sigma
Phosphocreatine	Sigma
Forskolin	Sigma
IBMX	Sigma
Rolipram	Schering AG (Berlin, Germany)
Aluminum oxide	Sigma
Dowex AG50W-X4 resin	Bio-Rad (Richmond, CA)
Polyprep columns	Bio-Rad
Glass columns, size B or size 21	Kontes (Vineland, NJ)

Stock Solutions and Columns

The following stock solutions should be prepared fresh on the day of the assay: 2.5 m*M* ATP, 2.5 m*M* cAMP, 40 mg/ml phosphocreatine, 16 mg/ml creatine phosphokinase, 1 m*M* GTP, 1 *M* Tris-HCl, pH 7.4, 1 *M* $MgCl_2$, 50 m*M* NaF, and 1 m*M* forskolin. Make all the above in 10 m*M* Tris, pH 7.4, except forskolin, which should be made up in dimethyl sulfoxide (DMSO) as a stock solution and then diluted with aqueous buffers prior to use. Also prepare stopping solution: 2% (w/v) Sodium dodecyl sulfate (SDS) containing 40 m*M* ATP and 12.5 m*M* cAMP, pH 7.4.

Make a 50% slurry of Dowex ion-exchange resin in water and pour into a series of polypropylene columns to a height of 3 cm. (Fig. 3). Make a

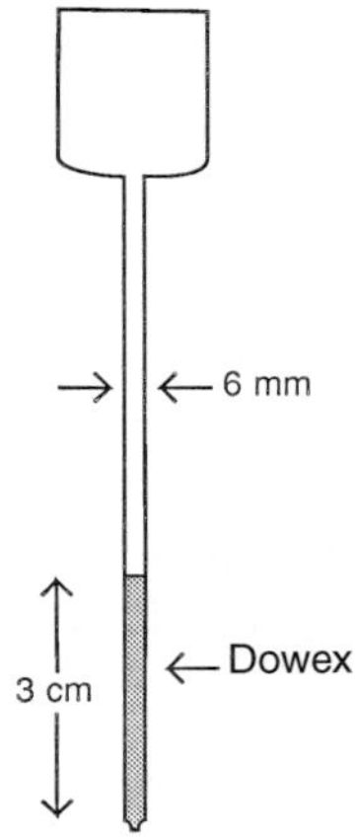

FIG. 3. Dowex ion-exchange columns used in the separation of ATP and cAMP.

similar slurry of aluminum oxide and pour into a series of polypropylene columns to a height of 2 cm.

Procedures

Assay

1. Mix together the following from the stock solutions described above: 100 μl ATP + 100 μl cAMP + 25 μl creatine phosphokinase + 25 μl phosphocreatine + 10 μl $MgCl_2$ + 100 μl Tris-HCl + 25 μl GTP + 20 μl [α-^{32}P]ATP (2 mCi/ml). This constitutes the cocktail.

2. Mix together 40 μl of cocktail + 20 μl ligand or 50 mM NaF or 10 μM forskolin or water + 40 μl of membranes (25 to 50 μg of protein). The assay mixture is incubated at 30° for 10 min and is terminated by the addition of 100 μl of stopping solution followed by 100 μl of [^{3}H]cAMP (add around 20,000 cpm) to measure the recovery of the radiolabeled cAMP. The preparation of membranes from cultured cells stably expressing chemokine receptors is described below under Preparation of Membranes.

3. Radiolabeled cAMP is separated from unreacted radiolabeled ATP and its radiolabeled reaction metabolites by a sequential two-step chromatography procedure described later under Isolation of Cyclic AMP by Two-Step Column Chromatography.

The stimulation of adenylate cyclase is measured in the presence of increasing amounts of agonist, and dose–response curves can be generated.

The inhibition of adenylate cyclase is usually carried out by measuring the ability of the test compound to inhibit the forskolin-induced enzyme activity. In addition, cyclase can be activated by stimulation with NaF as a control to demonstrate that the assay procedure is working correctly.

Preparation of Membranes

1. Cells transfected with chemokine receptors are resuspended to a final concentration of 2×10^7 cells/ml in 50 m*M* Tris-HCl buffer, pH 7.4, containing 5 μg/ml each of leupeptin and aprotinin, 0.1 m*M* phenylmethylsulfonyl fluoride (PMSF), 0.05 m*M* Pefabloc, and 1 m*M* EDTA (lysis buffer).

2. The cells are placed in a nitrogen cavitation chamber under 300 psi of pressure at 4° for 10 min.

3. The lysed cells are then centrifuged at 500 *g* for 20 min at 4°. The cell pellet, which consists of cell debris and nuclei, is discarded, and the supernatant is centrifuged at 48,000 *g* for 30 min. After centrifugation the pellet is removed and resuspended to a final concentration of 1 to 1.5 mg/ml in lysis buffer and stored at −80° until further use. Purified plasma membranes can be prepared from this total membrane preparation if necessary using sucrose density gradient centrifugation.

Preparation of Dowex and Alumina Columns

1. Either Kontes glass columns (with a tulip-shaped upper reservoir) or Bio-Rad polypropylene columns are plugged at their ends with glass wool, and around 3 ml of a suspension of the Dowex AG50W-X4 resin or aluminum oxide is poured into the columns (Fig. 3).

2. The packed resin bed volume should be at a height of 3 cm for the Dowex and 2 cm for the aluminum oxide using columns of 6 mm inner diameter.

3. To activate the Dowex columns wash them with 4 ml of 1 *M* HCl followed by 4 ml of water. To activate the alumina columns wash them twice with 4 ml of 0.1 *M* imidazole hydrochloride buffer pH 7.5. Alternatively, Tris-HCl buffer, pH 7.5, can also be used.

Regeneration of Dowex and Alumina Columns

1. To regenerate the columns rinse the columns with 10 ml of water.

2. Wash the Dowex columns with 10 ml of 1 *M* HCl, followed by 10 ml of water.

3. Wash the alumina columns with 10 ml of 0.1 *M* imidazole hydrochloride buffer, pH 7.5, followed by 10 ml of water.

Isolation of Cyclic AMP by Two-Step Column Chromatography

1. After the assay has been terminated, tip the contents of the tubes over the Dowex columns and then wash out the tubes with 1 ml of water and pour this onto the Dowex columns. Discard the eluates.

2. After the samples have run through the resin wash once more with 1 ml of water, discard the eluates, and place the Dowex columns over the alumina columns (Fig. 4).

3. Add 4 ml of water to the Dowex columns, allow this to drain into the alumina columns, and discard the eluate from the alumina column.

4. Arrange the alumina columns over the scintillation vials, add 4 ml of 0.1 *M* imidazole buffer, and collect the eluate in the vials (Fig. 4). Add 10 ml of scintillation fluid (Hydrofluor) to each of the vials, cap them, shake vigorously, and count using a dual-label ($^3H/^{32}P$) channel setting.

5. A typical separation of radiolabeled ATP and cAMP by sequential chromatography over Dowex and alumina columns is shown in Fig. 5. The procedure needs to be standardized for each new batch of columns used in the assay, and new columns should be calibrated before use. Calibration is important because there could be significant variations in the elution profiles of the radiolabeled nucleotides, and if this is not carefully controlled recoveries could be affected, which would affect the ultimate precision of the assay.

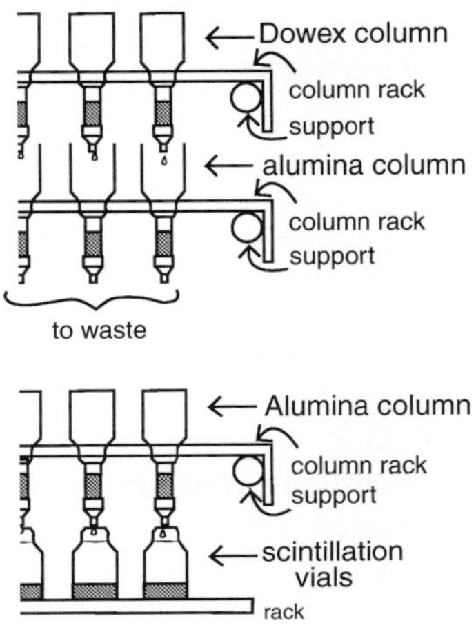

FIG. 4. Two-phase column chromatography setup used in the separation of cAMP from ATP.

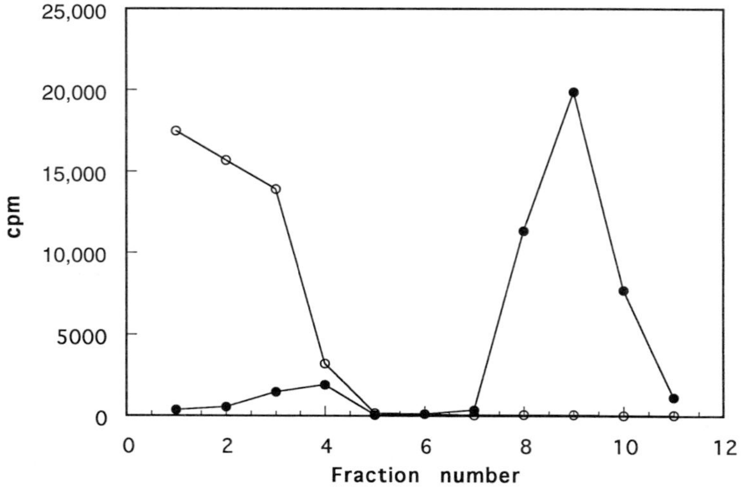

FIG. 5. Typical calibration of the Dowex and alumina columns used in the separation of ^{32}P-labeled ATP from cAMP. A mixture of [^{32}P]ATP (○) and [^{3}H]cAMP (●) standards were separated over the two-phase chromatography system described in the text. Fractions 1 to 6 represent water washes through the Dowex and alumina columns. Fractions 7 through 11 represent imidazole elution from the alumina column.

6. For each assay set up the following additional samples into scintillation vials:

Blank	4 ml of water
^{32}P standard	40 μl of 1:100 diluted cocktail + 4 ml water
^{3}H standard	100 μl of [^{3}H]cAMP + 4 ml water
Zero time	40 μl cocktail + 40 μl water + 20 μl ligand + 4 ml water

Sample Calculation

Data from a typical experiment, in which membranes prepared from cells stably transfected with the C-C chemokine receptor CCR1 were stimulated with forskolin in the presence and absence of increasing amounts of RANTES and MCP-1, is used to illustrate the calculation of adenylate cyclase activity (see Table I).

Assay conditions are as follows. The stock ATP concentration in this experiment is 2.5 mM; because 100 μl of this stock solution is used to prepare a cocktail of total volume 405 μl, the cocktail contains 250 nmol ATP per 405 μl. Because 40 μl of cocktail is used in the assay, this is equivalent to $40/405 \times 250 = 39{,}506$ pmol of ATP in a final incubation mixture of 100 μl. At the end of the incubation after the reaction is termi-

TABLE I
ASSAY TO DETERMINE ADENYLATE CYCLASE ACTIVITY: CONVERSION OF [^{32}P]ATP TO [^{32}P]cAMP

Sample	[^{32}P]cAMP formed (dpm)	[^{3}H]cAMP recovery (dpm)	[^{32}P]cAMP corrected (pmol/mg/min)
Water blank	52	14,712	—
Zero blank	45	14,835	—
Basal	1,782	14,442	55
Forskolin (10 μM)	9,692	14,820	295
Forskolin (10 μM) + RANTES (1 nM)	3,879	14,398	121
Forskolin (10 μM) + MCP-1 (1 nM)	9,357	14,133	299
Forskolin (10 μM) + MCP-1 (1 μM)	6,874	14,556	213

nated 18,980 dpm of [^{3}H]cAMP is added to the sample to measure the recovery following the two-step chromatography. The incubation time of the experiment is 10 min, and the protein concentration is 0.05 mg. The amount of [α-^{32}P]ATP added is 3.3×10^6 dpm.

The results are expressed as picomoles of cAMP per milligram protein per minute (pmol cAMP/mg/min) and are calculated as follows:

$$\text{pmol cAMP/mg/min} = \frac{\text{pmol ATP in sample} \times \text{dpm [}^{3}\text{H]cAMP added}}{\text{dpm }^{32}\text{P added} \times \text{time} \times \text{mg protein}} \times \frac{\text{dpm[}^{32}\text{P]cAMP formed} - \text{dpm zero}}{\text{dpm [}^{3}\text{H]cAMP recovered}}$$

Thus, basal adenylate cyclase activity expressed as pmol/mg/min is calculated as follows, using additional data from Table I:

$$\text{Basal activity} = \frac{39{,}506 \times 18{,}980}{3.3 \times 10^6 \times 10 \times 0.05} \times \frac{1782 - 45}{14{,}442} = 55 \text{ pmol/mg/min}$$

Samples are usually assayed in duplicate, and this, together with the addition of the [^{3}H]cAMP internal standard to monitor the recovery after the chromatography steps, ensures that this is a very precise and reproducible assay. Recovery of the assay is of the order of 70% or greater. The design of the column racks, which are made of Plexiglass, ensures that the Dowex columns fit precisely over the alumina columns so that the eluate from the former drips directly onto the latter (Fig. 4). The Plexiglass tube racks have been similarly configured to fit over standard 20-ml scintillation vials, which are arranged in a 2 × 10 array in tube racks (Fig. 4). The washing of the columns is largely automated by the use of large reservoirs of each of the wash solutions, which can be rapidly delivered to the columns by the use of an automatic syringe system. The columns can be generated and reused many times; typically, we replace the aluminum oxide columns after 30 to 40 uses.

Adenylate Cyclase Activity: Conversion of [^{3}H]Adenine to [^{3}H] cAMP

In the following method, cultured cells or tissue slices are incubated with [^{3}H] adenine, which effectively labels the adenine nucleotide pool of the cell. After washing away any unincorporated radiolabeled adenine, the cells are then stimulated with agonist that converts the radiolabeled ATP to [^{3}H]cAMP. The incubation is then terminated by the addition of trichloroacetic acid, and, after centrifugation of the precipitated cellular extracts, clarified supernatants are subjected to dual-column chromatography over Dowex and alumina, as described previously, to determine the amount of radioactivity incorporated into cAMP.

Prelabeling of Cells

1. Human embryonic kidney cells transfected with CCR1 receptors are grown in 24-well tissue culture plates until confluent and then labeled overnight with 2 μCi/ml of [^{3}H]adenine (10–30 μCi/ml) in Ham's F12/low glucose Dulbecco's modified Eagle's medium (GIBCO, Grand Island, NY) containing 10% fetal calf serum.

2. The cells are then washed in serum-free medium containing 0.1% bovine serum albumin (BSA), 10 μM Rolipram, and 1 mM IBMX. After washing the cells are resuspended in the same medium containing either chemokine alone, forskolin (10 mM) alone, or chemokine plus forskolin for 10 min at 30°.

3. At the end of the incubation the medium is aspirated and the reaction terminated by the addition of 1 ml of ice-cold 5% trichloroacetic acid containing 1 mM cAMP and 1 mM ATP. Addition of the trichloroacetic acid precipitates the proteins and essentially terminates the conversion of ATP to cAMP.

4. Following incubation at 4° for 30 min the acidified extracts are centrifuged (15,000 g for 5 min at 4°), and the clarified extracts are removed and neutralized with base.

5. Around 10,000 cpm of [^{14}C]AMP is added as a recovery marker, and the radiolabeled nucelotide pools are separated by the two-step se-

quential chromatography procedure over Dowex and alumina described above.

6. After the eluates are collected into scintillation vials, 10 ml of scintillant is added (Hydrofluor). The vials are shaken vigorously and then counted in a scintillation counter using a dual-label $^3H/^{14}C$ program.

7. For each assay set up the following additional samples into scintillation vials:

Blank	4 ml water
^{14}C standard	50 μl of [^{14}C]cAMP + 4 ml water
Total ^3H nucleotide	50 μl taken from 1 ml of trichloroacetic acid supernatant + 4 ml water

Sample Calculation

Data from a typical experiment, in which cells stably transfected with the CCR1 receptor are stimulated with forskolin in the presence and absence of increasing amounts of RANTES and MCP-1, are used to illustrate the calculation of adenylate cyclase activity (Table II). The amount of cAMP generated by addition of forskolin is taken as 100%, and the ability of the chemokines to inhibit this process is expressed as a percentage of the forskolin effect.

1. First calculate the total [^3H]ATP present in the cells, that is, disintegrations per minute (dpm) present in a 50-μl aliquot taken from the 1 ml of trichloroacetic acid supernatant multiplied by 20.

2. Then calculate the recovery of cAMP from each column. This is estimated from the amount of [^{14}C]cAMP recovered divided by the amount of [^{14}C]cAMP added.

TABLE II

ASSAY TO DETERMINE ADENYLATE CYCLASE ACTIVITY: CONVERSION OF [^3H]ADENINE TO [^3H]cAMP

Sample	[^3H]cAMP (dpm)	[^{14}C]cAMP (dpm)	[^3H]cAMP corrected for ^{14}C spillover (dpm)[a]	[^3H]cAMP corrected for recovery (dpm)	Total [^3H] ATP (dpm)	Conversion (%)
Basal	2,805	6,318	2,174	3,450	890,150	0.38
Forskolin	14,058	6,852	13,373	19,666	898,635	2.18
Forskolin + 1 n*M* RANTES	5,692	6,743	5,018	7,489	896,512	0.83

[a] 10,000 dpm of [^{14}C]cAMP was added.

3. The cAMP pool for each sample is then normalized to its own ATP pool, and data are expressed as percentage conversion of ATP to cAMP as follows:

$$\text{Percentage conversion} = \frac{[^3\text{H}]\text{cAMP} \times 100}{[^3\text{H}]\text{ATP}}$$

Thus for cells incubated in the absence of forskolin (basal) the percentage conversion of ATP to cAMP is

$$\text{Basal activity} = \frac{3450 \times 100}{890{,}150} = 0.38$$

A typical dose–response curve generated over a range of RANTES and MCP-1 concentrations is given in Fig. 6 along with results for other chemokines.

Adenylate Cyclase Activity: cAMP Production Measured by Radioimmunoassay or Enzyme Immunoassay

Cultured cells are incubated with agonist and the cAMP produced by the cells over a timed interval is measured by RIA or by enzyme EIA. A variety of kits are available from Amersham and New England Nuclear.

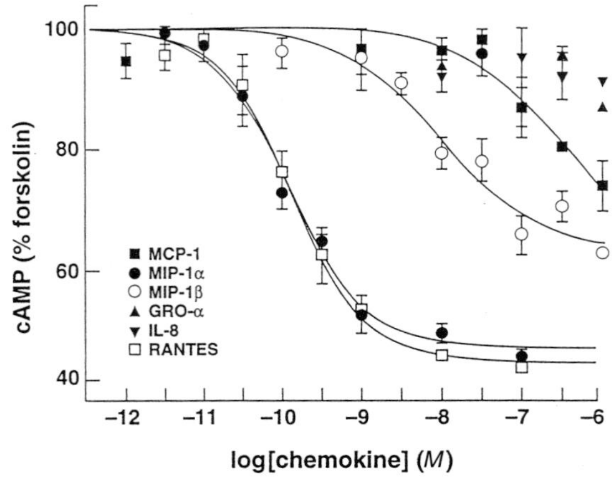

FIG. 6. Chemokine-mediated inhibition of adenylate cyclase by the CCR1 receptor. The inhibition of adenylate cyclase by chemokines acting through the CCR1 receptor expressed in human embryonic kidney cells was measured as described in the text. The data are expressed as the percentage inhibition of forskolin-induced cAMP by increasing concentrations of various chemokines. (Reproduced with permission from Myers *et al.*[6])

Basically the principle of the RIA is a competition between a radioactive and a nonradioactive antigen for a fixed number of antibody binding sites. The kit contains labeled antigen, usually ^{125}I-labeled cAMP, which competes for binding to the specific cAMP antibodies with the unlabeled cAMP that has been extracted from tissues or cells. The labeled antibody–antigen complex is separated from unbound label by immunoprecipitation with a second antibody.

The principle behind the EIA is very similar. A 96-well microtiter plate coated with anti-rabbit antibodies is used to capture rabbit antibodies to cAMP. To this antibody sandwich biological samples containing unlabeled cAMP and a fixed quantity of peroxidase-labeled cAMP are added. At the end of the incubation the substrate, which is a complex of tetramethylbenzidine and hydrogen peroxide, is added and the resultant blue color read at 630 nm in a microtiter plate spectrophotometer. The assay depends on a competition between the unlabeled cAMP in the biological samples and a fixed quantity of peroxidase-labeled cAMP for a limited number of binding sites on the rabbit anti-cAMP. The amount of cAMP in the biological samples is determined from a cAMP standard curve that is set up in parallel.

No further details of the principles behind either of these two methods are given because commercially available kits can be readily purchased from a number of different vendors. An assay procedure applicable to cultured cells is given below.

1. Human embryonic kidney cells transfected with CCR1 receptors are grown in 24-well tissue culture plates in Ham's F12/low glucose Dulbecco's modified Eagle's medium (GIBCO) containing 10% fetal calf serum.

2. After confluence the medium is replaced, and the cells are then incubated in serum-free medium containing 0.1% BSA, 10 μM Rolipram, and 1 mM IBMX. Cells are then stimulated by the addition of forskolin (10 μM) alone or chemokine plus forskolin for 10 min at 30°.

3. Reactions are stopped by the addition of boiling water or by the addition of ice-cold 65% (v/v) ethanol. In either case samples are clarified by centrifugation (15,000 *g* for 5 min) to remove precipitated proteins. If ethanol extraction is carried out, the ethanol must first be removed by drying the samples in a stream of nitrogen. The amount of cAMP produced is estimated either by RIA or by EIA.

[21] Analysis of Signal Transduction Following Lymphocyte Activation by Chemokines

By KEVIN B. BACON

Introduction

The chemokines are a superfamily of small molecular mass proteins related in primary structure by the conservation of a four-cysteine motif.[1,2] Presently, the superfamily is subdivided into three subgroups: the C-X-C family, typified by members such as interleukin-8 (IL-8), growth-related oncogene (GRO-α), and neutrophil-activating peptide-2 (NAP-2), retains all four N-terminal cysteines with the first two being separated by a single intervening amino acid. The C-C family, typified by members such as RANTES, macrophage inflammatory protein-1α (MIP-1α), MIP-1β, MIP-1γ, and macrophage chemotactic protein-1 (MCP-1), retains all four cysteines without any intervening amino acid between the first two cysteines. The third subfamily, represented by the unique member lymphotactin, has lost two of the cysteines (Cys^1 and Cys^3) but displays significant homology to the MIP-1 proteins.[3]

Biological classification of these molecules originally attempted an exclusive and reductionist cataloging, by characterizing the C-X-C members as neutrophil-activating proteins and the C-C members as monocyte-activating proteins. These classifications have not withstood the test of time, as very early analyses of IL-8 revealed a very potent T-lymphocyte activating capacity.[4–6] Although this is, even today, the center of controversy, the effects of IL-8 and certain other C-X-C chemokines on lymphocytes have been repeatedly observed,[2,7–9] as have its effects on B cells, eosinophils, endothe-

[1] M. Baggiolini, B. Dewald, and B. Moser, *Adv. Immunol.* **55,** 97 (1994).
[2] K. B. Bacon and T. J. Schall, *Int. Arch. Allergy Immunol.* **109,** 97 (1996).
[3] G. S. Kelner, J. Kennedy, K. B. Bacon, S. Kleyensteuber, D. A. Largaespada, N. A. Jenkins, N. G. Copeland, J. F. Bazan, K. W. Moore, T. J. Schall, and A. Zlotnik, *Science* **266,** 1395 (1994).
[4] C. G. Larsen, A. O. Anderson, E. Apella, J. J. Oppenheim, and K. Matsushima, *Science* **243,** 1464 (1989).
[5] K. B. Bacon, J. Westwick, and R. D. R. Camp, *Biochem. Biophys. Res. Commun.* **165,** 349 (1989).
[6] K. B. Bacon and R. D. R. Camp, *Biochem. Biophys. Res. Commun.* **169,** 1099 (1990).
[7] S. Qin, G. LaRosa, J. J. Campbell, H. Smith-Heath, N. Kassam, X. Shi, L. Zeng, E. C. Butcher, and C. R. Mackay, *Eur. J. Immunol.* **26,** 640 (1996).

0076-6879/97 $25.00

lial cells, and keratinocytes.[1,2] Equally important was the evidence demonstrating that certain (almost all) C-C family members were highly effective lymphocyte agonists[2] as well as potent stimulators of eosinophils and basophils.[2,10]

These observations, coupled with the (to date) exclusively T and natural killer (NK) cell-active lymphotactin (C-class), have raised important questions concerning the mode(s) of action of chemokines on the lymphocyte. Chemokines are known to operate through the common denominator of specific yet related seven-transmembrane (7-TM) G-protein-coupled receptors.[1,10] In a manner analogous to other agonists known to couple 7-TM receptors, ligand binding is known to transduce signals via heterotrimeric G proteins. The $\beta\gamma$ subunits of the G proteins are believed to activate phosphoinositide-specific phospholipase C, generating the metabolites inositol 1,4,5-trisphosphate (IP_3) and diacylglycerol (DAG) through hydrolysis of phosphoinositide bisphosphate.[11] The IP_3 binds specific receptors in the plasma membrane and on the endoplasmic reticulum (ER), inducing release of intracellular calcium, which in conjunction with DAG activates protein kinase C (PKC). The PKC then potentially activates a cascade of signal transduction events both intracytoplasmically and within the nucleus.

It is precisely at this "core" mechanism where most investigators have concentrated their experimentation, and its simplicity has proven extremely effective in facilitating receptor cloning and basic analyses of transfected receptors, as well as having the obvious advantages for screen development. However, application of these pathways and specific inhibitors thereof during in-depth investigation through biological assay has not always delivered the complete picture simply because the biological readout was highly specific and limited, not only in the context of general physiology, but also in the scope and limitations of the assay itself. It is becoming evident that, in a manner similar to mechanisms outlined in the literature for classic 7-TM agonists, there is a host of peripheral, but not necessarily any less important, signal transduction pathways activated by chemokines in lymphocytes. These include phosphorylation/dephosphorylation of major cytoplasmic proteins and activation of small GTP-binding proteins (Rho, Rac, cdc42Hs), mitogen-activated protein kinase (MAPK), and phospholipase D.[11–13] Because aspects of signal transduction including the activation of

---

[8] L. F. Santamaria Babi, B. Moser, M. T. Perez Soler, R. Moser, P. Loetscher, B. Villiger, K. Blaser, and C. Hauser, *Eur. J. Immunol.* **26,** 2056 (1996).

[9] P. Loetscher, M. Seitz, M. Baggiolini, and B. Moser, *J. Exp. Med.* **184,** 569 (1996).

[10] T. J. Schall and K. B. Bacon, *Curr. Opin. Immunol.* **6,** 865 (1994).

[11] G. M. Bokoch, *Blood* **86,** 1649 (1995).

heterotrimeric and low molecular weight G proteins,[13a] and cAMP pathways[13b] are covered in other chapters in this volume, along with calcium flux and phosphoinositide turnover,[13c] emphasis here is not placed on those methods.

Chemokine Activation of Lymphocytes

This section has been included to detail my experiences with lymphocytes and chemokines, and in doing so emphasize the fact that chemokines (C-X-C, C-C, and C) do activate T and B lymphocytes and NK cells in chemotaxis assays and beyond. Numerous investigators have acquired volumes of data linking the chemokines with the activation of T-lymphocyte migration and have attempted to show that the criteria for analyses have consistently been dependent on one major aspect of bioassay execution, namely, the preparation of cells. Every bioassay, irrespective of the cell type under analysis, is subject to intrinsic donor variability. In addition to this highly important characteristic is the inconsistency introduced during handling and preparation.

With respect to peripheral blood lymphocytes (PBL), mouse lymphocytes, human bone marrow cells, and thymocytes (murine and human), the best conditions for preparation and maintenance of signal transducing capacity are those performed rapidly and at 4°. Although there are schools of thought that maintain that preparation at room temperature is vital, within the time frame of cell preparation (Ficoll, wash three times, antibody labeling, and Dynal bead negative selection) the cells rapidly down-regulate and/or lose the capacity to respond to chemokines in both calcium flux assay and phosphorylation. Although this does not appear to impinge on the chemotactic response of lymphocytes to the chemokines, it makes it next to impossible to support chemotaxis data with assays of more direct ligand–receptor interaction (signal transduction).

This necessity for 4° preparation prior to immediate use is obviously likely to conflict with numerous phosphorylation-related assays that may require periods of culture in serum-free medium at 37° prior to cell activation and lysis. If the response is robust enough, however, this has not been a problem. On the whole, using freshly isolated PBL or thymocytes,

[12] C. Knall, S. Young, J. A. Nick, A. M. Buhl, G. S. Worthen, and G. L. Johnson, *J. Biol. Chem.* **271,** 2832 (1996).
[13] C. Laudana, J. J. Campbell, and E. C. Butcher, *Science* **271,** 981 (1996).
[13a] S. K. Beckner, *Methods Enzymol.* **288,** 309 (1997).
[13b] R. Horuk, *Methods Enzymol.* **288,** 326 (1997).
[13c] K. B. Bacon, *Methods Enzymol.* **288,** 362 (1997).

preparation of cells in serum-free medium or HBSS has facilitated the successful visualization of phosphorylated proteins.

The use of T-cell clones has proved to be an extremely useful alternative to PBL. The relatively straightforward generation of clones, of either CD4 or CD8 or Th0, Th1, or Th2 phenotypes, has allowed investigation using tools free of donor variability. The clones we have used have afforded a vast increase in the amount of data we have amassed on chemokine-induced signaling.[13c] Once again, however, the advantage is somewhat fortuitous because one needs to thoroughly screen numerous clones for responder populations (in a manner analogous to normal T cells). Once a responder population has been identified, it is always advisable to freeze early passage cells for future parental lines, as we have found that several clones lost their responsiveness with prolonged culture. The advantage with using clones, apart from the consistency of response, is obviously the ability to have cells ready without the time-consuming preparation. Although appearing as a disadvantage, the lengthy starvation periods in serum-free medium prior to analysis of phosphorylation is a useful quirk of the system. Because clones are exposed to antigen/mitogen and IL-2, they will necessarily be in an activated state, and this will include many of their phosphorylatable proteins; thus, the prolonged starvation is required. In addition, the signal-to-noise ratio is far greater in clones than for PBL or thymocytes.

Protein Kinase Activation

The activation of protein kinases represents a major component of the signal transduction machinery of most cells in response to numerous agonists. The resulting protein phosphorylation(s) is undoubtedly one of the major regulatory mechanisms for cellular activation, either via transduction/amplification of signals from exogenous stimuli or in the oncogenic transformation of cells via mutated or virally encoded forms of these kinase enzymes. The most extensively studied protein kinases are those that catalyze the phosphorylation of serine/threonine and tyrosine residues (including the dual-specificity kinases), although there exists a significant amount of literature concerning those affecting cysteine, aspartate/glutamate, and histidine, arginine, or lysine residues. The serine/threonine protein kinases can be roughly classified into cyclic nucleotide-dependent (cAMP, cGMP) kinases; calcium-dependent kinases (PKC, phosphorylase kinase, MLCK, calmodulin-dependent kinases); and messenger-independent kinases [β-adrenergic receptor kinases (β-ARK)], $p70^{S6}$ kinase, glycogen synthase kinase-3, and casein kinases I and II (CKI and CKII).

Ligands that elevate cAMP/cGMP, and activate the consequent signal tranduction pathways [e.g., prostaglandin E_2 (PGE_2)], have negative regulatory effects on lymphocytes (K. B. Bacon, unpublished). Although this is by no means a hard-and-fast rule, I have not performed assays rigorously investigating cAMP or cGMP in response to chemokines in lymphocytes and thus will refrain from further discussion; suffice it to say that in laboratories not dedicated to biochemical analysis of these kinases and pathways, there are numerous assay kits available from Amersham (Arlington Heights, IL). These kits (available in the Biotrak series) have the potential to measure metabolites in the picomolar range and thus may be useful for purposes of preliminary characterization in the nonspecialist laboratories.

However, the activation of PKC is a necessary consequence of G-protein-coupled 7-TM receptor ligation by chemokines, and, by extrapolation, so may be the activation of G-protein-coupled receptor kinases following ligand–receptor ligation (perhaps in a manner analogous to the β-ARK enzymes). Whether the activation of PKC determines the subsequent activation of other kinases and phospholipases should be determined in the respective assays. In addition, the activation of $p70^{S6}$ kinase has been suggested as a consequence of phosphatidylinositol 3-kinase (PI-3K) activation. As Turner *et al.*[14] have demonstrated the RANTES-induced activation of PI-3K in T lymphocytes, it is likely that the activation of serine/threonine kinases in lymphocytes following chemokine stimulation plays an important role in the activation mechanisms.

The tyrosine kinases (TKs) can be roughly subdivided into RTKs (e.g., growth factor receptor kinases), nonreceptor TKs (including the plasma membrane-associated kinases; *src* and *syk* family members), and cytoplasmic TKs (e.g., *abl*). Numerous methodological texts detail procedures for accurate measurement of these kinases; thus, the reader should also refer to these standards to appreciate the general principles involved.[15,16]

Assay of Protein Kinase C (Serine/Threonine Kinase)

Protein kinase C exists in several isoforms subdivided into three groups[17]; the classic or conventional enzymes (cPKC; α, βI, βII, γ) all require calcium as a coactivator, along with DAG and phosphatidylserine (PS); the new enzymes (nPKC; δ, ε, η, θ, μ) all require DAG and PS but not calcium; and the atypical enzymes (aPCK; ζ, λ) all activated (at least

[14] L. Turner, S. G. Ward, and J. Westwick, *J. Immunol.* **155,** 2437 (1995).

[15] D. G. Hardie, ed., "Protein Phosphorylation—A Practical Approach." Oxford Univ. Press, London, 1993.

[16] J. E. Coligan, A. M. Kruisbeek, D. H. Margulies, E. M. Shevach, and W. Strober, eds., "Current Protocols in Immunology," Chap. 11. Wiley, New York, 1994.

[17] Y. Nishizuka, *FASEB J.* **9,** 484 (1995).

the ζ form) by phosphatidylinositol trisphosphate (PIP_3), a product of the PI-3K enzyme, but not by DAG or phorbol esters.

The most straightforward assay is to measure histone H1 phosphorylation, although myelin basic protein (MBP) has been used as a substrate as well. Prior to assay, dissolve phosphatidylserine (PS; 1 mg/ml) and diolein (100 mg/ml) in chloroform, evaporate under a stream of nitrogen, and dissolve by sonication into 20 m*M* Tris-HCl, pH 7.5, for 5 min. Set up enough tubes (siliconized) for the specific assay and add 20 μl of assay mix [25 m*M* Tris-HCl, pH 7.5, 6.25 m*M* $MgCl_2$, 0.125 m*M* $CaCl_2$, 10 μ*M* [γ-^{32}P]ATP (1000–3000 Ci/mmol); 10 μg/ml PS and 1 μg/ml diolein; and 250 μg/ml histone H1]. Lyse cells in lysis buffer containing protease and phosphatase inhibitors but no EGTA or EDTA. Phosphatase inhibitors prevent the action of serine/threonine proteases, whereas the omission of divalent cation ion chelators is necessary due to the calcium dependence of PKC. Add 50 ml of supernatant, vortex, and incubate for 5–30 min at 30°. (The time course of assay must be established by comparison against a known concentration and specific activity of purified enzyme, or standard assay system.) Stagger addition of sample according to the size of the assay.

Following incubation, add 3 ml of 25% (w/v) ice-cold trichloroacetic acid (TCA) [or 1.5 ml of 50% (w/v) TCA, or 1 ml of 75% (w/v) TCA] to terminate the reaction and precipitate the phosphorylated protein. Leave samples on ice for 10–15 min. The precipitate can be collected by passing the suspension through a nitrocellulose filter using a suction device and three washes of 25% (w/v) TCA. Alternatively, many investigators simply centrifuge the suspension (2500 *g*; 15 min), remove the supernatant, and repeat the process three times by washing with TCA. Dried filters can then be transferred directly to scintillation vials, scintillation fluid added, and radioactivity counted. Using the alternative method, the precipitate must be redissolved in phosphate-buffered saline (PBS) and transferred for counting. This manipulation could feasibly cause errors in the final counts achieved; thus, the filter method is recommended.

Pharmacological Modification of Protein Kinase C Activity

To analyze the role(s) of PKC in lymphocyte function, many investigators have adopted the approach of pharmacological agonism or antagonism by use of specific (and sometimes nonspecific) drugs. A number of compounds, most notably the phorbol esters, have been used as direct PKC agonists, bypassing the need for receptor-mediated signals. 12-*O*-Tetradecanoylphorbol 13-acetate (TPA/PMA) and phorbol 12,13-dibutyrate have frequently been used as direct PKC activators in many leukocyte and nonleukocytic cells.[15] 1-Oleoyl-2-acetyl-*sn*-glycerol and 1,2-*sn*-dioctanoylglycerol, both DAG analogs, have also been shown to have some effect on PKC-dependent mechanisms. The (−)-enantiomer of indolactam V and the analog (−)-7-octylindolactam V have been shown to be highly effective PKC agonists (in the nanomolar range.)[18,19] These compounds essentially form the backbone structure of the naturally occurring teleocidins (from *Streptomyces* sp.) and are known to have PKC agonistic activity. In addition, the bryozoan *Bubula neritina* contains natural PKC-activating compounds (known as bryostatins 1 through 17).[20] These compounds, however, display dose-dependent agonistic, antagonistic, and partial agonistic effects on PKC, with the agonistic, antagonistic, and partial agonist effects being identified in assays using phorbol esters. Bryostatin 1 binds PKC in the low picomolar range (2.5 pM), indicating an extremely high affinity and hence potency of these compounds in modulating PKC activity.

With respect to PKC inhibitors, members of the synthetic naphthalenesulfonamide family H7 (K_i 6 μM) have been used for inhibition of chemokine effects.[6,21] Although these compounds have been reported as specific C-kinase inhibitors in many publications in the literature, their activity on A-kinase (cAMP-, cGMP-dependent protein kinase; K_i 3 μM), MLCK (K_i 90 μM), and PKG (K_i 6 μM) is now well established. The compound W-7 (a calmodulin antagonist) is often used as an indirect inhibitor of C-kinase activity. Significant inhibition of protein kinase C has been achieved using staurosporine (K_i in the low nanomolar range); however, like the isoquinoline compounds, this antibiotic affects many other serine/threonine and tyrosine kinases. The fungal metabolite calphostin C (K_i 1 mM) is one of the most selective inhibitors, affecting the α, β, and γ isozymes of PKC but not A-kinase or tyrosine kinases.[22] The potency of calphostin C does, however, appear to be light sensitive.[23] High selectivity for PKC comes from the compound chelerythrine chloride.[24] This naturally occurring alkaloid, although poorly soluble in aqueous solution, is noncompetitive with ATP and demonstrates relatively high potency (IC_{50} 600 nM), inhibiting PKC by binding to the catalytic domain (irrespective of the

[18] H. Fujiki and T. Suimura, *Adv. Cancer Res.* **49,** 223 (1987).

[19] J. Heikkila and K. Åkerman, *Biochem. Biophys. Res. Commun.* **162,** 1207 (1989).

[20] G. R. Pettit, *Prog. Chem. Org. Nat. Prod.* **57,** 153 (1991).

[21] K. B. Bacon, L. Flores-Romo, P. F. Life, D. D. Taub, B. A. Premack, S. J. Arkinstall, T. N. C. Wells, T. J. Schall, and C. A. Power, *J. Immunol.* **154,** 3654 (1995).

[22] E. Kobayashi, H. Nakano, M. Morimoto, and T. Tamaoki, *Biochem. Biophys. Res. Commun.* **159,** 548 (1989).

[23] R. F. Bruns, F. D. Miller, R. L. Merriman, J. J. Howbert, W. F. Heath, E. Kobayashi, I. Takahashi, T. Tamaoki, and H. Nakano, *Biochem. Biophys. Res. Commun.* **176,** 288 (1991).

[24] J. M. Herbert, J. M. Augereau, J. Gleye, and J. P. Maffrand, *Biochem. Biophys. Res. Commun.* **172,** 993 (1990).

regulatory domain). The bisindolylmaleimide compounds have been manufactured (based on the staursporine structure) as highly specific, cell-permeant PKC inhibitors that are competitive inhibitors of the ATP binding site (K_i 10 nM).[25,26]

Because the activation of PKC and its subsequent amplification of both nuclear signaling pathways and tyrosine kinase pathways are such integral parts of the biological reactions of cells in response to chemokines, we have found that accurate analysis of PKC activity has been critically dependent on the type of inhibitor used. Initially, one needs to ensure high lipophilicity enabling rapid uptake into cells for immediate binding either to the catalytic or DAG-binding sites of PKC. The shorter the time period required for preincubation, the less likely is the potential for nonspecific inhibition of both serine/threonine and tyrosine kinases. As with any pharmacological antagonist, the lower the toxicity and higher the potency, the greater is the applicability to multiple bioassay formats. We have used bisindolylmaleimide I {GF 109203X; 2-[1-(3-dimethylaminopropyl)-1*H*-indol-3-yl]-3-(1*H*-indol-3-yl)maleimide} at submicromolar concentrations and obtained greater than 50% inhibition of particular readouts, which were not comparable with use of H7 or calphostin C at similar concentrations.

Assay of Protein Tyrosine Kinase

Aside from specific assays (see later), general considerations in the analysis of protein tyrosine kinase activity are outlined here. As has been the case so far, chemokine stimulation of T lymphocytes has shown some protein phosphorylation as measured using whole cell lysates and antiphosphotyrosine antibody (4G10).[27] The studies have been far from exhaustive, and thus the identification of novel tyrosine kinase activity has so far not been forthcoming. As such, one has had to rely on characterization of chemokine effects on known kinases. Because availability of kinase substrates in high enough concentration and purity is usually not an option, one has had to resort to the use of synthetic amino acid polymers [e.g., MBP or poly(Glu-Tyr)]. Whereas these do not approach the physiological substrate (as many tyrosine kinases rely on secondary/tertiary protein structure as well as tyrosine residues to define the site of phosphorylation), these synthetic peptides are nonetheless extremely useful.

As a preliminary screen for the action of protein tyrosine kinase activation in response to chemokines, one can simply use specific inhibitors to

[25] D. Toullec, P. Pianetti, H. Coste, P. Bellevergue, P. T. Grand, M. Ajakane, V. Baudet, P. Boissin, E. Boursier, and F. Loriolle, *J. Biol. Chem.* **266,** 15771 (1991).

[26] Z. Kiss, H. Phillips, and W. H. Anderson, *Biochim. Biophys. Acta* **1265,** 93 (1995).

[27] K. B. Bacon, B. A. Premack, P. Gardner, and T. J. Schall, *Science* **269,** 1727 (1995).

pharmacologically modulate the suspected kinase activity. These assays have the advantage of giving a quick indication of potential kinase involvement and are facilitated by the large number of commercially available inhibitors.

Protein Tyrosine Kinase/Phosphatase Inhibitors

Prior to analysis of specific kinases, the pharmacological inhibition of tyrosine kinases in whole cells and lysates may be used to determine protein tyrosine kinase (PTK) activity. The erbstatin analog 2,5-dihydroxymethylcinnamate is highly cell permeant and has a K_i of 3.4 μM for inhibition of the epidermal growth factor (EGF) RTK catalytic site.[28] Genistein, although a competitive inhibitor of ATP binding, does not affect all TKs and has limited cell permeability.[29] Herbimycin A, a cell-permeant inhibitor of *src* tyrosine kinases, has an IC_{50} of approximately 900 nM.[30,31] Lavendustin A is also cell permeable (IC_{50} of 500 nM) and has little effect on PKC or protein kinase A (PKA).[32,33] The tyrphostins have IC_{50} values in the high nanomolar to low micromolar range and vary in their cell permeability. Bistyrphostin has an IC_{50} value in the range of 400 nM, and we have used this reagent effectively to inhibit tyrosine kinase-dependent calcium flux in T-cell clones (K. B. Bacon, unpublished, 1995).[34] Depending on the assay in question, we have preferred herbimycin A as a PTK inhibitor because its effects have been greater than those of the tyrphostins. In addition, there appear to be certain discrepancies concerning the length of incubation of cells with these inhibitors. We have opted for overnight incubation with doses near the IC_{50} of herbimycin A as these are less toxic and cells can remain in culture during the coincubation. We have used the less permeant inhibitors (genistein, tyrphostins) with permeabilized cells and thus have been able to limit the time required for permeabilization without destruction of the cells (30 min maximum).

Although much less experimentation has been performed using phosphatase inhibitors in the investigation of chemokine activity on lymphocytes, the identification and cloning of novel protein tyrosine phosphatases

[28] K. Umezawa, T. Hori, H. Tajima, M. Imoto, K. Isshiki, and T. Takeuchi, *FEBS Lett.* **260,** 198 (1990).

[29] K. Migita, K. Eguchi, Y. Kawabe, A. Mizokami, T. Tsukada, and S. Nagataki, *J. Immunol.* **153,** 3457 (1994).

[30] S. L. Weinstein, M. R. Gold, and F. E. De, *Proc. Natl. Acad. Sci. U.S.A.* **88,** 4148 (1991).

[31] D. J. Park, H. K. Min, and S. G. Rhee, *J. Biol. Chem.* **266,** 24237 (1991).

[32] J. M. Bishop, *Annu. Rev. Biochem.* **52,** 307 (1983).

[33] C.-Y. J. Hsu, P. E. Persons, A. P. Spada, R. A. Bednar, A. Levitski, and A. Zilberstein, *J. Biol. Chem.* **266,** 21105 (1991).

[34] A. Levitski and C. Gilon, *Trends Pharmacol. Sci.* **12,** 171 (1991).

indicate that the number of members of this family will approach that associated with the PTK family. Included in this class are membrane-spanning proteins (e.g., CD45 and LAR), as well as cytoplasmic phosphatases (e.g., PTP1B and PTP1C).[35] The complex issue of protein tyrosine phosphatase activation and measurement is not discussed here, but the reader is referred to the outstanding text by Tonks[36] in which the intricacies of substrate choice and preparation for *in vitro* assay are clearly and concisely described. A simple pharmacological approach in standard bioassays using a combination of inhibitors of phosphatase action, namely, vanadate (1 μM) and hydrogen peroxide (100 μM), has proved useful. I have primarily used these compounds in the analysis of phospholipase D (PLD) activity in T lymphocytes, demonstrating the dependence under certain conditions, on PTK activity, in response to chemokines.

Serine/Threoinine-Specific Protein Phosphatase Inhibitors

Serine/threonine-specific protein phosphatase (PP) inhibitors include the lipophilic compound okadaic acid, which has effects on protein phosphatases with relative potencies in the order PP2A > PP1 = PP2B.[37] Calyculin A is equipotent for PP2A and PP1, whereas PP2A activities have been selectively inhibited by cantharidin/cantharidic acids, endothall, and endothall thioanhydride.[38–40] For a simple bioassay okadaic acid can be added at a final concentration of up to 1 μM for 15 min prior to assay. Alternatively, one can analyze the phosphorylation status of cells prelabeled with [^{32}P] phosphate. Under these circumstances, lysis of cells should be performed in the presence of NaF and sodium pyrophosphate to ensure maximal phosphatase inhibition, as well as EGTA and EDTA to chelate the free Ca^{2+} and Mg^{2+} that may activate protein kinases and calcium-dependent phosphatases.

Phosphorylation of Proteins in Whole Cell Systems

Metabolic labeling of cells is efficiently achieved with ortho [^{32}P] phosphate. An important consideration when attempting analysis of phosphate

[35] S. M. Brady-Kalnay and N. K. Tonks, *Curr. Opin. Cell Biol.* **7,** 650 (1995).
[36] N. K. Tonks, *in* "Protein Phosphorylation: A Practical Approach" (D. G. Hardie, ed.), p. 231. Oxford Univ. Press, London, 1993.
[37] C. Bialojan and A. Takai, *Biochem. J.* **256,** 283 (1988).
[38] M. Suganuma, H. Fujiki, S. H. Furuya, S. Yoshizawa, S. Yasumoto, Y. Kato, N. Fusetani, and T. Sugimura, *Cancer Res.* **50,** 3521 (1990).
[39] Y. M. Li, C. Mackintosh, and J. E. Casida, *Biochem. Pharmacol.* **46,** 1435 (1993).
[40] R. E. Honkanen, *FEBS Lett.* **330,** 283 (1993).

incorporation into specific proteins is the equilibrium labeling of intracellular ATP pools. Because only a small percentage of exogenous radiolabel is actually incorporated into the ATP pool, one must be careful to not label for excessively long periods to prevent augmentation of the nonspecific signal. Many proteins contain phosphorylated species but do not respond to the exogenous stimuli under question; thus, to avoid excessive background labeling, the optimal time course must be established. In addition, many phosphoproteins are continually subject to the action of kinases and phosphatases, and many have multiple phosphorylation sites. In such instances, there may be a tight window of increase in phosphate content, the visualization of which will be critically determined by the labeling period. To maximize the labeling, it is advisable to minimize the exogenous phosphate pool by use of a phosphate-free medium.

Materials and Methods for Phosphorylation

Lymphocytes (PBL, clones) should be used at 10^6–10^7 cells per reaction. Lymphocytes and clones should be starved for at least 24 hr prior to analysis if they have been in culture. Freshly prepared PBL (in the cold) and thymocytes, normally prepared with ice-cold Hankes' balanced salt solution (HBSS), can be analyzed immediately. Cells are pelleted and resuspended at a final concentration of 10^8 cells/ml in phosphate-free RPMI 1640 (GIBCO Grand Island, NY) containing 10% dialyzed fetal calf serum (FCS, GIBCO). Sufficient ortho [^{32}P] phosphate is added to obtain a final concentration of 0.5 mCi/ml in a T-75 culture flask, and cells are incubated for 2–3 hr at 37° (Ensure use of a protective acrylic/Plexiglass covered box in the incubator.) After labeling, the cells are centrifuged (250 *g*; 10 min) and two times washed in excess PBS (250 *g*; 10 min), the supernatant is removed into an appropriate waste container, and then the cells are resuspended in phosphate-free medium at 10^7 cells/0.5 ml for reaction. All manipulations should be carried out using suitable protection, and further experimentation should be performed behind the necessary Plexiglass shielding.

Having prepared the sample tubes (Eppendorf, 1.5 ml) with respective agonists/antagonists, add the cells at appropriate time intervals (for time course experiments) and concentrations and incubate for the desired time in a shaking water bath (37°). Reactions can be stopped most rapidly by addition of ice-cold PBS containing phosphatase inhibitors, or by rapid centrifugation (2000 rpm, 4°, 3 min in an Eppendorf centrifuge). The cells can then be washed twice in ice-cold PBS prior to lysis. After the final wash ensure that the PBS is aspirated to the top of the pellet. Retain the cell pellet on ice for further analysis.

The phosphorylation status of the cell lysate can then be assessed by sodium dodecyl sulfate–polyacrylamide gel electrophoresis (SDS–PAGE).

Lyse cells by addition of 100 μl ice-cold lysis buffer [20 m*M* Tris (pH 7.9), 137 m*M* NaCl, 5 m*M* EDTA, 10% (v/v) glycerol, 1% (v/v) Triton X-100] containing protease and phosphatase inhibitors [1 m*M* phenylmethylsulfonyl fluoride (PMSF), 10 μg/ml each aprotinin and leupeptin, 1 m*M* sodium orthovanadate, 1 m*M* EGTA, 100 μ*M* β-glycerophosphate, 10 m*M* NaF, and 1 m*M* tetrasodium pyrophosphate]. The lysates are left on ice for 15 min with periodic vortexing. Following lysis the debris is centrifuged (14,000 rpm in an Eppendorf centrifuge; 4°; 10 min) to obtain supernatants and the pellet discarded. Supernatants can be used immediately, for SDS–PAGE or immunoprecipitation, or stored at −70° for future use.

For electrophoresis, at least 50 μg of total protein (as measured using the standard Bradford assay or the Pierce, Rockford, IL, BCA reagent) is loaded per lane. Generally, we have started with 12% Tris–glycine SDS gels or 4–20% Tricine gels to assess total protein phosphorylation profiles. If the gradient gels are overloaded, however, there may be insufficient resolution. Following initial analysis, the acrylamide percentage is adjusted depending on the species of interest. Following SDS–PAGE, the gels are fixed and dried according to standard protocols, and phosphorylation is assessed by autoradiography.

Immunoprecipitations

The reagents required for lysis are identical to those indicated above. In addition, protein A- or protein G-Sepharose (1 mg/ml; Pharmacia LKB, Piscataway, NJ) or (goat, anti-mouse, anti-rabbit, or anti-rat)-immunoglobulin G (IgG)-ararose (Sigma, St. Louis, MO) is used for immunoprecipitation in the absence of directly conjugated (commercially available) antibody.

Cell preparation and lysis are carried out as described above. Following lysis and incubation for 2 hr (4°) with primary antibody, protein A/protein G is added (approximately 25–50 μl) and the supernatant/slurry rotated for a minimum of 2 hr (4°). Alternatively, directly coupled antibody–agarose (20 μl) is added directly to the supernatant for 2 hr, with rotation at 4°. The immune complex is then centrifuged (14,000 rpm, 4 min, 4°) and the supernatant removed by aspiration. The immune complex is washed three times in cold lysis buffer, 1X SDS running buffer is added, and the sample is boiled, cooled on ice, recentrifuged, and then subjected to SDS–PAGE. Gels are fixed and dried as above, and the phosphorylated species are visualized by autoradiography.

Analysis of Protein Tyrosine Phosphorylation by Western Blot

Western blotting is a relatively straightforward method that allows analysis of proteins phosphorylated on tyrosine in the absence of radiolabeling. Because many cytoplasmic proteins are phosphorylated in lymphocytes following T-cell or B-cell receptor stimulation, the system can be adequately controlled for using cross-linked anti-CD3 or anti-IgM antibodies.

Cells are cultured in standard medium lacking serum (i.e., minus 0.5% FCS) for a minimum of 24 hr. Agonists are then added to cells (10^7 cells per reaction) in serum-free medium and the reactions incubated for a specific time depending on the assay. Gels are prepared and run according to methods outlined above. Following SDS–PAGE, Western transfers are performed (90 V, 1 hr; 70 V, 1.5 hr; 30 V, overnight) on methanol-activated Immobilon-P membranes (Millipore, Bedford, MA). [Ensure complete activation (wetting) of membranes in methanol to prevent loss of proteins during transfer due to lack of binding.] Following transfer, nonspecific sites are blocked 1 hr at room temperature or overnight at 4° using block buffer [Tris-buffered saline (TBS), 5% BSA or milk powder, 0.1% (v/v) Tween 20]. The Western blots are then stained using antiphosphotyrosine antibody [monoclonal, 4G10, UBI (UBI, Lake Placid, NY); PY20, Santa Cruz (Santa Cruz Biotechnology, Santa Cruz, CA)] preferably overnight (4°). The blots are then washed using TBS containing 0.5% (v/v) Nonidet P-40 (NP-40) (three to five times, 10 min each with rotation). Secondary horseradish peroxidase (HRP)-coupled species-specific anti-IgG antibody in block buffer (1 : 10,000, or according to manufacturer's instruction) is added to the filters for 1 hr at room temperature with gentle rotation. The filters are then washed in wash buffer again three to five times (10 min, with rotation). The filters are blotted slightly (Whatman, Clifton, NJ, 3M paper), and phosphorylated species are visualized using ECL (enhanced chemiluminescence) reagent (Amersham) or Supersignal CL-HRP (Pierce) and autoradiography according to the standard protocols.

Analysis of Protein Interaction by Coimmunoprecipitation

In a manner analogous to the stimulation of lymphocytes by antigen receptor ligation, and, similarly, integrin cross-linking, chemokines stimulate the coassociation of phosphorylated proteins with adapter molecules and kinase enzymes (K. B. Bacon, unpublished).[41] Physiologically, this may be relevant to the amplification of multiple signal transduction pathways within the same cell, eventually translating chemokine receptor ligation into cell activation and proliferation.[27] For this reason, one may be interested in analysis of the molecules involved, and this can be performed by coimmunoprecipitation assays in a manner similar to the immunoprecipitations performed above. Following immunoprecipitation and having an idea of the species under investigation, the acrylamide percentage of the gel is chosen to

[41] K. B. Bacon, M. C. Szabo, H. Yssel, J. B. Bolen, and T. J. Schall, *J. Exp. Med.* **184,** 873 (1996).

ensure complete resolution of the protein bands from the immunoglobulin heavy and light chains (55–60 kDa and approximately 28–35 kDa, respectively), if analysis is being performed under reducing conditions.

Analysis in Kinase Activity

The analysis of protein kinase activity is an extremely useful technique because a number of proteins that exhibit major functional activity in cell transformation are also those that become highly phosphorylated by chemokines (as assessed by whole cell lysate phosphotyrosine or exogenous substrate phosphorylation analyses). Although these kinase enzymes may represent some of the most important regulators of cell activation and transformation, one of the problems associated with simple antiphosphotyrosine Western blots of these proteins (assuming a protein tyrosine kinase) is that kinase activity is not always associated with increased phosphotyrosine staining. For precisely this reason a number of rapid and reproducible assays have been standardized to assess kinase activity of the precipitated enzyme. One must ensure that the antibody of choice for immunoprecipitation does not interfere with a catalytic or regulatory site on the kinase. Although there are certain anti-MAPK antibodies that do not bind the catalytic site of the enzyme (hence, immune complex kinase assays have been easy to develop), in the case of the S6k (a serine/threonine kinase) epitope-tagged S6k (HA) was generated for immunoprecipitation using anti-HA antibodies.[42]

Assay of Nonreceptor Tyrosine Kinase Activity: Focal Adhesion Kinase (*pp^{125FAK}*). The analysis of focal adhesion kinase (FAK) activity demonstrates a relatively typical assay that we have used involving an exogenous substrate. Essentially, phosphotyrosine incorporation into the exogenous substrate and the enzyme are measured.[41]

Cells (10^7 cells/reaction) are serum starved for at least 24 hr and stimulated with the appropriate agonist(s). Following stimulation of cells with the appropriate concentration of agonist and for appropriate times, the cells are lysed and supernatants retained as described previously. The lysis buffer consists of NP-40 instead of TritonX-100 (1% NP-40, 50 m*M* Tris-HCl, pH 8.0, 150 m*M* NaCl, 0.25% deoxycholate, 5 m*M* EDTA; containing the protease and phosphatase inhibitors 1 m*M* PMSF, 10 μg/ml each aprotinin and leupeptin, 1 m*M* sodium orthovanadate, 1 m*M* EGTA, 100 μM β-glycerophosphate, 10 m*M* NaF, and 1 m*M* tetrasodium pyrophosphate). Protein concentration of the centrifuged lysate is measured using standard methods, and two portions, 50 μg each, of the sample are incubated

[42] M. M. Chou and J. Blenis, *Cell (Cambridge, Mass.)* **85,** 573 (1996).

with antibody/antibody-Sepharose for 2 hr at 4° with rotation in separate Eppendorf tubes (2 hr with uncoupled antibody plus 1 hr with secondary antibody-coupled Sepharose). The immune complex is then washed twice with lysis buffer and once with excess kinase buffer (20 m*M* PIPES, pH 7.2, containing 5 m*M* each of $MgCl_2$ and $MnCl_2$). One sample is then incubated in kinase buffer containing 10 μCi [γ-^{32}P]ATP, 5 m*M* ATP (in a 20-μl volume) for 20 min at 30° in a shaking water bath. The other sample is incubated in kinase buffer containing 10 μCi [γ^{32}P]ATP, 5 m*M* ATP, and 0.4 mg poly(Glu-Tyr) in a 20-μl volume for 20 min at 30° in a shaking water bath.

The reaction of the first sample is stopped by rapid centrifugation to pellet the immune complex and remove the supernatant. Then 1X SDS sample buffer is added, and the sample is boiled for 5 min and then loaded onto an 8% polyacrylamide gel. Once the gel is dried, the FAK band is localized by comparison with Western-blotted nonradioactive lysate, excised, and counted by liquid scintillation to assess ^{32}P incorporation. The reaction containing exogenous substrate is stopped by addition of 10% ice-cold TCA, the precipitates rapidly pelleted, and the supernatant removed by aspiration or wash over filters. The radioactive phosphate incorporated into poly(Glu-Tyr) is then assessed by liquid scintillation counting.

Comment. Relevant to processes such as these is the growing appreciation that many kinase enzymes in the cell are stably or transiently associated with other kinases or phosphoproteins. These associated proteins may be responsible for some (or even a large part) of the phosphorylation observed both in exogenous substrates and the kinase of interest (not excluding autophosphorylation). At this level, careful stoichiometric analyses must be performed, as well as potential kinase assays using truncated mutants of the kinase to remove domains associated with autophosphorylation, SH2 or SH3 binding, or other association motifs. Additional considerations such as phosphoamino acid analysis and phosphopeptide mapping are not detailed here, and the reader is referred to standard and detailed texts on the subject.[15,16]

Assay of Mitogen-Activated Protein Kinase Activity

It has become apparent that the activation of MAPK and c-Jun amino-terminal kinase (JNK) pathways represents a critical mechanism in lymphocyte transformation in response to a variety of ligands.[43,44] The MAPK enzymes are serine/threonine-specific kinases (p42, p44, and p54 MAPK)

[43] P. Angel and M. Karin, *Biochim. Biophys. Acta* **1072,** 129 (1991).
[44] R. Seger and E. G. Krebs, *FASEB J.* **9,** 726 (1995).

regulated by dual phosphorylation of tyrosine and threonine.[45] The activation of these pathways may or may not be dependent on prior stimulation of the small GTP-binding proteins of the Ras family in conjunction with Raf.[44,46] There exist a number of simple protocols that facilitate the measurement of MAPK and JNK enzymes. Although there are a number of upstream regulatory kinase enzymes (MEKK, MEK), the reader is referred to specific publications for in-depth methodology of the measurement of such activity.[44]

Gel-Shift Assay of p42/p44 Mitogen-Activated Protein Kinase Activation

The analysis of phosphorylated MAPK can be demonstrated by SDS–PAGE and antiphosphotyrosine blotting. Stimulated cells (serum-starved) are lysed, and 50 μg of protein is loaded onto a 12.5% low cross-linker polyacrylamide gel.[46] The gel is run overnight until the 30-kDa molecular mass marker is run off the bottom of the gel. The region from the 30-kDa marker to 69-kDa marker is cut and proteins transferred to Immobilon-P membranes. Standard Western blotting procedure is followed to stain the phosphorylated MAPK with 4G10 (PY20). The blot is then stripped and restained with anti-ERK1/ERK2 monoclonal antibody (Zymed, South San Francisco, CA) to locate the position of MAPK and ensure equal protein loading. A similar method using precast 10% gels from NOVEX can also be employed; however, this has not always been as effective or reproducible as the larger 12.5% gels.

In general these methods are applicable to most lymphoid cells provided that there is a relatively large signal-to-noise ratio following stimulation. With respect to T-cell clones, the background phosphorylation may be high, and hence obtaining the best signal may be dependent on performing a time course experiment. In addition, the optimal concentration of agonist and time of stimulation will vary among the chemokines and different cell types. Of greatest importance is the serum starvation (and cytokines if T-cell clones are being used).

Phosphorylation of Myelin Basic Protein: Exogenous Substrate for Mitogen-Activated Protein Kinase

Full analysis of MAPK activity is aided by performing an assay to assess the extent of phosphorylation of myelin basic protein (MBP). In a similar manner to the assays outlined above for protein kinase activity measure-

[45] J.-H. Her, S. Lakhani, K. Zu, J. Vila, P. Dent, T. W. Sturgill, and M. J. Weber, *Biochem. J.* **296,** 25 (1993).

[46] M. L. Samuels, M. J. Weber, J. M. Bishop, and M. McMahon, *Mol. Cell. Biol.* **13,** 6241 (1993).

ments, this assay should be performed in conjunction with the gel shift assays.

Lysates are prepared from stimulated cells as before and kept on ice for immediate use. Protein determinations are vital to transfer 50 μg protein per assay. Mouse anti-human ERK-1/ERK-2-Sepharose (20 μl) is added to immunoprecipitate the MAPK enzymes (2 hr, 4°). Immune complexes are washed twice in lysis buffer, once in kinase buffer [25 mM Tris-HCl (pH 7.5), 137 mM NaCl, 40 mM $MgCl_2$, 40 mM HEPES (pH 7.4), 10% (v/v) glycerol]. Kinase buffer containing 200 μM ATP (per sample), MBP (20 μg per sample), and [γ-^{32}P]ATP (5 μCi per sample) is added to the pellet and incubated in a shaking water bath (30 min, 30°). The reaction is terminated by rapid centrifugation and removal of excess radioactivity (supernatant), and 1× SDS sample buffer is added. The sample is boiled for 5 min, cooled on ice, centrifuged to recapture condensation, and then loaded onto the gel (15%). Once the 14-, 21-, and 30-kDa markers have been adequately separated, the current is turned off. The bottom of the gel is cut off (it contains a high proportion of unincorporated radioactivity), and either the gel is dried or proteins are transferred to Immobilon-P membranes for autoradiography. MBP runs at around 18 kDa. Phosphorylated MBP can then be quantified by use of a phosphorimager.

c-Jun Kinase Activation

c-Jun is a vital component of the transcriptional activator AP-1 and is believed to be activated in TPA in T lymphocytes.[43] Although MAPK and CKII had been shown to negatively regulate the activity of c-Jun through phosphorylation, the kinase(s) responsible for activation of c-Jun by phosphorylation of amino-terminal serine residues has been identified as JNK (c-Jun amino-terminal kinase).[47]

The assay is performed in a similar manner to the immunoprecipitation of ERK1/2 except that cell lysates (from stimulated and unstimulated) cells are incubated with glutathione-agarose (GSH-agarose) containing 10 μg of either glutathione S-transferase (GST) or GST–c-Jun(1–223) fusion protein. The incubation mixture is rotated for 2–3 hr at 4° prior to centrifugation of the beads. The beads are then washed three times in excess HEPES buffer [20 mM HEPES (pH 7.7), 50 mM NaCl, 2.5 mM $MgCl_2$, 0.1 mM EDTA, 0.05% Triton X-100], and the pelleted beads are resuspended in 30 ml kinase buffer [20 mM HEPES (pH 7.6), 20 mM $MgCl_2$, 20 mM β-glycerophosphate, 20 mM p-nitrophenyl phosphate, 0.1 mM sodium vanadate, and 2 mM dithiothreitol] containing 20 μM ATP and 5 μCi

[47] H. Hibi, A. Lin, T. Smeal, A. Minden, and M. Karin, *Genes Dev.* **7,** 2135 (1993).

[γ-^{32}P]ATP. The kinase reaction is incubated for 20 min at 30° and terminated by washing in HEPES buffer. The phosphorylated proteins are subjected to SDS–PAGE using 10% gels. Phosphorylation of specific bands can then be determined by use of phosphorimagers or excision of the species of interest for liquid scintillation counting.

Derijard *et al.*[48] epitope-tagged JNK-1 (hemagglutinin or Flag) and immunoprecipitated JNK prior to incubation with GST–c-Jun(1–223). The in-gel kinase assay that involves the electrophoresis of c-Jun binding proteins, isolated using a similar GSH-agarose/GST–c-Jun immune-capture method, has also been used.[47,48] The proteins are resolved on 10% gels polymerized in the presence of 40 μg/ml GST–c-Jun. The gel is then washed (2 times for 30 min) in 100 ml 20% 2-propanol, 50 m*M* HEPES (pH 7.6) to remove SDS. Subsequently, the gel is washed twice for denaturation of proteins (30 min each time) in 100 ml buffer A [50 m*M* HEPES (pH 7.6), 5 m*M* 2-mercaptoethanol]. Renaturation was performed by incubation in 200 ml of buffer A containing 6 *M* urea (25°, 1 hr). The gel is then incubated in buffer A containing 0.05% Tween 20 with (serially) 3, 1.5, or 0.75 *M* urea. The gel is then washed (three times) in 100 ml buffer A containing 0.05% Tween 20 at 4°. This is followed by incubation in kinase buffer (as described above) containing 50 μM ATP and 5 μCi [γ-^{32}P]ATP for 1 hr at 30°. Termination of the reaction is performed by washing the gel in 100 ml 5% TCA and 1% sodium pyrophosphate at 25° (three times). The gel is then dried and phosphorylation assessed by autoradiography.

Phospholipase D Activation

Phospholipase D (PLD) is now recognized as a vital component of signal transduction pathways, vesicular trafficking, and cell transformation.[49,50] Numerous methods have been used to assess PLD activation, as PLD-induced hydrolysis of phosphatidylcholine produces phosphatidic acid (PA) and free choline. Alternatively, PA can be metabolized to DAG, which is then capable of activating PKC. Multiple forms of PLD exist, although the first example of a specific clone for PLD (PLD 1) was only described in 1995.[51] The exact mechanism(s) of agonist-induced PLD activation is unclear, but GTPγS-dependent, pertussis toxin-insensitive cofactors have been identified as the ADP-ribosylation factor [ARF; although which

[48] B. Derijard, M. Hibi, I.-H. Wu, T. Barrett, B. Su, T. Deng, M. Karin, and R. J. Davis, *Cell (Cambridge, Mass.)* **76,** 1025 (1994).

[49] M. M. Billah and J. C. Anthes, *Biochem. J.* **269,** 281 (1990).

[50] M. F. Roberts, *FASEB J.* **10,** 1159 (1996).

[51] S. M. Hammond, Y. M. Altshuller, T.-C. Sung, S. A. Rudge, K. Rose, J. Engebrecht, A. J. Morris, and M. A. Frohman, *J. Biol. Chem.* **270,** 29640 (1995).

one(s) is still unclear] and the small molecular weight G-protein RhoA.[52–56] Although some effects of PKC have been described (e.g., the down-regulation of PLD activity by chronic phorbol ester pretreatment), Singer and colleagues have demonstrated that membrane-derived PLD from brain required PKCα as a cofactor for ARF and Rho, but that this synergism was independent of kinase activity.[57] Different PLDs have been classified on the basis of subcellular location. The membrane form mPLD is associated with plasma, Golgi, and ER membranes. This PLD specifically catalyzes the hydrolysis of phosphatidylcholine (PC) and is dependent on Ca^{2+}, phosphatidylinositol 4,5-bisphosphate (PIP_2), GTP-binding protein(s) (RhoA and ARF), and PKCα. The form found in cytoplasm, cPLD, is effective on phosphatidylinositol (PI), phosphatidylethanolamine (PE), and PC. The activity is ARF-dependent. Li *et al.*[58] have demonstrated a GPI-linked, Ca^{2+}-dependent PLD (sPLD), whereas Balboa and Insel[59] have described a nuclear PLD that was Rho-dependent.

Some years ago it was noticed that PLD can stimulate a transphosphatidylation reaction in the presence of a primary alcohol; thus, instead of generation of PA and choline in the presence of ethanol, phosphatidylethanol was produced without any other metabolite. Although it was believed that this transphosphorylation reaction had no physiological significance, it nonetheless led to the establishment of an extremely rapid and reproducible assay for measurement of PLD activity.[60] It now has been shown that transphosphorylation may have a significant role *in vivo*, as this reaction results in significant attenuation of second mesenger production (DAG and PA) from both the PLC and PLD hydrolysis pathways.[61]

Cells (10^6–10^7 cells per reaction) do not necessarily have to be starved unless one is investigating phosphorylation interactions with PLD. Two hundred microliters of [^{3}H]oleic acid {[9,10,(*n*)-^{3}H]oleic acid (2–10 Ci/mmol); or [9,10,(*n*)-^{3}H]palmitic acid (40–60 Ci/mmol)} is added to a T-175 culture flask and the ethanol evaporated under a stream of nitrogen. Lymphocytes are resuspended at 10^7 cells/ml in Dulbecco's modified Eagle's

[52] S. Cockroft, G. M. H. Thomas, A. Fensome, B. Geny, E. Cunningham, I. Gout, I. Hiles, N. F. Totty, O. Truong, and J. J. Hsuan, *Science* **263,** 523 (1994).

[53] A. L. Boman and R. A. Kahn, *Trends Biochem. Sci.* **20,** 147 (1995).

[54] K. Oniguchi, Y. Banno, S. Nakashima, and Y. Nozawa, *J. Biol. Chem.* **271,** 4366 (1996).

[55] K. C. Malcolm, C. M. Elliott, and J. H. Exton, *J. Biol. Chem.* **271,** 13135 (1996).

[56] H. Kuribara, K. Tago, T. Yokozeki, T. Sasaki, Y. Takai, N. Morii, S. Narumiya, T. Katada, and Y. Kanaho, *J. Biol. Chem.* **270,** 25667 (1995).

[57] W. D. Singer, H. A. Brown, X. Jiang, and P. C. Sternweis, *J. Biol. Chem.* **271,** 4504 (1996).

[58] J.-Y. Li, K. Hollfelder, K.-S. Huang, and M. G. Low, *J. Biol. Chem.* **269,** 28963 (1994).

[59] M. A. Balboa and P. A. Insel, *J. Biol. Chem.* **270,** 29843 (1995).

[60] S. F. Yang, S. Freer, and A. A. Benson, *J. Biol. Chem.* **242,** 477 (1967).

[61] W. J. Van Blitterswijk and H. Hilkmann, *EMBO J.* **12,** 2655 (1993).

medium (DMEM)/Ham's F12 + 10% dialyzed FCS, 2 mg/ml fatty acid-free BSA, and 50 μg/ml gentamycin. Cells are labeled overnight at 37°. Lymphocytes are then washed twice (250 *g*, 10 min) in fresh medium to remove excess label (taking care to dispose of all waste in the required manner) and resuspended in a volume enough to allow 5×10^6 cells per 500 μl per reaction. (Ensure that you retain an unlabeled population of cells for background analysis.) Add 0.58% (v/v) (130 m*M*) ethanol (final concentration) to the final cell suspension and leave for 5–15 min to equilibrate. Reactions are all performed in 12 × 25 mm siliconized glass tubes to which agonists/antagonists have been added and allowed to proceed for a designated period of time at 37°. We have noticed a large difference in optimal time courses for experiments with lymphocytes and lymphocytic cell lines. Each cell should be tested with and without agonist for up to 1 hr to determine optimal PLD activity.

Cells are then rapidly pelleted (500 *g*; 5 min), and supernatants aspirated off. The cells are disrupted and lysed by addition of acidified methanol (100:1, v/v, methanol/HCl), then chloroform, then HCl (1:1:1, v/v). Immediately on addition of HCl an emulsion should form, which then disperses on vortexing. Add 40 μg of tracer 1,2-dioleoyl-*sn*-glycero-3-phosphoethanol (in chloroform; Avanti Polar Lipids, Birmingham, AL) and vortex vigorously for 15 sec. The samples are then centrifuged to separate the phases (1000 *g*; 20 min; 4°). Three phases should form: the top aqueous phase, a middle (insoluble material) phase, and a lower chloroform phase. The lower phase is removed, evaporated under nitrogen, and stored at −80° or redissolved in 20 μl of chloroform for assay. It is best to use glass Pasteur pipettes to remove the lower phase. One can first remove the upper phase, then rotate the tube to partition the insoluble "disc" onto the side of the tube before removing the chloroform phase. Alternatively, one can simply place the Pasteur pipette directly through the two upper phases and into the chloroform phase. Both methods are messy, and one should be absolutely sure not to aspirate any of the insoluble or aqueous materials along with the chloroform.

The thin-layer chromatography (TLC) plates (silica gel 60 TLC sheets; Merck, Gibbstown, NJ), heat-activated for 15 min (60°), should be marked with a horizontal line approximately 2 cm from the bottom and this line marked off at 2-cm intervals to delineate lanes. The 20-μl sample should then be carefully spotted at the center of each lane, using a stream of hot air to evaporate the chloroform. Once prepared the TLC plates can be run (two at a time) in appropriate TLC gel tanks using a solvent phase of ethyl acetate 2,2,4-trimethylpentaneacetic acid/water (13:2:3:10, v/v). Normal solvent migration takes approximately 1 hr 40 min to reach 1 cm from the top of the plate. Once resolved, the plates are air dried (in a chemical hood) and then placed in a TLC tank containing a beaker of iodine. The iodine vapor binds the unlabeled tracer (placed in a control/sample-free lane) to indicate the position of the [^{3}H]phosphatidylethanol, and this region is outlined with a pencil. The region containing the metabolite of interest is then scraped from the plate into scintillation vials, 5 ml of scintillant is added, and the samples are counted for ^{3}H incorporation. We have found it best to let the samples leach from the silica gel into the scintillant for at least 12 hr before counting.

Using this basic assay, one has the possibility of pharmacological manipulation using specific inhibitors of the numerous cofactors known (or believed) to be involved in the PLD stimulation. Inhibition of Rho proteins can be affected using C3 exoenzyme of *Clostridium botulinum,* which irreversibly ADP-robosylates Rho. Brefeldin A (a metabolite of *Eupenicillium brefeldianum*) has been used as an inhibitor of ARF proteins (hence, PLD activity). Brefeldin A causes disintegration of the Golgi network in certain cells, including T cells, thereby potentially preventing the translocation of proteins to the plasma membrane as well as the correct macromolecular complex associating for optimal PLD activation.[53] PKC inhibitors and PMA can also be used to determine the role of PKC (if any) in PLD activation in chemokine-stimulated lymphocytes. The latter two classes of inhibitors, being cell permeant, can simply be added to cells 20 min prior to agonist stimulation. The C3 exoenzyme, like GTPγS and vanadate/hydrogen peroxide, must be introduced either by electropermeabilization or cell permeabilization. We have opted for permeabilization using either β-escin, streptolysin O, or bacterial α-toxin. Of these treatments, streptolysin O is the most tried and tested; however, α-toxin has also provided a more "gentle" permeabilization because the pores rapidly reseal in the absence of toxin and presence of extracellular Ca^{2+}.[62,63] We have observed potent and robust stimulation of PLD activity in T cells in response to numerous chemokines. The important characteristic of these assays is the optimization, because not all chemokines stimulate to the same maximal phosphatidylethanol levels and at the same concentration.

Cell Permeabilization

The cell permeabilization protocols we have used are acute and allow rapid resealing. As with any permeabilization procedure, the extent and subsequent leakage should be checked by microscopy using the fluorescently labeled dextrans available from Molecular Probes (Eugene, OR). Incubation of cells for the short permeabilization periods is feasible using

[62] G. Ahnert-Hilger, S. Bhakdi, and M. Gratzl, *J. Biol. Chem.* **260,** 12730 (1985).
[63] M.-F. Bader, J.-M. Sontag, D. Thierse, and D. Aunis, *J. Biol. Chem.* **264,** 16426 (1989).

a glutamate medium (120 mM potassium glutamate, 20 mM potassium acetate, 3 mM $MgCl_2$, 20 mM Na–HEPES, pH 7.4, 1 mM EGTA, 1 mg/ml BSA).[64]

The choice of permeabilization agent as opposed to electropermeabilization/electroporation or scrape-loading is entirely dependent on one's experience. The electropermeabilization and scrape-loading may only be suitable for certain cell types and under certain conditions. Reports exist on the use of these methods for neutrophils (electropermeabilization) and fibroblasts with some success; however, much more evidence exists for the successful permeabilization of lymphoid cells using streptolysin O or α-toxin.[62,63] The process is relatively straightforward, although one must practice extreme caution in handling these agents. Incubate labeled cells in glutamate buffer containing α-toxin for 20 min at 37°. Centrifuge cells and remove supernatant. Resuspend cells in glutamate medium containing enough $CaCl_2$ to give a final concentration of 1 μM free Ca^{2+} and include ethanol for assay as previously described.

It should be noted that there are numerous reports in the literature detailing the loss of PLD activity in response to removal of cytosol. As an indication of the effective resealing of the plasma membranes after using α-toxin, the basal and stimulated levels of PLD activation in permeabilized cells is actually often higher than in unpermeabilized counterparts. Whether this is because permeabilization releases an inhibitory influence, simply increases the basal intracellular Ca^{2+}, or frees a greater concentration of low molecular weight G proteins and PIP_2 to act as cofactors is at present unclear; suffice it to say that the permeabilization of lymphocytes for analysis of PLD activity is a useful and viable assay.

Acknowledgment

DNAX is wholly funded by the Schering Plough Corporation.

[64] G. R. Dubyak, S. J. Schomisch, D. J. Kusner, and M. Xie, *Biochem. J.* **292,** 121 (1993).

[22] Calcium Mobilization and Phosphoinositide Turnover as Measure of Chemokine Receptor Function in Lymphocytes

By KEVIN B. BACON

Introduction

Despite the wealth of data defining the basic molecular characteristics of individual members of the chemokine superfamily, comparatively little is known of fundamental receptor-mediated signal transduction mechanisms that translate ligand binding into cellular responses such as chemotaxis, adhesion, and activation in lymphocytes. A limited number of G-protein-coupled seven-transmembrane (7-TM) receptors have so far been identified as specific chemokine receptors.[1] From the vast literature that exists on the action of ligands which bind 7-TM receptors, it is known that one of the signal transduction mechanisms stimulated by these receptors results in the activation of specific isoforms of phospholipase C (PLC) via GTP-binding proteins in a manner analogous to the activation of adenylate cyclase.[2] PLC activation results in the hydrolysis of membrane-associated phosphoinositol 4,5-bisphosphate [PI(4,5)P_2] to yield inositol 1,4,5-trisphosphate (IP_3) and diacylglycerol (DAG). IP_3 is firmly established as a second messenger linking the ligation of cell surface receptors to the release of intracellular calcium (Ca^{2+}i.c.).[3,4] This increase in calcium ions along with DAG is known to directly activate calcium-dependent isoforms of protein kinase C (PKC). Subsequently, IP_3 can be metabolized by a 5-phosphatase to generate I(1,4)P_2 or by a 3-kinase leading to formation of *myo*-inositol (1,3,4,5)-tetrakisphosphate (IP_4). IP_4 has also been proposed as a second messenger, although in lymphocytes it is postulated that this metabolite is involved in the control of extracellular Ca^{2+} flux through specific membrane channels.[5,6] In addition, PLC hydrolysis of PI(4,5)P_2 may result in the generation of *myo*-inositol 1:2-cyclic 4,5-trisphosphate (Ins-1:2,4,5P_3).

[1] C. A. Power and T. N. C. Wells, *Trends Pharmacol. Sci* **17,** 209 (1996).
[2] T. J. Schall and K. B. Bacon, *Curr. Opin. Immunol.* **6,** 865 (1994).
[3] C. P. Downes and C. H. MacPhee, *Eur. J. Biochem.* **193,** 1 (1990).
[4] M. J. Berridge, *Nature (London)* **361,** 315 (1993).
[5] A. H. Guse, E. Roth, and F. Emmrich, *Biochem. J.* **288,** 489 (1992).
[6] R. A. Wilcox, R. A. Chaliss, G. Baudin, A. Vasella, B. V. L. Potter, and S. R. Nahorski, *Biochem. J.* **294,** 191 (1993).

0076-6879/97 $25.00

Phosphorylation of phosphatidylinositol may also take place in the 3 position by specific phosphatidylinositol 3-kinase (PI-3K) activity.[7] This kinase appears to be highly regulated by phosphorylation, most commonly being activated as a consequence of classic growth hormone receptor ligation. The generation of $I(3,4)P_2$ and $I(3,4,5)P_3$ metabolites has raised questions as to their physiological role, as they are poor substrates for PLC. Of significance, however, are the findings that the PI-3K pathway and the Rho family G proteins Rac 1 and Cdc42 are implicated in the induction of the mitogenic response via the activation of $p70^{S6k}$.[8,9] Although PI-3K is not directly responsible for activation of $p70^{S6k}$, its metabolites $I(3,4)P_2$ and $I(3,4,5)P_3$ acting on calcium-independent PKC isoforms, PI_3P acting on the serine/threonine kinase Akt, or PI-3K activation of $p64^{PAK}$ have been proposed as direct stimulators.

In addition to PI-specific PLC, calcium release from intracellular stores in the endoplasmic reticulum (ER) *in vitro* has also been shown to be directly stimulated by products of the ceramide lipid pathway. Sphingosine 1-phosphate, a metablite of ceramide, is believed to be the physiological mediator of calcium release, but unequivocal proof has not been forthcoming. Sphingosylphosphorylcholine, although not yet identified *in vivo*, has, however, been shown to be a direct agonist.[10]

Three distinct receptors for IP_3 have been characterized, only some of which may be detectable in lymphocytes.[11] It is well documented that IP_3 mediates Ca^{2+} entry in mature T cells during the proliferative response to antigen and, potentially, during the apoptotic response in thymocytes.[12,13] IP_3 binds to specific IP_3 receptors (IP_3R) in the plasma membrane and ER and these cocap with the T-cell receptor (TCR)–CD3 complex during antigenic stimulation, indicating a major role for this metabolite in the activation of lymphocytes. However, unlike TCR ligation, which stimulates the phosphorylation of a specific kinase, ZAP-70, that binds PI-specific PLCγ via its SH2 domains[14] leading to PIP_2 hydrolysis and IP_3 generation, 7-TM receptors for chemokines appear to function via a different mechanism. It is postulated that receptor ligation results in activation of pertussis

[7] S. G. Ward, C. H. June, and D. Olive, *Immunol. Today* **17,** 187 (1996).

[8] M. M. Chou and J. Blenis, *Curr. Biol.* **7,** 806 (1995).

[9] M. M. Chou and J. Blenis, *Cell (Cambridge, Mass.)* **85,** 573 (1996).

[10] M. A. Beaven, *Curr. Biol.* **6,** 798 (1996).

[11] A. A. Khan, M. J. Soloski, A. H. Sharp, G. Schilling, D. M. Sabatini, S.-H. Li, C. A. Ross, and S. H. Snyder, *Science* **273,** 503 (1996).

[12] D. J. McConkey, P. Hartzell, J. F. Amador-Perez, S. Orrenius, and M. Jondal, *J. Immunol.* **143,** 1801 (1989).

[13] L. P. Kane and S. M. Hedrick, *J. Immunol.* **156,** 4594 (1996).

[14] J. B. Bolen, *Curr. Opin. Immunol.* **7,** 306 (1995).

toxin-sensitive heterotrimeric G proteins of the $G_i\alpha_2$ and $G_i\alpha_3$ subclasses,[15–18] the $\beta\gamma$ subunits then being responsible for activation of PLCβ isoforms. It is becoming clear, however, that another class of G proteins, namely, those of the pertussis toxin-insensitive or G_q family, which stimulate PLCβ to generate IP_3, represents the major stimulatory mechanism for PI-specific PLC.[15,16] Although we have shown that this pathway appears to operate in normal T lymphocytes and some T-cell clones,[18a] the exact significance is as yet unclear. There is thus a vast area of unchartered research terrain that awaits investigation with respect to calcium flux and IP_3 turnover, the significance of which will pertain to the resurgent question of redundancy in ligand number with respect to cloned receptors: What is the divergence in G-protein coupling to specific receptors in a given cell and the specific PLC isotype activated? With approximately 16 Gα, 4 β, and 7 γ subunits expressed in mammalian tissues, the 9 known isoforms of the PLC families, and the 3 IP_3R isoforms, this signal transduction pathway in isolation has the capacity for enormous specificity among different chemokines in the lymphocyte.

This chapter details methodology commonly used in the measurement of both calcium flux and PI turnover and how it becomes applicable to the study of chemokine action on lymphocytes.

Calcium Flux

The knowledge that mRNA for many of the known 7-TM receptors for chemokines are expressed in lymphocytes (C. Power, personal communication, 1995) leads us to speculate that surface expression of specific receptors may be central to the driving force behind much of the biological effects seen, such as chemotaxis and activation. This has led to certain controversy in initial studies of T-cell activation where calcium flux has been difficult to demonstrate. Binding analyses utilizing ^{125}I-labeled chemokine or specific antibody have, however, clearly demonstrated the presence of certain receptors, including, the interleukin-8 (IL-8) receptors IL-8RA and IL-8RB,[19–21]

[15] D. Wu, G. LaRosa, and M. I. Simon, *Science* **261,** 101 (1993).

[16] Y. Kuang, Y. Wu, H. Jiang, and D. Wu, *J. Biol. Chem.* **271,** 3975 (1996).

[17] P. M. Dubois, D. Palmer, M. L. Webb, J. A. Ledbetter, and R. A. Shapiro, *J. Immunol.* **156,** 1356 (1996).

[18] S. J. Myers, L. M. Wong, and I. F. Charo, *J. Biol. Chem.* **270,** 5786 (1995).

[18a] K. B. Bacon, abstract presented at the First Gordon Conference on Chemotactic Cytokines, June 1994.

[19] B. Moser, L. Barella, S. Mattei, C. Schumacher, F. Boulay, M. P. Colombo, and M. Baggiolini, *Biochem. J.* **294,** 285 (1993).

[20] K. B. Bacon, L. Flores-Romo, P. F. Life, D. D. Taub, B. A. Premack, S. J. Arkinstall, T. N. C. Wells, T. J. Schall, and C. A. Power, *J. Immunol.* **154,** 3654 (1995).

as well as potentially the C-C chemokine receptors CKR1,[22] CKR2,[22] CKR3 (D. Dairaghi and K. B. Bacon, unpublished, 1996), CKR4,[23] and Duffy[24] (D. Dairaghi and K. B. Bacon, unpublished, 1996) on T cells. Coupled with this finding is the original characterization of lymphocyte chemotactic responses to IL-8, RANTES, macrophage inflammatory protein-1α (MIP-1α), MIP-1β, macrophage chemotactic proteins (MCP-1 to MCP-4), lymphotactin, and CC-F18, the latter two, however, awaiting receptor assignments.[2,25,26] In addition, mRNA for $G_i\alpha_{2\text{-}3}$ and $G\alpha_q$ family members 11 and 16 (and potentially others) have all been identified in lymphocytes or lymphoid cell lines, so where does the problem of controversy fit in?

Primarily receptor–ligand interaction leading to increased calcium flux has been difficult (and sometimes impossible) to demonstrate. This apparently relates to the status of the lymphocyte.[20,21] Investigators appear to have particular idiosyncrasies concerning the method of preparation of lymphocytes; however, in our hands, maintaining cells at 4° throughout isolation and prior to assay has consistently resulted in reproducible chemokine-induced calcium flux with the best signal-to-noise ratios. Where it has become impractical to use freshly isolated lymphocytes, T-cell clones and certain T-cell lines have also proved adequate substitutes, although the prior stimulation with antigen or cytokines must be carefully controlled. As with any cell line, however, substantial screening for responder populations is a necessity, and once again, in our hands, lines such as Jurkat and Molt-4 have demonstrated chemokine-stimulated Ca^{2+} flux.

By far the most convenient methodology for analysis of Ca^{2+} flux in nonspecialist laboratories is fluorimetry. As opposed to the first generation calcium indicators such as Quin-2, preferred dyes in use today such as Fura-2 and Indo-1 (Molecular Probes, Eugene, OR) display stronger fluorescence and large shifts in fluorescence excitation or emission intensities when Ca^{2+} is bound. Such shifts can be readily observed when using standard calibration solutions containing different concentrations of calcium (see below). The ratio of fluorescence excitation or emission intensities at two

---

[21] S. Qin, G. LaRosa, J. J. Campbell, H. Smith-Heath, N. Kassam, X. Shi, L. Zeng, E. C. Butcher, and C. R. Mackay, *Eur. J. Immunol.* **26,** 640 (1996).

[22] P. Loetscher, M. Seitz, M. Baggiolini, and B. Moser, *J. Exp. Med.* **184,** 569 (1996).

[23] C. A. Power, A. Meyer, K. Nemeth, K. B. Bacon, A. J. Hoogewerf, A. E. I. Proudfoot, and T. N. C. Wells, *J. Biol. Chem.* **270,** 19495 (1995).

[24] K. Soo, personal communication, 1995.

[25] G. S. Kelner, J. Kennedy, K. B. Bacon, S. Kleyenstueber, D. A. Largaespada, N. A. Jenkins, N. G. Copeland, J. F. Bazan, K. W. Moore, T. J. Schall, and A. Zlotnik, *Science* **266,** 1395 (1994).

[26] T. Hara, K. B. Bacon, L. C. Cho, A. Yoshimura, Y. Morkawa, N. G. Copeland, D. J. Gilbert, N. A. Jenkins, T. J. Schall, and A. Miyajima, *J. Immunol.* **155,** 5352 (1995).

different wavelengths has therefore been used to calculate an accurate measure of the concentration of intracellular calcium ions. Similar techniques for fluorescence changes can be determined using Indo-1-loaded cells and fluorescence activated cell sorting (FACS). More specialist techniques including single-cell analysis by confocal microscopy, digital imaging, or patch clamp/electrophysiological analyses are not discussed here. Certain aspects of these techniques have been thoroughly covered.[27–30]

Cell Preparation

Normal Lymphocytes. As previously explained, our preferred methods for preparation of lymphocytes and thymocytes from both human and murine tissue have been those which minimize the manipulations and maintain the cells at 4°.

Following separation of lymphoid cells from whole blood or tissue by standard methodology using Ficoll, peripheral blood lymphocytes (PBL) can then be separated into T-cell and B-cell populations by standard negative selection techniques. FACS separations are to be avoided to prevent any potential activation that may occur during the antibody labeling. Peripheral blood monocytes (PBMC) are incubated with anti-CD14 and anti-CD19 Dynal beads (Dynal, Oslo, Norway), 30 min at 4° with rotation, and bound monocytes and B cells are removed by magnetic separation. Greatest purity is obtained by repeating this process at least once. Purity is then measured by FACS, or standard histochemical staining technique. Further separations if desired can be performed on the T-lymphocyte population using standard negative selection techniques, to obtain $CD4^+$, $CD8^+$ subpopulations and further to $CD45RO^+$ or $CD45RA^+$ populations of the $CD4^+$ and $CD8^+$ selections.

Thymocytes can be separated on the basis of similar principles using negative selection, although this is only useful in the separation of the double negative ($CD4^-/CD8^-$) population. During preparation of thymocytes, whether unseparated or negatively selected, keeping them at 4° is vital to observe any chemokine-stimulated Ca^{2+} flux (K. B. Bacon, unpublished, 1996).

Negatively selected T or B cells, or thymocytes, are washed and maintained on ice for no longer than 1 hr prior to use. These preparations are obviously standard, and the complexity of each cell separation is governed

27 E. D. W. Moore, P. L. Becker, K. E. Fogarty, D. A. Williams, and F. S. Fay, *Cell Calcium* **11,** 157 (1990).
28 A. Uto, H. Arai, and Y. Ogawa, *Cell Calcium* **11,** 29 (1990).
29 T. A. Ryan, P. J. Millard, and W. W. Webb, *Cell Calcium* **11,** 145 (1990).
30 T. Takematsu and W. G. Weir, *Cell Calcium* **11,** 111 (1990).

by the accessibility of the tissue sample under investigation. We have used standard methods for preparation of T and B cells from PBL, tonsilar B cells, splenic lymphocytes, unfractionated thymocytes, and the double negative thymocyte population. The standard protocols are clearly unsuited to preparations of intraepithelial gut lymphoid cells, gut and skin γ/δ T cells, or the individual subpopulations from thymus. The limitations of current protocols (which inevitably involve FACS) are thus translated into the experimental situation where one runs the risk of potentially desensitizing the cells during the separation procedure.

T-Cell Clones. An extremely useful tool, T-cell clones are likely to give the most consistent results. Under normal circumstances, the investigator simply maintains clones, once established, and thus the requirement of extensive preparation does not apply. There are two important considerations with T-cell clones: (1) addition of exogenous antigen and cytokines (IL-4 or IL-12 for polarization) and (2) cell cycle. As with most clones periodic addition of allogeneic "feeder" layers and phytohemagglutinin (PHA), for polyclonal stimulation, results in the presence of numerous attenuated cell types, as well as frantic activation and proliferative kinetics of the T-cell clone. One must familiarize oneself with the growth kinetics of the clone to allow use at a time when the majority of irradiated "feeders" have been removed from the culture. As with many clones, the periodic addition of IL-2, and possibly other cytokines to drive polarity, is a necessity for continued growth. Residual cytokine may affect calcium flux depending on the sensitivity of the cell to such stimuli and the chemokine receptor number. Should this be an important consideration, optimal reactivity is ensured by 24-hr starvation from cytokine prior to assay. This starvation protocol may also be relevant to serum because many chemokines can be found in desensitizing concentrations in serum.

Even when one has considered the exogenous stimuli, the cell cycle may play an important role in the responsiveness of the clones. On all the clones we have analyzed, there have been cycle-dependent changes in calcium flux profiles. As the clone reaches the plateau phase of its growth cycle, there tends to be a reduction in the reactivity toward chemokines, in terms of the magnitude of the calcium flux.

Lymphocyte Lines. Although few lines have proved responsive to chemokines unless transfected with specific receptors, lymphocyte lines would be by far the easiest cells to work with because the response would be either positive or negative and this would appear never to vary. In our hands, certain Jurkat sublines have shown Ca^{2+} flux responses to a dose range of RANTES, as have Molt-4 and HPB-ALL. We have by no means attempted a comprehensive survey of lymphoic cell lines for chemokine responses, however.

Transfected Lymphoid Cell Lines. The molecular characterization of many of the chemokine receptors (CKRs) has facilitated their transfection and hence expression in certain null backgrounds such as the 3T3 fibroblastic cell line or the HEK 293 epithelial cell line.[31–33] One of the criticisms associated with this system for analysis of CKR function was that the host cell was not representative of the correct "environment" as far as the lymphoid cell phenotype is concerned. In an attempt to overcome such limitations, a transfection system has been established in which a murine cell line L1/2 and Jurkat T cells have been transfected with human CKRs. This has provided a high-expressing line, free of endogenous CKRs, for analysis of both trafficking and signal transduction mechanism.[34] The ability to use a robust cell line, stably expressing high copy numbers of CKRs, will greatly facilitate research efforts in many areas of chemokine research.

Choice of Indicator

There are currently many fluorescent indicators on the market, the individual specifications of which relate to the concentration and type of ion that one wants to measure, the cell type, and, importantly, the instrumentation available for measurement. In using these dyes, one is looking for reasonable specificity of the dye for Ca^{2+} over other divalent cations and for analyses of chemokine function on leukocytes. This has been largely attained using Fura-2 or Indo-1. These indicators bind calcium in a 1:1 stoichiometry, making the calibration system very straightforward.

One of the most useful advances in the field was the development of indicators that could be introduced to cells noninvasively. This has been facilitated by the incorporation of an acetoxymethyl ester group that allows membrane permeability and once inside the cell is cleaved by nonspecific esterases, trapping the dye in the cytoplasm. Minor problems associated with these dyes, however, include incomplete deesterification, and hence leakage. Although this is certainly a consideration for precise imaging studies, the use of excitation/emission ratios for calculation of [Ca^{2+} i.c.] have negated this problem.[27]

[31] H. Deng, R. Liu, W. Ellmeier, S. Choe, D. Unutmaz, M. Burkhart, P. Di Marzio, S. Marmon, R. E. Sutton, C. M. Hill, C. B. Davis, S. C. Peiper, T. J. Schall, D. R. Littman, and N. R. Landau, *Nature* (*London*) **381,** 661 (1996).

[32] T. Dragic, V. Litwin, G. P. Allaway, S. R. Martin, Y. Huang, K. A. Nagashima, C. Cayanan, P. J. Maddon, R. A. Koup, J. P. Moore, and W. A. Paxton, *Nature* (*London*) **381,** 667 (1996).

[33] K. Neote, D. DiGregorio, J. Mak, R. Horuk, and T. J. Schall, *Cell* (*Cambridge, Mass.*) **72,** 415 (1993).

[34] J. J. Campbell, S. Qin, K. B. Bacon, C. R. Mackay, and E. C. Butcher, *J. Cell Biol.* **134,** 255 (1996).

Currently, one of the most widely used indicators, and our indicator of choice, is the acetoxymethyl ester of the BAPTA derivative Indo-1 (excitation λ = 355 nm). This indicator displays strong shifts between free (λ = 485 nm) and calcium-bound forms (λ = 405 nm). A particularly useful modification to this derivative has been its coupling to dextran. Although one must adopt invasive loading techniques for this indicator, the lack of compartmentalization, excessive leakage, or binding to intracellular proteins (common problems associated with the free dye) afford the investigator longer periods for experimentation. Indo-1 is our choice over Fura-2 because the latter is more subject to compartmentalization. Indo-1 can also be used with the less-flexible long-wavelength argon laser of a FACS machine and does not require the specialized quartz cuvettes that Fura-2 does. Fluo-3, a long-wavelength calcium indicator, is especially attractive for FACS owing to its almost nonexistent autofluorescence and huge increases in emission intensity once calcium is bound.

Cells in medium plus 10% (v/v) fetal calf serum (FCS) at 10^7 cells/ml can be loaded with between 1 and 10 μM Indo-1AM (we routinely use 3 μM). [Indo-1AM is dissolved in dimethyl sulfoxide (DMSO) to 1 mM, stored at $-20°$ in the dark.] We have obtained best loading at room temperature (this also prevents compartmentalization). Cells are kept in the dark to prevent photolysis of Indo-1AM, and they are gently rotated to enhance mixing. Loading is allowed to proceed for 45 min to 1 hr. Where cell lines or transfectants are used it is preferable to have split them 24 hr prior to experimentation, and to have them nearer 80% confluence as opposed to overconfluent. Although adequate loading is obtained in the presence of serum, enhanced loading is facilitated by use of Indo-1AM (10 μM final) and Pluronic (Molecular Probes)[35] at a final concentration of 0.025% (v/v). Loading is then left to proceed for 45 min to 1 hr with rotation at room temperature. Following loading, cells are pelleted to remove excess Indo-1AM and resuspended in physiological solution without phenol red or serum (as this obviously contributes to intrinsic fluorescence). We have consistently used Hanks' balanced salt solution (HBSS)/1% (v/v) bovine serum albumin (BSA) as the resuspension buffer. Cells are retained on ice prior to experimentation, to prevent excessive dye leakage and chemokine receptor down-regulation.

Fluorimetry

As explained above, Indo-1 is useful in both standard fluorimeters (as long as there are dual monochromators for measurement of emission inten-

[35] I. A. S. Drummond, A. S. Lee, E. Resendez, and R. A. Steinhardt, *J. Biol. Chem.* **262,** 12801 (1987).

sities) as well as cytometers (FACS). Individual fluorimeters and sorters are not discussed. Choice of application is therefore dependent on the scale and complexity of experiments. Routine analysis of single cell clones, lines, or unfractionated PBL for screening purposes is best suited to the fluorimeter, where large sample numbers can be handled over the space of 2–3 hr. Our progression to FACS has been in the analysis of rare populations where the separation procedures would irreversibly desensitize the cells to chemokine stimuli. A prime example of this has been in the analysis of double negative, double positive, $CD4^+$ and CD8 single positive thymocytes within a whole thymocyte population. The availability of three-color (or more) analyses allows simple staining and experimentation without the need for complex cell separation procedures. Despite the complexity of the procedure, which inevitably results in greater time frames of experimentation per sample, fewer samples, and longer analysis of data, FACS may be the only alternative one has for adequate measurement of calcium in these types of samples.

Analyses via FACS have essentially followed the methodology of Alexander *et al.*,[36] using a FACStar^{PLUS} flow cytometer. In the instance where the cell population is homogeneous, excitation is in the long-wavelength UV range (350–365 nm; multiple lines), recording emission of bound (405/10 nm bandpass) and free calcium (485/22 nm bandpass). The ratio of emission spectra for bound/free Indo is expressed for each event/cell and computed as a function of time. There are certain compensations necessary for situations requiring analysis of multiple cell types such as whole thymocyte populations, where phenotypic labeling is necessary with phycoerythrin (PE)- and PE–cytochrome-labeled anti-CD4 or anti-CD8 antibodies. The emission spectrum for bound Indo-1 is collected at 510/30 nm bandpass to avoid overlap into the 488 nm excitation wavelength for antibody–PE, and the free-Indo-1 is recorded at 395/25 nm bandpass. The limitation we have found with the FACS is the "dead" time, or time between injection of sample and recording of initial events, which is related to the sample module and whether one has to disassemble prior to sample addition. The system we have used involved direct injection of agonist while the sample is in place, using an established flow rate to limit the time of sampling to less than 10 sec.

Examples of Calcium Flux Profiles

Figures 1, 2, and 3 demonstrate a variety of Ca^{2+} flux profiles obtained from T cells, clones, and thymocytes and provides clear examples of the

[36] R. B. Alexander, E. S. Bolton, S. Koenig, G. M. Jones, S. L. Topalian, C. H. June, and S. A. Rosenberg, *J. Immunol. Methods* **148,** 131 (1992).

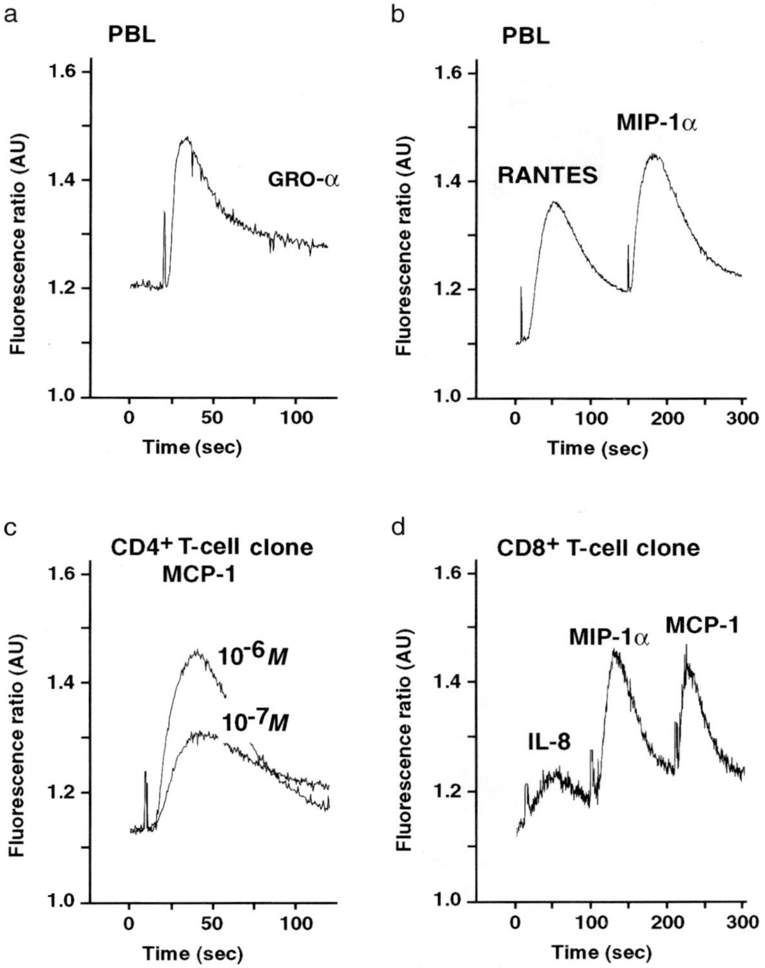

FIG. 1. (a–d) Chemokine-induced calcium flux profiles obtained using PBL (a, b) or T-cell clones (c, d). Results are expressed as fluorescence ratio over time (see text).

heterogeneity one must expect. Analysis of all profiles shows that the magnitude is never the same between the different lymphoid cell types. For example, the clones (both CD4$^+$ and CD8$^+$; Fig. 1c,d) as well as the line A3.2 (Fig. 2a) demonstrate larger magnitude fluxes that return rapidly to baseline. In contrast, the PBL response is smaller and there almost always appears to be a plateau phase (Fig. 1a,b). (These profiles are in stark contrast to that of the lines Jurkat and Molt-4, where the onset is

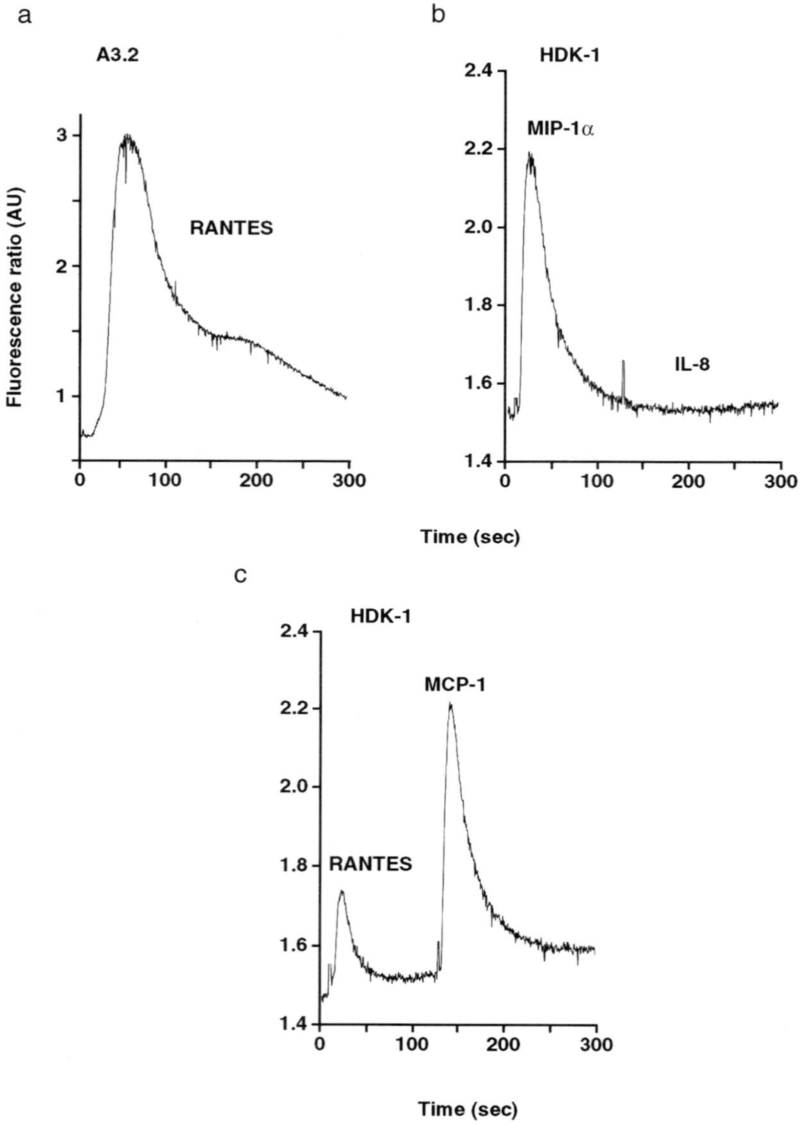

FIG. 2. (a) RANTES-induced calcium mobilization in the TCR$\alpha^+\beta^+$, $CD4^-$, $CD8^-$ T-cell hybridoma A3.2. (b) and (c) Chemokine-induced calcium mobilization on the murine T-cell clone ($CD4^+$) HDK-1.

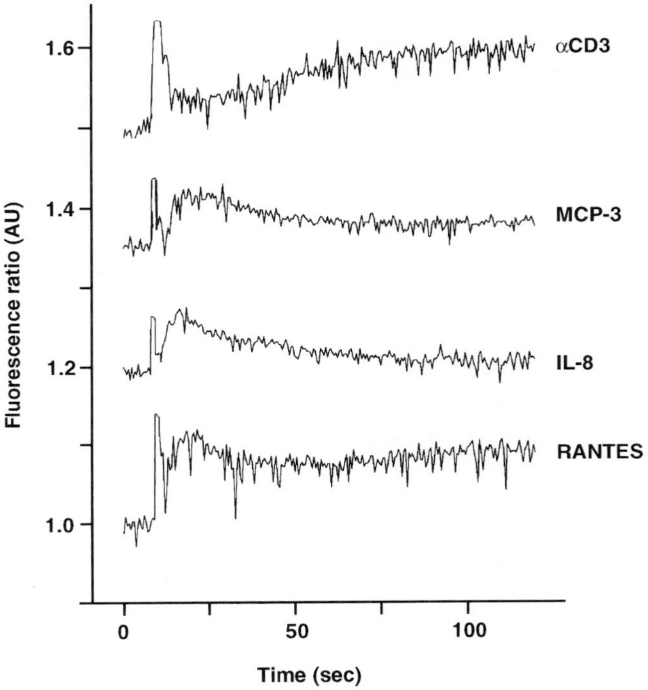

FIG. 3. Calcium flux profiles obtained in response to chemokine stimulation of freshly prepared human thymocytes, in comparison with anti-CD3.

delayed and the entire flux ratio remains elevated; data not shown.) The thymocytes (Fig. 3) display comparably smaller profiles than any other lymphoid cell we have analyzed. With any isolated lymphoid cell type (human or mouse) there has always been heterogeneity in the responses obtained and donor-specific reactivity, irrespective of the preparation and handling procedures. Not all PBL will respond to IL-8, for example, and not every thymocyte preparation will respond to chemokines.

There is an important consideration with these profiles. Other than the lines, each demonstrates a rapid transient onset consistent with the immediate IP_3-mediated release from stores induced by G-protein-coupled 7-TM receptors. In addition, the more robust responses, such as those exhibited by the clones and PBL, show that these cells may regulate the expression of receptors, or the coupling of intracellular signal transduction components. In any clone that is assayed (and normal PBL), there may be Ca^{2+} flux in response to, for example, IL-8, MCP-1, and MIP-1α, two of this combination, only one, or none at all. The same applies to MCP-3, RANTES, and growth-related oncogene-α (GRO-α).

The limitation observed with the fluorimetry system is one of sensitivity. Although current photomultipliers allow a relatively high degree of sensitivity, we have consistently observed flux profiles only at concentrations of chemokine greater than 1 n*M*. In most instances these concentrations are 10- to 1000-fold greater than the optimal concentration for chemotaxis and largely greater than the best K_d value for specific ligand–receptor binding. Whether this is simply because elicitation of the chemotactic response is independent of immediate Ca^{2+} flux transients, or whether these transients are preceded by low amplitude oscillations, remains to be determined. Alternatively, a threshold level of IP_3 generation may be required to induce higher magnitude and amplitude transients. Clearly, the optimal system for investigation of these questions is digital imaging/confocal microscopy, where individual cells can be analyzed.

Additional Considerations

Ensure cells are well dispersed. Large clumps interfere with the trace of the profile when the clump passes in the path of the light source.

We have measured flux in lymphocytes under a number of different conditions, including those involving permeabilization of cells to allow membrane-impermeant agonists or inhibitors to enter the cell. Basically, cells acutely permeabilized using standard reagents such as streptolysin O, β-escin, or α-toxin can subsequently be loaded with Indo-1AM. The leakage of the dye affords some instability of the trace; however, assuming that the permeabilization protocol has not been too harsh, the cell membrane seals well enough to retain the test compounds and the dye. Using this technique, we have visualized inhibition of chemokine-induced Ca^{2+} flux by inhibitory peptides corresponding to the C-terminal domain of G proteins and cell-impermeant tyrosine kinase inhibitors (K. B. Bacon, unpublished, 1996).

Measurement of Ca^{2+} flux parameters including release from intracellular stores and influx through membrane channels has been analyzed using the cell-permeant Ca^{2+}-ATPase inhibitor thapsigargin, in the absence of exogenous stimuli. This has helped, in part, to determine the magnitude and profile of Ca^{2+} release in cells (e.g., transfectants) where there is an absence of known exogenous stimuli or G-protein-coupled 7-TM receptors. An extension of these measurements is the contribution of Ca^{2+} influx to the total profile one observes. Most investigators simply remove extracellular calcium by adding 3 m*M* EGTA immediately prior to addition of agonist. In contrast, addition of 10$\times$ [Ca^{2+}] to the cuvette allows one to observe the contribution of calcium-induced calcium influx through membrane channels. In certain T lymphocytes, the addition of excess Ca^{2+} extracellularly causes an increase in the magnitude of the sustained phase of the flux, whereas in more sensitive lymphoid systems, one may observe an immediate "shutoff" or reduction to baseline of the secondary phase. This response is correlated with closing of membrane channels to further influx through an as yet unidentified mechanism.

Calibration Curves

Calibration curves are important in the quantitation of Ca^{2+} flux changes following receptor ligation both for standard screening purposes and in the establishment of measurement systems for different cell types (even within the lymphoid population). One will ultimately want to obtain absolute values for Ca^{2+} flux from resting intracellular concentrations of approximately 100 n*M* (depending on cell type). Again, the system is standard and widely referenced in the literature, including the protocols provided by the manufacturers of Indo-1AM and calibration solutions (e.g., Molecular Probes).

The necessary calibration curves for generation of an accurate dissociation constant (K_d) for calcium bound to Indo-1 allow calculation of absolute Ca^{2+} concentration.[37] Generation of emission spectra for Indo-1 at various concentrations of Ca^{2+} is performed with accurately titrated buffers containing EGTA and Ca^{2+}, solutions which can be easily made or purchased from Molecular Probes. This is a useful starting point because the fluorescence profiles of Fura-2 and Indo-1, hence the K_d, differ according to the instrumentation available for measurment.[28] The dissociation constant is also highly related to the cell, because the excitation/emission optima differ between *in vitro* calibration buffers and the viscosity of the cell cytoplasm. The ratio values and their corresponding [Ca^{2+} i.c.] can be tabulated and stored as reference tables for experiments where spectra can be compared. Again, the derivations of the formula for calculation are published elsewhere and so are not detailed extensively here.

Briefly, fluorescence ratios of free and bound calcium (F_1/F_2) following agonist (R) are equated with the calibrating ratios of fluorescence in Ca^{2+}-saturating (R_{max}) or Ca^{2+}-free (R_{min}) solution. The saturation ratios are often obtained by addition of ionomycin to loaded cells in high [Ca^{2+}] buffer, whereas EGTA at 10 m*M* is included to determine the calcium-free solution. Entering these values into the following equation allows a reasonably accurate determination of calcium concentration changes in response to agonist.

[37] G. Grynkiewicz, M. Poenie, and R. Y. Tsien, *J. Biol. Chem.* **260,** 3440 (1985).

$$[Ca^{2+}] = K_d[(R - R_{min})/(R_{max} - R)] \times S_{f2}/S_{b2}$$

In the equation, S_{f2}/S_{b2} is the fluorescence ratio of free and bound Indo at 405 nm, and the K_d of Indo-1 for Ca^{2+} (around 250 nM) has been determined for the specific application according to standard calibration procedures. Readers are referred to the original characterization of these indicators[37] as well as texts that have been written concerning these methods.[38]

Troubleshooting

Loading. Inefficient loading is usually not problematic given the 1-hr time frame. However, in instances where loading has not been optimal, this is easily recognized by the extremely high baseline fluorescence and instability of the trace. In our hands the problem associated with insufficient loading is the state of the cells. In most instances, this is a manifestation of dying cells. In addition, loading at 37° will be less than at room temperature. Given the photosensitivity of the dye, insufficient protection from light during the loading process will ultimately reduce the efficiency.

Cell Viability. Related to the loading problems, concerns about cell viability weighs heavily on the experiment at all levels. Assuming one is satisfied with the viability of the cells, optimal reactivity is related to optimal growth conditions. Thus, freshly isolated PBL, in our hands, require maintenance in the cold prior to analysis and will give varied results depending on use immediately or following culture for long periods. We have found that the time at room temperature allowed for Indo-1 loading does not affect the cell response, as long as it does not exceed 1 hr. This variation may be related to apoptosis of a responding (sub)population or down-regulation of chemokine receptors owing to natural cell cycle progression or the desensitization by chemokines and/or other cytokines in serum. Responsive cell lines, while being the most stable and reproducible, must be recently split to avoid overconfluence. In addition, and this applies for T-cell clones, the age of the culture (passage number) is a critical factor. Following generation of a cell line, or clone, it is ideal to have sufficient backup frozen samples because certain lines and numerous clones, in our experience, lose their flux capacity with time. Although this does not appear to affect the chemotactic response, for example, it does limit the number of informative and useful assays that can be performed on that particular cell.

Saturation of Fluorescence Curves. Saturation of fluorescence has been a problem that often leads investigators along the negative conclusion pathway in fluorimetry but is entirely related to the system settings (photo-

[38] A. M. Gurney, *in* "Receptor-Effector Coupling: A Practical Approach" (E. C. Hulme, ed.), p. 117. Oxford Univ. Press, London, 1990.

multiplier gain and monochromator settings) of the fluorimeter being used. Simply assuming that the cell number for analysis in cuvettes is standard is unacceptable. The cell number and loading determine the fluorescence intensity that may saturate the dual emission monochromators, indicating an almost perfect ratio of 1. Where this occurs, the solution is simply to lower the cell concentration until the wavelength intensities are considerably reduced and sufficiently separate to allow changes in fluorescence ratios.

Inositol Lipid and and Phosphoinositide Measurements

There are a number of methods that can be used in the analysis of inositol lipids and phosphoinositides. The complexity of the metabolites requires careful and controlled analysis, which is afforded by a limited number of effective experimental manipulations. Investigators may want to correlate chemokine receptor activation with the production of inositol phosphates (IPs) as a whole fraction or to quantitatively analyze individual metabolites. Such issues determine the use of radiolabeled inositol, addition of lithium chloride (LiCl), or extraction of metabolites from unlabeled samples for use in competitive binding assays for mass analysis. The use of LiCl has greatly enhanced the analysis of total IP because the salt is a specific inhibitor of the enzymes that degrade IP metabolites, namely, inositol polyphosphate 1-phosphatase and inositol monophosphatase.[39]

Labeling Cells with [^{3}H]inositol

The efficient labeling of phosphoinositide pools is often a vital consideration in the analysis of agonist-stimulated phosphatidylinositol (PI) turnover. With respect to lymphocytes, it is feasible to maintain the cells in culture for long enough periods to facilitate uptake of tritiated inositol to a level where the specific activity will give a reliable and reproducible indication of the metabolites formed following receptor stimulation. In such cases it is vital to maintain cultures in inositol-free medium and label for 24–48 hr to attempt adequate equilibrium loading. We have used Medium 199 (GIBCO, Grand Island, NY), although any custom-made physiological buffer or medium that allows normal culture may be used, as long as the media are inositol-poor or deficient. Additionally, the use of dialyzed heat-inactivated FCS minimizes the competition for equilibrium loading.

[39] L. M. Hallcher and W. R. Sherman, *J. Biol. Chem.* **255,** 10896 (1980).

[^{3}H]Inositol specific activity is an important consideration owing to the low K_m of the enzyme phosphatidylinositol synthase.[40] High radioactive concentrations are only useful if the specific activity is also high and the endogenous inositol is maintained at as low a level as possible. We have thus used labeling conditions of 2–5 μCi/ml (15 Ci/mmol) for 48 hr in Medium 199. There are different preparations available from Amersham (Arlington Heights, IL) that allow the choice of stabilizer (ethanol, or solid phase) as well as a huge difference in specific activity ranging from 10–20 to 80–120 Ci/mmol, affording the investigator the ability to dilute the preparations to the desired specific activity. Following labeling, cells are thoroughly washed to remove excess [^{3}H]inositol and equilibrated in medium containing 10 m*M* LiCl for 15 min prior to agonist stimulation.

Extraction Procedures

Whereas most investigators have been concerned with the quantitative analysis of IP_3 generated as a function of chemokine receptor stimulation, there are a number of assays that allow measurements of inositol lipids as well as the phosphates. Extraction of inositol lipids is usually measured by quantitation of the ^{32}P-labeled metabolites PI, glycerophosphoinositol phosphate (PIP), and PIP_2 by thin-layer chromatography (TLC). This is potentially more cumbersome than most assays and has been adequately described elsewhere.[41–43] A more simple and rapid assay involves measurement of the deacylated water-soluble derivatives [glycerophosphoinositol, glycerophosphoinositol phosphate (from PIP), and the bisphosphate (from PIP_2)], by simple anion exchange. Analysis of inositol phosphates is more readily performed using alternative extraction procedures, which include acidified chloroform/methanol, perchloric acid (PCA), or trichloroacetic acid (TCA) extraction, both of the latter two procedures being followed by neutralization of samples using concentrated hydroxide, freon/*n*-octylamine, or diethyl ether (see later). I do not describe the extraction or analysis of cyclic inositol lipids nor the CDP-diacylglycerol, as I have not employed these methods in my analyses; they have been adequately described elsewhere.[44]

[40] K. G. Oldham, *in* "Receptor–Effector Coupling: A Practical Approach" (E. C. Hulme, ed.), p. 99. Oxford Univ. Press, London, 1990.

[41] C. M. Liedtke, *Am. J. Physiol.* **262,** L183 (1992).

[42] E. Lee, K. Punnonen, C. Cheng, A. Glick, A. Dlugosz, and S. H. Yuspa, *Carcinogenesis* **13,** 2367 (1992).

[43] M. R. Hanley, D. R. Poyner, and P. T. Hawkins, *Methods Enzymol.* **197,** 149 (1991).

[44] P. P. Godfrey, *in* "Signal Transduction: A Practical Approach" (G. Milligan, ed.), p. 105. Oxford Univ. Press, London, 1992.

Lipid Extraction. As with any assay system, to ensure reproducibility and adequate comparison, it is necessary to treat all samples in an identical fashion, including the blanks, negative controls, and positive controls. In addition, "spiking" of samples with standards allows suitable comparison with experimental samples and an assessment of the recovery of radioactive metabolites during the extraction procedures. Immediately following stimulation, the sample reaction is terminated by addition of acidified chloroform/methanol. There are several different approaches to this method. Samples (cells) are extracted with 0.8 ml methanol–HCl (100 : 6, v/v), chloroform is added (1 : 1, v/v), and then the samples are centrifuged for separation of phases. Reextraction can be performed using chloroform/methanol/HCl (50 : 50 : 3, v/v). Alternatively, terminate the reaction using 4 ml of methanol. Add 6.3 ml of chloroform and 4 ml of 2.4 *M* HCl, then add a further 4 ml of chloroform. Reextract with 2 ml of chloroform three times, pooling the phases for sample analysis. A third procedure involves extraction with chloroform/methanol/HCl (40 : 80 : 1, v/v). Reextract with chloroform/0.1 *M* HCl (1 : 2, v/v).

In general, the preliminary extraction is allowed to proceed on ice for up to 30 min. The vortexed extract is then centrifuged (2000 *g*, 15 min), the lower (lipid) phase removed, the upper aqueous phase reextracted, and the lipid phases pooled and evaporated under nitrogen.

Water-Soluble Inositol Phosphate Extraction. Extraction using ice-cold perchloric or trichloroacetic acid have been widely used in the extraction process. We have used equal volumes of 10% PCA and cell suspension (usually 250 ml). The terminated sample is left on ice (10 min), and then precipitated material is removed by rapid centrifugation (2000 *g*, 5 min). Extracted supernatant is added (4 : 1, v/v) to EDTA (10 m*M*) for extraction of inositol phosphates. Alternatively, following PCA precipitation, 80% of the supernatant can be directly neutralized using 0.5 *M* KOH/9 m*M* sodium tetraborate. The disadvantage with this procedure is that the volume of hydroxide/tetraborate must be carefully monitored such that the pH of the supernatant reaches pH 8–9. Following alkalinization, the precipitated potassium perchlorate must be centrifuged out, and only then is the sample ready for analysis. We have used the neutralization according to Sharps and McCarl[45] by addition of an equal volume of tri-*n*-octylamine/1,1,1-trichlorotrifluoroethane (freon, 1 : 1, v/v). The neutralized sample is vortexed for 15 sec then centrifuged (2500 *g*, 15 min) where it separates into three phases. The top phase contains the aqueous extract of inositol phosphates entirely free of perchloric acid or neutralization reagents.

[45] E. S. Sharps and R. L. McCarl, *Anal. Biochem.* **124,** 421 (1982).

Trichloroacetic acid extraction proceeds in similar fashion. Samples are terminated with TCA (5:1, v/v) and the precipitated material removed by rapid centrifugation as above. Extracts are then neutralized with diethyl ether (1:4, v/v) in glass tubes. Tubes are placed in a freezing bath (methanol/dry ice). The ether phase is removed, and the sample is thawed and reextracted. Purified total phosphates are refrozen, vacuum desiccated, and stored for analysis. Inositol phosphates (radiolabeled) can then be analyzed using Dowex chromatography.

Deacylation of Inositol Lipids. The following procedure is useful in instances where one does not have the capacity for TLC analyses of lipid metabolites. Deacylation of the lipids leads to the formation of the glycerophosphoinositols. This protocol is taken from those of Jenkinson *et al.*[46] The dried lipid samples from initial lipid extraction are dissolved in 1.2 ml chloroform/methanol (5:1, v/v) and 0.4 ml of 0.5 *M* NaOH in methanol/water (19:1, v/v). Vortex samples for 20 min and then add chloroform (1 ml), methanol (0.6 ml), and water (0.6 ml); mix thoroughly and centrifuge 3000 *g* for 10 min. One milliliter of the upper phase is neutralized with boric acid (0.25 *M*) or by passage through a Dowex-50 (H^+ form) column. The column is washed with 1 ml water and the eluate adjusted to pH 7 with $NaHCO_3$. Alternatively, the methylamine method is also applicable.[47] The lipid residue is dissolved in 0.4 ml butanol and 1.6 ml methanol. Mix vigorously and sonicate, then add 2 ml of 40% methylamine solution followed by incubation (53°–56° for 1 hr). Evaporate under N_2 and then redissolve in 1 ml water (containing 1 mg/ml mannitol) plus 1.2 ml of butanol/petroleum ether/ethyl formate (20:4:1, v/v), mix, and centrifuge. The water phase is collected and the organic phase reextracted with 0.5 ml water/mannitol. The water/mannitol is back-extracted with butanol/petroleum ether/ethyl formate (20:4:1, v/v) and the aqueous phase evaporated under N_2. Samples are redissolved in water prior to addition to the columns. Radiolabeled standards are extracted in similar fashion to compare the recovery from chloroform extracts.

Column Separations of Glycerophosphoinositols

Samples from this lipid deacylation procedure can be added to Dowex AG1-X8 columns (formate form; 200–400 mesh) and eluted with 1 *M* formate/0.1 *M* formic acid for assesment of total inositol lipids by liquid scintillation counting of a fraction of the eluate.[40,44] Alternatively, load the mixture on a 1-ml Dowex AG1-X8 column and elute with 15 ml of 0.18

[46] S. Jenkinson, S. R. Nahorski, and R. A. J. Chaliss, *Mol. Pharmacol.* **46,** 1138 (1994).

[47] D. Button, A. Rothman, C. Bongiorno, E. Kupperman, B. Wolner, and P. Taylor, *J. Biol. Chem.* **269,** 6390 (1994).

M ammonium formate/5 m*M* sodium tetraborate. The combined eluate should contain the glycerophosphoinositol, the deacylated product of PI. Glycerophosphoinositol phosphate (PIP) is eluted with 20 ml of 0.4 *M* ammonium formate/0.1 *M* formic acid and glycerophosphoinositol bisphosphate (PIP_2) with 20 ml of 1 *M* ammonium formate/0.1 *M* formic acid. A fraction of all samples can then be subjected to scintillation counting.

myo-[^{3}H]Inositol Lipid Quantitation by Thin-Layer Chromatography

The dried lipid phase is resuspended in chloroform for analysis by TLC using heat-activated (80°, 20 min) potassium oxalate-impregnated silica TLC plates (plastic-backed).[41,42] Spot 10-μl lipid samples equidistant and approximately 1 cm from the base of the TLC plate. The plates are then developed in chloroform/acetone/methanol/acetic acid/water (40:15:13:12:8, v/v). Phospholipids are visualized on plates by exposure to iodine vapor and comparison to retardation of authentic standards. Phospholipids can then be scraped from the plates and counted by liquid scintillation.

Column Separation of Water-Soluble Inositol Phosphates

A stepwise formate gradient is used, as for the separation of glycerophosphoinositols.[42,48] Some investigators have added phytic acid hydrolysate to the aqueous fraction to enhance the recovery and separation. One milliliter of Dowex AG1-X8 (formate form, 200–400 mesh) is packed into plastic columns. Acid-extracted inositol phosphates (aqueous extract) are diluted with 10 ml of water for addition to the column. Elute free inositol with 20 ml of water; GPI can be eluted with 16 ml of 60 m*M* formate/5 m*M* sodium tetraborate, IP with 16 ml of 0.2 *M* formate/0.1 *M* formic acid, IP_2 with 16 ml of 0.4 *M* formate/0.1 *M* formic acid, IP_3 with 8 ml of 0.8 *M* formate/0.1 *M* formic acid, and IP_4 with 8 ml of 1.2 *M* formate/0.1 *M* formic acid. Then 2 *M* formate/0.1 *M* formic acid can be used for regeneration of the column. A similar method using chloride elution has also been employed with some success:[41] Dowex AG1-X8 (chloride form, 200–400 mesh) is used instead of the formate form for column packing. Elute [^{3}H]inositol, glycerophosphoinositol, and inositol phosphate with 10 ml of 30 m*M* HCl, IP_2 with 3 ml of 90 m*M* HCl, and IP_3 with 3 ml of 500 m*M* HCl. An important consideration in the scintillation counting of IP_3 and IP_4 eluates is the high salt concentration. Either add methanol or predilute the eluate with water to adequately partition the phases.

[48] G. B. Willars and S. R. Nahorski, *Br. J. Pharmacol.* **114,** 1133 (1995).

Chromatographic Separation of Inositol Phosphate Metabolites

High-performance liquid chromatography (HPLC) is unquestionably one of the most powerful techniques for analysis of individual metabolites. The disadvantages are that the method is time consuming, especially where conversion of radioactivity recovery is to be correlated with absolute concentration, and not every laboratory has access to the necessary (expensive) equipment. We have used the method according to Dean and Moyer,[49,50] although other well-established methods exist[42,51] Anion-exchange HPLC using a Partisil 10 SAX column, with ammonium phosphate elution gradients, is the most widely used method. The equipment required consists of a Partisil 10 SAX (strong anion exchange) column (4.6 mm $\times$ 25 cm), a precolumn (anion exchange pellicular packing), an automated gradient controller, and an on-line radioactivity detector. [^{3}H]Inositol polyphosphate standards are used as controls. Gradient elution of metabolites using aqueous ammonium phosphate (pH 3.8) is as follows: 0–30 min, 0.01–0.08 *M*; 30–60 min, 0.2–0.28 *M*; 60–90 min, 0.5–0.52 *M*. One-milliliter fractions are collected, and incorporated ^{3}H in 200 μl from the fractions is counted using liquid scintillant.

One of the best methods includes the following gradient of ammonium phosphate (pH 3.8) at 1 ml/min, using authentic standards as references: 0 to 0.3 *M* over 15–87 min (contains nominally GPI, IP, and IP_2); 0.3 to 0.7 *M* over 87–119 min (contains IP_3); 0.7 to 1 *M* over 119–125 min (IP_4); then isocratic elution at 1 *M* ammonium phosphate 125–145 min to elute IP_4 and IP_6. Lee *et al.*[42] have used similar HPLC conditions except that the elution program was a linear gradient of 0–60% ammonium phosphate (1.5 *M*, pH 4.1) at 1 ml/min. We have used the initial gradient system successfully in the analysis of IL-8-induced PI turnover.[49] In this investigation the presence of LiCl proved invaluable, and the recovery of IP, IP_2, and IP_3 indicated a direct role for IL-8 in the activation of T cells. The generation of IP_3 was further supported by use of the direct IP_3 mass assay (see below). Should the assessment of differences in signaling between selected T-cell-activating chemokines be required, HPLC would appear to be useful as a method to potentially analyze differences in PI turnover.

Mass Measurements of Inositol Tris- and Tetrabisphosphates

Mass measurements have been performed using preparations of binding proteins or the commercially available kits provided by Amersham. Prepa-

[49] K. B. Bacon, D. G. Quinn, J.-P. Aubry, and R. D. R. Camp, *Blood* **81,** 430 (1993).
[50] N. M. Dean and J. D. Moyer, *Biochem. J.* **242,** 361 (1987).
[51] E. W. Haeffner and U. Wittmann, *J. Lipid Mediators* **5,** 237 (1992).

ration of binding proteins (crude fractions containing receptors for IP_3 and IP_4 in adrenal glands and rat cerebellum, respectively) is described elsewhere.[44] For our purposes, the use of the commercially available binding kit for IP_3 has been the most easily adaptable and rapid to use for assay of large samples. These are suitably detailed in several references[40,52] as well as the manufacturer's protocols and thus are not further detailed here. Of note, however, is the necessity for adequate neutralization prior to assay using the binding protein. We have used the Amersham kit for analyses of both T- and B-lymphocyte responses to IL-8, MIP-1α, MIP-1β, RANTES, and MCP-1[21] (K. B. Bacon, unpublished).

Interestingly, stimuli as diverse as those mentioned above induced similar patterns of IP_3 production between T and B cells, with respect to both maximal levels obtained and the concentrations of chemokine used to induce these maximal levels. Although this is convenient for the screening of chemokine signal transduction in lymphocytes, it does indicate a potential field of research that requires more in-depth analysis. Is it possible that levels of IP_4 or PIP_2, important mediators for signal transduction pathways, including membrane channel opening (IP_4) and phospholipase D (PLD)/ARF/Rho activation (PIP_2),[53] are differentially regulated by chemokines in lymphocytes? What about the role of other metabolites, including the cyclic inositol phosphates? These and other questions will surely be clarified as more researchers become involved in chemokine research using newly available molecular tools (receptors) and the simple techniques outlined in this chapter.

Acknowledgment

DNAX Research Institute is wholly funded by the Schering Plough Corporation.

[52] R. A. J. Chaliss, I. H. Batty, and S. R. Nahorski, *Biochem. Biophys. Res. Commun.* **154,** 684 (1988).

[53] M. F. Roberts, *FASEB J.* **10,** 1159 (1996).

Author Index

Numbers in parentheses are footnote reference numbers and indicate that an author's work is referred to although the name is not cited in the text.

A

Abe, M., 165
Abood, M. E., 43
Abraham, W. M., 244, 264
Abrams, J. S., 242
Acah-Orbea, H., 185
Accavitti, M. A., 70
Ackerman, V., 245–246, 265
Adams, A. B., 166
Adams, D. H., 265
Adams, G. K., 245, 262(22)
Adams, R. R., 57, 64(15)
Adang, M. J., 40
Adcock, I. M., 258
Adelman, M., 188
Adermann, K., 72
Adler, H., 166
Aguirre, V., 259
Ahlstedt, S., 243
Ahmed, A., 264
Ahnert-Hilger, G., 360, 361(62)
Ahuja, M., 64, 150
Ahuja, S. K., 57, 73, 86, 109, 117, 117(12)
Aiuti, A., 198, 199(86)
Ajakane, M., 347
Akamizu, T., 58
Åkerman, K., 346
Akerman, K. E., 43
Alam, R., 241, 252(62), 253–255, 256(2)
al-Aoukaty, A., 313
Albelda, S. M., 215
Albertine, K. H., 162, 176(2), 177(2)
Aldovini, A., 119
Alexander, R. B., 370
Alexander, W., 4
Alexander, W. A., 125
Alfonso, L. F., 165
Ali, H., 4
Alizon, M., 119
Alkhatib, G., 85, 118, 119(3), 131(3)
Allavena, P., 86, 97(16)
Allaway, G. P., 85, 118–119, 119(4), 131(4), 368
Allen, R., 199(107), 200
Allen, R. M., 199(109, 111), 200
Allende, J., 110
Alphonso, M., 21
Al-Saady, N., 165
Altschul, S. F., 71
Altshuller, Y. M., 357
Alvarez, F. J., 165
Alvfors, A., 164
Amador-Perez, J. F., 363
Amara, A., 199
Amatruda, T. T., 313
Ameri, A., 163
Ancic, P., 243
Andalibi, A., 150
Anders, J., 43
Anderson, A. O., 340
Anderson, D. C., 264
Anderson, D. L., 164
Anderson, G., 151
Anderson, W. H., 347
Anderson-Walters, D. J., 255
Andrews, P. W., 34
Andrews, W. J., 57, 64(15)
Angel, P., 354, 356(43)
Angeletti, C. A., 197
Angiolillo, A. L., 195
Anraku, Y., 322
Ansiowicz, A., 199
Anthes, J. C., 357

Antonsson, J. B., 164
Aoki, J., 85
Appella, E., 199, 250, 252, 255, 272, 340
Apple, T., 265
Applebury, M. L., 316
Apuzzo, M., 97
Arai, H., 70, 81(16), 82(16), 83, 83(16), 366, 375(28)
Arai, I., 166
Araujo, D. M., 28
Araya, M., 169, 170(5)
Arenberg, D. A., 190, 195, 201, 201(58, 60), 202(60), 203(60), 204(60), 207(60), 210(60), 211(117, 118), 214(117), 217(118), 218(118)
Arenzana-Seidedos, F., 130, 199
Arisawa, M., 322
Arkinstall, S. J., 346, 364, 365(20)
Armatruda, T. T., 326
Arnaiz, A., 165
Aronica, S., 313
Arriza, J. L., 4
Arthur, A. K., 38
Arya, S. K., 130
Asano, T., 323
Aschauer, H., 250, 272
Ashorn, P. A., 119, 122(10)
Assoufi, B., 253
Atchison, R. E., 57, 64(12)
Atherton-Fessler, S., 39
Atkinson, L., 253
Aubry, J.-P., 382
Auerbach, R., 190, 196(2)
Augereau, J. M., 346
Aunis, D., 360, 361(63)
Ausprunk, D. H., 191
Avdi, N., 264
Axel, R., 113, 119
Axelsson, B. I., 164

B

Bach, M., 43
Bacherlerie, F., 199
Bacon, K. B., 85–86, 109, 114(14), 117(14), 241, 252, 254, 312, 340–341, 341(2), 342, 343(13c), 344, 346, 346(6), 347, 348, 352, 352(27), 353(41), 362, 364–365, 365(2, 20), 366, 368, 374, 382–383
Bader, M.-F., 360, 361(63)
Badwey, J. A., 3
Baganoff, M. P., 165
Baggiolini, M., 3, 15, 72, 130, 198–199, 199(79, 80, 84, 103, 104), 200, 202, 250, 252(61), 253, 255–256, 257(61, 92), 272, 282, 284(11), 340, 341, 341(1), 364–365
Bahl, O. P., 38
Bailey, O., 183
Baird, A., 192
Bajaj, M. S., 163
Bajaj, S. P., 163
Baker, J. B., 202, 251, 256, 265
Bakhle, Y. S., 261
Bakker-Woudenberg, I. A. J. M., 226
Balaraman, V., 165
Balboa, M. A., 358
Balch, W. E., 309
Baldwin, J. M., 56
Balentien, E., 3
Balis, J. U., 164–165
Balk, R. A., 164
Ban, T., 58
Banks, P. L., 164–165
Banna, P., 164
Banno, Y., 358
Bar, T. H., 191
Barak, L. S., 3–4, 4(2)
Barany, G., 17
Bardwell, L., 199
Barella, L., 364
Barkans, J., 253
Barker, J., 151, 199(110), 200
Barlett, R. H., 166
Barneon, G., 243
Barnes, P. J., 244, 247, 253
Barnwell, J., 17, 35, 150
Barrett, T., 357
Basha, M. A., 199(105, 106), 200
Basiripour, L., 121
Bates, M. E., 243
Battey, J. F., 38, 43
Batty, I. H., 383
Baudet, V., 347
Baudin, G., 362
Bauer, S. I., 195, 202, 202(57)
Baumruker, T., 246
Baxter, G. T., 87, 88(22), 90(22)
Bayse, G. S., 134, 137(2), 140(2)
Bazan, J. F., 85, 340, 365
Beaubien, B. C., 247

Beaven, M. A., 363
Becker, A. B., 244
Becker, F. S., 199(111), 200
Becker, P. L., 366, 368(27)
Beckmann, M. P., 26, 57, 58(5)
Beckner, S. K., 309, 311, 316–318, 319(35), 320(35), 321, 322(29), 342
Bednar, R. A., 348
Behrens, B. L., 244
Bellevergue, P., 347
Bellini, A., 245–246, 265
Ben-Baruch, A., 10, 15(30), 312
Bengali, K., 10, 15(30)
Benjamin, C. D., 264
Bennett, E. D., 165
Bennett, G. L., 18, 134
Ben-Nun, A., 187
Benovic, J. L., 4
Benson, A. A., 358
Bentley, A. M., 246
Benya, R. V., 43
Ben-Yaakov, M., 242
Berger, E. A., 85, 118–119, 119(2, 3), 120, 121(2), 122(10, 17), 123(17), 131(3)
Berger, J., 132
Berghuis, A. M., 321
Berkman, N., 247, 253, 258
Berliner, J. A., 150
Bernard, S., 151
Berridge, M. J., 302, 306, 307(2)
Berson, J. F., 57, 64(11), 70, 118, 120, 129, 130(29), 131
Berstein, G., 323, 324(67)
Berthiaume, Y., 176, 177(7)
Bessia, C., 199
Besson, G., 129
Bett, A. J., 239
Beume, R., 165
Beutler, B., 236
Bhakdi, S., 360, 361(62)
Bhardwaj, R., 196
Biache, G., 43
Bialojan, C., 349
Bianchi, G., 86, 97(16)
Bickel, C., 257
Bidlingmaier, F., 3
Billah, M. M., 357
Billiar, T. R., 166
Binns, O. A., 165–166
Birkenbach, M., 85
Birks, C. W., 26, 57, 58(5)
Birnbaumer, L., 315, 323
Bischoff, J., 191(20), 192
Bischoff, S. C., 252, 255, 257(84)
Bishop, D. H. L., 39–40
Bishop, J. M., 348, 355
Bjertnaes, L. J., 165
Black, D., 117, 165
Blakemore, W. F., 184
Blanchard, D., 64
Blank, J. L., 323, 324(67)
Blaser, K., 341
Blecher, M., 316–318
Blenis, J., 353, 363
Bleul, C. C., 198–199, 199(86)
Blocher, C. R., 164
Blomqvist, A. G., 85
Bloom, R. J., 164
Bloor, C. M., 192
Blumenfeld, W., 197
Blumenthal, R., 119, 132
Bochner, B., 257
Boichot, E., 244
Boissin, P., 347
Bokoch, G. M., 39, 58, 323, 341
Bolanowski, M. A., 165
Bolen, J. B., 352, 353(41), 363
Bolton, A. E., 134, 137(4, 5), 142(5)
Bolton, E. S., 370
Boman, A. L., 358, 360(53)
Bond, M. W., 246
Bongiorno, C., 380
Bonifacino, J. S., 121
Bonvini, E., 311
Boone, C., 58
Booth, T. F., 39
Borden, E. C., 195
Borg, T., 164
Borgeat, P., 308
Boring, L., 57
Borovetz, H. S., 166
Bosco, M. C., 199, 202(91)
Bottazzi, B., 86
Bouck, N. P., 191, 195, 197(9)
Boudin, H., 43
Boudy, S., 57
Bouillon, T., 320
Boulay, F., 364
Bourne, H. R., 83, 311, 325
Boursier, E., 347

Bousquet, J., 243
Bouvier, M., 43
Boyars, M., 241, 256(2)
Boyce, F. M., 40
Boyer, J. L., 323
Boylan, A. M., 169
Boyle, W. J., 9
Bozon, C., 43
Braciak, T. A., 254
Bradely, S., 39
Bradl, M., 183, 188
Brady-Kalany, S. M., 349
Bramson, J. M., 239
Brandt, D. R., 320, 323
Brann, M. R., 43
Braquet, P., 244
Brem, H., 193, 195, 200, 201(114)
Brem, S., 195, 195(44)
Brewer, G. J., 31
Brickey, D. A., 39
Brieland, J. K., 256
Brinkley, M., 151
Britton, B., 165
Broaddus, V. C., 161, 163, 169–170, 170(5), 173
Brockmann, D. C., 163
Brodde, O. E., 322
Broder, C. C., 57, 64(11), 85, 118–119, 119(2, 3), 120, 121(2), 122(17), 123(17), 129, 130(29), 131(3)
Broide, D. H., 262
Brooks, P. C., 191(21, 22), 192, 195(21, 22)
Brosius, J., 326, 327(3)
Brown, A. M., 310
Brown, H. A., 358
Brown, K. O., 323
Brown, L., 244
Brown, R. F., 3
Brown, T. T., Jr., 164
Brown, Z., 150, 199(108), 200
Browne, D., 40
Broxmeyer, H. E., 313
Brunner, C., 188
Brunner, T., 252, 252(61), 253, 255, 257(61, 84)
Bruns, R. F., 346
Buchanan, S. A., 165–166
Bucher, N. L., 40
Buck, C. A., 215
Buhl, A. M., 3, 341(2), 342
Buhl, M. A., 313
Buller, R., 119
Bulmen, D. E., 183
Burdick, M. D., 195, 199(112, 113), 200–201, 201(58, 60), 202(60), 203(60), 204(60), 207(60), 210(60), 211(117, 118), 214(116, 117), 217(118), 218(118)
Burgener, R., 198, 199(84)
Burgess, S., 39
Burkhart, M., 85, 118, 119(5), 368
Burnier, J., 17, 26(4), 64, 150
Burrows, L., 251
Burtonboy, G., 130
Buss, J. E., 39
Busse, W. W., 242
Butcher, E. C., 313, 340, 341(12), 342, 365, 368, 383(21)
Butterfield, J. H., 246
Button, D., 380
Byers, H. R., 202
Byrne, K., 164

C

Cabral, G. A., 43
Cahero, T. G., 3
Calayag, M. C., 114
Calhoun, W. J., 242–243
Camerato, T., 18
Camp, R. D. R., 254, 340, 346(6), 382
Campbell, A. M., 243
Campbell, E. M., 254
Campbell, J. J., 313, 340, 341(12), 342, 365, 368, 383(21)
Campbell, P. A., 222
Canfield, D., 186
Cao, Y., 195
Caon, A. C., 307–308, 308(16)
Caput, D., 252(61), 253, 257(61)
Car, B., 198, 199(84)
Cardon, M. C., 40
Carey, P. D., 164
Carley, W. W., 150
Caron, M. G., 3–4, 4(2), 43, 58, 326, 327(2)
Carr, M. W., 255
Carre, C., 244
Carrico, C. J., 166
Carroll, M., 251, 253

Carroll, P. R., 197
Carroll, R. G., 166
Carson, H. F., 195, 202(57)
Carstens, E. B., 40
Carver, J. M., 119
Casanas, S. J., 43
Casasnovas, J. M., 198, 199(86)
Casey, P. J., 4, 39, 319
Casida, J. E., 349
Castiello, A., 164–165
Cavanagh, S. L., 57
Cayanan, C., 85, 118, 119(4), 131(4), 368
Celestin, J., 261
Cen, Y.-H., 38, 68, 70
Ceradini, D., 129, 130(28)
Cerletti, C., 3, 86, 306
Cerretti, D. P., 26, 57, 58(5)
Cerrito, P. B., 155
Ceska, P., 199(108), 200
Cetin, Y., 72
Chaliss, R. A., 362, 380, 383
Chan, S. Y., 195, 201(60), 202(60), 203(60), 204(60), 207(60), 210(60)
Chan, S. D. H., 84
Chand, N., 262
Chanez, P., 243
Chang, C.-H., 119
Chantry, D., 28
Chapman, R. W., 245, 262(22)
Charcot, J., 182
Charo, I. F., 57, 64(12), 70, 72–73, 74(15), 77, 77(2), 79(2), 81(16), 82(16), 83, 97, 105(27, 28), 109, 313, 326, 338(6), 364
Chaverri-Almada, L., 150
Chazenbalk, G. D., 43
Cheever, F., 183
Cheifetz, I. M., 165
Chen, C., 124, 195
Chen, W., 325
Chen, W. Y., 38
Chen, X., 43
Chen, Y.-H., 3
Chen, Y. J., 70
Cheng, C., 378, 381(42), 382(42)
Cheng, Q. C., 3
Cheng, S. H., 38
Chenier, T. C., 253
Chensue, S. W., 199(105, 106, 108, 109, 111), 200, 243, 255
Cheresch, D. A., 191(21), 192, 195(21)
Cherish, D. A., 191(22), 192, 195(22)
Cherwinski, H., 246
Chesebro, B., 119
Chi, E. Y., 162, 164
Chidiac, P., 43, 325
Chihara, J., 253
Chio, C. L., 98
Chirgwin, J. M., 228
Cho, L. C., 365
Choe, H., 118, 119(7), 122(7), 199
Choe, S., 64, 85, 118, 119(5), 129, 130(28), 150, 368
Chou, M. M., 353, 363
Christianson, G., 195
Christman, J., 164
Christophers, E., 250, 252, 253(59), 266, 271–272, 272(2), 276, 280(2), 282, 284(12, 13), 285, 289(13), 294
Chung, K. F., 244, 247
Chuntharapai, A., 11, 15, 17, 26, 26(4), 64, 150, 169
Church, M. K., 251
Cima, C., 164
Cisek, L. J., 38
Clanton, T. L., 166
Clapham, P. R., 64, 119, 129, 150
Clark, R., 39
Clark, R. A., 191(21), 192, 195(21)
Clark, W. H., 197
Clarke, S. J., 256
Clark-Lewis, I., 3, 130, 199, 202, 254, 260(72)
Clayberger, C., 252
Cleveland, J. L., 311
Coalson, J. J., 166
Cobb, M. H., 309
Cocchi, F., 130
Cocci, F., 166
Cochrane, C. G., 41, 43, 47(34), 55(34), 57–58, 165–166
Cockett, A. T. K., 195
Cockroft, S., 358
Codina, J., 4
Coeffier, E., 262
Coffman, R. L., 246
Cognaux, J., 130
Cohen, I. R., 187
Cohn, S. M., 164–165
Cohn, Z., 199, 202(90)

Colapietro, A. M., 3, 4(2)
Colbran, R. J., 39
Colby, T., 26
Coleman, D. E., 321
Coligan, J. E., 344, 354(16)
Collins, H., 257
Collins, J., 164
Collins, J. A., 163
Collins, P. D., 241, 247, 249(40), 251, 254(40), 257(39, 40), 260(40), 261–262
Collman, R. G., 57, 64(11), 118, 119(6), 120, 120(6), 122(6), 129–130, 130(29)
Colman, A., 108, 109(2), 117(2)
Colombo, M. P., 364
Colotta, F., 192
Colpaietro, A.-M., 4
Combadiere, C., 73, 85–86, 118, 119(3), 131(3)
Condina, J., 323
Cone, R. D., 325
Conklin, B. R., 83, 311
Conlon, K., 254
Conn, H., 195
Connelly, A. J., 97, 105(27)
Connolly, A. J., 73, 109
Content, J., 114
Contreras, R., 114
Cook, D. G., 129
Coons, M. S., 166
Cooper, S., 313
Copeland, N. G., 85, 259, 340, 365
Corbin, R. S., 165–166
Corden, J. L., 38
Corrigan, C. J., 253
Cortes, A., 264
Coste, H., 347
Cotecchia, S., 58
Cotman, C. W., 28
Cotran, R., 190, 193, 196(39), 197(1)
Coughlin, S. R., 73, 80(11), 97, 105(27), 109
Coulson, A. R., 207
Courtneidge, S. A., 3
Couture, L., 43
Coyle, A. J., 253, 259
Coyle, T., 244
Craig, D. M., 165
Crawford, M., 236
Crawford, S., 39
Crespo, P., 3
Croitoru, K., 254
Cromwell, O., 243, 251
Crooke, S. T., 308
Crum, R., 195
Cullen, B. R., 132
Cullino, P. M., 197
Cunningham, E., 358
Cunningham, P., 254
Cuomo, A. J., 262
Cybulsky, M. I., 261

D

Dahinden, C. A., 252, 252(61), 253, 255, 257(61, 84)
Dahlback, C. M., 164
Dahms, T. E., 163
Dairaghi, D., 365
Dalgleish, A. G., 119
Daly, J. W., 326
Damiani, P., 166
D'Amore, P. A., 193
Danieli, T., 132
Daniels, J., 183
D'Apuzzo, M., 256
Darbonne, W. C., 265
Das, A., 264
Dauber, J. H., 164
Daugherty, B. L., 261
Davies, D. E., 251
Davies, E. V., 306
Davies, G. C., 166
Davies, M. P. A., 303
Davies, R. J., 253
Davis, C. B., 85, 118, 119(5), 368
Davis, M. D., 190, 197(7)
Davis, M. M., 252
Davis, N. G., 58
Davis, R. J., 357
Davis, R. L., 188
Davis Turner, J., 150
Davoes, D. E., 253
De, F. E., 348
Dean, N. M., 382
de Cordoba, S. R., 35
DeGregorio, D., 72
de Herreros, A. G., 38
Deinzer, M., 38
Dejana, E., 192
Del Canto, M., 183
Delner, G. S., 340
DeMartino, J., 57, 261

DeMartino, J. A., 57
DeMichele, S. J., 164–165
Deng, H., 85, 118, 119(5), 368
Deng, T., 357
Denis, M., 164
Dennis, M., 43
Dent, P., 355
Deppeler, C. L., 165
Der, C. J., 39
Derijard, B., 357
Deutsch, E., 199
Devalia, J. L., 253
Devergne, O., 150
DeVico, A. L., 130
Devrotes, P. N., 70
Dewald, B., 3, 15, 72, 198, 199(79), 202, 256, 257(92), 340, 341(1)
Diamantis, W., 262
Diamond, A. B., 20
Diaz, P., 243
Dieterich, K., 10
Dietsch, G. N., 28
Digilio, L., 57, 64(12)
Dignan, R. J., 164
DiGregorio, D., 73(3), 368
Di Marzio, P., 85, 368
Dimitrov, D. S., 119
Ding, J., 3
Dion, S. B., 4
DiPietro, L. A., 193, 196, 196(38), 201(70), 202(70)
Dixit, V. M., 192, 197, 199(110), 200
Dlugosz, A., 378, 381(42), 382(42)
Dobner, T., 85
Doerfler, W., 38
Dohlmam, H. G., 326, 327(2)
Dolhnikoff, M., 166
Domingo, D. L., 38
Doms, R. W., 57, 64, 64(11), 70, 118, 119(6), 120, 120(6), 122(6), 129, 130, 130(29), 131, 150
Dong, G. Z., 43
Donner, A. L., 195, 202(57)
Doranz, B., 70
Doranz, B. J., 57, 64(11), 118, 119(6), 120, 120(6), 122(6), 129–130, 130(29), 131
Dorinsky, P. M., 166
Dormehl, I. C., 163
Dornreiter, I., 38
Dörschner, A., 252, 253(59), 266, 282, 284(12)
Doucet, J. P., 320
Dower, S. K., 285
Downes, C. P., 362
Downey, W. E., III, 4
Dragic, T., 85, 118, 119(4), 131(4), 368
Drazen, J. M., 258, 259(100)
Dreesman, G. R., 39
Drenckhahn, D., 38
Drevets, D. A., 222
Drong, R. F., 98
Drott, P., 164
Drummond, I. A. S., 369
Du, J.-G., 38, 68, 70, 129
Dubois, G. C., 252, 255(55)
Dubois, P. A., 313
Dubois, P. M., 364
DuBose, R. A., 4
Dubyak, G. R., 361
Dumont, J. N., 109
Dunn, C. J., 243
Durell, S., 132
Durham, S. R., 243, 246–247, 253
Dyer, B. J., 252
Dzuiba, J., 195, 201(60), 202(60), 203(60), 204(60), 207(60), 210(60)

E

Earl, P., 120, 122(18), 123(18), 124(18)
Earley, K., 43
Easa, D., 165
Ebers, G. C., 183
Edelson, J. D., 166
Edgren, E. L., 165
Edinger, A. L., 118, 131
Egan, R. W., 245, 262(22)
Eguchi, K., 348
Ehrengruber, M. U., 3
Ehuja, M., 129
Eisenstein, R., 193, 195, 197(42)
Eisfelder, B. J., 313
Elder, J. T., 199(110), 200
Elhammer, A. P., 38
Elliott, C. M., 358
Elliott, G. A., 243
Ellmeier, W., 85, 118, 119(5), 368
Elner, S. G., 196, 199(104), 200–201, 201(70), 202(70)
Elner, V. M., 196, 199(104), 200–201, 201(70), 202(70)

Elwood, W., 244
Emilie, D., 150
Emmrich, F., 362
Emrich, T., 85
Enander, I., 243
Encinas, J., 184
Endres, M. J., 34, 64, 129, 150
Engebrecht, J., 357
Engerman, R. L., 190, 197(7)
Erikson, R. L., 39
Eriksson, M., 164
Eriksson, O., 164
Erturk, E., 195
Eskandari, M., 199(102), 200
Espinoza-Delgado, I., 199, 202(91)
Estes, M. K., 323
Etkin, L. D., 39
Evanoff, H. L., 199(109), 200, 256
Evans, R. L., 43
Evans, S. W., 311, 318, 319(35), 320(35), 321
Evege, E. K., 31
Exler, R., 166
Exton, J. H., 323, 324(67), 358
Exum, S., 58

F

Faccioli, L. H., 251, 257
Fahrenholz, F., 43
Falk, A., 163–164
Falk, J. L., 39
Falterman, K., 191
Fan, T. P. D., 201
Fan Chung, K., 253
Fang, Z. T., 150
Fanning, E., 38
Fantone, J. C., 256
Farber, C.-M., 130
Farber, J. M., 195, 199, 202(89), 207(89)
Farfel, Z., 83
Fargin, A., 43
Farrar, W. L., 311, 318, 319(35), 320(35), 321
Farzan, M., 118, 119(7), 122(7)
Farzen, M., 199
Fasolato, C., 302
Faubion, D., 252(62), 253
Faulkner, P., 40
Fay, F. S., 366, 368(27)
Feeney, W., 254
Feinberg, M. B., 121
Feng, Y., 85, 118, 119(2, 3), 121(2), 131(3)
Fenselau, A., 195
Fensome, A., 358
Ferdeghini, M., 166
Ferguson, G. P., 192
Ferguson, S. S. G., 3–4, 4(2)
Fernandez, X., 245, 262(22)
Ferrara, P., 252(61), 253, 257(61)
Feucht, P. H., 17
Fields, T. A., 319
Fiers, W., 114
Filippi, E., 166
Fingar, V. H., 148, 155
Finger, E., 261
Fink, M. P., 164
Finn, K. C., 165
Fischer, E., 251
Fisher, J. E., 164–165
Fisher, P. A., 57
Flanagan, R., 253
Flavahan, N. A., 243
Flax, J., F197
Fleisch, J. H., 164
Flores-Romo, L., 346, 364, 365(20)
Fluck, R., 164
Fogarty, K. E., 366, 368(27)
Fogelman, A. M., 150
Foley, J. J., 308
Folkesson, H. G., 163, 173
Folkinan, J., 192, 193(27)
Folkman, J., 190–191, 193, 193(5), 195, 195(17, 44), 196(5), 197, 197(1, 4), 200, 200(4), 201(4, 114)
Fong, T. M., 312
Fong, Y. L., 39
Fontanini, G., 197
Foran, S. K., 245, 262(22)
Forceille, C., 130
Forsgren, O., 164
Forsgren, P. E., 164
Forssmann, U., 72, 256, 257(92)
Forssmann, W.-G., 72
Forst, H., 166
Forsythe, P., 241, 252(62), 253–255, 256(2)
Foster, D. C., 17, 26(7), 27(7)
Foster, P. S., 244, 262(15)
Fournier, S., 320
Fraker, P. J., 134, 137(3), 139(3)
Franci, C., 57, 64(12), 73, 97, 105(27, 28), 109
Francke, U., 109, 117(11)

Francoeur, C., 164
Fraser, M. J., 40
Frazier, W. A., 195
Freedman, N. J., 4
Freer, S., 358
Freisewinkel, I., 39
Freissmuth, M., 321, 323
Frese, K., 38
Frew, A. J., 251
Frick, O. L., 244
Fridman, R., 192, 193(27)
Friend, D., 259
Frierson, H., 165–166
Fritsch, E. F., 113
Fritz, R., 187
Frohman, M. A., 357
Fuerst, T. R., 125–126
Fuhlbrigge, R. C., 198, 199(86)
Fuhrman, B. P., 165
Fujiki, H., 346, 349
Fujimoto, M., 245
Fujimura, M., 256
Fujisawa, T., 243
Fukuchi, Y., 163
Fukuda, K., 108, 195
Fung-Leung, W., 185
Fuortes, M., 164–165
Furukawa, T., 165
Furusho, K., 245
Furuya, S. H., 349
Fusetani, N., 349

G

Gadek, J. E., 166
Galanaud, P., 150
Galas, M. C., 3
Gallegos, C., 17, 150
Galleguillos, F. R., 243
Gallichan, W. S., 239
Gallin, E. K., 108
Gallo, R. C., 130
Gallup, J. Y., 195
Gamble, J. R., 191(19), 192
Gao, J.-L., 73, 85–86, 109, 117(11)
Garcia-Zepeda, E. A., 73, 261
Gardner, P., 86, 312, 347, 352(27)
Garnovskaya, M. N., 43
Garotta, G., 72, 256, 257(92)
Gärtner, I., 271
Gary, D., 185
Garzino-Demo, A., 130
Gastiasoro, E., 165
Gattai, V., 166
Gatto, L. A., 164
Gauldie, J., 239, 254
Gautan, N., 311
Gayle, R. B. III, 26, 57, 58(5)
Geiger, R., 132
Geiser, T., 3, 202, 252(61), 253, 257(61)
Genain, C. P., 186, 188
Genard, M., 199(102), 200
Geny, B., 358
George, S. R., 43
Georges, M., 130
Gerard, C., 118, 119(7), 122(7), 261, 313, 326
Gerard, N. P., 118, 119(7), 122(7), 261, 313, 326
Gerdin, B., 164
Germann, P. G., 165
Gerritsen, M. E., 150, 192
Gettys, T. W., 43
Geyer, H., 38
Geyer, R., 38
Ghavanian, M. N., 243
Giedlin, M. A., 17, 246
Gierschik, P., 10, 313, 320
Giladi, E., 117
Gilbert, C., 307, 308(16)
Gilbert, D. J., 365
Gilby, T., 253
Gill, G. S., 98
Gilligan, L., 166
Gilman, A. G., 309–310, 319, 321, 323
Gilon, C., 348
Gimbrone, M. A., Jr., 193, 196(39)
Gimpl, G., 43
Gish, W., 71
Glabinski, A. R., 35, 182
Glass, M. C., 201, 211(117, 118), 214(117), 217(118), 218(118)
Gleich, G. J., 242–243, 253
Glenny, R. W., 151
Gleye, J., 346
Glick, A., 378, 381(42), 382(42)
Godard, P., 243
Godfrey, P. P., 378, 380(44), 383(44)
Godiska, R., 28
Goebeler, M., 196
Goeddel, D. V., 252, 254

Goff, C. D., 165–166
Gold, M. R., 348
Gold, W. M., 244
Goldblum, S. E., 227
Golding, H., 119
Goldsmith, E. J., 309
Goldsmith, M. A., 57, 64(12)
Goldstein, I. M., 169
Goldstein, L. A., 256
Gonzalez, M. C., 243
Gonzalo, J.-A., 254, 256, 259, 260(72), 261
Good, D. J., 195
Goodman, R. B., 17, 26(7), 27(7), 162
Gordon, C. A., 17
Gorman, C., 19
Gosling, J., 57, 64(12)
Gotgone, G., 150
Goulbourne, I. A., 166
Gout, I., 358
Goverman, J., 184
Goya, T., 163, 165
Graff, R., 323
Graham, F. L., 239, 254
Grand, P. T., 347
Grant, D., 195
Grant, D. S., 201
Grant, J. A., 241, 252(62), 253–255, 256(2)
Gratzl, M., 360, 361(62)
Gray, G., 57, 58(6), 195, 202(57)
Gray, P. W., 28
Grazino, M. P., 323
Greeley, W. J., 165
Greenberg, S. M., 192, 195(34)
Greenwood, F. C., 134, 137(1)
Gregory, H., 282, 284(13), 289(13)
Gregory, T. J., 164–165
Griffith, B. P., 166
Griffiths, A. D., 27
Griffiths, C. E. M., 199(110), 200
Griffiths-Johnson, D. A., 241, 247, 249(40), 254(40), 257(39, 40), 258, 260(40), 261–262
Grimley, P. M., 125
Gröner, A., 38
Gross, J. L., 192
Grotjohnan, H. P., 165–166
Grynkievicz, G., 302, 303(3), 375, 376(37)
Gudermann, T., 56
Gundel, R. H., 244
Guo, H.-H., 38, 68, 70, 129, 148
Gurdon, J. B., 108
Gurevich, V. V., 4
Gurney, A. M., 376
Guse, A. H., 362
Gusella, G. L., 199, 202(91)
Guth, B. D., 151
Gutierrez-Ramos, J.-C., 254, 256, 259, 260(72), 261
Gutkind, J. S., 3

H

Haase, W., 43
Hachicha, M., 86
Haddara, W., 239
Hadley, T. J., 55, 57, 58(7), 64, 64(7), 150
Haeffner, E. W., 382
Hafeman, D. G., 87, 90(21)
Hafner, D., 165
Haga, T., 43, 108
Hague, N., 313
Haines, G. K., 199(112, 113), 200
Halks-Miller, M., 27
Hall, A., 309, 314(4), 321, 325(4)
Hallacher, L. M., 377
Haller, E. M., 164–165
Hallet, M. B., 306
Hamid, Q. A., 246–247
Hammond, M. E. W., 17
Hammond, S. M., 357
Han, J. H., 3
Hanahan, D., 197
Hancock, G., 199, 202(90)
Haney, J. A., 201
Hangen, D. H., 164
Hanley, M. R., 378
Hara, M., 251
Hara, T., 365
Harden, T. K., 3, 323
Hardesty, R. L., 166
Hardie, D. G., 344, 345(15), 354(15)
Hardin, M. J., 197
Haribabu, B., 4
Haribabu, D., 4
Harlan, J. M., 176
Harlow, E., 24
Harlow, L. A., 196, 199(112), 200, 201(70), 202(70)
Harris, C. C., 197
Harris, E. D., Jr., 196

Harrison, J. E., 262
Harrison, R., 252(62), 253
Harter, J. M., 313
Hartig, P. C., 40
Hartzell, P., 363
Hasegawa, N., 164
Hasel, K. W., 119
Haseltine, W., 121
Hassan, N., 313
Hattori, S., 316, 321, 322(29)
Hauber, R., 132
Haudenschild, C. C., 195
Hauschke, D., 165
Hauser, C., 38, 341
Hauser, S. L., 186, 188
Hawes, A. S., 251
Hawes, B. E., 3
Hawkins, P. T., 378
Hawkins, R. E., 27
Hayashi, S., 190
Hayashi, Y., 246
Hayashida, H., 166
Hayman, S., 164
Heard, J. M., 199
Heath, W. F., 346
Hébert, C. A., 17, 26, 26(4), 64, 150, 161, 163, 169, 173, 202, 251, 265
Hedrick, S. M., 363
Heikkila, J., 346
Heilker, R., 198
Hein, L., 73, 80(11)
Hellmich, M. R., 43
Henderson, R., 56
Hendey, B., 308
Henis, Y. I., 132
Henson, P. M., 264
Hepler, J. R., 323
Her, J.-H., 355
Herbert, C., 254
Herbert, J. M., 346
Herlyn, M., 197
Herman, A., 73, 97, 105(27), 109
Herman, A. G., 251
Hernan, L. J., 165
Herrera, R., 38
Hesselgesser, J., 27, 55, 57, 58(7), 64(7), 150
Heuser, L. S., 151
Heuz, G., 114
Hibi, H., 356, 357(47)
Hibi, M., 357
Hickey, W. F., 187
Hickling, K. G., 165
Higashijima, T., 323, 324(67)
Higashimoto, I., 253
Higgins, J. B., 4
Hilbert, M., 56
Hiles, I., 358
Hilkmann, H., 358
Hill, C. M., 85, 118, 119(5), 368
Hillman, N. D., 165
Hilt, S., 17
Hirano, H., 265
Hirose, T., 108
Hirschl, R. B., 166
Hirst, M., 84
Hirst, S. J., 247
Hitt, M., 239
Ho, A., 185
Ho, S. N., 73
Hoch, R. C., 165
Hoeffel, J. M., 169
Hofer, A. M., 307
Hoflack, J., 56
Hogan, S. P., 244, 262(15)
Hogeland, K., 38
Holgate, S. T., 246, 251, 253
Hollfelder, K., 358
Holloran, M. M., 199(113), 200
Holm, B. A., 165
Holman, M. J., 164–165
Holmes, R., 57, 64(15)
Holmes, W. E., 18
Holmgren, L., 195
Homandberg, G. A., 195
Honda, Z., 108
Honjo, T., 198
Honkanen, R. E., 349
Hood, L., 184
Hoogenboom, H. R., 27
Hoogewerf, A. J., 109, 114(14), 117, 117(14), 365
Hopp, T. P., 285
Hora, D., 254
Hori, T., 348
Hori, Y., 201
Horovitz, J., 166
Horton, R. M., 73
Horuk, R., 18, 26–28, 47, 55, 57, 58(7), 64(7), 72, 73(3), 129, 130(28), 134–135, 149–150, 153, 326, 342, 368

Hosaka, S., 199(113), 200
Hosey, M. M., 4, 43
Hosie, M. J., 122
Höss, A., 38
Hoth, M., 302
Houslay, M. D., 324
Howard, D. E., 132
Howarth, P. H., 251
Howbert, J. J., 346
Howe, P. H., 10
Hoxie, J. A., 64, 70, 122, 129, 150
Hsu, C.-Y. J., 348
Hsuan, J. J., 247, 249(40), 254(40), 257(40), 260(40), 358
Hu, D. E., 201
Hu, T., 191(22), 192, 195(22)
Huang, C., 43
Huang, K.-S., 358
Huang, S.-K., 247
Huang, Y., 85, 118, 119(4), 131(4), 368
Huber, N. E., 246
Hubscher, S., 265
Huet, J. C., 43
Huff, R. M., 98
Hulme, E. C., 43
Humphries, G. M. K., 84
Hunninghake, G. W., 242
Hunt, A. J., 195, 202(57)
Hunt, H. D., 73
Hunt, L. W., 253
Hunter, T., 9
Hunter, W. M., 134, 137(1, 5), 142(5)
Hurst, E., 182
Hyers, T. M., 163

I

Iannettoni, M. D., 201, 211(118), 217(118), 218(118)
Ichiyama, A., 108
Ida, N., 241, 256(2)
Idell, S., 166
Ikezu, T., 319
Imai, T., 130, 166
Imoto, M., 348
Ingber, D. E., 191, 195(17, 18), 197
Inglese, J., 4, 324
Inoue, S. B., 322
Insel, P. A., 322, 358
Irvin, C. G., 244
Irvine, G. B., 280
Ishai-Michaeli, R., 192, 193(27)
Ishii, K., 73, 80(11), 166
Ishioka, C., 245
Ishizaka, A., 164
Isshiki, K., 348
Ito, K., 108
Iwamota, P., 242

J

Jacob, T. D., 166
Jacobs, K. H., 10, 313, 317, 320
Jacobs, L., 164
Jacobs, R. F., 164
Jacobson, M. R., 247
Jaffe, B. M., 164–165
Jaffe, E., 192
Jakubicki, R., 262
Jambrosic, J., 197
James, K. K., 166
Jansen, J. R., 165–166
Jansson, C. C., 43
Janvier, D., 64
Jarvis, D. L., 38
Jay, M., 227
Jazin, E. E., 85
Jebson, P., 166
Jeffery, P. K., 242
Jenkins, N. A., 85, 259, 340, 365
Jenkinson, S., 380
Jenks, M. H., 70
Jensen, R. T., 43
Jessell, T. M., 113
Jewell, J. E., 40
Ji, I., 58
Ji, T., 58
Jia, G.-Q., 256, 259
Jiang, H., 325, 364
Jiang, X., 358
Jilek, P., 86
Jirik, F. R., 118
Johnson, G. L., 3, 313, 341(12), 342
Johnson, J., 189
Johnson, K., 166
Jolin, A., 165
Jolley, D., 57, 64(15)
Jondal, M., 363
Jones, C. T., 322
Jones, G. M., 370

Jones, M. L., 165, 256
Jones, S. A., 3, 312, 313(14)
Jones, T. L. Z., 324
Jongstra, J., 252
Jorgensen, J., 252
Jorquera, H., 110
Jose, P. J., 241, 247, 249(40), 254(40), 257(39, 40), 258, 260(40), 261–262
Josefson, K., 85
Joseph, A. E., 165
Joseph, D., 262, 264
Julius, D., 83, 113
June, C. H., 363, 370

K

Kadletz, M., 164
Kahn, B. A., 165
Kahn, R. A., 358, 360(53)
Kain, W., 199
Kajikawa, O., 176
Kakazu, T., 253
Kallen, R. G., 38
Kamal, G. D., 166
Kameyama, K., 43
Kameyoshi, Y., 252, 253(59), 266, 282, 284(12)
Kanaho, Y., 358
Kanazawa, M., 164
Kane, L. P., 363
Kaneko, T., 251
Kang, C. Y., 39
Kangawa, K., 108
Kanoff, M. E., 92
Kao, J., 303
Kaplan, A. P., 255, 257(86)
Kaplan, G., 199, 202(90)
Kapur, V., 165
Karabin, G. D., 199(110), 200
Kargman, S., 307
Karin, M., 354, 356, 356(43), 357, 357(47)
Karp, M., 43
Kasahara, K., 199(107, 111), 200
Kasahara, T., 246
Kasper, J., 195, 201(60), 202(60), 203(60), 204(60), 207(60), 210(60)
Kasprzyk, P. G., 38
Kassam, N., 254, 260(72), 340, 365, 383(21)
Katada, T., 320, 323, 358
Kato, K., 39
Kato, T., 166
Kato, Y., 349
Katz, H., 259
Kaufman, M. E., 57, 58(6)
Kawabe, Y., 348
Kawanishi, C. Y., 40
Kay, A. B., 243, 246–247, 251, 253, 257
Kehoe, M., 165
Keinanen, K., 43
Kelner, G. S., 85, 365
Kelvin, D. J., 254
Kenamore, C., 255
Kennedy, J., 85, 340, 365
Kennedy, P. E., 85, 118, 119(2, 3), 121(2), 131(3)
Kennedy, R. C., 39
Kermode, J. C., 145
Kern, F. H., 165
Key, M. L., 245
Khan, A. A., 363
Khouja, A., 322
Kieff, E., 85
Kiel, D. P., 164
Kilian, J. G., 163
Kilian, U., 165
Kim, B., 183
Kim, C. M., 4
Kim, J., 26, 53, 70, 264
Kim, J. Y., 70
Kim, K. J., 11, 15, 17, 21, 26(4), 64, 150, 169
Kim, Y. J., 313
Kimata, H., 245
Kimura, H., 187
Kingston, R. E., 124
Kirchgessner, T., 150
Kirchhoff, K., 72
Kirsch, G. E., 310
Kirshnan, V. L., 253
Kishimoto, W., 166
Kiss, Z., 347
Kist, A., 72
Kita, H., 242, 253
Kitamura, S., 246
Kjeldsberg, M. A., 58
Klagsbrun, M., 190, 192–193, 193(5, 27), 196(5)
Klass, D. J., 222
Klausner, R. D., 121
Kleimberg, J., 265
Klein, U., 43
Kleinman, H. K., 195

Klenk, H.-D., 38
Kleuss, C., 321
Kleyensteuber, S., 85, 365
Klier, G., 191(22), 192, 195(22)
Kloc, M., 39
Knall, C., 3, 341(12), 342
Knelder, A., 264
Knierim, M., 192
Knighton, D. R., 191
Kobayashi, E., 346
Kobayashi, T., 165
Kobilka, B., 73, 80(11)
Koch, A. E., 196, 199(112, 113), 200–201, 201(70), 202(70)
Koch, W. J., 3, 4(2)
Koenig, S., 370
Koh, D., 185
Köhler, G., 16
Kohn, L. D., 58
Kojima, M., 108
Kolakowski, L. F., Jr., 41, 87
Konerding, M. A., 192
Konteatis, Z. D., 57
Kopf, M., 259
Koprowski, H., 197
Kormos, C., 255
Kosek, J. C., 121
Kosugi, S., 58
Koup, R. A., 85, 118, 119(4), 129, 130(28), 131(4), 368
Kowallik, P., 151
Kozarsky, K., 121
Kozlosky, C. J., 26, 57, 58(5)
Kramer-Bjerke, J., 195
Krangel, M. S., 198, 199(83), 202(83), 246
Krans, H. M. J., 315
Krasznai, G., 165
Krebs, E. G., 354, 355(44)
Kreider, B., 72, 256, 257(92)
Kremer, L., 256, 261
Krensky, A. M., 252
Kreutner, W., 245, 262(22)
Krieger, M., 121, 252, 255, 257(84)
Krikorian, K., 182
Krishnaswamy, G., 247
Kron, I. L., 165–166
Kruisbeek, A. M., 344, 354(16)
Kruithoff, K. L., 164
Kruys, V., 114
Kuang, W.-J., 18
Kuang, Y., 325, 364
Kubbies, M., 72
Kubo, T., 108
Kuchroo, V., 184
Kuettner, K. E., 193, 197(42)
Kuhns, D. B., 109, 117(11)
Kuliszewski, M., 163
Kulmburg, P. A., 246
Kumai, M., 247
Kumar, M., 164–165
Kuna, P., 255, 257(86)
Kung, H. F., 199
Kunkel, S. L., 150, 190, 195–196, 198–199, 199(80, 84, 102–113), 200–201, 201(58, 60, 70), 202(60, 70, 92), 203(60), 204(60), 207(60), 210(60), 211(117, 118), 214(116, 117), 217(118), 218(118), 220, 243, 255–256
Kunkel, T. A., 207
Kupperman, E., 380
Kurachi, D., 253
Kurashima, K., 256
Kuratsu, J.-I., 252
Kuribara, H., 358
Kurihara, K., 85
Kuroda, K., 38
Kusner, D. J., 361
Kusui, T., 43
Kuusinen, A., 43
Kwan, C. Y., 306
Kwan, S. P., 20
Kwatra, M. M., 4

L

Labacq-Verheyden, A. M., 38
Labbe, O., 92, 107(25)
Labrecque, J., 43
Lachmann, B., 165
Lackey, M. N., 164
Lacoste, J. Y., 243
Lad, P. M., 314, 315(26)
Laemmli, U. K., 209
Lagente, V., 244
Lajiness, M. E., 98
Lakhani, S., 355
Lam, V., 15
Lameh, J., 84
Lamorte, G., 86, 97(16)

Landau, N. R., 64, 85, 118, 119(5), 129, 130(28), 150, 368
Lane, C. D., 108
Lane, D., 24
Lane, S. J., 256
Lane, W. S., 195
Lanford, R. E., 38–39
Lanfrancone, L., 192
Langer, R., 193, 195
Lansing, T. J., 3
Lapa-e-Silva, J. R., 264
Lapointe, G. R., 17
Lapouméroulie, C., 130
Largaespada, D. A., 85, 340, 365
Larhammar, D., 85
Larkin, S., 264
LaRosa, G., 57, 58(6), 118, 119(7), 122(7), 257, 313, 340, 364–365, 383(21)
Larsen, C. G., 252, 255(55), 340
Larsen, G. L., 244
Larson, L., 184
Larsson, A., 163–164
Lassman, H., 188
Latteri, S., 164
Lau, L. C. K., 251
Laudana, C., 313, 341(12), 342
Lauredo, I. T., 264
Laursen, R. A., 188
Lavi, E., 129
Lavu, S., 199
Lazarus, S. C., 244
Leach, C. L., 165
Lebacq, J. A., 38
LeBeau, M. M., 195
Lebwohl, D., 38
Ledbetter, J. A., 313, 364
Leder, P., 192, 195(34), 246, 258, 259(31, 100), 261
Lee, A., 193
Lee, A. S., 369
Lee, E., 321, 378, 381(42), 382(42)
Lee, J., 17–18, 26(4), 64, 150, 241, 256(2)
Lee, J. C., 57
Lee, K., 195
Lee, M. S., 39
Lee, T. H., 256
Lee, V. M.-Y., 28, 34(8)
Leeman, M., 165
Lee-Parritz, D., 186
Lees, M. B., 188
Lefkowitz, R. J., 3–4, 58, 324, 326, 327(2)
Lefort, J., 253
Legler, D. F., 199
Leibovich, S. J., 190, 193, 196, 196(6)
Leibovich, S. L., 193
Lellouch-Tubiana, A., 253
Lemons, R. S., 195
Lenaerts, J.-P., 256
Lennard, E. S., 164
Lenoir, G., 85
Leof, E. B., 10
Leonard, E. J., 199, 250, 252, 255, 272
Leone, D. R., 264
Lerman, M. I., 255
Lesperance, E., 222
Lett-Brown, M. A., 252(62), 253–255
Letts, L. G., 244
Letvin, N. L., 186, 188
Leung, Y. M., 306
Levasseur, S., 86
Levin, D. E., 322
Levitski, A., 348
Levitzki, A., 3
Levy, J. A., 118
Lew, W., 199
Lhiaubet, A. M., 43
Li, H., 72, 256–257, 257(92)
Li, J.-Y., 358
Li, S.-H., 363
Li, W. Z., 165
Li, X., 109, 117(11)
Li, Y., 39, 72, 256, 257(92)
Li, Y. M., 349
Liao, F., 195
Liao, N., 307, 308(16)
Liaw, L., 192
Libert, F., 57, 64(11), 120, 129–130, 130(29)
Lichtman, A., 259
Lichtwarck-Aschoff, M., 165
Lieberman, M., 121
Liedtke, C. M., 378, 381(41)
Liesnard, C., 129–130
Life, P. F., 346, 364, 365(20)
Lilly, C. M., 258, 259(100)
Lima, S. H., 72, 256, 257(92)
Lin, A., 356, 357(47)
Lin, G.-S., 259
Lin, M. C., 314, 315(26)
Lin, S. H., 43
Linden, J., 43

Linder, M. E., 319, 325
Lindley, I., 199(103, 104, 108), 200, 245, 251
Lindsay, T. F., 165
Ling, N., 192
Lingappa, V. R., 114
Linington, C., 183, 188
Lipman, D. J., 71
Lipp, M., 85
Lisowska, E., 64
Littman, D. R., 85, 118, 119(5), 368
Litwin, V., 85, 118–119, 119(4), 131(4), 368
Liu, M., 257
Liu, M. C., 247
Liu, R., 3, 85, 118, 119(5), 129, 130(28), 368
Lloyd, A. R., 254
Lloyd, C., 256
Lloyd, C. M., 261
Lobb, R. R., 264
Locati, M., 3, 86, 97(16), 306
Loetscher, M., 97, 199, 256, 257(92)
Loetscher, P., 72, 130, 256, 341, 365
Loew, D., 21
Logsdon, C. D., 114
Loh, T. T., 306
Lohse, M. J., 4
Lokerse, A. F., 226
Londos, C., 329
Long, D., 118, 131
Long, P. V., 201
Longo, D. L., 245
Lopez-Heredia, J., 165
Loriolle, F., 347
Lotvall, J. O., 244
Lotz, M., 262
Lou, Y.-C., 55, 57, 58(7), 64(7)
Low, M. G., 358
Lowe, P. N., 39
Lowry, S. F., 251
Lu, L., 256
Lu, Z., 188
Lu, Z.-H., 55–57, 58(7), 64, 64(7), 70, 148, 150
Luckow, V., 39
Luini, W., 3, 86, 97(16), 306
Lukacs, N. W., 243, 255–256
Lundkvist, K., 164
Luo, L., 39
Luo, W., 43
Luo, Z., 313
Lusis, A. J., 57, 150
Lusso, P., 130
Luster, A. D., 73, 192, 195(34), 199, 202(90), 246, 255, 258, 259(31, 100), 260(79), 261
Lustig, K. D., 83
Luther, E., 255
Luthin, D. R., 43
Luttrell, L. M., 3
Luttress, D. K., 3
Lutty, G. A., 195
Luxembourg, A., 43
Lynch, J. P., 199(102, 105, 106), 200

M

Macchiarini, P., 197
MacDonald, M. E., 129, 130(28)
MacDonald, R. J., 228
Machen, T. E., 307
MacIntyre, D. E., 254
Mackay, C. R., 28, 118, 119(7), 122(7), 254, 260(72), 261, 340, 365, 368, 383(21)
Mackay, I. S., 247
Mackintosh, C., 349
MacLean, J., 73
MacLean, J. A., 255, 260(79)
MacPhee, C. H., 362
Maddon, P. J., 85, 118–119, 119(4), 131(4), 368
Madri, J., 191, 195(16)
Maeda, A., 108
Maeda, S., 39
Maffrand, J. P., 346
Mägert, H.-J., 72
Maghazachi, A. A., 313
Maheshwari, S., 195
Maier, R. V., 164
Maione, T. E., 195, 202, 202(57)
Maiorana, A., 197
Mak, J., 368
Mak, J. Y., 41, 72, 73(3), 87
Mak, T., 185
Malcolm, K. C., 358
Malech, H. L., 108
Malkowitz, L., 261
Mallet, A. I., 252, 253(59), 266, 282, 284(12), 288, 293(16)
Mancianti, M. L., 197
Maniatis, T., 113
Manni, A., 265
Mansfield, P. J., 313
Mantel, C., 313

Mantovani, A., 3, 86, 97(16), 306
Manzana, W., 57, 64(15)
Maragoudakis, M. E., 195
Marbaix, G., 108
Marcello, M. F., 164
March, C. J., 285
Maree, M., 164
Marfaing-Koka, A., 150
Margulies, D. H., 344, 354(16)
Marini, J. J., 166
Marini, M., 245–246
Mark, R., 57, 58(6)
Marks, R. M., 192, 199, 199(102), 200
Markwell, M. A. K., 137, 139(7)
Marleau, S., 261–262
Marmon, S., 85, 118, 119(5), 368
Marriott, D., 195, 201(60), 202(60), 203(60), 204(60), 207(60), 210(60)
Marsh, D. G., 247
Marsh, M., 64, 129, 150
Marsh, W. R, 244
Marshall, M. S., 313
Martin, A. W., 150
Martin, G. R., 201
Martin, M. A., 121
Martin, S. R., 85, 118, 119(4), 129, 130(28), 131(4), 368
Martin, T. R., 17, 26(7), 27(7), 162, 176
Martinez, C., 256, 261
Martinez, R., 43
Martinez, S., 110
Martins, M. A., 166
Martonyl, C. L., 201
Marzio, P. D., 118, 119(5)
Massacesi, L., 186
Massion, P. R., 251
Matlow, A., 164
Matsouka, I., 85
Matsumoto, Y., 165
Matsuo, H., 108
Matsuse, T., 163
Matsushima, K., 15, 86, 198–199, 199(81, 82), 202(91), 250–252, 255(55), 256, 272, 340
Matsuura, N., 166
Matsuura, Y., 40
Mattei, S., 364
Mattera, R., 323
Matthaei, K. I., 244, 262(15)
Matthay, M. A., 162–163, 173, 176, 176(2), 177(2, 7)
Matthias, J., 247
Matthias, U., 191(19), 192
Mattoli, S., 245–246, 265
Matzger, W. J., 253
Mauser, P. J., 245, 262(22)
Maxfield, F. R., 308
Mayer, R., 195
Mazarakis, D. D., 199(112), 200
McAllister, P. K., 243
McCarl, R. L., 379
McCarthy, S. A., 303
McChesney, D. J., 114
McColl, S. R., 86, 301, 304, 307–308, 308(16)
McConkey, D. J., 363
McConnell, H. M., 87, 88(22), 90(21, 22)
McCormick, F., 39
McCune, J. M., 121
McDermott, A. B., 113
McDermott, D., 108–109, 117(11)
McDougal, J. S., 119
McKnight, A., 64, 129, 150
McMahon, M., 355
McMillen, M. A., 164–165
Meliones, J. N., 165
Mello, R. J., 195
Melvold, R., 183
Menard, L., 3–4, 4(2)
Mencia-Huerta, J. M., 244
Menconi, M. J., 164
Meng, Q., 253
Menger, M., 151
Merridge, M. J., 362
Merrifield, R. B., 17
Merriman, R. L., 346
Merritt, J. E., 303
Messing, J., 207
Messmer, K., 166
Mette, S. A., 215
Metz, C., 165
Metzger, W. J., 242, 244
Meurer, R., 254
Meyer, A., 109, 114(14), 117(14), 365
Meyer, G., 191(19), 192
Meyer, M., 72
Meyers, D. M., 165
Michel, M. C., 322
Michel, P.-B., 243
Michiel, D. F., 312
Michna, B., 164
Migita, K., 348

Mihm, F. G., 163
Mikami, A., 108
Miki, I., 108
Miko, I. J., 27
Mikulaschek, A., 166
Milan, R., Jr., 166
Millard, P. J., 366
Miller, D. L., 87, 88(22), 90(22)
Miller, F. D., 346
Miller, F. N., 151
Miller, L. K., 40
Miller, M. D., 198, 199(83), 202(83)
Miller, S., 183
Miller, W., 71
Milligan, G., 321–322, 322(49)
Milstein, C., 16
Mims, S. J., 125
Min, H. K., 348
Minami, M., 108
Minden, A., 356, 357(47)
Miniati, M., 166
Minta, A., 303
Minty, A., 252(61), 253, 257(61)
Mishina, M., 108
Mita, S., 246
Mitra, R. S., 197, 199(110), 200
Mitsuyama, T., 165
Miura, Y., 246
Mixon, M. B., 321
Miyajima, A., 38, 365
Miyamoto, T., 108
Miyata, T., 163
Mizer, L. A., 166
Mizokami, A., 348
Moarefi, I., 38
Modig, J., 164
Moghtader, R., 10
Mohler, M. A., 265
Moldawer, L. L., 251
Molino, M., 3, 86, 306
Molkowitz, L., 57
Mollereau, C., 92, 107(25)
Mong, S., 308
Monick, M., 242
Monteclaro, F. S., 57, 64(12), 70, 74(15), 77
Montefort, S., 251
Montesano, R., 192
Montgomery, A. M. P., 191(22), 192, 195(22)
Monti, S., 166
Montour, J. L., 192
Moore, E. D. W., 366, 368(27)
Moore, J. P., 85, 118, 119(4), 131(4), 368
Moore, K. W., 85, 340, 365
Moore, S. K., 255
Moores, K. E., 303
Moqbel, R., 247, 249(40), 251, 253, 254(40), 257, 257(40), 260(40)
Mori, T., 85, 166
Morii, N., 358
Morikawa, S., 39
Morimoto, M., 346
Morishita, K., 199
Morisu, M., 165
Morita, E., 250, 271, 272(2), 280(2)
Moriuchi, H., 166
Morkawa, Y., 365
Moro, O., 84
Moroyama, E. K., 166
Morris, A. J., 323, 357
Morris, S. B., 201, 211(118), 217(118), 218(118)
Morris, S. J., 119, 132
Morrison, M., 134, 137(2), 140(2)
Moscatelli, D., 192
Moseley, P., 242
Moser, B., 3, 15, 130, 199, 202, 312, 313(14), 340, 341, 341(1), 364–365
Moser, R., 341
Moses, M., 195
Moses, M. A., 193
Mosmann, T. R., 246
Moss, B., 119–120, 122, 122(10, 18), 123(18), 124(18), 125–126
Moss, J., 325
Moss, R. F., 165
Mouton, J. W., 226
Moyer, J. D., 382
Mrowietz, U., 250, 271, 272(2), 280(2), 285
Muceniece, R., 43
Mueller, S. G., 3, 6(11), 7(11), 8(11, 12), 11(11), 15
Mukaida, N., 15, 198, 199(82), 256
Mulheron, J. G., 43
Mullen, J. B., 163–164, 166
Mullen, P. G., 164
Mullenbach, G., 17
Mulligan, M. S., 165
Munoz, M., 43

Munson, P. J., 24, 76
Murabito, R., 164
Murakami, H., 166
Murata, A., 166
Murayama, Y., 319
Murphy, K. R., 244
Murphy, P. M., 57, 73, 85–86, 108–109, 109(4), 117, 117(4, 11, 12), 118, 119(3), 131(3), 312
Murphy, W. J., 245
Murray, M. J., 164–165
Musser, J. M., 165
Musso, T., 199, 202(91)
Mustard, R. A., 164
Mutsuda, T., 256
Muyldermans, G., 130
Myers, E. W., 71
Myers, S. J., 72, 73, 77(2), 79(2), 97, 105(27), 109, 313, 326, 338(6), 364
Myklebust, R., 165
Mythen, M. G., 165

N

Naccache, P. H., 301, 304, 307–308, 308(16)
Nadel, J. A., 244, 251
Nagase, T., 163
Nagashima, K. A., 85, 118–119, 119(4), 131(4), 368
Nagataki, S., 348
Nahorski, S. R., 325, 362, 380–381, 383
Nahum, A., 166
Nakagawa, N., 164
Nakajima, S., 253
Nakamura, H., 164
Nakamura, M., 108
Nakano, H., 346
Nakano, M., 166
Nakano, T., 198
Nakao, A., 166
Nakashima, S., 358
Nakayama, D. K., 166
Nakhosteen, J. A., 256
Nariuchi, H., 195
Narumiya, S., 358
Nasman, J., 43
Navarro, J., 57, 58(6), 117
Nebigil, C. G., 43
Neil, J. C., 122
Nelson, N., 26, 57, 58(5)
Nemeth, K., 109, 114(14), 117(14), 365
Neopolitan, C., 193, 197(42)
Neote, K., 41, 72, 73(3), 86–87, 368
Neubig, R. R., 313
Newman, W., 118, 119(7), 122(7), 254, 257, 260(72), 261
Newton, R., 253
Ng, G. Y., 43
Nguyen, M., 191(20), 192
Nguyen, M. H., 188
Nichols, M. E., 17, 35, 150
Nick, J. A., 3, 341(12), 342
Nicklen, S., 207
Nickoloff, B. J., 197, 199(110), 200
Nicosia, S., 314, 315(26)
Nield, G. H., 199(108), 200
Nielson, J. B., 165
Nieman, G. F., 164
Niles, E. G., 126
Nishimoto, I., 319, 324(37)
Nishio, I., 166
Nishiyama, A., 188, 189(25)
Nishizuka, Y., 344
Nori, T., 86
Norin, A., 166
Northrup, J. K., 319, 323
Noso, N., 253, 257, 294
Notvall, L., 39
Nourshargh, S., 251, 257, 264
Novak, R. F., 195
Nozawa, Y., 358
Numa, S., 108
Nuoffer, C., 309
Nürnberg, B., 56
Nussbaum, O., 120, 122(17), 123(17)

O

Obata, T., 164
Obayashi, Y., 199
Oberlin, E., 199
O'Conner, B., 253
O'Dowd, B. F., 43
Odumeru, J., 164
Ogata, E., 319
Ogawa, Y., 366, 375(28)
Ogg, S., 39
O'Hanley, P., 163–164

Ohya, Y., 322
Okado, H., 108
Okajima, F., 320
Okamoto, T., 319, 324(37)
Oker-Blom, C., 43
Okoyama, H., 124
Olate, J., 110
Oldham, K. G., 378, 380(40), 383(40)
Olive, D., 363
Olsen, R., 165
Olson, N. C., 164
Olson, W. C., 119
Omann, G. M., 313
O'Neill, S., 222
Oniguchi, K., 358
Onorato, J. J., 4
Ooi, B. G., 40
Oostveen, J. A., 243
Opdenakker, G., 256
Oppenheim, J. J., 10, 15, 15(30), 198–199, 199(81, 82), 250, 252, 254–255, 255(55), 257(86), 272, 312, 340
Oppermann, M., 4
Op't Holt, T. B., 166
Orci, L., 192
O'Reilly, M. S., 195
Orrenius, S., 363
Orringer, M. B., 201, 214(116)
Ostrowski, J., 58
O'Sullivan, B. P., 164
Overton, H. A., 40
Owicki, J. C., 87–88, 88(22), 90(22)
Ownbey, R., 255, 260(79), 261

P

Paczan, P. R., 165
Page, C., 28, 34(8), 244, 253
Page, M. J., 39
Pajot-Augy, E., 43
Palmer, D., 313, 364
Pang, I., 323
Panoutscaopoulou, N., 195
Papo, M. C., 165
Pappenheimer, A., 183
Parce, J. W., 87, 88(22), 90(21, 22)
Pardigol, A., 72
Parent, A., 166
Park, D. J., 348
Parker, E. M., 43
Parker, L. L., 39
Parmentier, M., 57, 64(11), 92, 107(25), 118, 119(6), 120, 120(6), 122(6), 129–130, 130(29)
Parolin, C., 199
Parulekar, M., 320
Paskanik, A. M., 164
Passaniti, A., 201
Patalano, F., 265
Paterson, H., 39
Paterson, J. F., 164–165
Patz, A., 195
Pauly, R. R., 201
Pavia, J., 43
Paweletz, N., 192
Paxton, W. A., 85, 118, 119(4), 129, 130(28), 131(4), 368
Pearl, R., 188
Pease, L. R., 73
Pedersen, S. E., 323
Peel, R. L., 166
Pei, G., 4
Peiper, S. C., 38, 47, 55–57, 58(7), 64, 64(7), 68, 70, 85, 118, 119(5, 6), 120(6), 122(6), 129, 148–150, 368
Peitzman, A. B., 166
Pelaprat, D., 43
Pelletier, S. L., 132
Penman, M., 121
Penner, R., 302
Pepinsky, B., 264
Peppel, K. D., 236
Pepper, M. S., 192
Pereira, P. M., 166
Perez, H. D., 57, 64(15)
Perez Soler, M. T., 341
Perkins, J. A., 247, 251
Permentier, M., 57
Pernollet, J. C., 43
Persons, P. E., 348
Persoon, N., 272
Peterson, H. I., 195
Petro, J., 195, 202(57)
Pettit, D. A., 43
Pettit, E. J., 306
Pettit, G. R., 346
Peveri, P., 250, 272
Pfaffenenbach, D., 190, 197(7)

Pfister, A., 253
Phan, S. H., 199, 199(102, 109), 200
Pheng, L. H., 164
Philips, G. D., 191
Phillips, H., 347
Pianetti, P., 347
Picone, A., 164
Pilewski, J., 215
Pili, R., 201
Pillar, J., 262
Pistolesi, M., 166
Pistorese, B. P., 162
Pitcher, J. A., 4
Pitchford, S., 84, 87, 88(22), 90(22)
Pitman, A., 245, 262(22)
Piwnica-Worms, H., 38–39
Pleasure, S. J., 28, 34(8)
Poenie, M., 302, 303(3), 375, 376(37)
Pohl, S. L., 315
Polentarutti, N., 86, 97(16), 192
Pollock, T. W., 164–165
Polverini, P. J., 190, 193, 193(3), 195–196, 196(3, 38, 39), 197, 201, 201(58, 60, 70), 202(60, 70), 203(60), 204(60), 207(60), 210(60), 211(117, 118), 214(116, 117), 217(118), 218(118)
Ponath, P. D., 118, 119(7), 122(7), 254, 257, 260(72), 261
Pope, R. M., 199(112, 113), 200
Portier, A., 150
Portis, J., 119
Possee, R. D., 40
Post, L. E., 38
Post, M., 163, 166
Post, T. W., 261
Poston, R. N., 256
Potter, B. V. L., 362
Potts, B. J., 121
Pouyssegur, J., 3
Power, C., 364
Power, C. A., 64, 109, 114(14), 117, 117(14), 129, 150, 258, 346, 362, 364–365, 365(20)
Poyner, D. R., 378
Pratt, B., 191, 195(16)
Preis, I., 195
Premack, B. A., 86, 312, 346–347, 352(27), 364, 365(20)
Premont, R. T., 324
Pretolani, M., 244, 264
Prevec, L., 239
Price, P. J., 31
Probst, W. C., 326, 327(3)
Proost, P., 73, 97, 105(28), 256–257
Prossnitz, E. H., 58
Prossnitz, E. R., 41, 43, 47(34), 55(34), 57
Proudfoot, A. E. I., 109, 114(14), 117, 117(14), 254, 365
Przybyca, A. E., 228
Przybylski, M., 38
Ptasienski, J., 4
Pullen, J. K., 73
Punnonen, K., 378, 381(42), 382(42)
Purcell, P., 110
Pytela, R., 169
Python, C. P., 322
Pyun, H. Y., 117

Q

Qadota, H., 322
Qin, S., 254, 260(72), 261, 340, 365, 368, 383(21)
Quan, J. M., 17, 26(7), 27(7)
Quehenberger, O., 41, 43, 47(34), 55(34), 57
Quilliam, L. A., 39
Quinlan, M. F., 164–165
Quinn, D. G., 382

R

Rabin, L. B., 121
Racik, P., 33
Raffin, T. A., 163–164
Raida, M., 72
Raiford, C., 199(102), 200
Rampart, M., 251
Ramsay, A. J., 244, 262(15)
Rana, S., 129–130
Ranchod, M., 164
Rankin, C., 40
Ransohoff, R. M., 28, 35, 182
Rapoport, B., 43
Rastinejad, F., 195
Raum, M. G., 38
Ravenscraft, S. A., 166
Raw, A. S., 321
Raymond, J. R., 3, 43

Re, F., 192
Reagan, J. D., 43
Reaman, G. H., 195
Reddigari, S. R., 255, 257(86)
Reddy, B., 39
Reilander, H., 43
Reinecke, M., 72
Reisfeld, R. A., 191(22), 192, 195(22)
Remick, D. G., 199, 199(102), 200
Remy, J. J., 43
Resendez, E., 369
Revak, S. D., 166
Reyes, G. R., 121
Rhee, S. G., 348
Rhodes, S., 39
Ribeiro, S. P., 163
Rice, G. C., 18, 265
Richards, C. D., 254
Richards, I. M., 243
Richardson, R. M., 4, 43
Richerson, H. B., 242
Richmond, A., 3, 6(11), 7(11), 8(11, 12), 11(11), 15, 199
Rieppi, M., 86, 97(16)
Riethmacher, D., 39
Rifkin, D. B., 192
Riggi, M., 164
Riley, D. T., 98
Rinaldo, J. E., 164
Ringler, D. J., 254, 260(72)
Rink, T. J., 302
Rinken, A., 43
Risau, W., 196
Risberg, B., 163–164
Rivers, T., 182
Roberts, J. D., 207
Roberts, M. F., 357, 383
Roberts, T. M., 38
Robertson, B., 165
Robeva, A. S., 43
Robichaud, A., 247
Robinson, D., 246
Robinson, E. A., 250, 252, 255, 272
Robishaw, J. D., 323
Roczniak, S., 195, 201(60), 202(60), 203(60), 204(60), 207(60), 210(60)
Rodbard, D., 24, 76
Rodbell, M., 314–315, 315(26), 326, 327(4), 329
Rodeck, U., 197
Rodriguez, J. L., 166
Rodriguez de Cordoba, S., 17, 150
Rogers, R. M., 164
Rolfe, M. W., 199(109, 111), 200
Rollins, B., 118, 119(7), 122(7)
Rollins, B. J., 255
Rollins, T. E., 57
Romagnani, S., 246
Romaschin, A., 165
Roof, D. J., 316
Roomi, M. W., 164
Rooney, S., 262
Rose, K., 357
Rose, S. S., 255
Rosen, H., 254
Rosen, J., 57
Rosen, O. M., 38
Rosenberg, G. B., 17, 26(7), 27(7)
Rosenberg, S. A., 370
Rosenblum, E. N., 125
Rosenfeld, M., 191(22), 192, 195(22)
Rosenfield, R. E., 17, 150
Rosenthal, K. L., 239
Rosenthal, R. A., 195
Ross, C. A., 363
Ross, E. M., 320, 323, 324(67), 325
Rossi, A. G., 247, 257(39)
Rostene, W., 43
Rot, A., 149, 252, 264–265
Roth, E., 362
Roth, S. J., 255
Rothenberg, M. E., 246, 258, 259(31, 100), 261
Rothman, A., 380
Rothschild, H. R., 164
Rott, R., 38
Rousset, D., 130
Rouzer, C. A., 307
Rozenblatt, S., 122
Rubenstein, R. C., 325
Rubinstein, P., 17, 35, 150
Rubinstein, R. E., 35
Rucinski, D., 255, 257(86)
Rucker, J., 57, 64(11), 70, 118, 119(6), 120, 120(6), 122(6), 129–130, 130(29), 131
Rudge, S. A., 357
Rüegg, C., 169
Ruffie, C., 264
Russell, M., 313
Rutter, W. J., 228
Ryan, T. A., 366

Ryan, U. S., 165
Ryder, A., 119

S

Sabatini, D. M., 363
Sabharwal, A. K., 163
Sadée, W., 84
Sadick, M., 169
Sadovnick, A. D., 183
Sagara, N., 195
Sage, E. H., 195
Sager, R., 199
Saggio, A., 164
St. John, R. C., 166
Saito, A., 253
Sakamaki, F., 164
Saldeen, T., 164
Saldiva, P. H. N., 166
Salesse, R., 43
Salomon, Y., 329
Salon, J. A., 85, 88
Samama, P., 4
Sambrook, J., 113
Samson, M., 57, 64(11), 92, 107(25), 118, 119(6), 120, 120(6), 122(6), 129–130, 130(29)
Samuels, M. L., 355
Samuelsson, B., 307
Samuelsson, T., 164
Sanderson, C. J., 261
Sandin, R., 164
Sanger, F., 207
Santamaria Babi, L. F., 341
Sapsford, R. J., 253
Sarafi, M. N., 73
Saragosti, S., 130
Sarau, H. M., 308
Sarkar, D., 119, 132
Sarma, V., 192, 199(110), 200
Sarmonika, M., 195
Sarnelli, R., 166
Sasaki, T., 358
Sasse, J., 192, 193(27)
Sato, K., 164
Sato, N., 195
Sato, T., 85
Sautel, M., 43
Savola, J. M., 43
Sayama, K., 164
Scalea, T. M., 164–165
Schade, A., 151
Schägger, H., 279, 285(9)
Schall, T. J., 27, 28(1), 41, 72, 73(3), 85–87, 118, 119(5), 148, 150, 241, 252, 254, 312–313, 340–341, 341(2), 346–347, 352, 352(27), 353(41), 362, 364–365, 365(2, 20), 368
Scharff, M. D., 20
Scheer, B. J., 246
Scheidtmann, K. H., 38
Scherle, P. A., 322
Schertler, G. F. X., 56
Schilling, G., 363
Schilling, W. P., 43
Schioth, H. B., 43
Schleimer, R., 257
Schmelzer, C. H., 21
Schnyder-Candrian, S., 199, 202(92)
Schomisch, S. J., 361
Schondorf, M., 3
Schouten, B. D., 164
Schraufstätter, I. U., 165–166
Schraw, W. P., 3, 6(11), 7(11), 8(11, 12), 11(11), 15
Schreiber, R. E., 58
Schreurs, J., 38
Schrier, D., 165
Schröder, J.-M., 250, 252–253, 253(59), 256–257, 266, 271–272, 272(2), 276, 280(2), 282, 284(12, 13), 285, 288, 289(13), 293(16), 294
Schubert, B., 310
Schultz, G., 56
Schultze-Osthoff, K., 196
Schulz, R., 151
Schulz-Knappe, P., 72
Schumacher, C., 364
Schuster, D. I., 326, 327(3)
Schuster, D. P., 165
Schutzer, K. M., 163–164
Schwartz, D., 150
Schwartz, L. B., 242
Schwartz, L. W., 164–165
Schwartz, O., 199
Schwartz, S. M., 192
Schwenkter, F., 182
Schwiebert, L. M., 257
Scott, C. F., 202
Scott, C. W., 323

Sealfon, S. C., 326, 327(3)
Seamon, K. B., 326
Sedgwick, J. B., 242–243
Seger, R., 354, 355(44)
Sehmi, R., 251, 257
Seibel, H. R., 192
Seidman, B., 108, 117(3)
Seitz, M., 341, 365
Self, A. J., 321
Selvan, R. S., 246
Semple, J. P., 197
Severini, M., 43
Seyama, Y., 108
Sgadari, C., 195
Shah, M. R., 199(113), 200
Shah, N. S., 166
Shaklee, C. L., 255
Shanafelt, A. B., 190, 195, 201(60), 202(60), 203(60), 204(60), 207(60), 210(60)
Shanley, T. P., 165
Shapiro, J., 163
Shapiro, R., 195
Shapiro, R. A., 313, 364
Shapiro, R. S., 166
Sharar, S. R., 176
Sharp, A. H., 363
Sharpe, R. J., 195, 202, 202(57)
Sharps, E. S., 379
Sharron, M., 70
Shaw, C., 280
Shaw, S., 265
Shayevitz, J. R., 166
Sheikh, S. P., 43
Shelley, S. A., 164–165
Shepherd, D., 243
Sheppard, D., 169
Sherman, W. R., 377
Shevach, E. M., 344, 354(16)
Shi, X., 254, 260(72), 340, 365, 383(21)
Shields, P. G., 197
Shields, T. S., 325
Shih, T. Y., 316, 321, 322(29)
Shimizu, T., 108
Shing, Y., 195
Shirozu, M., 198
Shockey, K. S., 165–166
Shockley, M. S., 84
Sholly, M. M., 192
Showalter, V. M., 43
Showell, H. J., 199, 199(102, 107), 200, 251
Shows, T. B., 192
Sibbald, W. J., 165
Siciliano, S. J., 57, 261
Sidiropoulos, D., 313
Sidky, Y. A., 195
Sidman, R. L., 33
Siebenlist, U., 265
Siegelman, M. H., 261
Sielaff, T. D., 164
Sielczak, M. W., 264
Sillard, R., 72
Sim, T., 241, 256(2)
Simon, M. I., 311, 313, 322–323, 326, 364
Simon, M.-T., 253
Simons, R. K., 164
Simony-Lafontaine, J., 243
Singer, W. D., 358
Sironi, M., 192
Sirontino, A., 261
Sjoerdsma, K., 242, 244
Sjostrand, U. H., 165
Skeyensteuber, S., 340
Skinner, R. H., 39
Sleath, P. R., 26, 57, 58(5)
Slepak, V. Z., 322–323
Slightom, J. L., 98
Slutsky, A. S., 163, 166
Smeal, T., 356, 357(47)
Smigel, M. D., 319
Smilowitz, H. M., 166
Smith, D., 251
Smith, D. R., 201, 214(116)
Smith, G. E., 40
Smith, H., 254, 260(72)
Smith, M., 26
Smith, P. D., 92
Smith, P. K., 165
Smith-Heath, H., 340, 365, 383(21)
Smrcka, A. V., 323, 324(67)
Smyth, R. J., 118, 119(6), 120, 120(6), 122(6), 129–130
Snyder, L. A., 326, 327(3)
Snyder, S. H., 363
Snyderman, R., 4
Sobel, R., 188
Soble, L. W., 193, 197(42)
Soderling, T. R., 39
Sodroski, J., 118, 119(7), 121, 122(7), 199
Sofia, R. D., 262
Soler, D., 257

Soloski, M. J., 363
Solski, P. A., 39
Sontag, J.-M., 360, 361(63)
Soo, K., 365
Sood, S. L., 165
Soreq, A., 108, 117(3)
Sorg, C., 196
Sorgente, N., 193, 197(42)
Sousa, A. R., 256
Sozzani, S., 3, 86, 97(16), 306
Spada, A. P., 348
Spain, D. A., 166
Speck, J. C., Jr., 134, 137(3), 139(3)
Spiegel, A. M., 326
Spindel, E. R., 117
Sprague, G. F., Jr., 58
Sprang, S. R., 321
Springer, M. H., 57
Springer, M. S., 57, 254, 261
Springer, T. A., 198–199, 199(86), 255, 264
Squartini, F., 197
Srinivason, S., 26, 57, 58(5)
Stabel, S., 39
Stafford, S., 241, 252(62), 253–254, 256(2)
Standiford, T. J., 199(105–107, 109, 111), 200, 220
Staub, N. C., 176, 177(7)
Steinberg, F., 192
Steinbrink, K., 196
Steinhardt, R. A., 369
Steinhorn, D. M., 165
Stellato, C., 257
Stenman, G., 199
Sterne-Marr, R., 4
Sternweis, P. C., 316, 323, 324(67), 358
Stevens, J. H., 163–164
Sticherling, M., 294
Stoebenau-Haggarty, B., 64, 129, 150
Stork, P. J., 325
Storm, G., 226
Strader, C. D., 312
Strathmann, M. P., 311, 322
Straub, C., 10
Streffer, C., 192
Strieter, R. M., 150, 190, 195–196, 198–199, 199(84, 102–109, 111–113), 200–201, 201(58, 60, 70), 202(60, 70, 92), 203(60), 204(60), 207(60), 210(60), 211(117, 118), 214(116, 117), 217(118), 218(118), 220, 243, 255–256
Strober, W., 344, 354(16)
Strubel, N. A., 191(20), 192
Stuart, G., 182
Studier, F. W., 126
Stuhlmann, H., 129, 130(28)
Sturgill, T. W., 355
Su, B., 357
Su, M.-C., 247
Suchard, S. J., 313
Suda, T., 246
Sudderick, R. M., 247
Sugama, Y., 246
Suganuma, M., 349
Sugerman, H. J., 164
Sugimura, T., 349
Suimura, T., 346
Sullivan, N., 118, 119(7), 122(7)
Sullivan, R., 192, 193(27)
Summers, M. D., 38–40
Sum-Ping, S. T., 166
Sun, Y., 118, 119(7), 122(7)
Sunderkotter, C., 196
Sung, T.-C., 357
Sur, S., 253
Sutterwala, S. S., 64, 129, 150
Sutton, R. E., 85, 118, 119(5), 368
Svoboda, M., 38
Swanson, H. T., 164
Sweet, L. J., 39
Swenson, K. I., 39
Sydenham, M., 39
Symreng, T., 166
Syrbu, S., 166
Szabo, M. C., 352, 353(41)
Szabo, S., 195
Szardenings, M., 43

T

Taborda-Barata, L., 253
Tachmes, L., 166
Tada, H., 198
Tago, K., 358
Tai, H. H., 43
Tait, A. R., 166
Tajima, H., 348
Takagi, H., 166
Takahashi, H., 108
Takahashi, I., 346
Takai, A., 349

Takai, Y., 358
Takatsu, K., 246
Takematsu, T., 366
Takeuchi, T., 348
Tamaoki, T., 346
Tanaka, N., 166
Tanaka, S., 250, 252, 255
Tanaka, Y., 265
Tang, W. J., 310
Tani, M., 35, 182
Tannock, I. F., 190
Tasaka, S., 164
Tashiro, K., 165, 198
Taub, D. D., 195, 245, 254, 346, 364, 365(20)
Taub, D. T., 201, 211(118), 217(118), 218(118)
Taylor, F. B., Jr., 163
Taylor, H. E., 43
Taylor, L., 57, 58(6), 117
Taylor, P., 380
Taylor, R. M., 201
Taylor, S., 195
Tazelaar, H. D., 164–165
Tekamp-Olson, P., 17, 150
Ten Kate, M. T., 226
Teran, L. M., 251, 253
Terashima, T., 164
Terrell, T. G., 251
Theiler, M., 183
Thelen, M., 3, 72, 256, 257(92), 312, 313(14)
Therrien, S., 307, 308(16)
Thiele, C., 43
Thierse, D., 360, 361(63)
Thies, S. D., 165–166
Thomas, G. M. H., 358
Thomas, H. G., 3, 199
Thomas, J. F., 64, 129, 150
Thomas, K. M., 57, 58(6), 117
Thompson, D. C., 195
Thompson, R., 182
Thomsen, D. R., 38
Thorner, J., 326, 327(2)
Thornton, A. J., 199(103), 200
Thrall, R. S., 166
Tiffany, H. L., 73, 86, 108–109, 109(4), 117(4, 11)
Tighe, D., 165
Timmermans, V. E., 84
Toda, H., 166
Toews, G. B., 199(106), 200
Toh, H., 108, 163
Tolsman, S. S., 195
Tomeczkowski, J., 72
Tomhave, E. D., 4
Tominaga, A., 246
Tonks, N. K., 349
Tonneson, M. G., 264
Tooley, R., 166
Topalian, S. L., 370
Torisu, M., 163, 246
Torricelli, J. R., 31
Torry, D. J., 254
Tosato, G., 195
Tota, M. R., 312
Totty, N. F., 247, 249(40), 254(40), 257(40), 260(40), 358
Toullec, D., 347
Touvay, C., 244
Toy, K. I., 252
Trapp, B., 188, 189(25)
Traynor, J. R., 325
Tribble, C. G., 165–166
Tricomi, S. M., 163
Trifaro, J. M., 320
Trooskin, S. Z., 166
Trowbridge, I. S., 38
Trumpp Kallmeyer, S., 56
Truong, O., 247, 249(40), 254(40), 257(40), 260(40), 358
Tsien, R. Y., 302–303, 303(3, 6), 375, 376(37)
Tsou, C. L., 57
Tsukada, T., 348
Tucker, A., 191, 195(16)
Tuohy, V., 35, 182, 187–189, 189(25)
Turcovski-Corrales, S. M., 245
Turner, C. R., 164–165
Turner, J. D., 64, 70, 122, 129
Turner, L., 344
Twort, C. H. C., 247

U

Uda, K. I., 166
Uejima, Y., 163
Uguccioni, M., 72, 97, 256, 257(92)
Ui, M., 320, 323
Umezawa, K., 348
Unanue, E. R., 193, 196(39)
Underwood, D., 312
Underwood, D. J., 57
Ungerleider, R. M., 165

Unutmaz, D., 85, 118, 119(5), 368
Urano, R., 164
Uto, A., 366, 375(28)
Uyehara, C. F., 165

V

Vacanti, J., 195
Valente, A. J., 265
Vallee, B. L., 195
Valls-i-Soler, A., 165
Van Ackern, C., 192
Van Biesen, T., 3
Van Blitterswijk, W. J., 358
Vance, P. J., 64, 129, 150
Van Damme, J., 73, 86, 97, 105(28), 195, 201(60), 202(60), 203(60), 204(60), 207(60), 210(60), 251, 256–257
Vanden Bos, T., 26, 57, 58(5)
van der Geer, P., 9
Van der Heijde, R. M., 165–166
VanDongen, A. M. J., 310
Van Heyningen, V., 23, 24(16)
Vanhoutte, P. M., 243
Van Oberghen-Schilling, E., 3
Van Riper, G., 57, 254
Van Wyke Coelingh, K., 38
Van Zee, K. J., 251
Varani, J., 197
Varesio, L., 199, 202(91)
Vargaftig, B. B., 244, 253, 262, 264
Varney, V. A., 247
Vasella, A., 362
Vasquez, A., 164
Vassali, G., 246
Vassart, G., 57, 64(11), 92, 107(25), 120, 129–130, 130(29)
Vassa Ui, J. D., 192
Vasudevan, S., 43
Vaughan, M., 325
Vecchi, A., 3, 86, 306
Venge, P., 243
Verhofstede, C., 130
Versprille, A., 165–166
Vignola, A. M., 243
Vik, T. A., 39
Vila, J., 355
Vilander, L. R., 57, 64(15)
Villa, C., 56
Villar, J., 163, 166
Villiger, B., 341
Virelizier, J.-L., 130, 199
Vitangcol, R. V., 202
Vittori, E., 245–246
Vlahos, C. J., 3
Vlodavski, I., 192, 193(27)
Volovitz, B., 242
Volpert, O. V., 195
von Jagow, G. V., 279, 285(9)
von Rueker, A. A., 3
Von Tscharner, V., 252(61), 253, 257(61)
Vrtis, R. F., 243

W

Wada, H. G., 84, 87, 88(22), 90(22)
Wagenvoort, C. A., 165–166
Waki, Y., 164
Waldo, G. L., 323
Walker, B., 119
Walker, C., 163
Walker, M. W., 85
Walker, P., 43
Walker, P. M., 165
Walsh, D. T., 247, 249(40), 254(40), 257(40), 260(40)
Waltmann, P., 244
Walz, A., 195, 198–199, 199(79, 80, 84, 112), 200, 201(60), 202(60, 92), 203(60), 204(60), 207(60), 210(60), 250, 255, 272, 282, 284(11)
Wang, H., 164
Wang, J., 254, 260(72), 261
Wang, J. H., 253
Wang, J. M., 10, 15(30)
Wang, L. H., 43
Wang, M., 4
Wang, Z.-X., 38, 47, 55–57, 58(7), 64, 64(7), 68, 70
Ward, P. A., 165, 199
Ward, S. G., 344, 363
Wardlaw, A. J., 251
Warren, J. S., 256
Warsop, H., 322
Washlestedt, C., 85
Wasniowska, K., 64, 150
Watanabe, T., 108, 322
Waterfield, M. D., 247
Wathelet, M., 114
Watson, K., 197

Watson, M. L., 251, 254
Waugh, J. B., 166
Webb, A. R., 165
Webb, M. L., 313, 364
Webb, W. W., 366
Weber, M. J., 355
Wechsler, A. S., 164
Wedegaertner, P. B., 325
Weerawarna, K. S., 26, 57, 58(5)
Weg, V. B., 251, 264
Wegenius, G., 164
Wegner, C. D., 244
Wehrly, K., 119
Weidner, N., 197
Weiner, H., 184
Weiner, L., 184
Weinstein, S. L., 348
Weir, W. G., 366
Weisbrode, S. E., 166
Weisman, D. M., 190, 196(6)
Weiss, R. A., 119
Weissman, I. L., 121
Wekerle, H., 187
Welch, W. R., 197
Weller, P. F., 246, 264
Wells, J. W., 325
Wells, T. N. C., 64, 114(14), 117, 117(14), 129, 150, 254, 258, 346, 362, 364–365, 365(20)
Weng, G., 85
Wershil, B. K., 256
Westwick, J., 150, 199(108), 200, 251, 254, 340, 344
Wheeldon, E. B., 164–165
White, J., 257
White, J. M., 132
White, J. R., 3, 15
White, M., 57, 58(6)
Whitehead, R. A., 191
Whitmore, T., 17, 26(7), 27(7)
Whyte, R. I., 201, 211(118), 214(116), 217(118), 218(118)
Wichelhaus, D. P., 322
Wiehle, S. A., 155
Wieland, T., 317
Wieman, T. J., 155
Wiener-Kronish, J. P., 162, 176(2), 177(2)
Wikberg, J. E., 43
Wilcox, R. A., 362
Wilke, C. A., 201, 214(116)
Wilkie, T. M., 322–323
Willars, G. B., 381
Willett, B. J., 122
Willey, R. L., 121
Williams, D. A., 366, 368(27)
Williams, F. M., 257
Williams, J. A., 114
Williams, N. G., 38
Williams, T. J., 241, 247, 249(40), 251, 254(40), 257, 257(39, 40), 258, 260(40), 261–262, 264
Wilson, J. D., 192
Wilson, P. T., 325
Windsor, A. C. J., 164
Winfield, J., 166
Winkler, J. D., 308
Winn, R. K., 176
Winter, G., 27
Witt, C., 247
Witt, D., 57, 58(6)
Wittmann, U., 382
Wolf, A., 184
Wolf, F. W., 192
Wolf, I., 85
Wolff, J. R., 191
Wollert, P. S., 164
Wolner, B., 380
Woloszyn, T. T., 166
Wong, G., 39
Wong, L. M., 72–73, 77(2), 79(2), 97, 105(28), 313, 326, 338(6), 364
Wood, W. I., 17–18, 26(4), 64, 150
Woodland, H. R., 108
Woodle, M. C., 226
Woods, A., 184
Woodard, R., 43
Worthen, G. H., 313
Worthen, G. S., 3, 341(12), 342
Wrann, M., 246
Wu, D., 313, 325, 364
Wu, I.-H., 357
Wu, K. M., 227
Wu, L., 118, 119(7), 122(7), 261
Wu, Y., 325, 364
Wyatt, L. S., 122

X

Xia, C., 253
Xiao, H.-Q., 247
Xie, M., 361

Xing, Z., 254
Xu, L. L., 10, 15(30)
Xu, N. Z., 3

Y

Yalamanchili, R., 85
Yamada, K., 165
Yamaguchi, Y., 246
Yamamoto, T., 176, 253
Yang, B., 164
Yang, S. F., 358
Yasumoto, S., 349
Ye, R. D., 41, 43, 47(34), 55(34), 57–58
Yee, F., 85
Yelton, D. E., 20
Yi, Y., 57, 64(11), 118, 119(6), 120, 120(6), 122(6), 129–130, 130(29)
Ying, S., 246–247, 253
Yokozeki, T., 358
Yoo, H., 85
York, J., 241, 256(2)
Yoshida, A., 245
Yoshimura, A., 365
Yoshimura, T., 199, 250, 252, 254–256, 258, 272
Yoshizawa, S., 349
Young, I. G., 244, 262(15)
Young, J., 282, 284(13), 289(13)
Young, J. S., 165–166
Young, P. R., 10, 15(30)
Young, S., 3, 341(12), 342
Yousem, S. A., 166
Youssef, M. E., 164
Yssel, H., 352, 353(41)
Yu, M., 188–189, 189(25)
Yuhki, N., 255
Yuizono, T., 166
Yuspa, S. H., 378, 381(42), 382(42)

Z

Zachariae, C. O., 15, 198, 199(82)
Zajchowski, D., 199
Zakour, R. A., 207
Zaller, D., 184
Zanetti, A., 192
Zavala, D., 242
Zeng, L., 340, 365, 383(21)
Zhang, T.-Y., 56, 64, 70
Zhou, D., 86, 97(16)
Zieg, P. M., 165
Zilberstein, A., 348
Zimmerberg, J., 119
Zlotnik, A., 85, 340, 365
Zonnekyn, L., 251
Zu, K., 355
Zwissler, B., 166

Subject Index

A

Acid-injury rabbit model, *see* Lung injury
Adenylate cyclase
adenine conversion to cyclic AMP, tritium radioassay
calculations, 337–338
cell prelabeling, 336–337
ATP conversion to cyclic AMP, phosphorous-32 radioassay
ATP regenerating system, 329
calculations, 334–336
chromatography
column preparation, 330–332
cyclic AMP isolation, 333–334
incubation conditions, 331
materials, 330
membrane preparation, 332
principle, 329
enzyme immunoassay, 338–339
forskolin activation, 327, 332
radioimmunoassay, 338–339
regulation by G proteins, 326–327
Angiogenesis
angiogenic factors, 193–196
angiostatic factors, 193–196
associated diseases, 196
C-X-C chemokine role
angiogenic activity, 201–202
angiostatic activity, 202–204, 207
assays, 200–201
ELR motif, 202–203, 207–210
endothelial cell chemotaxis, 202–203, 210
phases, 191–192
tumor angiogenesis
non-small cell lung cancer
enzyme-linked immunosorbent assay of chemokines, 213, 216–217
interleukin-8 role, 214–216
IP-10 angiostatic activity, 216–219
mouse model for study, 211–212
tumor analysis, 212–213
vessel density assay, 216, 219
VLA-2 expression detection by flow cytometry, 215
vascular mass, 197
wound healing regulation, 190–191
Animal models, *see* Asthma; Lung injury; Multiple sclerosis animal models; Pneumonia, bacteria
Antibody, *see* Monoclonal antibody, chemokine receptor
Asthma
allergic mechanisms and proinflammatory cytokines, 245–247
animal models
dogs, 244
guinea pig, 243–244
mice and rats, 244
primates, 244–245
rabbits, 244
chemokine purification from inflammatory lavage fluids, 247, 249
chemokine role in leukocyte trafficking
adhesion molecule expression, 263–264
endothelial cell interactions, 264
experimental induction, 242
histology, 242–243
prevalence, 242
selective cell recruitment role
eotaxin, 257–262
interleukin-5, 261–262
interleukin-8, 250–251
macrophage inflammatory protein-1α, 254–255
macrophage inflammatory protein-1β, 254–255

monocyte chemotactic protein-1, 255–257
monocyte chemotactic protein-2, 256–257
monocyte chemotactic protein-3, 256–257
RANTES, 252–254
symptoms, 242

B

Bacterial pneumonia, *see* Pneumonia, bacteria
Bacteriorhodopsin, structure, 56
Baculovirus–insect cell expression system
baculovirus features, 39–40
chemokine receptors
expression level quantification, 47, 52–53, 55
insect cell culture, 45–46
optimization of expression, 52
recombinant virus purification, 46–47
transfection, 45–46
transfer vector construct preparation, 43–44
insect cell lines, 45
polyhedron gene promoter, 40, 55
posttranslational processing of proteins, 38–39
serpentine receptor superfamily proteins, examples, 41–43
transposon-based system for preparing recombinant baculoviruses, 55
Bolton–Hunter reagent, *see* Iodination
Brain
fetal cell culture, 29–31, 33
hNT neuron culture, 34
neuroblast migration in cortex formation, 33

C

Calcium flux
calcium sources in chemokine activation, 86–87
channels, 301–302
chemokine receptor desensitization assay, 308–309
chemokine receptor expression assay in *Xenopus* oocytes
aequorin assay, 117
calcium-45 release assay
agonist response, 115
incubation conditions, 114–115
CXCR2 analysis, 8–9
fluorescent dye assay
calibration and quantification, 304–305, 375–376
cell loading, 303–304
fluorescence detection, 304, 369–370
manganese quenching, 306
principle, 302–303, 365–366
saturation, 376–377
T cell calcium flux analysis
cell preparation, 366–367
dye selection, 368–369
flux profiles of various cell types, 370–371, 373–374
loading of cells, 376
lymphocyte lines, 367
T cell clones, 367
transfected cell lines, 368
viability of cells, 376
inhibition
calcium chelators, 305–306, 374–375
channel blockers, 306, 374
intracellular calcium level
concentration before flux, 301
modulation, 306–307
RANTES response phases, 86
sensitivity compared to chemokine chemotaxis assays, 308, 374
CC-CKR1, *see* CCR1
CC-CKR2, *see* CCR2
CC-CKR5, *see* CCR5
CCR1
CCR1–CCR2B chimeric receptor
construction, 73, 75
enzyme linked immunosorbent assay of expression, 80
epitope tagging, 73
G protein-coupled domain identification, 80–81, 83–84
inositol phosphate formation assay, 80
monocyte chemotactic protein-1 binding assay, 75–77
DARC–CCR1 chimeric receptor
construction, 65–68
detection of expression by flow cytometry, 68–69

expression in baculovirus–insect cell system, 68–69
expression in developing human brain, 36–37
G protein coupling, 105, 313, 325
sequence homology with CCR2, 71–73
transient transfection in Chinese hamster ovary cells, 92–93, 101, 103, 339
CCR2
CCR1–CCR2B chimeric receptor
construction, 73, 75
enzyme linked immunosorbent assay of expression, 80
epitope tagging, 73
G protein-coupled domain identification, 80–81, 83–84
inositol phosphate formation assay, 80
monocyte chemotactic protein-1 binding assay, 75–77
G protein coupling, 313
isoforms, 71
monocyte chemotactic protein-1 binding assay, 75–77
sequence homology with CCR1, 71–73
transient transfection of CCR2B in Chinese hamster ovary cells, 92–93, 101, 103
CCR5
cell-surface expression analysis, 129–130
expression
baculovirus–insect cell system, 52–53, 55
developing human brain, 36–37
human immunodeficiency virus binding, 85, 126–127
human immunodeficiency virus env-mediated fusion assays, 70
monoclonal antibody generation and binding, 53, 55
transient transfection in Chinese hamster ovary cells, 92–93, 101, 103, 107
Chemokine classification, 84–85, 149, 197–198, 340
Chemotaxis assay
chemokine assays in skin disease
desensitization assay, 273–274
eosinophils
Boyden chamber assay, 272
β-glucuronidase assay, 272
isolation from human blood, 271
polymorphonuclear leukocytes
Boyden chamber assay, 269–270
β-glucuronidase assay, 270–271
neutrophil isolation, 268–269
CXCR2 receptor, analysis of truncated mutants, 10–11, 14–15
endothelial cells in angiogenesis, 202–203
Chimeric chemokine receptor
CCR1–CCR2B chimeric receptor
construction, 73, 75
enzyme linked immunosorbent assay of expression, 80
epitope tagging, 73
G protein-coupled domain identification, 80–81, 83–84
inositol phosphate formation assay, 80
monocyte chemotactic protein-1 binding assay, 75–77
DARC–CCR1 chimeric receptor
construction, 65–68
detection of expression by flow cytometry, 68–69
expression in baculovirus–insect cell system, 68–69
expression detection with monoclonal antibodies, 63–65
gene construction
fusion of restriction endonuclease fragments, 59–60, 63
ligation polymerase chain reaction, 59, 61–63
overlap polymerase chain reaction, 59–61, 65–67, 73, 75
host and graft selection, 58–59
junction design, 59
rationale for structure–function analysis, 56–58
selection of receptor pairs, 71–73
Chloramine-T, *see* Iodination
Cholera toxin, identification of G protein-coupled receptors, 316–317
Chorioallantoic membrane assay, angiogenesis, 201
Confocal microscopy, chemokine receptors in developing human brain, 37–38
Cornea micropocket assay, angiogenesis, 201, 210
CXCR2
calcium fluorimetry with Fura-2, 8–9
chemotaxis assay, 10–11

degradation analysis, 8
expression analysis in human brain
confocal microscopy imaging, 37–38
fetal brain
cell culture, 29–31, 33
fixation and sectioning, 34
hNT neuron culture, 34
immunohistochemical staining, 35–37
expression in baculovirus–insect cell system, 52–53
GTP binding assay, 10
melanoma growth stimulating activity binding assay, 6
metabolic labeling with [35S]methionine, 8
monoclonal antibody preparation, 34–35
phosphorylation assay
overview of assays, 11, 14
phosphoamino acid analysis, 9–10, 14
in vivo, 7, 14
sequestration assay, 11
serine phosphorylation, 4
site-directed mutagenesis, 5–6
stable transfectant generation, 4, 6
truncated mutants
effect on chemotaxis, 14–15
generation, 5
Western blot analysis, 7–8, 14
CXCR4
cell-surface expression analysis, 129–130
expression directed by recombinant baculovirus, 55
human immunodeficiency virus binding, 85, 126–127
Cyclic AMP, *see* Adenylate cyclase
Cytosensor Microphysiometer System
capsule cup seeding of cells, 88–89
cell preparation for analysis
immune cells, 92
THP-1 cells, 92
chemokine effects
chemokine receptor transient transfection in Chinese hamster ovary cells, 92–93, 101, 103
neutrophil response to interleukin-8, 101
T cell and monocyte response, 101
THP-1 cells
antibody blocking studies, 98, 101
desensitization studies, 97–98
interleukin-8, 95, 97–98, 101
macrophage inflammatory protein-1α, 95, 97–98, 101
methylisobutylamiloride effects on response, 98
monocyte chemotactic protein-1, 95, 97–98
RANTES, 95, 97–98
extracellular acidification measurement, 87–90
light-addressable potentiometric sensor, 87, 90
reagents, 93–94
running medium, 93
sensor chamber, 89

D

DARC
biological functions, 150
DARC–CCR1 chimeric receptor
construction, 65–68
detection of expression by flow cytometry, 68–69
expression in baculovirus–insect cell system, 68–69
expression in baculovirus–insect cell system, 52–53, 55
signal transduction, 41
Duffy chemokine receptor, *see* DARC

E

EAE, *see* Experimental autoimmune encephalitis
ECAR, *see* Extracellular acidification rate
ELISA, *see* Enzyme-linked immunosorbent assay
ENA-78, endothelial cell chemotaxis, 202–203
Enzyme-linked immunosorbent assay
adenylate cyclase assay, 338–339
cell surface expression assay of chemokine receptors, 80
interleukin-8, 171–172, 181, 213, 287–288
IP-10, 213, 216–217
macrophage inflammatory protein-2, 229–231
monoclonal antibody
affinity assay, 23–24
hybridoma screening, 21–22

Eosinophil
chemotaxis assay, *see* Chemotaxis assay
chemotaxis in asthma, 251–255, 259–262, 265–266
Eotaxin
discovery, 257
gene cloning, 258–259
selective cell recruitment in asthma, 259–262
sequence homology
chemokines, 257–258
species differences, 259, 261
tissue distribution, 259
Experimental autoimmune encephalitis, *see* Multiple sclerosis animal models
Extracellular acidification rate
chemokine effects
chemokine receptor transient transfection in Chinese hamster ovary cells, 92–93, 101, 103
neutrophil response to interleukin-8, 101
T cell and monocyte response, 101
THP-1 cells
antibody blocking studies, 98, 101
desensitization studies, 97–98
interleukin-8, 95, 97–98, 101, 105
macrophage inflammatory protein-1α, 95, 97–98, 101, 105
methylisobutylamiloride effects on response, 98
monocyte chemotactic protein-1, 95, 97–98, 105
RANTES, 95, 97–98, 105
measurement by Cytosensor Microphysiometer System, 87–90

F

FAK, *see* Focal adhesion kinase
Flow cytometry
calcium flux analysis in T cells, 370
chemokine receptor expression quantification in baculovirus–insect cell expression system, 52–53, 55, 68–69
monoclonal antibody hybridoma screening, 22
VLA-2 detection in non-small cell lung cancer, 215
Focal adhesion kinase, assay, 353–354
Fura-2
calibration and quantification, 304–305
cell loading, 303–304
CXCR2 analysis of calcium flux, 8–9
fluorescence detection, 304
manganese quenching, 306
principle of calcium assay, 302–303
Fusin, *see* CXCR4

G

GARK, *see* G protein-activated receptor kinase
Gene therapy, murine model of *Klebsiella* pneumonia
interleukin-12, 238–240
macrophage inflammatory protein-2, 240–241
Genistein, protein tyrosine kinase inhibition, 348
β-Glucuronidase
chemotaxis assay, 270–272
degranulation assay, 272–273
G protein
CCR1–CCR2B chimeric receptor, G-protein-coupled domain identification, 80–81
chemokine receptor coupling, 105, 312–314, 325
families, 310–311
GTP
binding assay, 317–320
GTPase assay, 320–321
identification of coupled receptors
affinity modification by G proteins, 314–315
reporter gene assay, 325
toxin effects, 310, 315–317
reconstitution with receptors, 322–324
signaling, *see also* Adenylate cyclase; Phosphatidylinositol 3-kinase; Phospholipase C; Phospholipase D
mechanisms, 309, 314, 326, 341
regulation, 311–312, 324–325
subunits, 309–311, 341
Western blot analysis, 321–322
G protein-activated receptor kinase, signal transduction for chemokine receptors, 4
GRO-α, *see* Growth related oncogene-α

Growth related oncogene-α
isoforms, 293
purification from lesional scale material, 289–293
GTP
CXCR2 binding assay, 10
G protein
binding assay, 317–320
GTPase assay, 320–321

H

Heparin affinity chromatography, chemokine purification from lesional scale material, 289, 294
Herbimycin A, protein tyrosine kinase inhibition, 348
High-performance liquid chromatography, *see also* Reversed-phase high-performance liquid chromatography
chemokine purification from inflammatory lavage fluids, 247, 249
chemokine purification from lesional scale material
cation-exchange chromatography, 276, 278, 282, 291–293, 295
size-exclusion chromatography, 274–276, 279–280
inositol phosphate metabolites, anion-exchange chromatography, 382
HIV, *see* Human immunodeficiency virus
Human immunodeficiency virus
CD4 receptor, 118–119
chemokine receptors
CCR5 binding, 85, 126–127
cell–cell fusion assays
antibody blocking studies, 130–131
detection of fusion, 119–120
effector cell requirements, 120–122
env-encoding plasmid construction and transfection, 127–128
env-mediated fusion assays, CCR5 in, 70
fusion reaction, 126
infection of HeLa cells by env-encoding vaccinia virus, 124–125
kinetic analysis, 131–132
reporter gene detection, 120, 123, 126, 132–133
signal optimization, 132–133
target cell requirements, 122–123
transfection of QT6 cells, 123–124
cell-surface expression analysis, 129–130
CXCR4 binding, 85, 126–127
overview, 118–119, 122
tropism of utilization, 119, 126–127
virus infection assays, overview, 119

I

ICAM-1, *see* Intercellular adhesion molecule-1
IL-2, *see* Interleukin-2
IL-5, *see* Interleukin-5
IL-8, *see* Interleukin-8
IL-12, *see* Interleukin-12
Immunohistochemical staining, chemokine receptors in developing brain
CCR1, 36–37
CCR5, 36–37
CXCR2, 35–37
Indo-1, calcium flux analysis in T cells, 369–370
Inositol phosphate
binding protein assays, 382–383
extraction
lipids, 379
monitoring, 378
water-soluble compounds, 379–380
glycerophosphoinositols
deacylation of inositol lipids, 380
ion-exchange chromatography, 380–381
quantification by thin-layer chromatography, 381
ion-exchange chromatography of water soluble compounds and metabolites, 381–382
metabolic labeling with tritiated inositol, 377–378
metabolism, 362–363
receptor formation assay, 80
receptor types, 363
signal transduction of chemokine receptors, 362
Intercellular adhesion molecule-1, cytokine effects on expression, 264
Interleukin-2, G protein assays
GTPase, 320
GTP binding, 317–319

Interleukin-5, selective cell recruitment in asthma, 261–262
Interleukin-8
CXCR1 receptor, transient transfection in Chinese hamster ovary cells, 107
depletion strategies in mice, 214–215
effects on polymorphonuclear leukocytes
complement expression, 223
immunoglobulin receptor expression, 223
microbicidal activity, 224
phagocytosis, 223
endothelial cell chemotaxis, 202–203
enzyme-linked immunosorbent assay, 171–172, 181, 213, 287–288
expression and purification of glutathione *S*-transferase fusion protein, 152–153
expression in brain, 28
extracellular acidification rate response
neutrophils, 101
T cells and monocytes, 101
THP-1 cells, 95, 97–98, 101, 105
fluorescent microspheres and receptor visualization by intravital microscopy
control experiments, 157–158
linkage, 153–154
receptor visualization, 156–157
videotape analysis, 158
G protein coupling of receptor, 313
lung injury model analysis
acid-injury rabbit model, 173–175
pleurisy rabbit model
neutralization with monoclonal antibody, 168–171
Western blot analysis, 171
enzyme-linked immunosorbent assay, 171–172, 181
monoclonal antibody preparation, 167–168
non-small cell lung cancer angiogenesis role, 214–216
purification from lesional scale material, 284–289
receptor A, antibody generation
affinity assays
enzyme-linked immunosorbent assay, 23–24
radioimmunoprecipitation assay, 23–24
fusion, 20–21
hybridoma screening
enzyme-linked immunosorbent assay, 21–22
flow cytometry, 22
immunization of mice, 20
ligand receptor binding assay for selection of blocking antibodies, 22–26
synthetic peptides as immunogens, 17, 24–27
transfected cell receptors as immunogens, 18–19, 24–27
receptor B, *see* CXCR2
selective cell recruitment in asthma, 250–251
sequence homology between receptors, 26
site-directed mutagenesis of ELR motif
angiogenic activity, 210
fusion protein expression and purification, 209–210
mutagenesis protocol, 207, 209
Interleukin-12, murine *Klebsiella* pneumonia
intratracheal gene therapy, 238–240
survival effects, 238
Intravital microscopy
fluorescent microsphere visualization
interleukin-8
control experiments, 157–158
linkage, 153–154
receptor visualization, 156–157
videotape analysis, 158
rat preparation for injection, 155–156
size, 151
principle, 151–152
resolution, 151
Iodination
bindability evaluation of chemokines, 145–146
Bolton–Hunter reagent
benzene evaporation, 142
incubation conditions, 143
preparation, 142
specificity of iodination, 134–135, 142
chloramine-T iodination
incubation conditions, 137, 139
reaction rate and quenching, 138
conformational effects, 135
interleukin-8
Bolton–Hunter reagent, 143

lactoperoxidase, 141
Iodogen method, 139–140
lactoperoxidase method, 140–141
macrophage inflammatory protein-1α with chloramine-T, 139
melanoma growth stimulating activity with chloramine-T, 6
monocyte chemotactic protein-1 with Bolton–Hunter reagent, 75–76
optimization of specific activity, 136–137
oxidation of proteins, 135–136
purity evaluation of modified chemokines
gel filtration, 146
mass spectrometry, 147
polyacrylamide gel electrophoresis, 145
reversed-phase high-performance liquid chromatography, 136, 146–147
trichloroacetic acid precipitation, 143–145
selection of technique, 134–135
specific activity calculation, 147–148
Iodogen, *see* Iodination
IP-10
angiostatic activity, 203–204, 207, 216–219
enzyme-linked immunosorbent assay, 213, 216–217
induction by cytokines, 199, 216

J

c-Jun kinase, assay, 356–357

K

Klebsiella pneumonia, *see* Pneumonia, bacteria
Knockout models, limitations in cytokine studies, 161

L

Lactoperoxidase, *see* Iodination
Lung cancer, *see* Angiogenesis
Lung injury
acid-injury rabbit model
blood gas analysis, 175
catheter placement, 174
extravascular lung water determination, 177
instillation, 174
interleukin-8 neutralization with monoclonal antibody, 173–175
lung vasculature permeability measurement, 175–176
mechanical ventilation, 175
sample collection
bronchoalveolar lavage, 181
edema fluid, 177
statistical analysis, 181
animal models for cytokine study
duration of experiments, 162
neutralization of cytokines, 166–167
outcome variable selection, 162
overview, 162–163
pleurisy rabbit model
endotoxin induction, 168, 170
immune cell quantification, 172–173
interleukin-8
enzyme-linked immunosorbent assay, 171–172, 181
neutralization with monoclonal antibody, 168–171
Western blot analysis, 171
pleural catheter placement, 169–170

M

Macrophage inflammatory protein-1α
extracellular acidification rate response
T cells and monocytes, 101, 105
THP-1 cells, 95, 97–98, 101, 105
radioiodination with chloramine-T, 139
receptor, *see* CCR1
selective cell recruitment in asthma, 254–255
Macrophage inflammatory protein-1β, selective cell recruitment in asthma, 254–255
Macrophage inflammatory protein-2
effects on polymorphonuclear leukocytes
complement expression, 223
immunoglobulin receptor expression, 223
microbicidal activity, 224
phagocytosis, 223
murine model of *Klebsiella* pneumonia
antibody blocking studies, 232–238

bacterial clearance role, 234–235
enzyme-linked immunosorbent assay, 229–231
immunolocalization in lung, 231–232
inflammation mediation, 233–234
intratracheal gene therapy, 240–241
messenger RNA analysis, 228–229
proinflammatory cytokine regulation, 235–236
survival effects, 236–238
MAPK, *see* Mitogen-activated protein kinase
MCP-1, *see* Monocyte chemotactic protein-1
MCP-2, *see* Monocyte chemotactic protein-2
MCP-3, *see* Monocyte chemotactic protein-3
Melanoma growth stimulating activity
CXCR2 binding assay, 6
radioiodination with chloramine-T, 6
Methylisobutylamiloride, effects on extracellular acidification rate response of chemokines, 98
MGSA, *see* Melanoma growth stimulating activity
Microinjection, *see* *Xenopus* oocyte
MIG
angiostatic activity, 203–204, 207
site-directed mutagenesis of ELR motif
angiogenic activity effects, 210
glutathione *S*-transferase fusion protein expression and purification, 208–209
mutagenesis protocol, 207–208
MIP-1α, *see* Macrophage inflammatory protein-1α
MIP-1β, *see* Macrophage inflammatory protein-1β
MIP-2, *see* Macrophage inflammatory protein-2
Mitogen-activated protein kinase
assays
gel-shift assay, 355
myelin basic protein as substrate, 355–356
signal transduction for chemokine receptors, 3, 341, 354–355
Monoclonal antibody, chemokine receptor
binding sites, 64
CCR5, antibody generation and binding, 53, 55
chimeric chemokine receptors, 63–65
CXCR2, antibody generation, 34–35
difficulty in generation, 16
interleukin-8 receptor A, antibody generation
affinity assays
enzyme-linked immunosorbent assay, 23–24
radioimmunoprecipitation assay, 23–24
fusion, 20–21
hybridoma screening
enzyme-linked immunosorbent assay, 21–22
flow cytometry, 22
immunization of mice, 20
ligand receptor binding assay for selection of blocking antibodies, 22–26
synthetic peptides as immunogens, 17, 24–27
transfected cell receptors as immunogens, 18–19, 24–27
Monocyte chemotactic protein-1
expression in brain, 28
extracellular acidification rate response
T cells and monocytes, 101, 105
THP-1 cells, 95, 97–98, 105
radioiodination with Bolton–Hunter reagent, 75–76
radioiodination, 75–76
receptor, *see* CCR2
selective cell recruitment in asthma, 255–257
Monocyte chemotactic protein-2, selective cell recruitment in asthma, 256–257
Monocyte chemotactic protein-3, selective cell recruitment in asthma, 256–257
MPO, *see* Myeloperoxidase
Multiple sclerosis animal models
experimental autoimmune encephalitis
chronic-relapsing form
immunization, 188–189
monitoring, 189–190
mouse maintenance, 188
pathogenesis, 187–188
guinea pigs, 185
immunization induction, 182–183, 186
myelin antibody role, 188

passive transfer, 186–187
primates, 185–186
rats, 185
transgenic mice, 184–185
toxic demyelination, 184
virus-induced central nervous system demyelination, 183
Myeloperoxidase
degranulation assay, 273
lung leukocyte influx assay, 227–228, 233

N

Neutrophil
chemokine effects in bacterial pneumonia
complement expression, 223
immunoglobulin receptor expression, 223
microbicidal activity, 224
phagocytosis, 223
chemotaxis assay, *see* Chemotaxis assay
isolation from human blood, 268–269, 271
Non-small cell lung cancer, *see* Angiogenesis

P

PCR, *see* Polymerase chain reaction
Pertussis toxin, identification of G protein-coupled receptors, 318
PF4, *see* Platelet factor-4
Phagocytosis, assay of chemokine effects on polymorphonuclear leukocytes, 222–223
Phosphatidylinositol 3-kinase, signal transduction for chemokine receptors, 3, 344, 363
Phosphodiesterase, inhibitors, 327
Phospholipase C, signal transduction for chemokine receptors, 3, 362–364
Phospholipase D
classification, 358
inhibitors, 360
signal transduction, 357–358
transphosphatidylation assay
cell permeabilization effects, 360–361
cell preparation, 358–359
extraction, 359
thin-layer chromatography, 359–360
Phosphorylation, protein, *see also specific kinases*
CXCR2 phosphorylation assay
overview of assays, 11, 14
phosphoamino acid analysis, 9–10, 14
in vivo, 7, 14
denaturing polyacrylamide gel electrophoresis, 350–351
immunoprecipitation, 351–353
metabolic labeling in whole cells, 349–350
protein interaction analysis by coimmunoprecipitation, 352–353
Western blot analysis of tyrosine phosphoproteins, 351–352
PKC, *see* Protein kinase C
Platelet factor-4, angiostatic activity, 203–204, 207
Pleurisy rabbit model, *see* Lung injury
Pneumonia, bacteria
chemokine effects on polymorphonuclear leukocytes
complement expression, 223
immunoglobulin receptor expression, 223
microbicidal activity, 224
phagocytosis, 223
cytokine mediation of host defense, 221
murine model of *Klebsiella* pneumonia
advantages of model, 224–225
inoculation, 225–226
interleukin-12
intratracheal gene therapy, 238–240
survival effects, 238
leukocyte influx assays, 226–228, 233
macrophage inflammatory protein-2
antibody blocking studies, 232–238
bacterial clearance role, 234–235
enzyme-linked immunosorbent assay, 229–231
immunolocalization in lung, 231–232
inflammation mediation, 233–234
intratracheal gene therapy, 240–241
messenger RNA analysis, 228–229
proinflammatory cytokine regulation, 235–236
survival effects, 236–238

pneumonia susceptibility of mouse strains, 225
treatment, 220
Polymerase chain reaction, *see also* Reverse transcription–polymerase chain reaction
chimeric chemokine receptor construction
fusion of restriction endonuclease fragments, 59–60, 63
ligation polymerase chain reaction, 59, 61–63
overlap polymerase chain reaction, 59–61, 65–67
Protein kinase C
agonists, 345–346
assays, 345
chemokine receptor signal transduction, 344
inhibitors, 346–347
isoforms, 344–345
Protein phosphatase, inhibitors, 348–349
Protein tyrosine kinase
assays, 347–348, 353–354
classification, 344
inhibitors, 348–349
Psoriasis
amino-terminal sequencing of chemokines, 283
angiogenesis, 196
chemokine purification
C-C chemokines, 293–294
cation-exchange high-performance liquid chromatography, 276, 278, 282, 291–293, 295
extraction from lesional scale material, 267–268, 274, 286, 294
gel filtration, 274–276, 279–280
growth related oncogene-α, 289–293
heparin affinity chromatography, 289, 294
interleukin-8, 284–289
polyacrylamide gel electrophoresis, 279, 288
RANTES, 294–296
reversed-phase high-performance liquid chromatography, 278–279, 281–283, 286, 288–291, 294–296
chemotaxis assays of chemokines
desensitization assay, 273–274
eosinophils
Boyden chamber assay, 272
β-glucuronidase assay, 272
isolation from human blood, 271
polymorphonuclear leukocytes
Boyden chamber assay, 269–270
β-glucuronidase assay, 270–271
neutrophil isolation, 268–269
polymorphonuclear leukocyte enzyme release in chemokine bioassay
desensitization assay, 273–274
β-glucuronidase assay, 272–273
myeloperoxidase, 273
PTK, *see* Protein tyrosine kinase

R

Radioiodination, *see* Iodination
RANTES
calcium flux response, 86
discovery, 252
expression in brain, 28
extracellular acidification rate response
monocytes, 105
THP-1 cells, 95, 97–98, 105
purification from lesional scale material, 294–296
selective cell recruitment in asthma, 252–254
Reverse transcription–polymerase chain reaction, macrophage inflammatory protein-2 messenger RNA analysis, 228–229
Reversed-phase high-performance liquid chromatography
chemokine purification
inflammatory lavage fluids, 247, 249
lesional scale material, 278–279, 281–283, 286, 288–291, 294–296
iodinated chemokines, 136, 146–147
RT–PCR, *see* Reverse transcription–polymerase chain reaction

S

Seven-transmembrane domain, chemokine receptor structure, 3, 16, 149

Site-directed mutagenesis
CXCR2, 5–6
interleukin-8 ELR motif
angiogenic activity effects, 210
fusion protein expression and purification, 209–210
mutagenesis protocol, 207, 209
MIG ELR motif
angiogenic activity effects, 210
glutathione *S*-transferase fusion protein expression and purification, 208–209
mutagenesis protocol, 207–208

T

Thapsigargin, calcium channel blocking, 306, 374
Theiler's murine encephalomyelitis virus, animal model of multiple sclerosis, 183
Thin-layer chromatography
inositol phosphate lipids, 381
phospholipase D assay, 359–360
TLC, *see* Thin-layer chromatography
T lymphocyte
calcium flux analysis by fluorescence
cell preparation, 366–367
dye selection, 368–369
fluorescence detection, 369–370
flux profiles of various cell types, 370–371, 373–374
loading of cells, 376
lymphocyte lines, 367
T cell clones, 367
transfected cell lines, 368
viability of cells, 376
chemokine activation assays, 342–343
chemokine receptor types, 364–365
donor variability, 343
Tumor angiogenesis, *see* Angiogenesis
Tyrosine kinase, *see* Protein tyrosine kinase

V

Vascular cell adhesion molecule-1, cytokine effects on expression, 264
VCAM-1, *see* Vascular cell adhesion molecule-1

W

Western blot analysis
CXCR2, 7–8, 14
G proteins, 321–322
interleukin-8, 171
tyrosine phosphoproteins, 351–352

X

Xenopus oocyte
anatomy, 109–110
calcium flux assay of chemokine receptor expression
aequorin assay, 117
calcium-45 release assay
agonist response, 115
incubation conditions, 114–115
messenger RNA injection of chemokine receptors
advantages of system, 116
animal husbandry, 111
comparison to mammalian cell systems, 117
defolliculation of oocytes, 112
disadvantages of system, 116–117
materials, 110–111
microinjection technique, 112–113
oocyte removal, 111–112
RNA preparation, 113–114
staging of oocytes, 109–110
orphan cloning and chemokine receptor subtypes, 109

ISBN 0-12-182189-7